I0894534

Handbook of Research on Ergonomics and Product Design

Juan Luis Hernández Arellano
Autonomous University of Ciudad Juárez, Mexico

Aide Aracely Maldonado Macías
Autonomous University of Ciudad Juárez, Mexico

Juan Alberto Castillo Martínez
University of Rosario, Colombia

Porfirio Peinado Coronado
Autonomous University of Ciudad Juárez, Mexico

A volume in the Advances in Chemical and Materials Engineering (ACME) Book Series

IGI Global
DISSEMINATOR OF KNOWLEDGE

Published in the United States of America by
IGI Global
Engineering Science Reference (an imprint of IGI Global)
701 E. Chocolate Avenue
Hershey PA, USA 17033
Tel: 717-533-8845
Fax: 717-533-8661
E-mail: cust@igi-global.com
Web site: http://www.igi-global.com

Copyright © 2018 by IGI Global. All rights reserved. No part of this publication may be reproduced, stored or distributed in any form or by any means, electronic or mechanical, including photocopying, without written permission from the publisher. Product or company names used in this set are for identification purposes only. Inclusion of the names of the products or companies does not indicate a claim of ownership by IGI Global of the trademark or registered trademark.

Library of Congress Cataloging-in-Publication Data

Names: Hernandez Arellano, Juan Luis, 1979- editor.
Title: Handbook of research on ergonomics and product design / Juan Luis Hernandez Arellano, Aide Aracely Maldonado Macias, Juan Alberto Castillo Martinez, and Porfirio Peinado Coronado, editors.
Description: Hershey, PA : Engineering Science Reference, [2018] | Includes bibliographical references.
Identifiers: LCCN 2017037491| ISBN 9781522552345 (h/c) | ISBN 9781522552352 (eISBN)
Subjects: LCSH: New products. | Product design. | Human engineering.
Classification: LCC TS171.4 .H36325 2018 | DDC 658.5/75--dc23 LC record available at https://lccn.loc.gov/2017037491

This book is published in the IGI Global book series Advances in Chemical and Materials Engineering (ACME) (ISSN: 2327-5448; eISSN: 2327-5456)

British Cataloguing in Publication Data
A Cataloguing in Publication record for this book is available from the British Library.

All work contributed to this book is new, previously-unpublished material. The views expressed in this book are those of the authors, but not necessarily of the publisher.

For electronic access to this publication, please contact: eresources@igi-global.com.

Advances in Chemical and Materials Engineering (ACME) Book Series

J. Paulo Davim
University of Aveiro, Portugal

ISSN:2327-5448
EISSN:2327-5456

Mission

The cross disciplinary approach of chemical and materials engineering is rapidly growing as it applies to the study of educational, scientific and industrial research activities by solving complex chemical problems using computational techniques and statistical methods.

The **Advances in Chemical and Materials Engineering (ACME) Book Series** provides research on the recent advances throughout computational and statistical methods of analysis and modeling. This series brings together collaboration between chemists, engineers, statisticians, and computer scientists and offers a wealth of knowledge and useful tools to academics, practitioners, and professionals through high quality publications.

Coverage

- Nanomaterials
- Statistical methods
- Coatings and surface treatments
- Computational methods
- Heat Treatments
- Industrial chemistry
- Electrochemical and Corrosion
- Chemical Engineering
- Fatigue and Creep
- Thermo-Chemical Treatments

IGI Global is currently accepting manuscripts for publication within this series. To submit a proposal for a volume in this series, please contact our Acquisition Editors at Acquisitions@igi-global.com or visit: http://www.igi-global.com/publish/.

The Advances in Chemical and Materials Engineering (ACME) Book Series (ISSN 2327-5448) is published by IGI Global, 701 E. Chocolate Avenue, Hershey, PA 17033-1240, USA, www.igi-global.com. This series is composed of titles available for purchase individually; each title is edited to be contextually exclusive from any other title within the series. For pricing and ordering information please visit http://www.igi-global.com/book-series/advances-chemical-materials-engineering/73687. Postmaster: Send all address changes to above address. Copyright © 2018 IGI Global. All rights, including translation in other languages reserved by the publisher. No part of this series may be reproduced or used in any form or by any means – graphics, electronic, or mechanical, including photocopying, recording, taping, or information and retrieval systems – without written permission from the publisher, except for non commercial, educational use, including classroom teaching purposes. The views expressed in this series are those of the authors, but not necessarily of IGI Global.

Titles in this Series

For a list of additional titles in this series, please visit: www.igi-global.com/book-series

Mechanical Properties of Natural Fiber Reinforced Polymers Emerging Research and Opportunities
Sarah S. Gebai (International University of Beirut, Lebanon) Ali M. Hallal (International University of Beirut, Lebanon) and Mohammad S. Hammoud (International University of Beirut, Lebanon)
Engineering Science Reference • copyright 2018 • 228pp • H/C (ISBN: 9781522548379) • US $165.00 (our price)

Advanced Solid Catalysts for Renewable Energy Production
Sergio González-Cortés (Oxford University, UK) and Freddy Emilio Imbert (Univsersidad de Los Andes, Venezuela)
Engineering Science Reference • copyright 2018 • 520pp • H/C (ISBN: 9781522539032) • US $225.00 (our price)

Production, Properties, and Applications of High Temperature Coatings
Amir Hossein Pakseresht (University of Tehran, Iran & Materials and Energy Research Center, Iran)
Engineering Science Reference • copyright 2018 • 557pp • H/C (ISBN: 9781522541943) • US $235.00 (our price)

Energetic Materials Research, Applications, and New Technologies
Rene Francisco Boschi Goncalves (Federal University of Para, Brazil) José Atilio Fritz Fidel Rocco (Aeronautics Institute of Technology, Brazil) and Koshun Iha (Aeronautics Institute of Technology, Brazil)
Engineering Science Reference • copyright 2018 • 367pp • H/C (ISBN: 9781522529033) • US $225.00 (our price)

Chemical Compound Structures and the Higher Dimension of Molecules Emerging Research and Opportunities
Gennadiy Vladimirovich Zhizhin (Russian Academy of Natural Sciences, Russia)
Engineering Science Reference • copyright 2018 • 221pp • H/C (ISBN: 9781522541080) • US $175.00 (our price)

Emergent Research on Polymeric and Composite Materials
R. Somashekar (University of Mysore, India) and Thejas Urs G. (University of Mysore, India)
Engineering Science Reference • copyright 2018 • 341pp • H/C (ISBN: 9781522530237) • US $235.00 (our price)

3D Printing and Its Impact on the Production of Fully Functional Components Emerging Research and Opportunities
Petar Kocovic (Union – Nikola Tesla University, Serbia)
Engineering Science Reference • copyright 2017 • 115pp • H/C (ISBN: 9781522522898) • US $150.00 (our price)

Advanced Applications of Supercritical Fluids in Energy Systems
Lin Chen (Tohoku University, Japan & Japan Society for the Promotion of Science (JSPS), Japan) and Yuhiro Iwamoto (Nagoya Institute of Technology, Japan)
Engineering Science Reference • copyright 2017 • 682pp • H/C (ISBN: 9781522520474) • US $235.00 (our price)

IGI Global
DISSEMINATOR OF KNOWLEDGE

701 East Chocolate Avenue, Hershey, PA 17033, USA
Tel: 717-533-8845 x100 • Fax: 717-533-8661
E-Mail: cust@igi-global.com • www.igi-global.com

List of Contributors

Table of Contents

Detailed Table of Contents

Work-related musculoskeletal disorders (WMSDs) prevention has become a global phenomenon and nowadays is one of the main challenges that ergonomics and work safety professionals face. The evaluation of risk factors exposition related to the WMSDs is one of the main activities that are performed for prevention. In order to accomplish it, many ergonomic assessment methods have been developed. In this chapter, an ergonomic observational individual risk assessment (ERIN) method is presented, which is distinguished for being easy to learn, apply, and it needs short training time. Results on the reliability, validity, and usability studies of ERIN are included too, as well as an example of how it can be used in the workplace intervention process.

Development of musculoskeletal disorders caused by awkward postures are present in tasks performed in small companies. Consequently, this chapter evaluates the postural risk level associated to the task of changing brake pads of a forklift, and proposes the design and the implementation of an ergonomic bench that helps to reduce this level of postural risk. As methods, postural risk level corresponding to the task was determined using REBA, whereas anthropometric analysis was developed, and axiomatic design and TRIZ methodologies were used to develop the proposed design. Results showed a REBA score of 8 (high-risk level). The final design is done of nylamid, and it contains some TRIZ principle, such as copying, extraction, and dynamicity. Respect axiomatic design, the bench meets some functional requirements, such as support the user's weight, avoid unsafety conditions, and be adjustable. With the TRIZ principles and functional requirements, this bench will help diminish postural risk level.

For public hospitals, the patients are generally placed on a waiting list prior to having their surgery; it means that they have an elective surgery. The elective surgery is what can safely be delayed for more than 24 hours, but this usually is moved when some external factors are present, one of these is attributable to medical skills. For this reason, it is necessary to analyze the behavior of the last one to determine its relationship with the elective surgery performance and how it increases the surgical deferral. For purposes of the desired analysis, it is necessary to use one of the analysis complex system techniques, in this case the modeling and simulation agent-based.

The population pyramid is being inverted and the designers are facing challenges in the way they design products, spaces, and services for elder people. Although the design is composed of both cultural and social aspects, the functional aspect takes relevance for autonomy and empowerment of the person that gets older. To design products centered on the physical capacities of the elder population, the aging suit has being used as a tool for the gerontodesign. In countries like Mexico, its application is limited by the manufacturing cost. For this reason, as part of a design methodology for elder people, a low-cost suit was developed focus on the design students and professionals located in Latin America.

In this chapter, the data collection is presented, where the base is the study of a group of elderly people in eight nursery houses and two hospitals in Bogota, Colombia. The objective of this research is to look through industrial design and the analysis of existing products, in which way the designer could offer an improvement to the products and with that how to increase the life expectancy. The user perception method is applied, therefore, and the interview was performed as a sequence of questions, where the

attributes of use and methods to design visuals. Specifically, flow diagrams, procedures lesson, and one-point lessons as visuals are studied in this chapter. The theoretical departure covered diverse areas, and activity theory served as conceptual framework. So, the overall purpose of this chapter is to (1) explore the defined specifications for design visuals, (2) propose a checklist to determine the levels of usability, and (3) evaluate the visuals utilized in two manufacturing companies serving as cases of study. Finally, the results have permitted the convenience of using the proposed checklist for checking the design and utilization of the visuals in manufacturing processes to be proven.

The use of novel technologies can be complicated for some people; even products designed for everyday use can present problems. If a product is difficult to use, it causes frustration and prevents usage. The objectives of this study are to conduct a usability test, assess the mental workload, and identify potential human errors during the process of listening to music using wireless earphones from a mobile phone device to detect which operations were the hardest in order to suggest modifications to the design and to bring a better experience to the user. During the usability test some difficulties among users were detected and a hierarchical task analysis (HTA) was developed. NASA-TLX and SHERPA methods were applied. The use of these earphones resulted in an intermediate mental workload, and four types of human errors were identified. The methods used in this study helped to detect those difficult tasks and subtasks for the users.

An analysis of different objects of daily use was made, specifically four models of lemon squeezers, three models of can openers, and four models of clothespins. Such analysis is made focusing on the strength needed to complete the task and the dimensional relation between the hands and object grips that impact both in ease of use and user well-being itself, identifying the gap between young and older adults characteristics and capabilities and the features of the object. This comparison was supported by a dynamometric study to identify the maximum capabilities of strength application in two different grip forces very common in daily life. These two grip forces are the ground of the interaction with the analyzed products. In conjunction with the force, the registry made an anthropometric assessment of different anatomical points in the hands.

Sustainable supply chain management (SSCM) practices in thermal power plants is dependent on mostly three pillars: social factor, economic factor, and environmental factor. So, in this chapter, sustainable supply chain management of Indian thermal power sector is evaluated. Artificial neural network (ANN) method is implemented to measure whether the benefits of sustainable supply chain management are achieved after practices of sustainable supply chain management in Indian thermal power sector. This chapter also designs a framework by QFD (quality function deployment) method to find solution for some unsatisfactory measures (inputs in sustainable factors) that are not achieved against outputs. As sustainable supply chain management practices in thermal power plants are influenced by a significant number of interrelated enablers and barriers, the drivers or enablers of SSCM are taken as the design requirement to improve SSCM in thermal power industries, and the most important driver is prioritized against the unsatisfied measurands in thermal power sector.

Foreword

The book *Handbook of Research on Ergonomics and Product Design* edited by Doctors Hernández Arellano, Maldonado Macías, Castillo-Martínez and Peinado Coronado comes as an important contribution to the ergonomics and product design field. With 21 chapters contributed by authors from 14 universities in four different countries, it is arranged in five sections: Ergonomic Analysis, Ergonomic Product Design, Anthropometrics, Usability and User Experience, and,finally, Additive Manufacturing and New Technologies.

The nature of the topics covered in the materials is a good reflection of the broadly-aimed spectrum set at the title, and certainly constitutes the main strength of this book. From design for elderly people to additive manufacturing, the reader will find proposals and reports related to anthropometry (data and devices), new methods for ergonomic evaluation, design of vehicles for special populations and for recreational purposes, applications to industrial processes, and methods for usability and users experience. Anyone interested in or practicing ergonomics and product design will greatly benefit from adding Handbook of Research on Theory, Methods and Applications in Ergonomics and Product Design to their collection.

Certainly, in Mexico this book sets a breaking point, being the very first that puts together theoretical proposals and research findings in both areas, ergonomics and product design. In fact, the book will be a useful tool for students, professors and researchers, making it a valuable addition to libraries in all the academic institutions dealing with these topic areas.

J Nieves Serratos Pérez
Universidad de Guanajuato, Mexico

Preface

INTRODUCTION

The International Ergonomics Association (IEA, 2017) has defined ergonomics or human factors as "the scientific discipline concerned with the understanding of interactions among humans and other elements of a system, and the profession that applies theory, principles, data, and methods to design in order to optimize human well-being and overall system performance". In this context, product design is an important field where ergonomics and human factors should be applied.

From this definition, we have highlighted the word "design" as the starting point where ergonomics and product design work together to develop better products, better experiences, and better usability with the unique objective of improving the quality life of users. However, this statement is not always accomplished, and every day plenty of new products are launched to the market, but they disappear rapidly. Why? The reason is simple: the acceptability of the product does not occur, the strongest reason is the difficulty in its use, and the use is the key to the integration of a product into the functional universe of the individuals, the product is not usable, and people forget or discard the product. From the perspective of ergonomics, the integration of objects, processes and procedures occurs when a user can perform the integration of these elements in two ways: the first occurs in the physical world through the identification of patterns of interaction in accessible sequences; the second is cognitive, this implies that actions can be executed because the element is accessible in terms of codification - decoding; also because the learning process is embedded in the knowledge base of the user, or facilitates the use of knowledge and experiences already installed in individuals.

In this order of ideas, it is important to understand that one of the scientific bases of design is the integration of ergonomics, both physical and cognitive, to the development process; this implies to integrate the user in the specifications, exploration, test and put in use of the possible solution. Designers usually focus on the object, leaving the user as a hypothetical element, this implies the use of ergonomics as an additive to the process; this misconception is the cause of bad designs or pathogenic designs (they cause damage to the health of the individuals). For this reason it is important to understand that ergonomics is at the base of a pyramid of decision, criteria and principles that lead to the introduction of a solution in the form of a product; It is important to understand that an ergonomic design contributes to facilitate or expand the number of options for users, so the designs that hinder, restrict or limit, are solutions devoid of ergonomics.

In this manner, the content of this book has been structured in five sections: (1) Ergonomic Analysis, (2) Ergonomic Product Design, (3) Anthropometrics, (4) Usability and User Experience, and (5) Additive Manufacturing and Technology. These sections will be described briefly in the following paragraphs. Afterwards, a complete description is given for each chapter.

Before starting any design process or ergonomic project, it is necessary to develop an ergonomic analysis (also known as Task Analysis; Gómez-Bull, Hernández-Arellano, & Ibarra-Mejía 2015). Typically, a design process and/or ergonomic project fails because the most critical and initial analysis is not properly conducted. Like this, the first chapter of the ergonomic analysis section of proposes a novel approach to developing the assessment of workplace risks for the identification of work-related musculoskeletal disorders. The second chapter proposes the ergonomic design of a bench to decrease postural risk Level. The last chapter of this section proposes a modeling simulation based Agents for reducing operative deferral in oncology in a virtual scenario using ergonomics.

The core of this book is precisely the ergonomic product design. Maybe, the most characteristically designed product is the chair. However, how it was designed? How was it tested? How was it validated? These questions are more complex when the design is focused on elderly people. To explore a low-cost and proper way to design, test and validate several products, the first chapter of this book presents an aging suit as an accessible and low-cost design tool for gerontodesign. The second chapter presents an environmental analysis to identify the design requirements of an ergonomic chair for elderly people. The third presents the exploration of the physical and functional characteristics required to design school furniture, the fourth and fifth chapters present the design of two new vehicles to offer new experiences to persons with disabilities and young people, respectively. The sixth chapter presents the incorporation of ergonomics, the analysis of user habits, trials and simulators into the design process.

One of the most important problems about anthropometrics is the lack of information/data to be incorporated into the design process. The case of Mexico and Latin America is very particular because, although anthropometric data does exist, it cannot be representative of the whole country (Hernandez-Arellano et al., 2016). In addition, product design methodologies do not include, as a relevant step, the inclusion and treatment of the anthropometric data and considerations to develop the product. This book includes three chapters related with anthropometrics, one of them presents a comparative study of different devices for data acquisition, the second one presents an ergonomic analysis of working postures on construction site workers and the third and last chapter of this section presents an application for data acquisition from a body measurement scanner.

Today, it is very common to purchase new devices as cell phones, computers, and electronic tablets, among others. Nevertheless, what happens when we start using a new device? The usability and user experience could be different for each user. For example, in 2017 new wireless earphones (The AirPods®) were launched by Apple (2017); products like these have been studied from a cognitive ergonomic approach to evaluating its usability and user experience (Stanton & Baber, 2002). Therefore, new indices with these concepts have been created. The first chapter of this section presents a checklist to determine the usability levels of standard work visuals. The second chapter presents a usability test and cognitive analyses during the use of wireless earphones. The third chapter presents the analysis of different objects of daily use such as lemon squeezers, and clothespins. The fourth chapter presents the comprehension of figurative food illustrations and the effectiveness of ergonomic guidelines. The fifth chapter describes the application of space design for finding solutions to social issues through the generation of social innovations. The last chapter presents the activities and results from a didactic strategy to generate conceptual designs of objects to be manufactured.

With the introduction of the additive manufacturing and new technologies the process of creating a rapid prototyping is changing every day. Now, products that are difficult to shape with traditional manufacturing techniques are now being printed with the different techniques and materials used by additive manufacturing. The first chapter of this section presents a methodology of design for remanufacturing products in the development of new products. The second chapter presents the transformation of a virtually produced gear to a physical product using rapid prototyping technique. The third and last chapter presents the evaluation of sustainability for the supply chain management of the Indian thermal power sector.

ORGANIZATION OF THE BOOK

This book includes regular research papers as contributions, which have been edited according to the norms and guidelines of IGI Global. The call for chapters was distributed among the lists of the field for researchers to submit their works to this project. As a result, 34 chapter proposals were received from countries as Colombia, India, United States, and Mexico. However, abstracts were subject to a screening process to ensure their clarity, authenticity, and relevance to this book. After the screening process, 22 proposals were invited to submit full versions, and two reviewers were assigned to every work to ensure the peer review process. In the end, 21 chapters were accepted for their publication after corrections requested by reviewers and editors were addressed.

Section 1: Ergonomic Analysis

The first section of this book is entitled "Ergonomic Analysis" and comprises three chapters described as follows:

Chapter 1 entitled "Individual Risk Assessment ERIN: Method for the Assessment of Workplace Risks for Work-Related Musculoskeletal Disorders" is proposed by author Yordan Rodriguez from National School of Public Health, University of Antioquia (Colombia). This chapter presents an ergonomic and observational Individual Risk Assessment (ERIN) method which is distinguished for being easy to learn, apply and it needs short training time. Results on the reliability, validity, and usability are included too, as well as an example of how the intervention process can be used in workplaces.

Chapter 2 entitled "Ergonomic Bench to Decrease Postural Risk Level on the Task of Changing Forklift's Brake Pads: A Design Approach" is proposed by authors Arturo Realyvásquez, Guadalupe Hernández-Escobedo from Instituto Tecnológico de Tijuana (Mexico) and Aide Aracely Maldonado Macías from Autonomous University of Ciudad Juárez (Mexico). This chapter addresses the evaluation of the postural risk level associated to the task of changing brake pads of a forklift and proposes the design and the implementation of an ergonomic bench that helps to reduce this level of postural risk. As methods, postural risk level corresponding to the task was determined using REBA, whereas anthropometric analysis was developed, and axiomatic design and TRIZ methodologies were used to develop the proposed design

Chapter 3 named "Modeling Simulation-Based Agents for Reducing Operative Deferral in Oncology in a Virtual Scenario Using Ergonomics" is proposed by authors Adriana Garcia from Universidad Politécnica de Aguascalientes (Mexico), Alberto Ochoa-Zezzatti and Juan Luis Hernández Arellano from Autonomous University of Ciudad Juárez (Mexico). This chapter analyzes the behavior of the last one to determine its relationship with the elective surgery performance and how it increases the surgical deferral. For purposes of the desired analysis is necessary to use one of the analyses of complex system techniques, in this case, the modeling and simulation agent-based.

Section 2: Ergonomic Product Design

The second section of this book is entitled "Ergonomic Product Design" and comprises six chapters described as follows:

Chapter 4 named "Aging Suit: An Accessible and Low-Cost Design Tool for the Gerontodesign" is proposed by author Annika Maya Rivero from Universidad Autónoma del Estado de México (Mexico). This chapter addresses the design, developing and testing of a low-cost suit focused on the design students and professionals located in Latin-America to design products centered on the physical capacities of the elder population the aging suit has been used as a tool for the gerontodesign. In countries like Mexico, its application is limited by the manufacturing cost.

Chapter 5 is named "Product and Space Analysis Using the Product Perception Method" and proposed by authors Mario Osmar Moreno and Juan Luis Hernandez from Autonomous University of Ciudad Juárez (Mexico). This chapter addresses the identification and analysis of the design requirements of a chair for elderly people. After the study, it was determined the necessity of designing an ergonomic chair that allows performing activities of personal hygiene in the shower, the requirements for the designing of the product were identified.

Chapter 6 is named "Pedagogy as a Fundamental Part of School Furniture Design" proposed by author Natalia Anaya Echeverría of the Universidad de Guadalajara (Mexico). This chapter addresses the exploration of the physical and functional characteristics required by the design of school furniture to be a facilitator of pedagogical and didactic practices to make a proposal for furniture design that meets both the pedagogical demands and the anthropometric variability of schoolchildren.

Chapter 7 named "Experimental Design to Analyze a Novel Stabilization Design of a Three-Wheel Vehicle" is proposed by authors Luis Elias Davila-Leyva and Ludovico Omar Soto Nogueira from Autonomous University of Ciudad Juárez (Mexico). This chapter addresses the analysis of a novel design to stabilize a three-wheel vehicle capable to be used without hands for people with disabilities or for simple recreation. The conclusion after the analysis resulted that the function of the mechanisms was optimal.

Chapter 8 named "Design of Recreational Vehicles for Young and Adult People as an Alternative to Physical Activation in Open Spaces" is proposed by authors Mayra Ivette Peña, Cesar Omar Balderrama Armendáriz and David Cortés Sáenz from Autonomous University of Ciudad Juárez (Mexico). This chapter addresses the analysis of the problems of sedentary lifestyle to propose a user-focused alternative that meets the necessary requirements for healthy energy consumption.

Chapter 9 named "Trials and Simulators Research in User Analysis and Sustainable Packaging Development" is proposed by authors Berthana Ma Salas Dominguez and Silvia Ana María Oropeza Herrera from Universidad Autónoma Metropolitana, Xochimilco (Mexico). This chapter addresses the incorporation of ergonomics, user habits analysis, trials and simulators in the design process and making users aware of the need of taking environmental actions can lead to a packaging design that promotes actions for a more sustainable world. The development of research methods in this area will lead to better design products.

Section 3: Anthropometrics

The third section is titled "Anthropometrics" and it is integrated by three chapters described as follows:

Chapter 10 named "Anthropometry Devices: A Comparative Study" is proposed by authors Julian Israel Aguilar-Duque, Guillermo Amaya, Victor M Juarez, Diego Tlapa, Ulises J. Tamayo from Universidad Autónoma de Baja California (Mexico) and Ana Y. Tovar-Hernández from the Autonomous University of Ciudad Juárez (Mexico). This chapter presents a comparative research of different anthropometric equipment designs with patent registration, comparing analogic devices versus digital devices, with the aim to define the relevance and status of the anthropometric devices.

Chapter 11 named "An Ergonomic Analysis on Working Postures of Construction Site Workers: A Framework for Construction Site Workers" is proposed by the author Suchismita Satapathy from KIIT University (India). This chapter presents an investigation to rank the maximum body movement range and body parts for different types of the workflow system. It may also help to develop policies for occupational safety and hazards in the workplace.

Chapter 12 named "Body Scanner Measurements for Apparel Design in Mexican Women" is proposed by authors Lilia Roselia Prado-León Carlos Aceves-Gonzalez from Universidad de Guadalajara (Mexico). This chapter aims to report data on body dimensions and proportions for apparel design in Mexican women and to analyze changes in dimensions and proportions among groups according to the age. Measurements of 1131 female subjects aged from 15 to 65 years from the Metropolitan Zone of Guadalajara, Mexico were obtained using the TC2 3D Body Scanning System. Mean, Standard Deviation, and percentiles 5, 50, and 95 were calculated for 28 dimensions of six age groups. Body proportion was calculated for 26 dimensions with respect to stature, using mean data of those groups.

Section 4: Usability and User Experience

The fourth section is related with "Usability and User Experience" and comprises six chapters described as follows:

Chapter 13 named "Usability on Standard Work Visuals in Manufacturing" is proposed by authors Guadalupe Hernández-Escobedo from Instituto Tecnologico de Tijuana (Mexico), Karina Cecilia Arredondo-Soto from Universidad Autónoma de Baja California (Mexico), Carlos Aceves-Gonzalez from Universidad de Guadalajara (Mexico) and Arturo Realyvásquez from Instituto Tecnologico de Tijuana (Mexico). This chapter presents a proposed checklist to determine the usability levels of standard work visuals (SWVs) used in manufacturing processes. The checklist considers attributes in terms of easy to use and methods facilitating their design. It is because there were not enough studies, which address both attributes of use and methods to design visuals.

Chapter 14 named "Usability Test and Cognitive Analyses During the Task of Using Wireless Earphones" is proposed by authors Nora G. Bustamante, Adrian A. Durán, Aide Aracely Maldonado Macías, Juan Carlos Ortiz Nicolás from the Autonomous University of Ciudad Juárez (Mexico) and Andres R. Quiñones from University of Texas at El Paso (USA). This chapter developed a usability test, assess the mental workload, and identify potential human errors during the process of listening to music using wireless earphones from a mobile phone device to detect which operations were the hardest in order to suggest modifications to the design and to bring a better experience to the user. During the usability test, some difficulties among users were detected and a Hierarchical Task Analysis (HTA) was developed.

Chapter 15 named "User Characteristics and Ergonomic Properties for Daily Objects Design" is proposed by authors John A. Rey-Galindo, Elvia Luz González-Muñoz and Alicia Libertad Rizo-Corona from Universidad de Guadalajara (Mexico). This chapter addresses the analysis of different objects of daily use, specifically four models of lemon squeezers, three models of can openers and four models of clothespins. Such analysis is made focusing on the strength needed to complete the task and the dimensional relation between the hands and object grips, that impact both in ease of use and user well-being itself, identifying the gap between young and older adult's characteristics and capabilities and the features of the object.

Chapter 16 entitled "The Comprehension of Figurative Images of Food Items: The Effect of Ergonomic Guidelines in Graphic Design" is proposed by authors Lilia Roselia Prado-León, Carlos Diaz de Leon Zuloaga, and Adrian Antonio Cisneros Hernandez from Universidad de Guadalajara (Mexico). This chapter addresses the assessing, within four stages, the comprehension of the figurative food illustrations and the effectiveness of ergonomic guidelines when used in the design process of these illustrations. The results showed that the familiarity and comprehension of some foods are limited due to the nutritional or cultural habits of the users. When comparing the graphic styles that use or lack of ergonomic guidelines, it is clear that the use of these guidelines helps the user to comprehend and identify better the graphical information presented.

Chapter 17 named "Understanding the Context of Design for Social Innovations: A Methodological Case Study" is proposed by authors Juan Carlos Ortiz Nicolás from the Autonomous University of Ciudad Juárez (Mexico) and Thomas Harrison from Imperial College London (UK). This chapter describes the application of designers to the space of finding solutions to social issues through the generation of social innovations. Based on a literature review, it was evident that design requires new methods to apply to the resolution of social issues and this chapter presents a new method that supports designers in the development of solutions within the Design for Social Innovation framework.

Chapter 18 named "A Creative Teaching Strategy to Generate Concept Designs and Their Possible Application in Tourism" is proposed by authors Juan Manuel Madrid Solórzano, Ercilia Loera Anchondo, Porfirio Peinado Coronado, Ludovico Soto Nogueira from the Autonomous University of Ciudad Juárez (Mexico). This chapter presents the activities and results from a didactic strategy developed to generate concept designs that can be perceived as possible objects to be manufactured. Similarly, this document analyzes the importance of creating concept designs for promoting touristic destinations through a logo.

Section 5: Additive Manufacturing and New Technologies

The fifth and last section of this book is titled "Additive Manufacturing and Technology." It is integrated by three chapters described as follows:

Chapter 19 named "Methodology to Apply Design for Remanufacturing in Product Development" is proposed by authors Karina Cecilia Arredondo-Soto from Universidad Autónoma de Baja California (Mexico), Rosa María Reyes-Martínez, Jorge De la Riva-Rodríguez, Humberto Hijar-Rivera, and Jaime Sánchez-Leal from Instituto Tecnológico de Ciudad Juárez (Mexico). This chapter discusses the definitions of Environmental Design, Design for Remanufacturing (DfR), as well as a case study to demonstrate the convergent point between these topics. The case study analyzes a product that was not originally designed as re-manufacturable, the decision was made to develop a reconstruction process that fulfilled the characteristics of remanufacturing.

Chapter 20 named "Selection of Prototyping Process and Part Orientation for Virtually Manufactured Gears" is proposed by authors Divya Zindani from NIT Silchar (India) and Kaushik Kumar from Birla Institute of Technology (India). This chapter addresses the transformation of a virtually produced gear to a physical product using rapid prototyping technique. The same can be segregated into two parts.

Chapter 21 named "Thermal Power Sector Sustainability: A Framework for Sustainable Supply Chain Management" is proposed by author Suchismita Satapathy from KIIT University (India). This chapter addresses the evaluation of the Sustainable Supply Chain Management of Indian thermal power sector. Artificial Neural Network (ANN) method is implemented to measure whether the benefits of sustainable supply chain management are achieved after practices of sustainable supply chain management in Indian thermal power sector.

TARGET AUDIENCE

The target audience of this book includes several professionals such as product engineers, industrial designers, researchers, industry practitioners, research scientists, academics involved in product design, product engineers, industrial and manufacturing engineers, graduate, and postgraduate students, among others. Also, the book could be used and consulted for any person with interest in ergonomics and product design.

REFERENCES

Apple. (2017). *AirPods*. Retrieved from https://www.apple.com/mx/shop/product/MMEF2BE/A/airpods

Gómez-Bull, K. G., Hernández-Arellano, J. L., & Ibarra-Mejía, G. (2015). A Proposed Methodology for Task Analysis in Ergonomic Evaluations. *Procedia Manufacturing*, *3*, 4756–4760. doi:10.1016/j.promfg.2015.07.573

Hernandez-Arellano, J. L., Talavera-Aguirre, G., Serratos-Perez, J. N., Maldonado-Macias, A. A., & Garcia-Alcaraz, J. L. (2016). Anthropometrics of University Students in Northern Mexico. *Open Journal of Safety Science and Technology*, *6*(4), 143–155. doi:10.4236/ojsst.2016.64011

IEA. (2017). *What Is Ergonomics?* Retrieved from http://iea.cc/whats/index.html

Stanton & Baber. (2002). Error by Design: Methods for Predicting Device Usability. *Design Studies, 23*(4), 363–84.

Acknowledgment

Our deepest acknowledgment to the researchers, professors, and students that sent their contributions to this project. At the beginning, we never thought to have the answer we have. Now, this book has 21 chapters from four countries (Colombia, India, United States, and Mexico) and from 14 universities. Thank you for trusting in this project!

This book would not have been possible without the support of Alma Lorelei, Jessica Pasaret, Mariana Salas, Pablo Muñoz, and Ariadna Vazquez. All of them students at the Universidad Autónoma de Ciudad Juárez, Mexico. Thank you for your effort and time during this project.

Similarly, editors would like to express special gratitude to IGI Global Publishing and associate editors for the opportunity to edit this book and for providing valuable comments to improve the selection of research works.

Section 1
Ergonomic Analysis

Chapter 1
Individual Risk Assessment (ERIN):
Method for the Assessment of Workplace Risks for Work-Related Musculoskeletal Disorders

Yordán Rodríguez
University of Antioquia, Colombia

ABSTRACT

Work-related musculoskeletal disorders (WMSDs) prevention has become a global phenomenon and nowadays is one of the main challenges that ergonomics and work safety professionals face. The evaluation of risk factors exposition related to the WMSDs is one of the main activities that are performed for prevention. In order to accomplish it, many ergonomic assessment methods have been developed. In this chapter, an ergonomic observational individual risk assessment (ERIN) method is presented, which is distinguished for being easy to learn, apply, and it needs short training time. Results on the reliability, validity, and usability studies of ERIN are included too, as well as an example of how it can be used in the workplace intervention process.

INTRODUCTION

Work-Related Musculoskeletal Disorders (WMSDs) prevention has become one of the main focus to the Ergonomics and Occupational Safety and Health fields. It is accepted and acknowledged that these diseases emergence is due to the existence of workstations/tasks that do not meet ergonomic principles, exposing workers to risk factors related to these diseases (NIOSH, 1997). On consequence, from a few years ago, many professionals and researchers have been interested in refining the evaluation of the exposure to risk factors related to the WMSDs, resulting in the creation of new methods/tools and the update of some existent ones (Takala et al., 2010). These can be grouped into three fundamental categories: direct measurement, self-report and observational (S. Bao, Howard, Spielholz, & Silverstein, 2007; David, 2005; ISO 11228-3, 2007; Li & Buckle, 1999; Neumann, 2006).

DOI: 10.4018/978-1-5225-5234-5.ch001

Copyright © 2018, IGI Global. Copying or distributing in print or electronic forms without written permission of IGI Global is prohibited.

Direct measurement methods usually use sensors placed over the individual's body (electrogoniometers, inclinometers, lumbar motion monitor, among others), and they offer a high accuracy on the measures if used correctly (Stephen Bao, Howard, Spielholz, Silverstein, & Polissar, 2009; N. Corlett, 1995; David, 2005; Li & Buckle, 1999). They are used for simulated tasks study and allow registering body coordinates in real time with high precision. As the main disadvantage, it can be mentioned that the sensors placed on the subject's body can cause discomfort and changes in its behavior, it requires time for the analysis and data interpretation, instrument calibration and workgroups with high qualification and experience. Due to technological limitations, such as the number of entry channels available on an instrument and its space for information storage, these methods are usually limited to evaluate a reduced number of people and body parts (Stephen Bao et al., 2009). Furthermore, the main disadvantage is that the related costs to direct measurement are usually very high, which makes them unpractical in large epidemiological studies on WMSDs or for most ergonomics internships.

Other advanced measure techniques have been developed for posture variation analysis on highly dynamic activities. For this, information is registered as from filming the individual on real time for a representative work time lapse, registering relevant information for its later analysis on specific programs. The analysis can include the use of biomechanical models that use anthropometric and posture information and manipulated load to calculate the momentum and forces between the body segments. The complexity of these models varies from static dimensions to dynamic tridimensional. These techniques are very expensive, require of highly qualified personnel, and are mostly used for workstation and objects design (cars, machinery, plane cabins, etc.), they do not have a practical usage for evaluating existing workstations and are better for the analysis of simulated tasks (Bubb et al., 2006; Chang & Wang, 2007).

Self-report methods are used to collect data on the exposure to physical and psychosocial factor on the workstation, these include work diaries, interviews, and questionnaires (Neville Anthony Stanton, Hedge, Brookhuis, Salas, & Hendrick, 2004), in which the usage of human body maps and varied punctuation scale are frequently used (E. N. Corlett & Bishop, 1976; Sinclair, 1995; Straker, 1999). These methods have the apparent advantage of easy usage, applicable to a broad spectrum of work situations and allowance to show large samples at relatively low costs. As the main disadvantage, it can be mentioned the necessity of large amounts of samples to achieve representativity and that the information obtained through the worker's perception is generally imprecise and unreliable (David, 2005), this approach of self-report has proven to have low validity and reliability levels related to the ergonomic interventions necessities (Li & Buckle, 1999). Nonetheless, ergonomics practitioners frequently use these types of methods on workplace assessments, given the apparent mentioned advantages.

On the other hand, observational methods are of practical use on a broad range of workstations, allow the study of large samples at relatively low costs and their usage usually does not require interrupting work. As main disadvantages, it could be mentioned that score systems are hypothetical, require knowledge and experience from the observer and the possibility of its bias when the evaluation is being executed. Observational methods have been considered a practical tool with a reasonable reliability among epidemiological investigations on WMSDs, for which they have gained popularity amongst safety and health professionals, since they respond better than direct methods and the self-report ones, which usually require time and limited resources and are in need of guides and intervention priority criteria settling (David, 2005; Takala et al., 2010).

So far, observational methods have had the biggest acceptance among ergonomists, despite having been mostly created for experts to use them (Li & Buckle, 1999; Takala et al., 2010). Furthermore, it is acknowledged that, for the correct application of these methods, specific ergonomics knowledge, experience, and an adequate training are required (Takala et al., 2010); these reasons have restrained its massive usage on the evaluation of workstations/tasks as part of WMSDs prevention.

For example, in Latin-American countries, the amount of experts on the usage of these methods is very limited. However, each time, normative, laws, and government institutions demand and encourage the perform ergonomic assessments of workstations/tasks. From this point of view, it would be worth to ask: How correctly are the ergonomic evaluations being done with the observational methods?

The results from a recent study, which involved 290 professionals from 20 different countries, with a notable representation of Latin-American countries, quantified and confirmed the suspicions that the author of this chapter had about the correct use of these methods. 30% out of the 442 reviewed evaluations presented mistakes (Diego-Mas, Alcaide-Marzal, & Poveda-Bautista, 2017). This study shows a clear alert: We have to take several actions to improve the workplace ergonomic evaluation process, in order to increase the efficiency of occupational WMSDs prevention.

In this case, on an attempt to contribute to attenuate the previous outlook, this chapter's author along with other investigators, starts many years ago the development of an observational method which took into account the necessities, limitations and capacities of non-experts, attempting to answer the workstations ergonomic evaluations demands on real contexts, whilst maintaining acceptable levels of reliability and validity. As a result, the Individual Risk Evaluation (ERIN) method was obtained (Rodríguez, 2001; Rodríguez, Viña & Montero, 2013a, 2013b), which is presented in this chapter, along with reliability, validity and usability method results, as well as a procedure on how the ERIN method can be used during the evaluation and intervention of workstations process, illustrated through an example.

INDIVIDUAL RISK ASSESSMENT: ERIN

Background

ERIN is an observational easy-to-use method, developed so non-expert and low-trained personal can evaluate individuals exposed to risks associated with WMSDs. It was conceived through existent methods, the epidemiologic evidence on WMSDs and the practical needs of non-expert users on the evaluation of WMSDs risk factors exposure on workstations (Rodríguez, 2011; Rodríguez et al., 2013a, 2013b).

ERIN evaluates the trunk, arm, wrist, and neck's posture and their movement frequency; the rhythm, given by the work's speed and the effective duration of the task; the intensity of the effort, result of the perceived effort by the evaluator and its frequency, and the self-evaluation/perception of the referred stress on the subject about the task it is performing (Appendix). On the method, many actions and risk levels are recommended according to the total risk (Table 1), which is calculated summing the seven evaluated variables risks. The used additive model allows to easily identify the influence of each factor and to locate which elements must be intervened to lower the risk level.

The brief time dedicated to the process evaluated with ERIN allows studying a large amount of workstations/tasks on different economy fields, at relatively low costs and without interrupting work. Its application only requires a sheet and pencil, although the previous training and knowledge of the observer will influence its effectiveness.

ERIN can be used to evaluate, design, and redesign static and dynamic tasks. Its score system allows establishing low criteria to evaluate the impact of the changes that have been made (before and after), assuming that low scores correspond to more favorable conditions.

Table 1. Risk levels and recommended action according to the total ERIN risk

Zone	Total Risk	Risk Level	Recommended Action
Green	6-14	Low	No changes are required
Yellow	15-24	Medium	Further investigations are needed and changes may be required
Orange	25-34	High	Investigations and changes are required soon
Red	≥ 35	Very high	Investigations and changes are required immediately

The ERIN results can guide the personal dedication to the management and prevention of occupational risks on which changes must be made and towards where. Once the changes are made, the same approach can be used to quickly show to what extent the risks have lowered and the work conditions have improved.

Development Process of ERIN

The ERIN method was developed in three stages: (1) Prototype building, (2) Improvement of the prototype, and (3) Conformation of the ERIN method (see Figure 1).

On the first stage a prototype of the method is made and tested, and the aspects that should be changed and improved, parting from the obtained results from the experiments and the evaluator's feedback are identified. The second stage was about improving the identified aspects of the prototype regarding the variable, score system, field sheet design and the procedure. This improved prototype was tested on the field and analyzed with the purpose of establishing the elements to be improved. During the third and last stage, the ERIN method conformation is made as a result of the changes done to the improved prototype and the addition of graphic design elements. The main results of the reliability, validity and usability evaluation that were made are presented at the end of the chapter.

ERIN Guideline of Application

In Figure 2, the ERIN method application process is shown.

Figure 1. ERIN method development process diagram
Source: Taken from (Rodríguez, 2011).

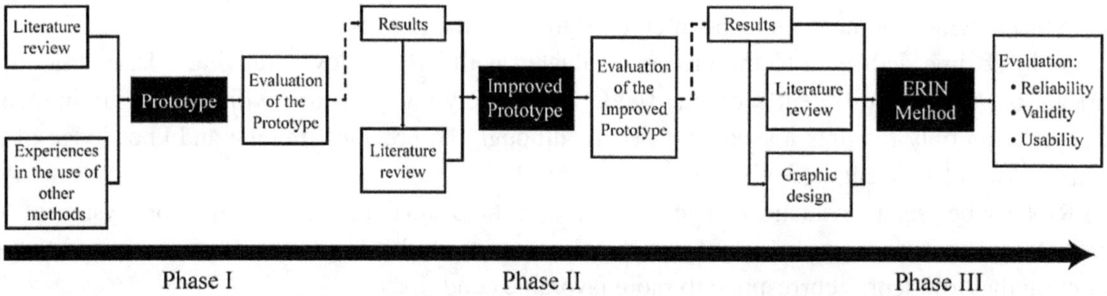

Figure 2. ERIN method process application

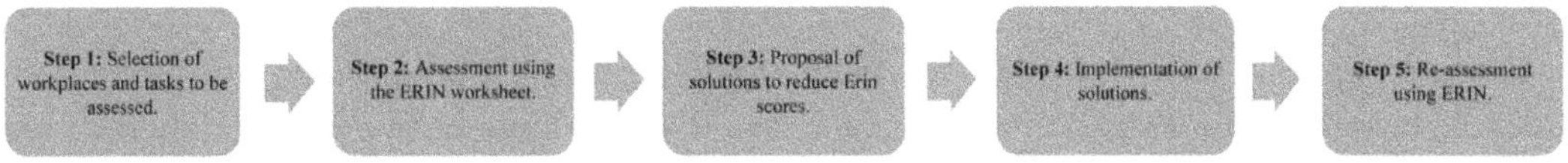

Step 1: Select the Workstations/Tasks That Will Be Evaluated Within the Company

Although ERIN was conceived to perform massive evaluations on the company, it is very convenient to establish a priority order. For the selection of the workstations to be evaluated, it must be considered the WMSDs statistics, the musculoskeletal symptoms, the worker's opinions, area chiefs, supervisors, and workstation impact on the productive system, among others. The experience and knowledge of the person conducting the research is a factor that favors a correct selection.

Step 2: Evaluate According to the ERIN Field Sheet

In this step, the workstations/tasks of greater risk and that need to be intervened to reduce the WMSDs risk factors are identified. In this case, ERIN is a tool that has been used to set priorities. If during the evaluation non-considered ERIN risk factors are identified (e.g. Vibration, awkward posture of lower extremities), these must be registered for their later study. In brief, the steps within the beginning of the ERIN field sheet (Appendix).

1. Observe the worker and select the critical posture for each body part evaluated (help yourself with text and figures).
2. Add an adjustment in case it corresponds to obtain the postural load.
3. Determine the variable risk given by the postural load and the body part movement; write it down in the corresponding box.
4. Determine the risk value for rhythm, effort and self-evaluation variables according to each table; write it down in the corresponding box.
5. Sum the risk values to obtain the Total Risk.
6. Determine the corresponding Risk Level.

Step 3: Propound Solutions Aimed to Reduce the ERIN Scores

The proposals (engineer, organizational and behavioral) that are performed are based on the evaluation obtained with ERIN, looking to reduce the given scored, at least on the most critical variables. The proposals must be accompanied by the before and after ERIN scores. It is possible that the changes in the variables scores do not lead to reduce the risk level. Nonetheless, a movement focused on improving the work conditions from an ergonomic point of view is created. It is highly recommended that the workers participate in the proposals and that these are certificated by a cost-benefit analysis.

Step 4: Implement the Solutions

Within the intervention process, this step has been the most complicated, for the solutions are often not implemented because of the lack of needed resources, or they have not been appropriately justified since the personnel sometimes use the ERIN method but do not possess the right education to justify the proposals in business terms. Fortunately, many times, the cost of the proposals can be ignored, or do not imply any cost at all, which undoubtedly has helped its implementation (Rodríguez & Pérez, 2013a, 2013b).

Step 5: Reevaluate Using ERIN

Once the workstation/task has been modified, it is reevaluated with ERIN. This will allow proving if the changes made were effective in the reduction of WMSDs risk factors exposition.

Annotations for the Correct Use of the ERIN Method

- Sometimes the work is executed with great speed, making difficult to observe some postures. In this case, the observer can ask the worker to simulate the posture for this can be observed in detail.
- In the given case that the body parts evaluated with ERIN (trunk, arm, wrist, and neck) maintain a low variability posture in time, it is recommended to evaluate the posture that is repeated the most for each body part and compare the results with critical postures.
- To evaluate the intensity's effort, two approaches are proposed. The first one is directed to ask the worker/s about the required effort to perform the task one time. This option is highly recommended when the personnel that performs the evaluation possess very to little experience. The second approach is to use the evaluator's criteria. The efficacy of the delegated perception of the effort, performed by non-expert personal has been reported (Dury et al., 2006). Nevertheless, given the high subjectivity required to evaluate this variable, it is recommended to integrate both approaches.
- In case that the task requires performing many types of effort, to sort this aspect out, the highest one performed must be selected and quantified to obtain the effort's intensity variable risk. Since the average's estimation of the effort can be though for non-experts and high force peaks could remain hidden.
- The movement frequency for each body part evaluated must be determined by the number of times said part moves and not by the times the critical posture repeats. This must be estimated by the time the performed work is done with the most intensity.
- The recording of the evaluated tasks is generally suggested, especially for the newcomers or low-experienced in these types of ergonomic evaluations. The use of this technique facilitates the estimation of postures and movement frequency, for the freezing of images and low motion camera use is very easy in video players nowadays. Besides, it allows for the evaluation to be performed at some other time and group. However, the advantages that the recording technique provides must be used carefully to avoid problems like the lack of image stability, insufficient time and cycles, poor workstation overview, frequent zoom-in's and out's, view-blocking objects that difficult the observation, and filming from positions that make the observation of body angles difficult (Cochran, Stentz, Stonecipher, & Hallbeck, 1999).

- It is possible that, when evaluating arm and wrists, it will not be evident which side (right or left) is the most affected, in this case, it is recommended to evaluate both sides and chose the one with more postural load, according to ERIN. This meaning that the posture is the aspect that determines which body part is the critical one, regardless of the other body part performing movements with more frequency.
- Sometimes, many people work at a workstation and it is necessary to assign a risk level to the workstation. In this case, it is recommended to select the trend for each of the evaluated variable. For example, let's suppose that seven workers that perform the same task on an assembly line are evaluated. When analyzing the trunk variable posture, two of them adopt a critical posture of 0-20º, for 20-60º and one above 60º. In this case, the critical trunk posture would be determined by the modal value (20-60º). Similarly, this would apply for the rest of the variables, forming the workstation score.
- Frequently, a worker must execute multiple tasks in one work shift. In these cases, the worker's risk is calculated using the following equation:

$$Multi\ Task\ ERIN = \sum TR_i \times P_i$$

where:

Multi-Task ERIN: Estimated risk for when a worker performs multiple tasks during a work shift.
TR_i: Total i task risk according to ERIN.
P_i: Proportion of the i task effective time, regarding the total tasks time.

If all of the evaluated tasks belong to the same workstation, the calculated risk corresponds to the workstation risk.

In Table 2, it is explained how to perform the multi-task analysis with ERIN.

EVALUATION AND INTERVENTION OF A WORKSTATION USING ERIN METHOD

The following example shows how can the ERIN method be used on the evaluation and improvement of a workstation, following the procedure described previously.

Table 2. Example of the multi-task analysis with ERIN

Task	ERIN Total Risk	Effective Task Duration (Hours)	Total Time Proportion	Equivalent Task Risk
A	25	4	4/8 = 0.5	25 x 0.5=12.5
B	30	2	2/8=0.25	30 x 0.25=7.5
C	28	2	2/8=0.25	28 x 0.25=7
Total		**8**	**1**	**27**

Selection and Description of the Box-Washing Task (Step 1)

On the workstation, the plastic boxes from the production area of a food product company are gathered. The task is cyclic and consists of the following activities: 1) move a dirty boxes pile to the washing machine, 2) insert three boxes through the washing machine entrance slit, taking one at a time from the pile and inserting them likewise, 3) gather the three clean boxes that come out of the exit canal from below the washing machine, and pile them, one at a time; lastly, 4) move the clean boxes pile to a storage located a few meters away to reincorporate them to the company's productive process. The floor will be permanently wet, for which the pile movement will be accomplished by sliding them. The plastic boxes have an approximate weight of 2.3 kg. Approximately 3 cycles per minute are performed, which is equivalent to 9 boxes per minute.

Evaluation Using the ERIN Field Sheet (Step 2)

Figure 3 shows the evaluation of the plastic boxes cleaning tasks delimited according to the ERIN field sheet. As it is explained at the beginning of the field sheet, the next step is to select the critical posture of the trunk, arm, wrist and neck adopted by the worker during the task cycle and the movement frequency of each body part. On the wrist and arm's case, the observer must choose which side, left or right, represents the highest postural risk. In case of any doubts, both sides are evaluated and the highest risk, according to ERIN, is chosen.

On Figure 4 it can be observed that the trunk's critical posture occurs the moment the worker has to reach the clean boxes coming out of the washing machine. The trunk is severely flexed (over 60º), which belongs to a score of (3), and it is also rotated, meaning an adjustment of (+1). Therefore, the trunk's postural load is 3+1=4. The trunk's movement during the task is "very frequent", more than 10 times/min. obtaining a score of 9 for the *Trunk*.

Figure 5 shows that the shoulder/arm critical posture occurs when the worker takes the dirty boxes placed on the piles upper side to introduce them in the washing machine, being the right arm the highest riskier. A similar situation is presented when the worker places the clean boxes on the upper side of the pile. The arm is severely flexed (over 90º), which belongs to a score of (3) and it is abducted, meaning an adjustment of (+1). Thus, the arm's postural load is 3+1=4. The arm's movement during the task is "very frequent" (almost a continuous movement), obtaining a score of 9 for the *Arm*.

Figure 6 shows that the shoulder/arm critical posture occurs when the worker is inserting the dirty boxes in the washing machine. The right wrist is severely extended (over 45º) which belongs to a score of (2) and is deviated (+1), plus the hand is holding the boxes (static object press) for longer than the cycles 50% of the time (+1). It must be mentioned that there are other moments during the task execution where the wrist performs flexo/extensions higher than 45º. Therefore, the wrist's postural load is 2+1+1=4. The right wrist's movement is "very frequent", more than 20 times per minute, obtaining a score of 6 for the *Wrist*.

Figure 7 shows that the neck's critical posture occurs when the worker is reaching for the clean boxes coming out of the washing machine. The neck is severely extended (over 20º), which belong to a score of (2), plus it is rotated, meaning an adjustment of (+1). Therefore, the neck's postural load is 2+1=3. The neck's movement during the task execution is "constant", obtaining a score of 7 for the *Neck*.

To accomplish the assigned load during the workday, the worker executes the activities at a "normal work speed" and requires approximately 7 hours of effective work, obtaining a score of 3 for the *Rhythm*.

The intensity of the effort, given by the muscular effort required to execute the task one time (taking a box from the pile and putting it in the washing machine/ taking a box from the machine's output and pile it), is considered by the worker as "something hard" and has a frequency higher than 10 times per minute, obtaining a score of 6 for *Effort's intensity*.

On this workstation, only one worker labors and he has perceived the task as "little stressing", which belongs to a score of (1) for *Self-evaluation*. Finally, the variable risks scores are added and a total risk of 41 is obtained, corresponding to a "very high" level risk. This indicates that it is required to make immediate changes.

Improvement Proposals Based on the ERIN Evaluation (Step 3)

With the purpose of reducing the exposition to WMSDs related risk factors, parting from the ERIN evaluation, three improvement proposals are made.

Proposal 1: Redesign the actual washing machine. The redesign would imply to modify the clean boxes output' slit height, rising it to the same level of the input slit (Figure 8 and Figure 9). This modification would avoid that the worker flexes the trunk (over 60º) and the rotation/inclination during the clean boxes pick up. This proposal implies the hiring of another worker to pick up the clean boxes coming out of the output slit, which implies an additional cost. This way, the workload is divided into two workers and two sub-tasks would be created, where similar activities would be executed (same risk exposure).

Proposal 2: In case the washing machine modification is not possible (Proposal 1), then it is recommended to hire another worker to pick up the clean boxes coming out of the output slit, located on floor's level (actual situation), but on the contrary side to where they are introduced to the washing machine. On this case, two sub-tasks would be created, each one executed by a worker: A) input of the dirty boxes on the washing machine and B) pick up of the clean boxes. This way, the workload would be divided between two workers, making a rotation between sub-tasks possible.

Proposal 3: Consists of modifying the work method. Actually, the worker stashes around 30 boxes, which causes the pile's height to notably rise, causing the worker to labor over the shoulder's height (shoulder/arm flexion over 90º), therefore, it is recommended to work with piles of 15 boxes, reducing the height to half of it and the shoulder/arm's flexion angle to 45-90º- The implementing of this proposal does not require an additional cost.

Re-Evaluating With ERIN to Measure the Proposals' Impact (Step 4)

Table 3 shows the variable ERIN method risk for the actual situation, and the projected for the integrated proposals.

As for the trunk's frequency movement, both of the integrated proposals group would reduce it from very frequent (over 10 times/min.) to frequent (5-10 times/min.) for both sub-tasks. Given the approximate production rhythm of 9 boxes/min., just one worker would perform around 18 flexions/min., for two trunk flexions are performed for the box, one when the dirty box is taken from the pile to insert it and another when the clean box is picked up from the washing machine.

Figure 3. Evaluation of the box-washing task with ERIN method

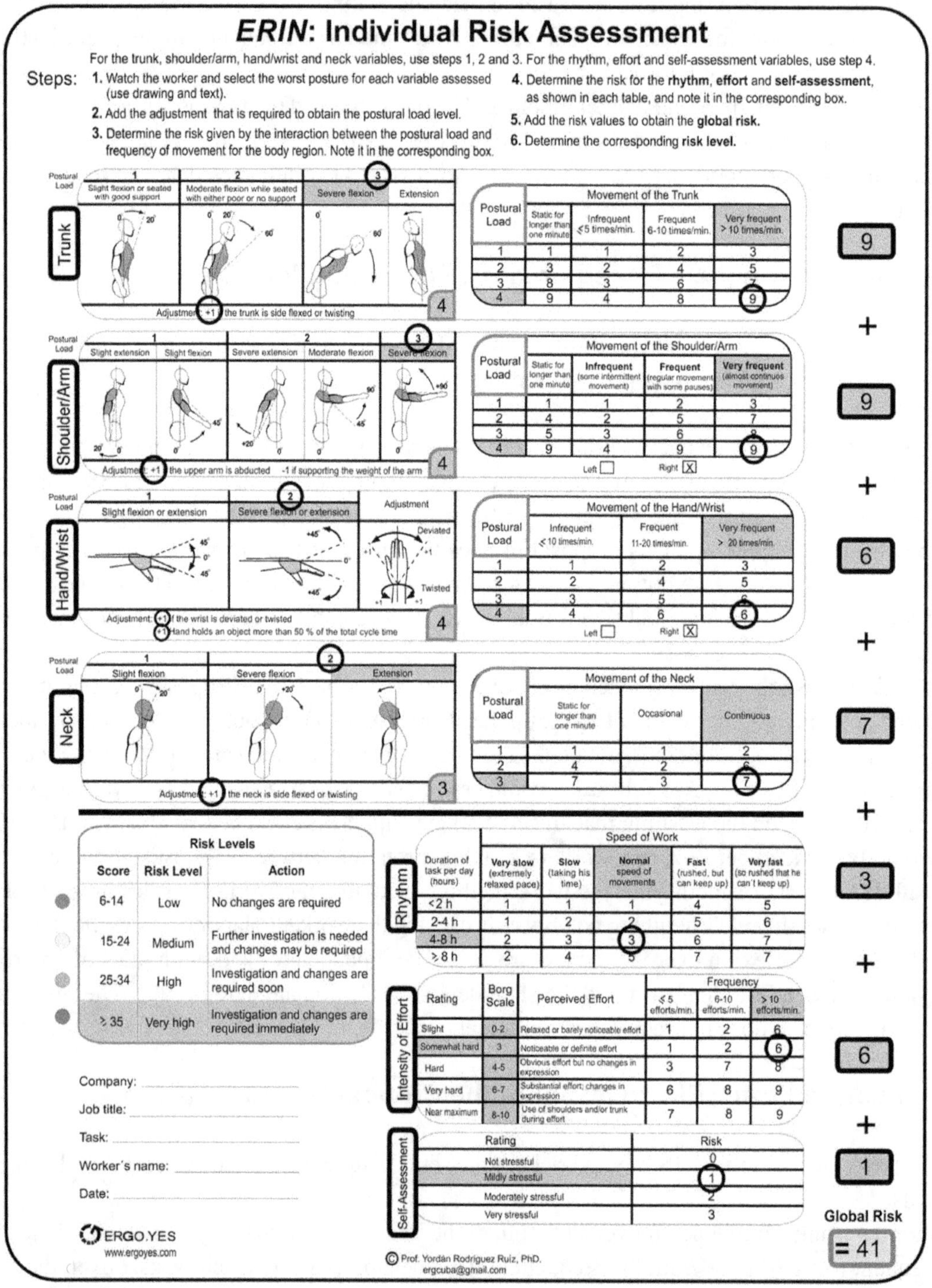

ERIN: Individual Risk Assessment

For the trunk, shoulder/arm, hand/wrist and neck variables, use steps 1, 2 and 3. For the rhythm, effort and self-assessment variables, use step 4.

Steps:

1. Watch the worker and select the worst posture for each variable assessed (use drawing and text).
2. Add the adjustment that is required to obtain the postural load level.
3. Determine the risk given by the interaction between the postural load and frequency of movement for the body region. Note it in the corresponding box.
4. Determine the risk for the **rhythm**, **effort** and **self-assessment**, as shown in each table, and note it in the corresponding box.
5. Add the risk values to obtain the **global risk**.
6. Determine the corresponding **risk level**.

Trunk — Postural Load: 1 Slight flexion or seated with good support; 2 Moderate flexion while seated with either poor or no support; 3 Severe flexion; Extension. Adjustment: +1 the trunk is side flexed or twisting. **4**

Postural Load	Static for longer than one minute	Infrequent ≤5 times/min.	Frequent 6-10 times/min.	Very frequent > 10 times/min.
1	1	1	2	3
2	3	2	4	5
3	8	3	6	7
4	9	4	8	9

Trunk risk: **9**

+

Shoulder/Arm — Postural Load: 1 Slight extension, Slight flexion; 2 Severe extension, Moderate flexion; 3 Severe flexion. Adjustment: +1 the upper arm is abducted; -1 if supporting the weight of the arm. **4**

Postural Load	Static for longer than one minute	Infrequent (some intermittent movement)	Frequent (regular movement with some pauses)	Very frequent (almost continuos movement)
1	1	1	2	3
2	4	2	5	7
3	5	3	6	8
4	9	4	9	9

Left ☐ Right ☒

Shoulder/Arm risk: **9**

+

Hand/Wrist — Postural Load: 1 Slight flexion or extension; 2 Severe flexion or extension. Adjustment: Deviated, Twisted. Adjustment: +1 if the wrist is deviated or twisted; +1 Hand holds an object more than 50 % of the total cycle time. **4**

Postural Load	Infrequent ≤ 10 times/min.	Frequent 11-20 times/min.	Very frequent > 20 times/min.
1	1	2	3
2	2	4	5
3	3	5	6
4	4	6	6

Left ☐ Right ☒

Hand/Wrist risk: **6**

+

Neck — Postural Load: 1 Slight flexion; 2 Severe flexion; Extension. Adjustment: +1 the neck is side flexed or twisting. **3**

Postural Load	Static for longer than one minute	Occasional	Continuous
1	1	1	2
2	4	2	6
3	7	3	7

Neck risk: **7**

+

Risk Levels

Score	Risk Level	Action
6-14	Low	No changes are required
15-24	Medium	Further investigation is needed and changes may be required
25-34	High	Investigation and changes are required soon
≥ 35	Very high	Investigation and changes are required immediately

Rhythm — Speed of Work

Duration of task per day (hours)	Very slow (extremely relaxed pace)	Slow (taking his time)	Normal speed of movements	Fast (rushed, but can keep up)	Very fast (so rushed that he can't keep up)
<2 h	1	1	1	4	5
2-4 h	1	2	2	5	6
4-8 h	2	3	3	6	7
≥ 8 h	2	4	5	7	7

Rhythm risk: **3**

+

Intensity of Effort

Rating	Borg Scale	Perceived Effort	Frequency ≤5 efforts/min.	6-10 efforts/min.	> 10 efforts/min.
Slight	0-2	Relaxed or barely noticeable effort	1	2	6
Somewhat hard	3	Noticeable or definite effort	1	2	6
Hard	4-5	Obvious effort but no changes in expression	3	7	8
Very hard	6-7	Substantial effort; changes in expression	6	8	9
Near maximum	8-10	Use of shoulders and/or trunk during effort	7	8	9

Intensity of Effort risk: **6**

+

Self-Assessment

Rating	Risk
Not stressful	0
Mildly stressful	1
Moderately stressful	2
Very stressful	3

Self-Assessment risk: **1**

Global Risk = 41

Company:

Job title:

Task:

Worker's name:

Date:

ERGO.YES www.ergoyes.com

© Prof. Yordán Rodríguez Ruiz, PhD. ergcuba@gmail.com

Figure 4. Trunk's critical posture on the box-washing task

Figure 5. Shoulder/arm's critical posture on the box-washing task

In the two integrated proposals groups, the division of tasks causes a reduction of effective time, reducing it from 7 hours to 3.5, which implies a reducing of the Rhythm variable score from 3 to 2 points, for the work speed, would still be "normal".

Proposal 3 solely affects the arm's postural load, but not its movement frequency. This reduces the arm's score from 9 to 7 points.

As Table 3 shows, the sub-task B risk is higher than sub-task A, for the severe flexion movements (over 60°), the trunk rotation, and the neck's extension and rotation remain at the time of the clean boxes pick up of the output slit on floor's level.

Figure 6. Wrist's critical posture on the box-washing task

Figure 7. Neck's critical posture on the box-washing task

In regard to the trunk's posture, it must be mentioned that, even though the proposals cut out the sever flexion (over 60º) and the trunk's rotation during the clean boxes pick up, the worker continues to perform trunk rotations over 60º during the pick up and stashing of the boxes from floor to pile. For said motive, the trunk's score only reduces from 9 to 6 points, meaning an aspect to improve with other intervention proposals.

Table 3. Proposals impact projection on ERIN method's scale

ERIN Method Variables	Actual Situation	Proposal 1 + Proposal 3	Proposal 2 + Proposal 3	
			Sub-task A	Sub-task B
Trunk	9	6	6	9
Arm	9	7	7	7
Wrist	6	6	6	6
Neck	7	6	6	7
Rhythm	3	2	2	2
Effort's intensity	6	2	2	2
Self-evaluation	1	1	1	1
Total risk	41	30	30	34
Risk level	**Very high**	**High**	**High**	**High**

Implementation and Tracing (Step 5)

It is left to the company's decision-makers' consideration which actions to implement. In these cases, the ergonomist or professional in charge must qualitatively and quantitatively offer the potential benefits that are generated on the health, security, productivity, and quality indicators. Involving from the beginning of the intervention process the different interested actors (ej. Workers and directives), facilitates the implementing process, for they feel involved in the proposed solutions.

Additional Commentaries

As it was previously showed (Table 3), despite the reduction of the total 41 points score (very high-risk level) to 30 points (high-risk level), the task remains a high risk, according to ERIN. This proves that the improvement process of the workstations/tasks is gradual, and how it gets difficult and the costs rise when the ergonomic principles are not taken into account from the design phase (Hendrick, 2008). Finally, it must be highlighted the importance of multiple measures integration, in order to reduce the exposition, for in many cases just one measure does not manage to positively impact on every aspect of the workstation/task to intervene, and thus reduce the exposition of WMSDs related risk factors.

RELIABILITY, VALIDITY, AND USABILITY

Many studies have been executed to evaluate the reliability, validity, and usability of the ERIN method. Following, the main results are described and presented (Rodríguez, 2011).

Figure 8. Washing machine redesign proposal

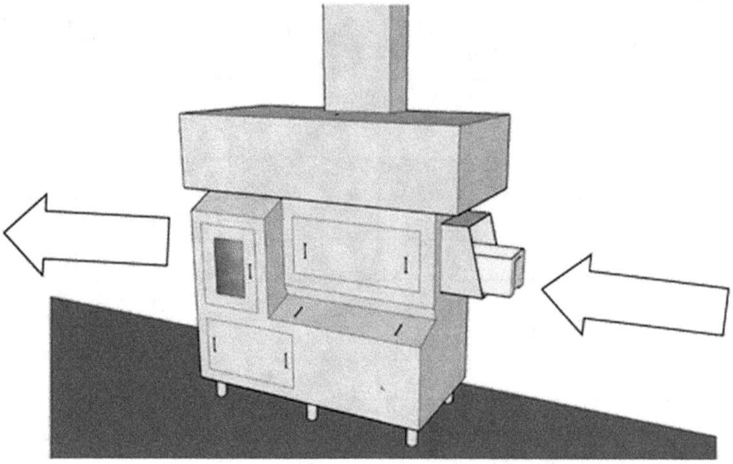

Figure 9. Washing machine redesign proposal

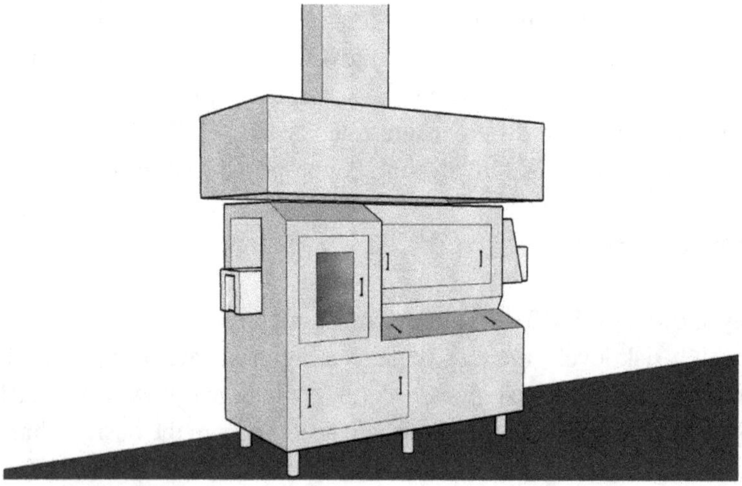

Inter-Observer Reliability

Study Description

A group of 42 subjects evaluated with videos and the ERIN's field sheet, seven tasks selected from the industry. In the selection, it was evaluated the variety of the tasks, and the representativity of different industry fields, filming quality, corporal angles visibility from different perspectives, standing, and seated tasks, and repressing of complete work cycles. The video projection was random, one at a time and the projection was repeated until each observer was done with the evaluation, in order to proceed to the next video. It must be mentioned that the evaluators could not stop the video. The participants were classified as non-experts, for they had not performed WMSDs risk evaluations before.

With the purpose of analyzing the training time influence, half of the evaluators received 3 training hours, whilst the other half received 6.

Methodology

For the dichotomic variables, the statistic ICC (KR-20) was used. For the remaining variable, which scale is ordinal, the statistic ICC (2.1) was used employing the individual measures and complete agreement (McGraw & Wong, 1996; Shrout & Fleiss, 1979). The results were interpreted as follows: if the ICC < 0140, the reliability is poor; if the ICC is between 0.40 and 0.75, the reliability is moderated to good; and if the ICC > 0.75 the reliability if excellent (Stephens, Vos, Stevens, & Steven Moore, 2006; Stevens, Vos, Stephens, & Steven Moore, 2004). Additionally, in the ordinal scale variable, it was used as an agreement measure the Kendall's concordance coefficient. This coefficient can vary from 0 (no association) to 1 (perfect association). The agreement levels measured with W are interpreted as follows: poor association $0.0 < W < 0.6$; moderate association $0.6 < W < 0.7$; good association $0.7 < W < 0.8$ and strong association $0.8 < W < 1.0$ (Lee & Ferreira, 2003).

Results and Analysis

It was obtained that the dichotomic variables (trunk adjustment, arm adjustment, wrist posture, neck adjustment and neck posture), presented the value of ICC (KR-20) over 0.75, indicating an excellent agreement among the evaluators. On the wrist adjustment evaluation (deviation or rotation) the agreement was poor according to the ICC (KR-20) = 0.081.

The agreement values varied more on the ten ordinal variables (trunk's posture and movement, arm's posture and movement, wrist's movement, neck's movement, work speed, effort, movement frequency and risk levels) than in the dichotomics.

In the trunk's posture evaluation, the agreement was excellent, according to the ICC (2.1) = 0.768, and the Kendall's coefficient of 0.807. In the variables evaluation: trunk's movement, arm posture, and effort, the agreement level is moderated to good since the ICC (2.1) values vary between 0.574 and 0.728, and the Kendall ones between 0.638 and 0.755. In the arm, wrist, and frequency movement, effort and work speed evaluation a poor agreement coefficient was obtained. The ICC (2.1) values vary from 0.088 to 0 0.298, and the Kendall's coefficient from 0.134 to 0.359.

In the total risk variables (obtained when summing the risk per variable) and the risk level (obtained when categorizing in one out of four levels the total risk value), the agreement was from moderate to good, with ICC (2.1) values of 0.465 and 0.471 respectively. When establishing two categories to evaluate the risk level: the first was a result of grouping up the risk levels "low" and "medium" in the category "low risk", and the second a result of grouping risk levels "high" and "very high" on the category "considerable risk", with the purpose of analyzing the agreement obtained when using the ERIN method to solely detect if there is a risk at all, it gives out that the ICC (KR-20) is 0.944, indicating an excellent reliability, and thus that the method is adequate for this purpose.

The agreement indexes were usually higher for the evaluator's group that received a higher training time, although in this study the differences were not big enough to give out definitive criteria on the influence of this factor. Nonetheless, the obtained results support the idea that if it is desired to raise the reliability, the training time must be raised as well (Takala et al., 2010).

Overall, the inter-observer reliability was moderated to good, since out of the 17 variables studied, 7 showed an excellent agreement, 5 moderate to good, and 5 poor. If in the variables where ICC (2.1), among which the five poorly agreed reported ones are, had used the average indexes, the evaluator's mistake impact would have been lower regarding the reliability indexes. As it was expressed, the average ICC's measures provide higher reliability indexes than individual indexes. A study that uses the average measures, must provide higher indexes than the calculated on this research (Shrout & Fleiss, 1979).

Lastly, it should be noted that if the evaluators had had the opportunity to stop the video or use the slow motion during the evaluations, as it is usually done nowadays, it is possible that higher agreement levels would have been reported in this study.

Intra-Observer Reliability

Study Description

A group of 17 subjects using the ERIN field sheet evaluated on two occasions a total of seven industry-selected tasks. The elapsed time between the first and second evaluation was three weeks.

Methodology

The statistics ICC (KR-20) and Cohen's Kappa were used for the dichotomic variables. The statistics ICC (2.1) and Spearman correlation coefficient were used for the ordinal scale variables. The interpretation of the ICC categories is the same as the explained for the inter-observer reliability analysis, and for the Cohen's Kappa values it is based on the Landis and Koch criteria cited by (Burt y Punnett, 1999), poor $K < 0.0$; light $0.0 < K < 0.2$; regular $0.2 < K < 0.4$; moderated $0.4 < K < 0.6$; substantial $0.6 < K < 0.8$, and almost perfect from $0.8 < K < 1$. The interpretation of the Spearman correlation values was: over 0.7 strong relation, between 0.3 and 0.7 a medium strength relation, and below 0.3 a weak relation (Li, Rosenthal et al, 1996).

Results and Analysis

It was obtained that for the dichotomic variables (trunk's adjustment, wrist posture, neck's adjustment, neck's posture) parting from the ICC (KR-20) and Cohen's Kappa statistics, showed a reliability level from moderate to good, and moderated, respectively.

On the wrist adjustment variable, the agreement was poor according to the statistical methods, coinciding on the obtained results on the inter-observer reliability study and other reported in the literature about the low-reliability levels on the smaller body areas evaluations, as is the wrist's case.

Out of the studied ordinal variables, three are classified as of excellent reliability, considering the ICC (2.1) values and the Spearman correlation coefficient, which varied between 0.786 and 0.833, and between 0.784 and 0.83, respectively. Five showed a moderated to good agreement, according to the ICC (2.1), with values between 0.416 and 0.676, and a medium according to the Spearman coefficient with values between 0.401 and 0.677.

It was only found a poor or low agreement on two variables. The first case was the wrist's movement, with a 0.256 value, according to the ICC (2.1), and 0.22, according to Spearman, coinciding with the expressed about the complexity of observation on this body part. In the case of the neck's movement, the agreement was poor, according to the Spearman values of 0.305, and an ICC (2.1) of 0.31.

Overall, the intra-observer reliability behaved moderated to good, for out of the 16 studied variables, 10 presented a moderate to good agreement, 3 excellent, and 3 poor.

Validity

Study Description

Study 1: The evaluation of seven tasks taken from videos by 42 observers, with three experts defined pattern were compared. These experts had the possibility of stopping the videos, freezing the images to observe the postures in detail, interchange criteria and reduce the speed of the film's projection. This procedure was established to reduce the present mistake on the observational processes.

Study 2: The evaluation of two workstations realized out of videos by 8 health field professionals (doctors and traumatology, physiology and rehab specialists), with three experts defined pattern were compared. It must be noted that these professionals had no experience on the usage of ergonomic evaluation methods, but they did on related subjects. The evaluators received an hour of training on the usage of the ERIN method.

Study 3: The obtained risk levels when evaluation 32 workstations with ERIN, RULA, and REBA methods were compared (Rodríguez, 2011). The RULA and REBA methods were selected among the existents because of the coincidences regarding the evaluated body zones, and ERIN related risk factors that propose a final risk level, result of the combination and consideration of the evaluated risk factors that have been used since many years ago on the workstations evaluation. Two observation strategies were employed for the evaluation: the critical posture, and the most common.

Methodology

Study 1 and Study 2: The total percent agreement was calculated, as well as the percentage with one and two or more different categories between the expert-given value, and the non-expert evaluators, for each of the 17 analyzed variables.

Study 3: Four levels were defined (low, medium, high, and very high). On the REBA method case, five risk level classification categories (Hignett & McAtamney, 2000), for which, in order to establish the comparison between the methods, categories 1 and two were grouped in the low level, remaining four categories as ERIN and RULA. To study the agreement level, according to the different methods, ICC (2.1) and percent agreement indexes were calculated.

Results and Analysis

Study 1: Out of the studied variables, 12 presented percent agreement above 60%. Out of these, five obtained values above 70%. Five variables presented percent agreement values lower than 60%, out of which four are related to the movement frequency estimation of the body parts (trunk, arm, wrist, and neck). These results correspond to the low-reliability indexes reported on these variables. Therefore, it is evident that the determination of the body parts movement frequency in real time is an aspect that the non-expert evaluators find difficult. The cases where the evaluations differed in more than one category were low, between 1 and 6%. Mostly coinciding with the lower total percent agreement variables previously referred.

Study 2: All of the variables, except the wrist and neck movement with percentages of 63%, presented total percent agreement above the 70%. Out of the 17 studied variables, 12 presented values above the 80%, and out of these, four were superior to 90%. No variable differed on more than one category. All of the above indicates that expert personnel with less than an hour of training is capable of offering validate and consolidate results when employing ERIN in workstations evaluation.

Study 3: The agreement between ERIN and RULA methods was, for the critical posture, of an ICC (2.1) = 0.403, and a percent agreement of 56%, as for the most common posture had an ICC (2.1) = 0.457, and a percent agreement of 56%. The agreement between the ERIN and REBA methods for the critical posture was of an ICC (2.1) = 0.554, and a percent agreement of 66%, as for the most common posture had an ICC (2.1) = 0.556, and a percent agreement of 59%. It was proposed that, regardless of the employed observation strategy, when using ERIN instead of RULA or REBA, the agreement regarding the final risk level is moderated to good. Nevertheless, the characteristics, strengths, and weaknesses must be analyzed in each of these methods when selecting them to evaluate a given situation (e.g. Personal that will use it, a dynamic or static task, involved body parts, among others).

On ERIN's predictive validity, until now no studies have been reported. To date, ERIN has been applied to the ergonomic evaluation of dissimilar workstations in different countries (Brazil, Canada, Cuba, Colombia, Mexico, Panama, Ecuador, India, China, Peru, among others). The evaluators, workers, and other interested subjects that have used it have considered it a useful and practical tool. This translates as ERIN having a good preliminary validity "face validity", as well as in contexts where many ergonomics-related situations must be solved, like in developing countries (Latin-Americans), the preliminary validity gains importance.

Usability

Study Description

Study 1: To a 42-subject group, classified as non-experts, after evaluating 7 video-projected tasks with the ERIN method, an instrument to evaluate ERIN's usability was provided. This instrument is a modification of one that was developed to measure the usability of informatics programs (Neville A Stanton & Young, 1999). As part of the usability study, the time that each subject took on each task evaluation was registered too.

Study 2: Like in the study described above, a group of 8 health field professionals was provided with an instrument designed to evaluate ERIN's usability, after evaluating two tasks. The evaluators received an hour of training previous to the ERIN method's usage.

The evaluation of two workstations realized out of videos by 8 health field professionals (doctors and traumatology, physiology and rehab specialists), with three experts defined pattern were compared. It must be noted that these professionals had no experience on the usage of ergonomic evaluation methods, but they did on related subjects. The evaluators received an hour of training on the usage of the ERIN method.

Methodology

Study 1 and Study 2: Table 4 shows the used tool which consists of 10 questions focused on exploring the following aspects: easy learning, little training time, needed time to realize the evaluation and applicability on a varied group of tasks (dynamic and statics). Starting from the tool, the usability index was obtained, varying from 0 to 100 points. This index is calculated when subtracting the value one to each answer of the odd questions and subtracting the value five of the answer to the even questions; next, the ten obtained values are summed and multiplied by 2.5.

Results and Analysis

Study 1: The usability index varied among 60 to 100 with an average of 80, and a deviation of 10. The results show that the practical value of the method, according to the user's criteria is high. Additionally, the agreement between the evaluators was analyzed, when responding the questions related to usability. To analyze the agreement among the subjects when answering said questions, the ICC (2.1) was calculated with 0.642 individual indexes and 0.987 averages. Indicating a regular to god agreement for each individual and excellent for the group (Stevens et al., 2004).

The average evaluation time per task was 6.5 minutes, with a tendency to reduce as more evaluations were being done. This was presumably due to the familiarization of the evaluators with the method.

Study 2: The usability index varied among 62.5 and 92.5 with an average of 84 and a deviation of 12. The results show that the method's practical value is high. To analyze the agreement among the professionals when answering the asked questions, the ICC (2.1) was calculated with an individual

Table 4. Format used to measure the ERIN's method usability

In case of the answer you consider taking into account that the analysis is different for even questions (2,4,6,8,10) and odds (1,3,5,7,9).					
	Total Disagreement				**Total Agreement**
1. ERIN is easy to use.	1	2	3	4	5
2. ERIN's evaluation takes too much time.	1	2	3	4	5
3. ERIN allows evaluating a broad range of tasks.	1	2	3	4	5
4. ERIN requires the help of specialized personnel.	1	2	3	4	5
5. Most people will learn to use ERIN quickly.	1	2	3	4	5
6. I need to learn many things to use ERIN.	1	2	3	4	5
7. I feel safe when using ERIN.	1	2	3	4	5
8. I find the tool difficult to use.	1	2	3	4	5
9. The ERIN's field sheet design facilitates the usage of the tool.	1	2	3	4	5
10. The terms used in ERIN are confusing.	1	2	3	4	5

Source: Built from (Neville A Stanton & Young, 1999)

index of 0.747 and 0.947 average. Indicating an agreement very close to excellent for each individual and excellent for the group. Therefore, the obtained agreement for the usability usage was excellent.

ERIN SOFTWARE

To ease the ERIN method's application, a free-access online application has been developed (Figure 10 and Figure 11), in English as well as in Spanish, and the ERIN 1.7 software (Figure 12 and Figure 13), that allows storing the evaluations, generating reports and making recommendations according to the evaluation's results.

FUTURE RESEARCH DIRECTIONS

Nowadays there is a higher interest for the approximation of a real evaluation on the exposition of real people to WMSDs risk. This is an extremely complex phenomenon given the multi-factor origin of these diseases, and how hard it becomes to control all of the risk factors present in the work environment, as well as outside of it. Some authors have presented other paradigms, where they highlight the necessity of a broaden vision to act beyond the workstation, on the hope of a bigger impact (Punnett et al., 2009). In my opinion, this philosophy will gain strength over the years, although I must point out that it will find

Figure 10. Online application to evaluate using the ERIN method
Source: www.ergoyes.com

Figure 11. Online application to evaluate using the ERIN method
Source: www.ergoyes.com

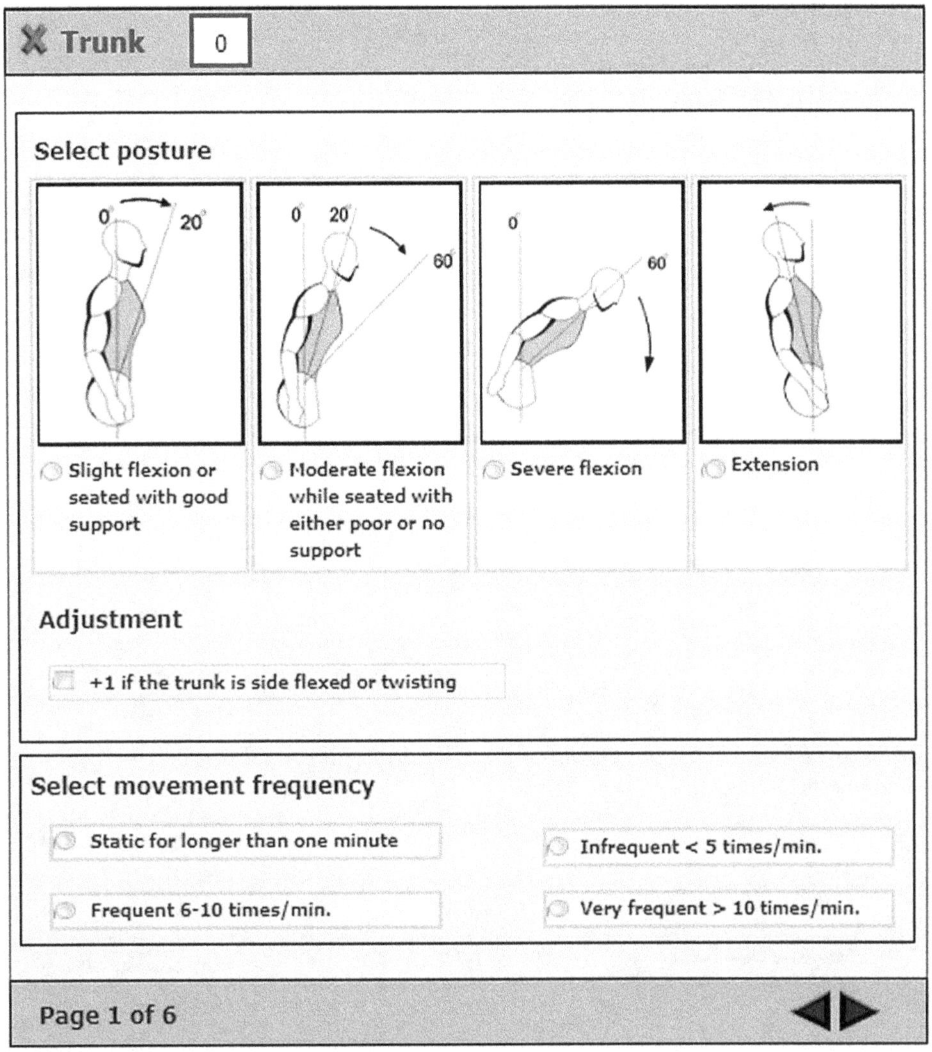

many barriers, mainly on sub-developed countries, where there are more urgent political, economical, and social problems to solve.

Although the race for the risk factor's evaluation perfectionism must not stop and will surely be an impulse by the incessant technological developments, the professionals and ergonomic-field researchers must increase their attempts to improve the effectiveness of the intervention and transformation process of the workstations and systems, under a systemic ergonomic perspective. As of today, it is very common to find workstation evaluation courses, but very few on how to intervene the workstations and systems. Expressing it differently, only by evaluating we are not going to change the actual overview that we have in regard of WMSDs, we must step into actions. In order to do so, the future development and perfecting of the ergonomic evaluation methods must find a balance between the effort dedicated to evaluating and the utility of the intervention results.

Figure 12. Software ERIN 1.7 for the ergonomic evaluation of the workstations
Source: www.ergoyes.com

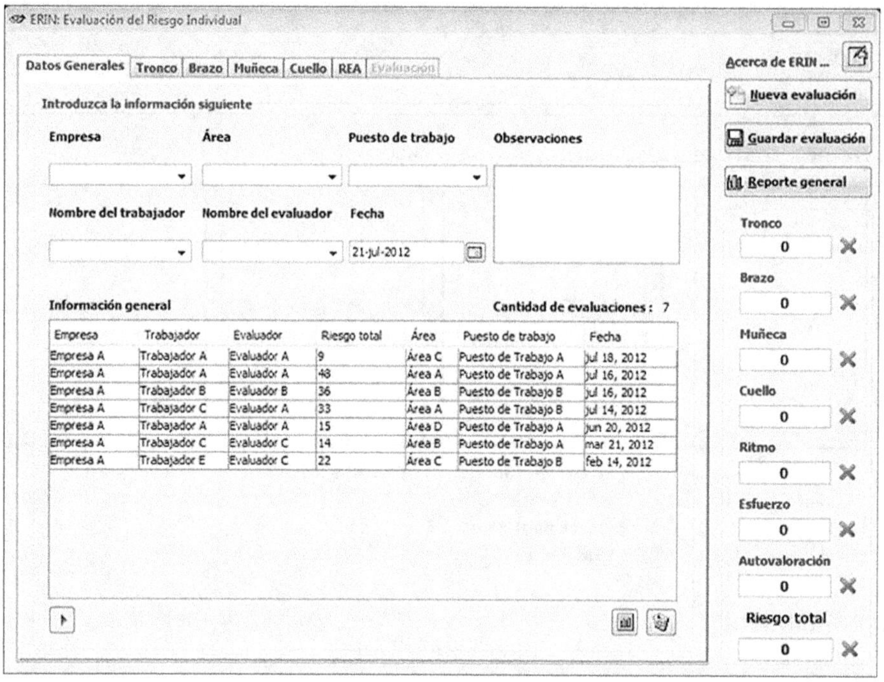

Figure 13. Software ERIN 1.7 for the ergonomic evaluation of the workstations
Source: www.ergoyes.com

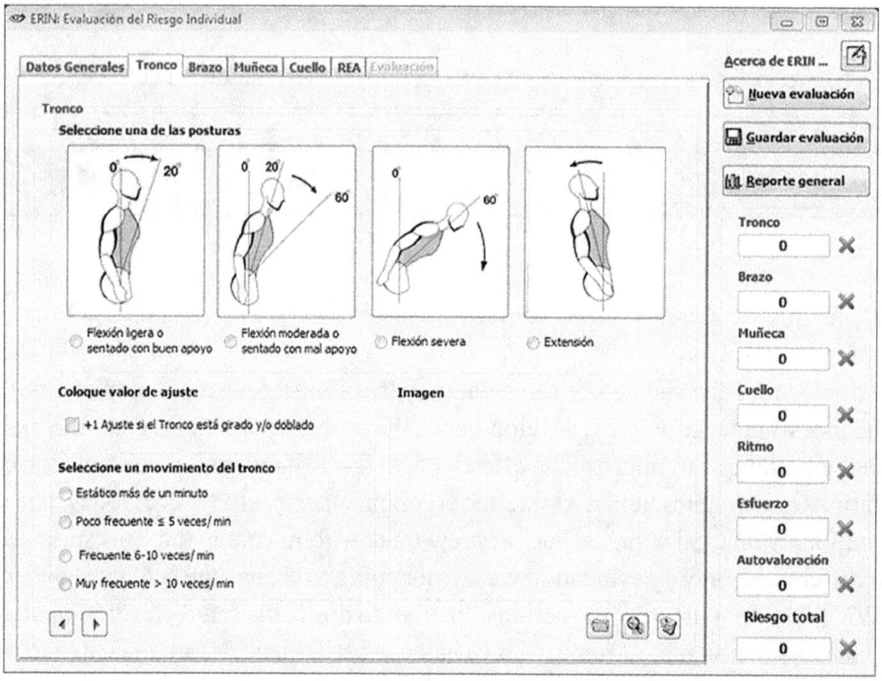

CONCLUSION

The ergonomic methods available to evaluate the exposure to WMSDs risk factors have been mostly developed for specialized personnel that also has experience in said activities can use them. This has made the massive evaluation of workspace/tasks very difficult in many countries as part of the primary preventive actions of these diseases since the existence of this personnel is limited. For said reason, the creation of tools and practical methods is necessary, for them to be correctly used by non-expert personal and to favor the evaluation-intervention process. With ERIN it is hoped to lower this problem, given the fact that it has proven to have an acceptable level of reliability and validity, as well as a great acceptance among professionals that perform ergonomic evaluations on real contexts since its use requires low training, sources and time.

ACKNOWLEDGMENT

Gratitude to professors Silvio Viña Brito and Ricardo Montero Martínez for their company during the ERIN method development process, and to Elizabeth Pérez Mergarejo for her valuable input on the conforming of this chapter.

REFERENCES

Bao, S., Howard, N., Spielholz, P., & Silverstein, B. (2007). Two posture analysis approaches and their application in a modified Rapid Upper Limb Assessment evaluation. *Ergonomics*, *50*(12), 2118–2136. doi:10.1080/00140130701458230 PMID:17852373

Bao, S., Howard, N., Spielholz, P., Silverstein, B., & Polissar, N. (2009). Interrater Reliability of Posture Observations. *Human Factors: The Journal of the Human Factors and of Ergonomics Society*, *51*(3), 292–309. doi:10.1177/0018720809340273 PMID:19750793

Bubb, H., Engstler, F., Fritzsche, F., Mergl, C., Sabbah, O., Schaefer, P., & Zacher, I. (2006). The development of RAMSIS in past and future as an example for the cooperation between industry and university. *International Journal of Human Factors Modelling and Simulation*, *1*(1), 140–157. doi:10.1504/IJHFMS.2006.011686

Chang, S. W., & Wang, M. J. J. (2007). Digital human modeling and workplace evaluation: Using an automobile assembly task as an example. *Human Factors and Ergonomics in Manufacturing & Service Industries*, *17*(5), 445–455. doi:10.1002/hfm.20085

Cochran, D. J., Stentz, T. L., Stonecipher, B. 1., & Hallbeck, M. S. (1999). Guide for Videotaping and Gathering Data on Jobs Analysis of Risks of Musculoskeletal Disorders. In W. Karwoski & W. S. Marras (Eds.), *The Occupational Ergonomics Handbook* (pp. 511–524). Boca Ratón, FL: University of Nebraska-Lincoln.

Corlett, E. N., & Bishop, R. P. (1976). A Technique for Assessing Postural Discomfort. *Ergonomics*, *19*(2), 175–182. doi:10.1080/00140137608931530 PMID:1278144

Corlett, N. (1995). The evaluation of posture and its effects. In J. r. Wilson & E. N. Corlett (Eds.), *Evaluation of human work. A practical ergonomics methodology* (2nd ed.; pp. 662–713). Nottingham, UK: Academic Press.

David, G. (2005). Ergonomic methods for assessing exposure to risk factors for work-related musculoskeletal disorders. *Occupational Medicine, 55*(3), 190–199. doi:10.1093/occmed/kqi082 PMID:15857898

Diego-Mas, J.-A., Alcaide-Marzal, J., & Poveda-Bautista, R. (2017). Errors Using Observational Methods for Ergonomics Assessment in Real Practice. *Human Factors,* (0). doi: 10.1177/0018720817723496

Drury, C. G., Atiles, M., Chaitanya, M., Lin, J.-F., Marin, C., Nasarwanji, M., ... Sunm, M. (2006). Vicarious perception of postural discomfort and exertion. *Ergonomics, 49*(14), 1470–1485. doi:10.1080/00140130600774372 PMID:17050388

Hendrick, H. W. (2008). Applying ergonomics to systems: Some documented "lessons learned". *Applied Ergonomics, 39*(4), 418–427. doi:10.1016/j.apergo.2008.02.006 PMID:18374303

Hignett, S., & McAtamney, L. (2000). Rapid Entire Body Assessment (REBA). *Applied Ergonomics, 31*(2), 201–205. doi:10.1016/S0003-6870(99)00039-3 PMID:10711982

ISO 11228-3. (2007). *Ergonomics -Manual handling- Part 3: Handling of low loads at high frequency* (Vol. 11228).

Lee, D., & Ferreira, J. (2003). *Reliability and usability evaluation of the Manual handling Assessment Charts (MAC) for use by non-regulatory professionals.* Health & safety laboratory.

Li, G., & Buckle, P. (1999). Current techniques for assessing physical exposure to work-related musculoskeletal risks, with emphasis on posture-based methods. *Ergonomics, 42*(5), 674–695. doi:10.1080/001401399185388 PMID:10327891

McGraw, K. O., & Wong, S. P. (1996). Forming Inferences About Some Intraclass Correlation Coefficients. *Psychological Methods, 1*(1), 30–46. doi:10.1037/1082-989X.1.1.30

Neumann, P. (2006). *Inventory of Tools for Ergonomic Evaluation: Research programme: Strategies and methods for Management of the Work Environment.* Academic Press.

NIOSH. (1997). Musculoskeletal Disorders and Workplace Factors: A Critical Review of Epidemiologic Evidence for Work-Related Musculoskeletal Disorders of the Neck, Upper Extremity, and Low Back. Cincinnati, OH: DHHS (NIOSH).

Punnett, L., Cherniack, M., Henning, R., Morse, T., Faghri, P., & Team, C.-N. R. (2009). A conceptual framework for integrating workplace health promotion and occupational ergonomics programs. *Public Health Reports, 124*(4_suppl1), 16-25.

Rodríguez, Y. (2011). ERIN: método práctico para evaluar la exposición a factores de riesgo de desórdenes músculo-esqueléticos (Doctor en Ciencias Técnicas Tesis de Doctorado). Instituto Superior Politécnico "José Antonio Echeverría", La Habana, Cuba.

Rodríguez, Y., & Pérez, E. (2013a). Implementación de un proceso de intervención ergonómica en una imprenta cubana. *Ação Ergonômica. Revista da Associação Brasileira de Ergonomia, 8*(2), 1–13.

Rodríguez, Y., & Pérez, E. (2013b). Introducción de mejoras ergonómicas. Beneficios para la organización. *Revista Técnica de Ingeniería de la Universidad del Zulia*, *36*(2), 1–10.

Rodríguez, Y., Viña, S., & Montero, R. (2013a). ERIN: A practical tool for assessing work-related musculoskeletal disorders. *Occupational Ergonomics*, *11*(2-3), 15.

Rodríguez, Y., Viña, S., & Montero, R. (2013b). A method for non-experts in assessing exposure to risk factors for work-related musculoskeletal disorders-ERIN. *Industrial Health*, *51*(6), 5. doi:10.2486/indhealth.2013-0008 PMID:24077444

Shrout, P. E., & Fleiss, J. L. (1979). Intraclass Correlations: Uses in Assessing Rater Reliability. *Psychological Bulletin*, *86*(2), 420–428. doi:10.1037/0033-2909.86.2.420 PMID:18839484

Sinclair, M. A. (1995). Subjective assessment. In J. R. Wilson & E. N. Corlett (Eds.), *Evaluation of Human Work. A practical ergonomics methodology* (pp. 69–100). London: Taylor and Francis.

Stanton, N. A., Hedge, A., Brookhuis, K., Salas, E., & Hendrick, H. W. (2004). *Handbook of human factors and ergonomics methods*. CRC Press. doi:10.1201/9780203489925

Stanton, N. A., & Young, M. S. (1999). Ergonomics Methods in the Design of Consumer Products. In W. Karwowski & W. S. Marras (Eds.), *The occupational ergonomics handbook* (pp. 741–760). Boca Raton, Florida.

Stephens, J.-P., Vos, G. A., Stevens, E. M. Jr, & Steven Moore, J. (2006). Test-retest repeatability of the Strain Index. *Applied Ergonomics*, *37*(3), 275–281. doi:10.1016/j.apergo.2005.07.007 PMID:16182231

Stevens, E. M., Vos, G. A., Stephens, J.-P., & Steven Moore, J. (2004). Inter-Rater Reliability of the Strain Index. *Journal of Occupational and Environmental Hygiene*, *1*(11), 745–751. doi:10.1080/15459620490521142 PMID:15673095

Straker, L. M. (1999). Body discomfort assessment tools. In W. Karwowski & W. S. Marras (Eds.), *Handbook of Occupational Ergonomics* (pp. 1239–1251). Boca Ratón, FL: Academic Press.

Takala, E.-P., Pehkonen, I., Forsman, M., Hansson, G.-Å., Mathiassen, S., Neumann, W., ... Winkel, J. (2010). Systematic evaluation of observational methods assessing biomechanical exposures at work. *Scandinavian Journal of Work, Environment & Health*, *36*(1), 3–24. doi:10.5271jweh.2876 PMID:19953213

KEY TERMS AND DEFINITIONS

Ergonomic Intervention: Changes directed towards the improvement of a task, workstation, or work system from the ergonomic perspective.

Ergonomics Assessment: Testing on the fulfillment of the principles and approaches that ergonomics promotes.

Inter-Observer Reliability: Consistency grade reached for various observers when evaluating a given situation using the same method.

Intra-Observer Reliability: Consistency grade reached by an observer when evaluating at least twice, given a situation, with the same method.

Linguistic Descriptors: Word used to describe the movement range and the frequency of the body parts movement.

Task: Group of actions that are executed to achieve a determined goal.

Usability: Magnitude in which a method is considered simple, easy to use and learn, and applicable to varied situations in real contexts.

Validity: Grade in which a method or tool measures what it is supposed to measure.

Workplace: Workspace where many worker-assigned tasks are executed.

APPENDIX

Figure 14 shows the ERIN worksheet.

Figure 14. ERIN worksheet

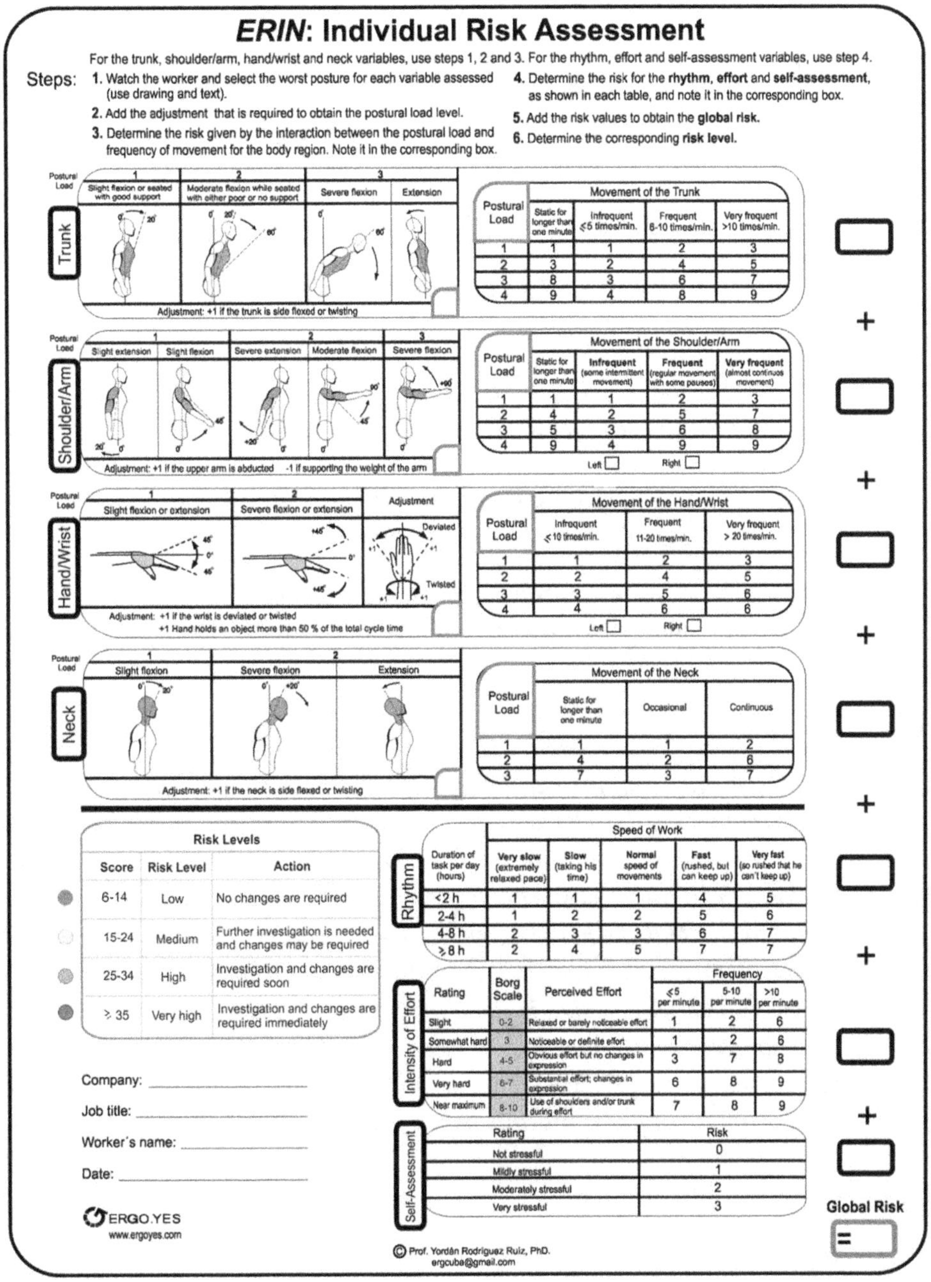

ERIN: Individual Risk Assessment

For the trunk, shoulder/arm, hand/wrist and neck variables, use steps 1, 2 and 3. For the rhythm, effort and self-assessment variables, use step 4.

Steps:

1. Watch the worker and select the worst posture for each variable assessed (use drawing and text).
2. Add the adjustment that is required to obtain the postural load level.
3. Determine the risk given by the interaction between the postural load and frequency of movement for the body region. Note it in the corresponding box.
4. Determine the risk for the **rhythm, effort** and **self-assessment**, as shown in each table, and note it in the corresponding box.
5. Add the risk values to obtain the **global risk.**
6. Determine the corresponding **risk level.**

Trunk — Postural Load: 1 = Slight flexion or seated with good support; 2 = Moderate flexion while seated with either poor or no support; 3 = Severe flexion, Extension

Adjustment: +1 if the trunk is side flexed or twisting

Postural Load	Movement of the Trunk: Static for longer than one minute	Infrequent ≤5 times/min.	Frequent 6-10 times/min.	Very frequent >10 times/min.
1	1	1	2	3
2	3	2	4	5
3	8	3	6	7
4	9	4	8	9

Shoulder/Arm — Postural Load: 1 = Slight extension, Slight flexion; 2 = Severe extension, Moderate flexion; 3 = Severe flexion

Adjustment: +1 if the upper arm is abducted -1 if supporting the weight of the arm

Postural Load	Movement of the Shoulder/Arm: Static for longer than one minute	Infrequent (some intermittent movement)	Frequent (regular movement with some pauses)	Very frequent (almost continuos movement)
1	1	1	2	3
2	4	2	5	7
3	5	3	6	8
4	9	4	9	9

Left ☐ Right ☐

Hand/Wrist — Postural Load: 1 = Slight flexion or extension; 2 = Severe flexion or extension; Adjustment: Deviated, Twisted

Adjustment: +1 if the wrist is deviated or twisted
+1 Hand holds an object more than 50 % of the total cycle time

Postural Load	Movement of the Hand/Wrist: Infrequent ≤10 times/min.	Frequent 11-20 times/min.	Very frequent > 20 times/min.
1	1	2	3
2	2	4	5
3	3	5	6
4	4	6	6

Left ☐ Right ☐

Neck — Postural Load: 1 = Slight flexion; 2 = Severe flexion, Extension

Adjustment: +1 if the neck is side flexed or twisting

Postural Load	Movement of the Neck: Static for longer than one minute	Occasional	Continuous
1	1	1	2
2	4	2	6
3	7	3	7

Risk Levels

Score	Risk Level	Action
6-14	Low	No changes are required
15-24	Medium	Further investigation is needed and changes may be required
25-34	High	Investigation and changes are required soon
≥ 35	Very high	Investigation and changes are required immediately

Rhythm

Duration of task per day (hours)	Speed of Work: Very slow (extremely relaxed pace)	Slow (taking his time)	Normal speed of movements	Fast (rushed, but can keep up)	Very fast (so rushed that he can't keep up)
<2 h	1	1	1	4	5
2-4 h	1	2	2	5	6
4-8 h	2	3	3	6	7
≥8 h	2	4	5	7	7

Intensity of Effort

Rating	Borg Scale	Perceived Effort	Frequency: ≤5 per minute	5-10 per minute	>10 per minute
Slight	0-2	Relaxed or barely noticeable effort	1	2	6
Somewhat hard	3	Noticeable or definite effort	1	2	6
Hard	4-5	Obvious effort but no changes in expression	3	7	8
Very hard	6-7	Substantial effort; changes in expression	6	8	9
Near maximum	8-10	Use of shoulders and/or trunk during effort	7	8	9

Self-Assessment

Rating	Risk
Not stressful	0
Mildly stressful	1
Moderately stressful	2
Very stressful	3

Global Risk =

Company: ______________

Job title: ______________

Worker's name: ______________

Date: ______________

ERGO.YES
www.ergoyes.com

© Prof. Yordán Rodríguez Ruiz, PhD.
ergcuba@gmail.com

Chapter 2
Ergonomic Bench to Decrease Postural Risk Level on the Task of Changing Forklift's Brake Pads:
A Design Approach

Arturo Realyvásquez
Instituto Tecnológico de Tijuana, Mexico

Guadalupe Hernández-Escobedo
Instituto Tecnológico de Tijuana, Mexico

Aide Aracely Maldonado Macías
Autonomous University of Ciudad Juárez, Mexico

ABSTRACT

Development of musculoskeletal disorders caused by awkward postures are present in tasks performed in small companies. Consequently, this chapter evaluates the postural risk level associated to the task of changing brake pads of a forklift, and proposes the design and the implementation of an ergonomic bench that helps to reduce this level of postural risk. As methods, postural risk level corresponding to the task was determined using REBA, whereas anthropometric analysis was developed, and axiomatic design and TRIZ methodologies were used to develop the proposed design. Results showed a REBA score of 8 (high-risk level). The final design is done of nylamid, and it contains some TRIZ principle, such as copying, extraction, and dynamicity. Respect axiomatic design, the bench meets some functional requirements, such as support the user's weight, avoid unsafety conditions, and be adjustable. With the TRIZ principles and functional requirements, this bench will help diminish postural risk level.

DOI: 10.4018/978-1-5225-5234-5.ch002

Copyright © 2018, IGI Global. Copying or distributing in print or electronic forms without written permission of IGI Global is prohibited.

INTRODUCTION

Most of the ergonomic assessment methodologies adopted in the manufacturing industry are implemented to assess ergonomic hazards in the production area (Tompa, Dolinschi, de Oliveira, Amick, & Irvin, 2010), yet other processes, such as machinery maintenance/repairs, material transportation, and/or plant cleaning, require similar ergonomic evaluations to ensure employee safety. To address this gap, this research takes place in a company that repairs and sells industrial machinery.

It is well known that maintenance and repairing tasks force employees to adopt awkward postures (Barros, Marçal, & Soares, 2015; Teymourian, Seneviratne, & Galar, 2016) and apply excessive force. In this sense, the maintenance technicians of the company wherein this study was conducted have claimed to experience discomfort when they change forklift brake pads. This task takes approximately one hour per brake, and to complete it, the technicians usually adopt awkward and long-lasting postures. According to the employees, the main musculoskeletal complaints experienced are in the back, the ankles, and the knees. Even though knee pads are worn, the employees claim these are not an effective solution to pain and discomfort. Figure 1 depicts the typical posture adopted by the employees as they change forklift brake pads.

According to the literature, awkward postures have a negative effect on employee health and performance. They can cause soft tissue injuries that favor the accumulation of metabolites and accelerate disc degeneration, which can ultimately lead to disc herniation (Roffey, Wai, Bishop, Kwon, & Dagenais, 2010). Also, awkward workplace postures can increase the risk of suffering from musculoskeletal disorders (MSDs) – such as osteoarthritis (da Costa & Ramos-Vieira, 2009; Gallagher, Pollard, & Porter, 2011) and fibromyalgia (da Costa & Ramos-Vieira, 2009) – and are strongly associated with low back pain (LBP) and whole-body vibration (Vandergrift, Gold, Hanlon, & Punnett, 2012).

As regards employee performance, awkward postures are less efficient than postures that keep joints near to a neutral posture (Gangopadhyay, Ghosh, Das, Ghoshal, & Das, 2010). Also, it has been argued that awkward postures force employees to apply more force than necessary to accomplish a particular task. Such excessive force in turn affects muscle loading and compressive forces on the internal vertebral disc.

The literature shows the need to improve the working conditions of those who adopt awkward and unsuitable postures to do their jobs. In the context of our research, one alternative to addressing this need is to design a bench, as an ergonomic work tool, to ensure that the technicians can change the brakes safely and comfortably. The importance of this bench relies on its advantages as an ergonomic intervention. Ergonomic interventions do not only reduce workplace hazards and increase employee safety; from a corporate point of view, they provide economic benefits, since they prevent companies from incurring additional costs (e.g. employee insurance) and increase employee retention, since workers feel and are safe.

LITERATURE REVIEW

REBA

The Rapid Entire Body Assessment (REBA) method is a highly sensitive tool for postural analysis. It is employed to assess unpredictable postures within the work environment. According to several researchers, REBA can be effectively applied to evaluate MSDs, since it provides the following advantages (Hignett & McAtamney, 2000; Maldonado-Macías, Realyvásquez, Hernández, & García-Alcaraz, 2015):

Figure 1. Posture adopted by employees during the task of changing forklift's brake pads
Source: Prepared by the authors (2017).

1. It develops a sensitive postural analysis system for MSDs risks in a variety of tasks.
2. It divides the body into segments that are individually encoded.
3. It provides a scoring system for muscular activity caused by static, dynamic, rapidly changing, and unstable postures.
4. It provides an action level with an urgency indication to find the kind of ergonomic interventions that must be implemented.
5. It is not time-consuming or resources consuming.

As mentioned, REBA has an action level system to support decision making after performing the postural analysis. This system is detailed in Table 1.

REBA is a popular method for postural analysis and MSDs risk assessment. In their research, (Lee, Yu, Choi, & Han, 2011) used this method to evaluate the task of large glass panels installation in the construction industry. Similarly, (Maldonado-Macías et al., 2015) relied on REBA to analyze various

Table 1. REBA's action level system

Action Level	Score	Risk Level	Required Corrective Actions
0	1	None	Corrective actions and further assessments are not required.
1	2-3	Low	Corrective actions and further assessment may be required.
2	4-7	Medium	Corrective actions and further assessments are required.
3	8-10	High	Corrective actions and further assessments are required as soon as possible.
4	11-15	Very high	Corrective actions and further assessments are immediately required.

tasks of the computer repairing process and identify levels of MSDs risk. As results, the authors found high levels of MSDs risks in every task and proposed a series of ergonomic interventions to address the ergonomic issues. Likewise, a most recent study performed by (Shirzaei, Mirzaei, Khaje-Alizade, & Mohammadi, 2015) evaluated ergonomic factors and profession-related postures with dental students. The authors directly observed the students' postures while treating patients. The analyzed postures were scored using REBA, and the results showed a medium level of risk for most subjects (84%).

Anthropometry

Anthropometry is the study of human body measurement and is used in the engineering sciences to ensure maximum product benefit and capability (Robinette, 2012). The field of application of anthropometry is broad and includes the description, comparison, evaluation, and identification of individuals or groups of individuals. Usually, anthropometry is used to describe the morphology of an individual or a sample of individuals. In addition, it can be used as a comparison basis of a population's sample to other samples or comparisons of the same sample over the time. For instance, the growth status of children that participate in specific sports competitions (Malina, 1995).

Anthropometry has a variety of applications in the design area. For instance, (Madriz-Quirós, Ramírez-Coretti, & Serrano-Montero, 2008) performed an anthropometric study to design ergonomic furniture for school age children in Costa Rica. The study included a sample of 1,293 children from all around the country, and 12 anthropometric measures were necessary to design the furniture. Similarly, (Azadeh, Fam, & Garakani, 2007) proposed a total ergonomic design approach to enhance productivity in complicated control systems. The proposal included performing an anthropometric study. In another work, (Gouvali & Boudolos, 2006) used anthropometry to find the match between school furniture dimensions and children anthropometry. Likewise, (Herrera-Ruiz, 2006) designed an amphibious wheelchair for hydrotherapy. For its design, the author used data from an already published anthropometric table.

Since anthropometric data follow a normal distribution, when a product is designed based on anthropometric data, percentiles and Z-scores are used to determine product dimensions (Hernández-Arellano, Aguilar-Duque, & Gómez-Bull, 2018). Percentiles are interpreted in the following way: if the *nth* percentile is determined for a specific body dimension, it means that *n%* are shorter than this percentile in this body dimension (Pheasant & Haslegrave, 2016). For instance, if the 95th percentile is calculated for anthropometric data of the height, it means that 95% of people have a height below than the 95th percentile value. To design the ergonomic bench, the authors had to know the anthropometry measures of the users (i.e. technicians) as well as the most appropriate design material and design methodology.

Nylamid

The bench to be designed must be robust and ergonomic. That is, it must be resistant enough to support the weight of the technicians, but and at the same it must be comfortable, easy to use, and easy to handle. The material that best satisfies these proprieties of the bench is Nylamid®. This material has a breaking strain of 142,559,000 N/m^2 (Martínez-Nopala, 2015), which is above the pressure that any person may exercise; also, it is two to eight times lighter than metals. Similarly, Nylamid® is rigid and has a desired mechanical, chemical, and thermal resistance (Herrera-Ruiz, 2006; Ramírez-Ibanez, 2004). A wide range of products are designed and manufactured using Nylamid®, such as rollers, chains, wear guides, impact parts, gears, pulleys, chairs (Herrera-Ruiz, 2006), vehicles (Martínez-Nopala, 2015), electromagnetic positioning systems, electrical power systems, and wind generators (Magdaleno-Adame, Olivares-Galvan, & Fofana, 2014).

Axiomatic Design

In engineering, Axiomatic Design (AD) sets the scientific and systematic basis of design processes. The main goals of AD are to increase creativity, reduce the random search process, minimize the iterative trial-and-error process, and determine the best designs among a set. Also, AD implies a continuous interplay between what designers/customers want to achieve and how they want to achieve it (Hirani & Suh, 2005; Suh, 1990, 2001). AD is a methodology that support the design of a wide variety of systems (manufacturing systems, software, products, and organizations, for instance) and establishes the customer needs (CNs) that the system must satisfy. Then, the functional requirements (FRs) and constraints (Cs) of the system to be designed are determined to meet such CNs. Next, the FRs must be matched to the design parameters (DPs). This step allows designers to identify and choose the DPs that are necessary to design the system. Once the DPs are chosen, designers must go to the process domain and identify the process variables (PVs) based on the creation of a new process or the use of an existing process (Suh, 1998).

The AD theory is composed of two key axioms, theorems, and corollaries. The first axiom sustains that the independence of the FRs must be always preserved. When there are two or more FRs, one of them should not affect the other. In this sense, the designer has to select the proper DPs to satisfy the FRs independently. This is done one by one in a mapping process between one FR and its corresponding DP (Won & Joon, 2005). Accordingly, the application of this axiom determines coupled, decoupled, and uncoupled designs. An ideal design is an uncoupled design, where one DP can satisfy each FR independently. However, these designs are commonly difficult to achieve, and decoupled designs can be acceptable designs as well. Both, uncoupled and coupled designs satisfy the independence axiom. On the other hand, a coupled design exists when DPs and FRs cannot be satisfied independently. In this case, designers must work creatively to change the product to a decoupled or uncoupled design.

Once the first axiom is accomplished, the second axiom is also required for better designs. The information axiom explains that, in all good designs, information content must be the minimum. This axiom establishes the idea that the success of a design is determined by the probability associated with achieving the FRs. This probability decreases with the amount of information necessary to fulfill the FRs (Lenz & Cochran, 2000).

The AD theory has been applied successfully in many areas. In their work, (Kulak, Cebi, & Kahraman, 2010) presented a literature review of AD applications and identified five application areas: product design (Ferrer, Rios, & Ciurana, 2009; Gumus, Ertas, Tate, & Cicek, 2008; Tang, Zhang, & Dai, 2009), system design (Bang & Heo, 2009; Helander, 2007; Lo & Helander, 2007), manufacturing system design (Durmusoglu & Kulak, 2008; Nakao, Kobayashi, Hamada, Totsuka, & Yamada, 2007; Schnetzler, Sennheiser, & Schönsleben, 2007), software design (Lindkvist & SÖDerberg, 2003; Togay, Dogru, & Tanik, 2008; Yi & Park, 2005), and decision making (Celik, 2009a, 2009b; Celik, Cebi, Kahraman, & Er, 2009). However, despite its multiple applications, some authors claim that AD by itself is not enough to solve design problems, and thus requires the support of techniques and methodologies for FRs determination and creative solutions. As a result, designers now combine AD with tools such as Failure Mode Error Analysis (FMEA), Design for Six Sigma (DFSS), the theory of inventive problem solving (TRIZ), and decision support multiattribute methods (Dickinson, 2006).

TRIZ

TRIZ is a Russian acronym for the Theory of Inventive Problem Solving. It was proposed by Altshuller and originated from several technical and patent information studies. The latter showed that only 1% of the solutions were truly pioneering inventions; the rest represented the use of previously known ideas and concepts but in a novel way (Altshuller, 1984; Yang & Zhang, 2000). This led to the conclusion that the idea of a design solution to a problem might be already known. But where and how this idea can be found? Based on a systematic view of the technical world, TRIZ provides different techniques and tools that help designers create a new design idea and avoid numerous trials and errors during the problem solving process (Yang & Zhang, 2000).

According to several authors, TRIZ is a powerful systematic ideation methodology for technical problem solving that leads to the enhancement of existent techniques (Kremer et al., 2012; Savransky, 2000). Moreover, TRIZ is a human-oriented knowledge-based systematic methodology of inventive problem solving (Savransky, 2000). In TRIZ, every factor that affects a system can be defined as a parameter, and there is a dependency relationship among system parameters. TRIZ structures a problem into a "contradiction statement" and derives solutions that address the problem statement both from technical and system perspectives. Hence, the ideality of the design increases as a parameter is improved without worsening the other parameter. In this sense, TRIZ demonstrates its capability as a support tool for original idea creation.

Despite the success of TRIZ in aiding idea generation, if implemented alone, TRIZ falls short in selecting the most appropriate idea. Therefore, using it in unison with appropriate tools – such as AD, in this study – is strongly recommended (Kremer et al., 2012).

TRIZ comprises 40 inventive principles to guide the TRIZ practitioner in developing useful concepts of solution for inventive situations (Yang & Zhang, 2000). Each solution represents a recommendation to make a specific change in a system and eliminate technical contradictions. The literature provides a contradiction table that recommends which principles should be considered in solving approximately 1,250 contradictions. Five of these 40 inventive principles are stated below (Altshuller, 2002; Technical Innovation Center, 2013):

- **Extraction:** Disturbing parts must be removed from an object only if it is necessary.
- **Counterweight:** Compensate for the weight of an object by combining it with another object that provides a lifting force.
- **Cushion in Advance:** Compensate for the relatively low reliability of an object with emergency measures prepared in advance.
- **Dynamicity:** The characteristics of an object or outside environment must be altered to provide optimal performance at each stage of an operation. Then, if an object is immobile, make it mobile, make it interchangeable, and divide the object into elements capable of changing their position relative to each other.
- **Copying:** It is better to use a simplified and inexpensive copy in place of a fragile original object, or an object that is inconvenient to operate.

In the last decade, various works have illustrated the advantages of combining TRIZ and AD and have reported how these techniques complement each other. AD analyzes the problem and structures it in the most suitable way, whereas TRIZ can solve the minimum number of design conflicts that are intrinsically present in a case study (Borgianni & Matt, 2016). That said, when TRIZ is combined with AD, it can analyze a complex problem and deduce the most relevant conflicts, since TRIZ uses AD to appropriately structure the problem. Similarly, AD is useful for analyzing the system and its requirements, but it completely lacks means to identify technical solutions. These solutions, according to the axioms that guide the design process, should be appropriately characterized by satisfying performances for each requirement and high controllability of the system. In conclusion, both methods – TRIZ and AD – can be applied jointly thanks to the not contradictory objectives they have.

METHODOLOGY

To design the bench from an ergonomic approach, authors propose a cross-sectional, non-experimental methodology that comprises six stages. The following sections review the used instruments and thoroughly discuss the methodology stages.

Instruments

Authors relied on the following instruments to design the bench from an ergonomic approach:

- **Computer Equipment:** Laptop Sony Vaio®
- **REBA, Online Software**
- **Design Software:** Authors initially decided to use SolidWorks® software, but due to a lack of experience, they finally used PowerPoint and Paint. Both programs encourage user creativity.

Method

As previously mentioned, the methodology followed to design the ergonomic bench comprises six stages as depicted in Figure 2.

Figure 2. Methodology proposed to design the ergonomic bench
Source: Prepared by the authors (2017).

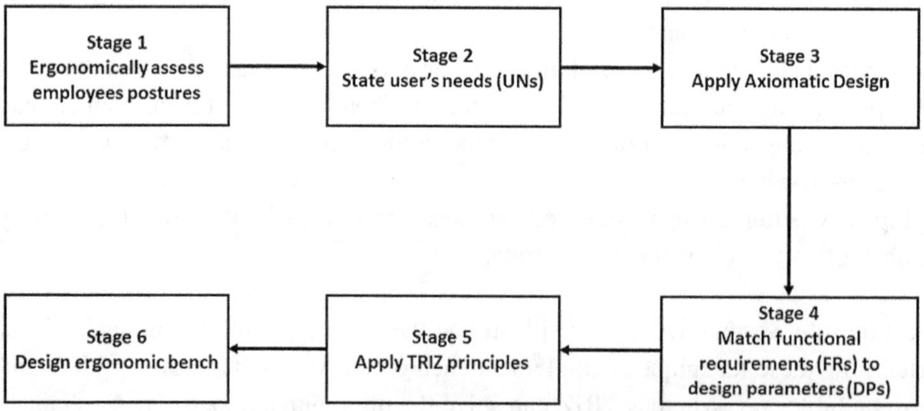

Stage 1 - Ergonomically Assess Employee Postures: The REBA method was implemented to ergonomically evaluate the postures adopted by the technicians in the task of changing forklift brake pads. To this end, authors relied on REBA online software, available on the *ergonautas* platform (Ergonautas.upv.es, 2017).

Stage 2 - State User Needs (UNs): Authors defined the technicians' needs that had to be satisfied by the ergonomic bench.

Stage 3 - Apply Axiomatic Design: Once the UNs were identified, authors defined the FRs of the design that had to satisfy such UNs. Also, authors set the constraints that would limit the ergonomic design of the bench.

Stage 4 - Match FRs to DPs: Each FR was assigned a DP that would allow the bench to meet the said FR. Likewise, authors relied on anthropometry to set the measures of the bench. Since the company has only eight employees, authors used the anthropometric data presented by (Ávila-Chaurand, Prado-León, González-Muñoz, & Muñoz, 2007), which correspond to a sample of 369 male industrial employees, ranging from 18 to 65 years old. That said, authors only used the data of those body segments that would interact with the bench. When designing for extremes, the dimensions of the ergonomic bench were defined by applying the 5th and 95th percentiles, calculated by means of equation (1) (Hernández-Arellano et al., 2018):

$$P_n = \overline{x} + Z\sigma \tag{1}$$

where P_n represents the value obtained for the *nth* percentile, $\overline{x}$ is the average value of the anthropometric data for a specific body segment, Z is the score of the accumulated normal distribution corresponding to the *nth* percentile, and σ is the standard deviation of the anthropometric data. Table 2 shows the corresponding Z-scores of each percentile.

Stage 5 - Apply TRIZ Principles: Once authors defined the UNs, FRs, Cs, and DPs, they searched for an existing design that could meet our FRs after performing the necessary modifications on it, following the TRIZ principles.

Stage 6 - Design the Ergonomic Bench: Once the percentiles of the bench were defined, authors built the piece of furniture using Nylamid as the design material.

RESULTS

The results of the proposed design methodology are discussed below:

Table 2. Percentiles and corresponding Z-scores

Percentile	Z-Score
5th	-1.645
95th	1.645

Stage 1 Results

Figure 3 shows the REBA score, the level of MSDs risk, and the action level obtained for the brake pads changing task. As can be observed, the score (8 points) implies a high level of MSDs risk for this task. Therefore, corrective actions and further assessments are required as soon as possible (Hignett & McAtamney, 2000; Maldonado-Macías et al., 2015).

Stage 2 Results

According to the results of stage 1, four UNs were identified: comfort, long-term health, short-term safety, and efficiency.

Results for Stage 3

The FRs that meet the identified UNs are:

- $\mathbf{FR_0}$ = Efficiency
- $\mathbf{FR_1}$ = Support user weight
- $\mathbf{FR_2}$ = Help maintain correct posture
- $\mathbf{FR_{21}}$ = Adjustable
- $\mathbf{FR_3}$ = Avoid unsafe conditions

FR_1 will help employees perform the task comfortably, since they no longer will have to support their own weight with the thighs. On the other hand, FR_2 will allow the technicians to adopt a healthier posture if compared with the one they currently adopt. That said, the bench can guarantee a correct posture if its design is anthropometrically adjustable – i.e. the design comprises FR_{21}. Both FR_2 and FR_{21} can improve employee health as they prevent MSDs occurrence. Finally, FR_3 will provide safer working conditions for the task of changing forklift brake pads.

Figure 3. REBA (Rapid Entire Body Assessment) score for the task of changing forklift brake pads
Source: Adapted from (Ergonautas.upv.es, 2017).

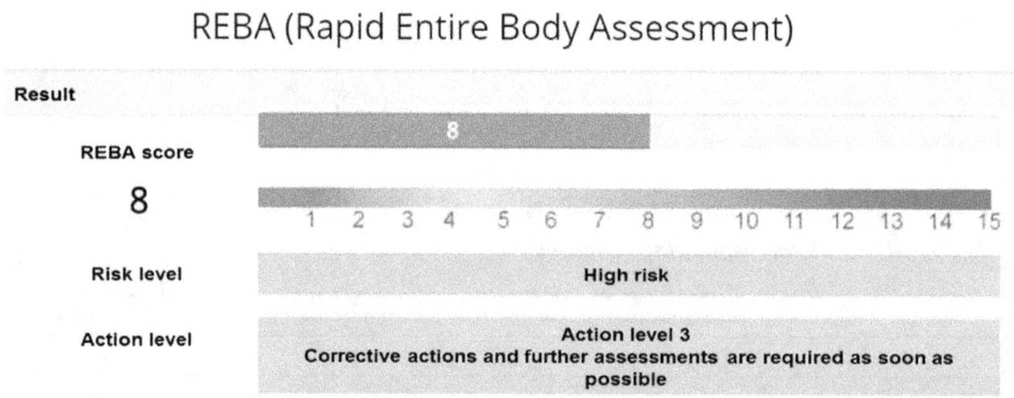

Together, FR_1, FR_2, FR_{21}, and FR_3 achieve FR_0 = efficiency, as they provide comfortable and safe postures for the technicians to reach components without extending the upper limbs. However, designs also imply constraints. In this work, authors identified two constraints that may prevent the design from being carried out: a lack of experience in using design software and a lack of material (i.e. Nylamid). In this sense, Figure 4 shows the mapping between the UNs (customer domain) and the FRs (functional domain) and the design constraints.

Results for Stage 4

Since the observed technicians support most of their weight with their thighs when they change forklift brake pads (i.e. they squat), FR_1 – support user weight – was related to DP_1 = create a seat. This seat will support most of the technicians' weight. On the other hand, since the employees flex their trunk when they change forklift brake pads, FR_2 – help maintain correct posture – was associated with DP_2 = Create a tilted support for the trunk. This DP_2 will allow employees to flex their trunk without supporting a force to maintain this position. Then, FR_{21} – Adjustable seat – was matched to DP_{21} = Design the seat and the trunk support with an adjuster. The adjuster will help employees fit the seat and the support according to their anthropometric measures. Finally, to meet FR_3 – avoid unsafety conditions – authors considered DP_3 = Design the seat and the trunk support with cotton pads covered with rubber. Cotton pads will cover the seats and the trunk support to offer comfort and protection.

Figure 4. Design constraints and mapping from user (or customer) domain to functional domain
Source: Prepared by the authors (2017).

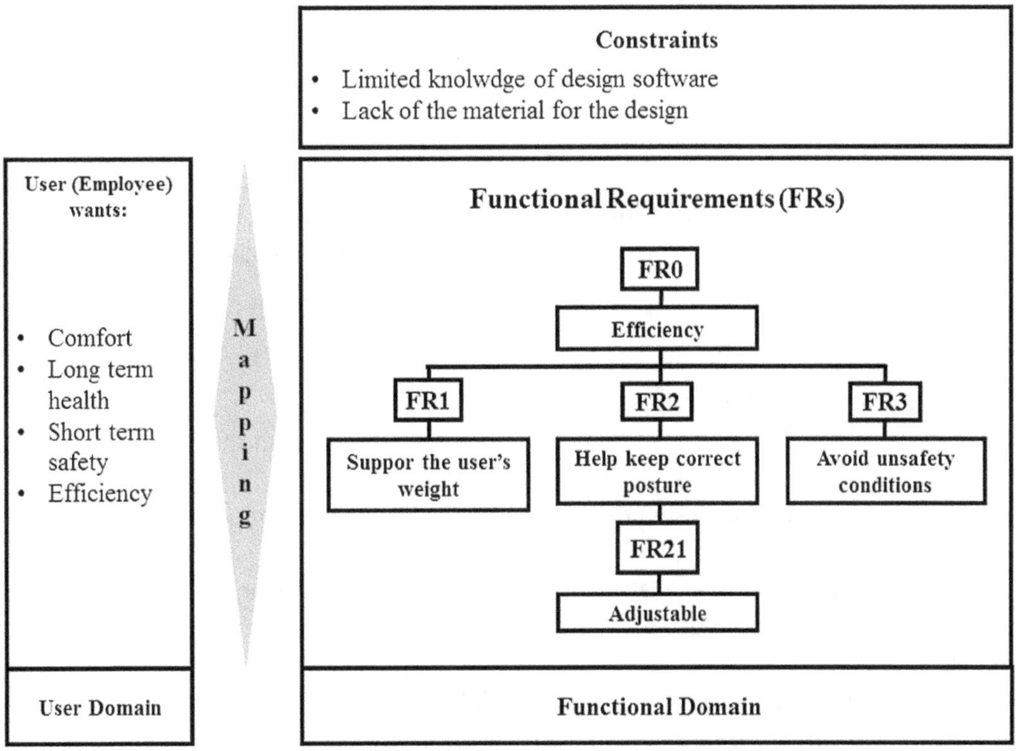

Then, the design matrix is given by equation (2) and equation (3):

$$\begin{Bmatrix} FR_1 \\ FR_2 \\ FR_3 \end{Bmatrix} = \begin{bmatrix} A_{11} & 0 & 0 \\ 0 & A_{22} & 0 \\ 0 & 0 & A_{33} \end{bmatrix} \times \begin{Bmatrix} DP_1 \\ DP_2 \\ DP_3 \end{Bmatrix} \quad (2)$$

$$\{FR_{21}\} = [B_{11}] \times \{DP_{21}\} \quad (3)$$

where A_{ij} and B_{11} represent non-zero elements and 0 represents a zero element.

The design equations for the matrix stated above are given by equation (4), equation (5), equation (6), and equation (7) as follows:

$$FR_1 = A_{11} \times DP_1 \quad (4)$$

$$FR_2 = A_{22} \times DP_2 \quad (5)$$

$$FR_{21} = B_{11} \times DP_{21} \quad (6)$$

$$FR_3 = A_{44} \times DP_3 \quad (7)$$

The design is uncoupled since the design matrix is diagonal. Therefore, the information content is zero, and it does not violate the information axiom (Cochran, Eversheim, Kubin, & Sesterhenn, 2000; Kim & Cochran, 2000). Hence, to obtain the best design, the sequence must be the following:

1. Create a seat.
2. Create a tilted trunk support.
3. Design the seat and the trunk support with an adjuster.
4. Design the seat and the trunk support with cotton pads covered with rubber.

The body dimensions used to design the ergonomic bench were the following:

- **Weight:** This dimension was taken into account since the technicians adopt a squatting posture when changing the forklift brake pads. Therefore, the bench is intended to support the technicians' weight.
- **Popliteal Height:** This dimension was used to define the seat's height.
- **Sitting hip Width:** This dimension was used to define the seat's width.

- **Buttock-Popliteal Length:** This dimension was used to define the length of the seat.
- **Seated Shoulder Height:** This dimension was used to define the height of the trunk support.
- Transverse diameter of the thorax.

Table 3 shows the average $\overline{x}$, the standard deviation (SD), and the 5th and 95th percentiles obtained from the anthropometric data for each body dimension that would interact with the bench. The percentiles were calculated by means of equation (1). It is necessary to clarify that these dimensions do not consider tolerances for clothing and footwear.

Figure 5 presents the bench components and their corresponding dimensions. Note that these dimensions correspond to the indicated percentiles of Table 3. The width of the trunk support was designed with the 5th percentile of the transverse diameter of the thorax to allow employees to freely move their arms and easy access their work area.

Stage 5 Results

After carefully reviewing pre-existing designs, authors selected the one depicted in Figure 6 as the starting point to develop our design. To make sure the pre-existing design could help meet the abovementioned FRs, authors applied the TRIZ principle of copying. Then, our ergonomic bench was designed with the anthropometric dimensions presented in Figure 5.

As can be observed in Figure 6, the height of the original support trunk is not adjustable; thus, authors redesigned this component according to the anthropometric dimensions presented in Figure 5c. However, the support is adjustable with respect to the angle taken by the trunk. Authors kept this DP, since it would help the technicians reach the forklift components comfortably. Then, they applied the dynamicity principle to the original design by adjusting the height and inclination of the trunk support and the height of the seat. Next, they implemented the extraction principle by removing those parts that would interfere with the employees' motions and their work area. Figure 6.b shows the removed parts circled in black. The base was positioned closer to the seat to reduce uncomfortable distances between the technicians and the forklifts. Likewise, the counterweight principle was applied to make sure that the technicians' weight could be supported by the ergonomic bench. Finally, authors implemented the cushion in advance principle by adding cotton pads for the seat and the trunk support. These pads provide a soft and comfortable contact between the users (i.e. technicians) and the bench parts.

Table 3. Body dimension, bench dimension, and corresponding percentiles

Body Dimension	Bench Dimension	$\overline{x}$	SD	Percentiles	
				5th	95th
Weight (Kg)	Material for seat and support for trunk	73	12.33	55.31	97.30*
Popliteal height	Seat height	412	25.65	374*	453*
Sitting hip width (mm)	Seat width	374	31.26	328	423*
Buttock-popliteal length (mm)	Seat length	476	28.92	432	526*
Seated shoulder height (mm)	Trunk support height	581	27.63	535*	638*
Transverse diameter of the thorax (mm)	Trunk support width	342	34.12	293*	398

* Percentiles selected to design the ergonomic bench.

Figure 5. Bench components and dimensions: a) seat, b) seat height, c) support for trunk
Source: Prepared by the authors (2017).

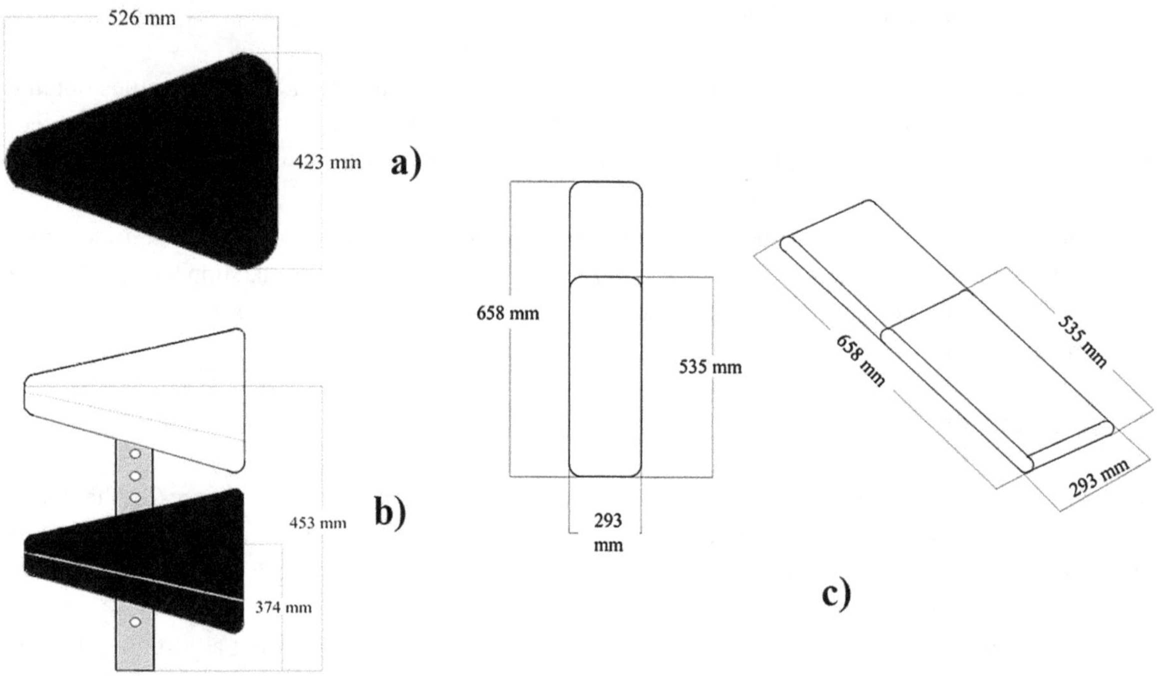

Figure 6. Starting point for the design of an ergonomic bench: a) original design, b) removed parts
Source: (Ebay, 2016).

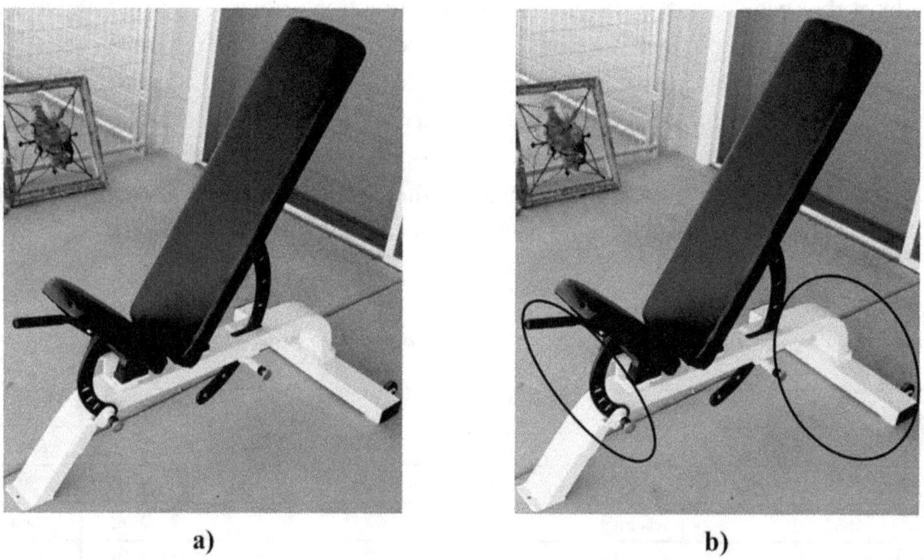

Stage 6 Results

Figure 7 shows the final design of the ergonomic bench. The colored lines represent the design dimensions based on the percentiles introduced in Table 3. Also, note that the trunk support is now height and angle adjustable, whereas the seat is only height adjustable. The solid part is made of Nylamid, which is a light and resistant material (Estrada-López, Vilchis-Martínez, Gutiérrez-Osorio, Arellano-Carmona, & Borja-Ramírez, 2013). Finally, note that the drawing shown in Figure 7 is not necessarily to scale. For instance, the maximum height of the seat is not on scale with the maximum height of the trunk support.

FUTURE WORK

As future work, authors will seek to actually construct the ergonomic bench and implement it. During the implementation process, research will aim to test the effects of the bench characteristics on factors such as employee performance, MSDs incidence, and manual handling. Then, the obtained results could contribute to a better understanding of employee safety, performance, and comfort. Moreover, since the level of postural risk was calculated by using REBA, it is convenient to bear in mind that to define this variable, REBA takes in into account whether a seat supports user weight, but it does not consider whether the weight is supported by a seat and a trunk support, as in this case. This means that, in the future, if a REBA analysis is conducted in a setting wherein this new ergonomic bench is implemented, the analysis would only take into account the trunk posture, but it would not consider whether the weight of the trunk is supported by the bench, which decreases the postural load on the person. For this reason, authors believe there might be an opportunity to further explore the shortcomings and limitations of REBA and propose an update.

Figure 7. Final ergonomic bench design
Source: (Ebay, 2016).

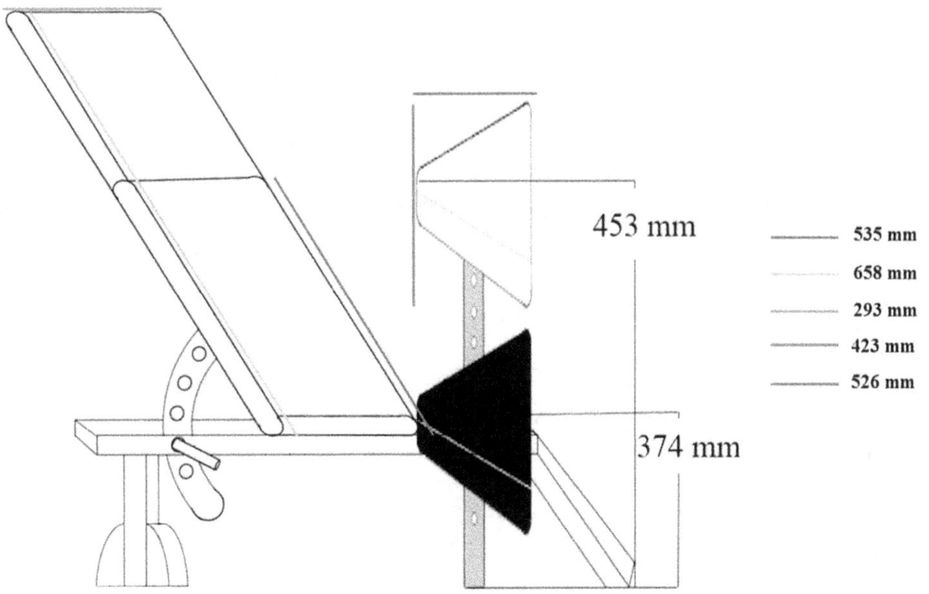

**For a more accurate representation see the electronic version.*

Another ergonomic solution for the task of changing forklift brake pads is an underground elevator, yet the final decision on whether implementing the bench or the elevator might depend on cost-related factors in terms of construction, implementation, and maintenance. That said, the elevator would allow employees to select a comfortable height and restful postures to change the brake pads. Also, the elevator would allow the company to save space, but it must be designed according to the anthropometric measures of its employees. Finally, authors encourage small companies from all the industrial sectors to research on and implement ergonomic tools that can enhance their performance while offering employees health and safety.

CONCLUSION

This research argues that anthropometry, combined with AD and TRIZ, is an excellent design tool, not only from an ergonomic approach, but in general. The importance of anthropometry lies in its ability to provide designers with precise design dimensions derived from the very specific characteristics of a sample. This makes anthropometry not only appealing, but also time-saving and resource-saving, since it prevents companies from implementing equipment devised from the characteristics of their employees. As for AD, authors conceive it as a tool that helps designers stay on the right track to reach their design goals. User needs, functional requirements, and design parameters serve as guidelines to the design process. Similarly, when implemented from an ergonomic point of view, TRIZ increases creative problem solving. Finally, as regards the ergonomic bench, authors expect that its implementation could diminish the level of MSDs risk in the task of changing forklift brake pads.

REFERENCES

Altshuller, G. (1984). *Creativity as an exact science : the theory of the solution of inventive problems*. New York: Gordon and Breach Science Publishers. Retrieved from http://cds.cern.ch/record/450367

Altshuller, G. (2002). *40 principles : TRIZ keys to innovation* (L. Shulyak & S. Rodman, Eds.; 3rd ed.). Worcester, MA: Technical Innovation Center. Retrieved from https://books.google.com.mx/books?id=mqlGEZgn5cwC&dq=triz&lr=&hl=es&source=gbs_navlinks_s

Ávila-Chaurand, R., Prado-León, L. R., González-Muñoz, E. L., & Muñoz, G. (2007). *Dimensiones antropométricas de población latinoamericana*. Guadalajara: Universidad de Guadalajara. Retrieved from https://www.researchgate.net/profile/Rosalio_Avila-Chaurand/publication/31722433_Dimensiones_antropometricas_de_la_poblacion_latinoamericana_Mexico_Cuba_Colombia_Chile_R_Avila_Chaurand_LR_Prado_Leon_EL_Gonzalez_Munoz/links/55490e8d0cf2ebfd8e3ad6bc/Dimensi

Azadeh, A., Fam, I., & Garakani, M. (2007). A Total Ergonomic Design Approach to Enhance the Productivity in a Complicated Control System. *Information Technology Journal*, *6*(7), 1036–1042. doi:10.3923/itj.2007.1036.1042

Bang, I. C., & Heo, G. (2009). An axiomatic design approach in development of nanofluid coolants. *Applied Thermal Engineering*, *29*(1), 75–90. doi:10.1016/j.applthermaleng.2008.02.004

Barros, R. Q., Marçal, M. A., & Soares, M. M. (2015). Ergonomic Analysis of the job of assembly and maintenance in an electronic equipment company. *Procedia Manufacturing*, *3*, 6542–6549. doi:10.1016/j.promfg.2015.07.953

Borgianni, Y., & Matt, D. T. (2016). Applications of TRIZ and Axiomatic Design: A Comparison to Deduce Best Practices in Industry. In Procedia CIRP (Vol. 39). Academic Press. doi:10.1016/j.procir.2016.01.171

Celik, M. (2009a). A hybrid design methodology for structuring an Integrated Environmental Management System (IEMS) for shipping business. *Journal of Environmental Management*, *90*(3), 1469–1475. doi:10.1016/j.jenvman.2008.10.005 PMID:19038488

Celik, M. (2009b). Establishing an Integrated Process Management System (IPMS) in ship management companies. *Expert Systems with Applications*, *36*(4), 8152–8171. doi:10.1016/j.eswa.2008.10.022

Celik, M., Cebi, S., Kahraman, C., & Er, I. D. (2009). An integrated fuzzy QFD model proposal on routing of shipping investment decisions in crude oil tanker market. *Expert Systems with Applications*, *36*(3), 6227–6235. doi:10.1016/j.eswa.2008.07.031

Cochran, D. S., Eversheim, W., Kubin, G., & Sesterhenn, M. L. (2000). The application of axiomatic design and lean management principles in the scope of production system segmentation. *International Journal of Production Research*, *38*(6), 1377–1396. doi:10.1080/002075400188906

da Costa, B. R., & Ramos-Vieira, E. (2009). Risk factors for work-related musculoskeletal disorders: A systematic review of recent longitudinal studies. *American Journal of Industrial Medicine*, *53*(3), 285–323. doi:10.1002/ajim.20750 PMID:19753591

Dickinson, A. L. (2006). Integrating Axiomatic Design Into a Design for Six Sigma Deployment (DFSS). *Design*, 2–7.

Durmusoglu, M., & Kulak, O. (2008). A methodology for the design of office cells using axiomatic design principles☆. *Omega*, *36*(4), 633–652. doi:10.1016/j.omega.2005.10.007

Ebay. (2016). *Precor Icarian Super Banco | eBay*. Retrieved July 20, 2017, from http://www.ebay.com/itm/Precor-Icarian-Super-Banco-/131881903124?_ul=AR

Ergonautas.upv.es. (2017). *Método REBA Rapid Entire Body Assessment*. Retrieved January 4, 2017, from http://www.ergonautas.upv.es/metodos/reba/reba-ayuda.php

Estrada-López, U., Vilchis-Martínez, S., Gutiérrez-Osorio, A., Arellano-Carmona, A., & Borja-Ramírez, V. (2013). Desarrollo de un Equipo Auxiliar para Rehabilitación de Pacientes con Lesiones en las Rodillas de bajo Costo. In *Memorias del XIX Congreso Internacional Anual de la Somim* (pp. 139–148). Pachuca, Mexico: SOMIM. Retrieved from https://www.researchgate.net/profile/Sean_Rodolfo_Vilchis_Martinez/publication/273772276_Desarrollo_de_un_equipo_auxiliar_para_rehabilitacion_de_pacientes_con_lesiones_en_las_rodillas_de_bajo_costo/links/550c53230cf2ac2905a3cc8b/desarrollo-de-un-equipo-au

Ferrer, I., Rios, J., & Ciurana, J. (2009). An approach to integrate manufacturing process information in part design phases. *Journal of Materials Processing Technology*, *209*(4), 2085–2091. doi:10.1016/j.jmatprotec.2008.05.009

Gallagher, S., Pollard, J., & Porter, W. L. (2011). Electromyography of the thigh muscles during lifting tasks in kneeling and squatting postures. *Ergonomics*, *54*(1), 91–102. doi:10.1080/00140139.2010.535025 PMID:21181592

Gangopadhyay, S., Ghosh, T., Das, T., Ghoshal, G., & Das, B. (2010). Effect of Working Posture on Occurrence of Musculoskeletal Disorders among the Sand Core Making Workers of West Bengal. *Central European Journal of Public Health*, *18*(1), 38–42. Retrieved from http://search.proquest.com/openview/6db3fd86bba430f5c0e2054d5b580110/1?pq-origsite=gscholar&cbl=54116 PMID:20586229

Gouvali, M., & Boudolos, K. (2006). Match between school furniture dimensions and children's anthropometry. *Applied Ergonomics*, *37*(6), 765–773. doi:10.1016/j.apergo.2005.11.009 PMID:16442494

Gumus, B., Ertas, A., Tate, D., & Cicek, I. (2008). The Transdisciplinary Product Development Lifecycle model. *Journal of Engineering Design*, *19*(3), 185–200. doi:10.1080/09544820701232436

Helander, M. G. (2007). Using design equations to identify sources of complexity in human–machine interaction. *Theoretical Issues in Ergonomics Science*, *8*(2), 123–146. doi:10.1080/14639220601092442

Hernández-Arellano, J. L., Aguilar-Duque, J. I., & Gómez-Bull, K. G. (2018). Methodology to Determine Product Dimensions Based on User Anthropometric Data. In J. L. García-Alcaraz, G. Alor-Hernández, A. A. Maldonado-Macías, & C. Sánchez-Ramírez (Eds.), New Perspectives on Applied Industrial Tools and Techniques (pp. 373–385). Cham, Switzerland: Springer International Publishing. doi:10.1007/978-3-319-56871-3_18

Herrera-Ruiz, M. (2006). *Silla de Ruedas Anfibia para Hidroterapia*. Instituto Politécnico Nacional. Retrieved from http://itzamna.bnct.ipn.mx/bitstream/handle/123456789/13911/Silla de ruedas anfibia para hidroterapia.pdf?sequence=1

Hignett, S., & McAtamney, L. (2000). Rapid Entire Body Assessment (REBA). *Applied Ergonomics*, *31*(2), 201–205. doi:10.1016/S0003-6870(99)00039-3 PMID:10711982

Hirani, H., & Suh, N. (2005). Journal bearing design using multiobjective genetic algorithm and axiomatic design approaches. *Tribology International*, *38*(5), 481–491. doi:10.1016/j.triboint.2004.10.008

Kim, Y.-S., & Cochran, D. S. (2000). Reviewing TRIZ from the perspective of Axiomatic Design. *Journal of Engineering Design*, *11*(1), 79–94. doi:10.1080/095448200261199

Kremer, G., Chiu, M.-C., Lin, C.-Y., Gupta, S., Claudio, D., & Thevenot, H. (2012). Application of axiomatic design, TRIZ, and mixed integer programming to develop innovative designs: A locomotive ballast arrangement case study. *International Journal of Advanced Manufacturing Technology*, *61*(5-8), 827–842. doi:10.100700170-011-3752-1

Kulak, O., Cebi, S., & Kahraman, C. (2010). Applications of axiomatic design principles : A literature review. *Expert Systems with Applications*, *37*(9), 6705–6717. doi:10.1016/j.eswa.2010.03.061

Lee, S., Yu, S., Choi, J., & Han, C. (2011). A methodology to quantitatively evaluate the safety of a glazing robot. *Applied Ergonomics*, *42*(3), 445–454. doi:10.1016/j.apergo.2010.09.002 PMID:20870215

Lenz, R. K., & Cochran, D. S. (2000). The application of axiomatic design to the design of the product development organization. Academic Press.

Lindkvist, L., & SÖDerberg, R. (2003). Computer-aided tolerance chain and stability analysis. *Journal of Engineering Design, 14*(1), 17–39. doi:10.1080/0954482031000078117

Lo, S., & Helander, M. G. (2007). Use of axiomatic design principles for analysing the complexity of human–machine systems. *Theoretical Issues in Ergonomics Science, 8*(2), 147–169. doi:10.1080/14639220601092475

Madriz-Quirós, C., Ramírez-Coretti, A., & Serrano-Montero, R. (2008). Estudio antropométrico para el diseño de mobiliario para niños de edad escolar en Costa Rica. *Tecnología en Marcha, 21*(4), 17–28. Retrieved from http://revistas.tec.ac.cr/index.php/tec_marcha/article/viewFile/221/218

Magdaleno-Adame, S., Olivares-Galvan, J., & Fofana, I. (2014). Electric Field Distribution to Analyze Possible Manufacturing Defects of a Large Preventive Autotransformer. In *International Conference on High Voltage Engineering and Application* (p. 4). Retrieved from http://s3.amazonaws.com/academia.edu.documents/34736938/final_tornillo_dielectrico_09_June_2014.pdf?AWSAccessKeyId=AKIAIWOWYYGZ2Y53UL3A&Expires=1498110339&Signature=emThzCWMeIUSTqLpL9GpmoiGWnA%3D&response-content-disposition=inline%3Bfilename%3DElectri

Maldonado-Macías, A., Realyvásquez, A., Hernández, J., & García-Alcaraz, J. (2015). Ergonomic assessment for the task of repairing computers in a manufacturing company: A case study. *Work (Reading, Mass.), 52*(2), 393–405. doi:10.3233/WOR-152118 PMID:26409363

Malina, R. (1995). Antropometría. *PubliCE Standard, 1.16*. Retrieved from http://g-se.com/es/articulos/articulo/pdf/antropometria-718

Martínez-Nopala, S. (2015). *Diseño y Construcción de un Vehículo Terrrestre para la Expulsión de Fauna Voladora en los Aeropuertos*. Escuela Superior de Ingeniería Mecánica y Eléctrica.

Nakao, M., Kobayashi, N., Hamada, K., Totsuka, T., & Yamada, S. (2007). Decoupling Executions in Navigating Manufacturing Processes for Shortening Lead Time and Its Implementation to an Unmanned Machine Shop. *CIRP Annals - Manufacturing Technology, 56*(1), 171–174. 10.1016/j.cirp.2007.05.041

Pheasant, S., & Haslegrave, C. (2016). *Bodyspace: Anthropometry, Ergonomics and the Design of Work* (3rd ed.). Boca Raton, FL: CRC Press.

Ramírez-Ibanez, M. (2004, December 14). *Aplicacion del Nylamid XL en un sistema de potencia*. Universidad de las Américas Puebla. Retrieved from http://catarina.udlap.mx/u_dl_a/tales/documentos/lim/ramirez_i_m/portada.html

Robinette, K. (2012). Anthropometry for Product Design. In G. Salvendy (Ed.), *Handbook of Human Factors and Ergonomics* (4th ed.; p. 1752). Wiley & Sons. doi:10.1002/9781118131350.ch11

Roffey, D. M., Wai, E. K., Bishop, P., Kwon, B. K., & Dagenais, S. (2010). Causal assessment of awkward occupational postures and low back pain: Results of a systematic review. *The Spine Journal, 10*(1), 89–99. doi:10.1016/j.spinee.2009.09.003 PMID:19910263

Savransky, S. (2000). *Engineering of creativity : introduction to TRIZ methodology of inventive problem solving*. Boca Raton, FL: CRC Press. doi:10.1201/9781420038958

Schnetzler, M. J., Sennheiser, A., & Schönsleben, P. (2007). A decomposition-based approach for the development of a supply chain strategy. *International Journal of Production Economics*, *105*(1), 21–42. doi:10.1016/j.ijpe.2006.02.004

Shirzaei, M., Mirzaei, R., Khaje-Alizade, A., & Mohammadi, M. (2015). Evaluation of ergonomic factors and postures that cause muscle pains in dentistry students' bodies. *Journal of Clinical and Experimental Dentistry*, *7*(3), e414–e418. doi:10.4317/jced.51909 PMID:26330941

Suh, N. (1990). *The principles of design. Oxford series on advanced manufacturing: 6.* New York: Oxford University Press.

Suh, N. (1998). Axiomatic Design Theory for Systems. *Research in Engineering Design*, *10*(4), 189–209. doi:10.1007001639870001

Suh, N. (2001). *Axiomatic Design: Advances and Applications.* Oxford University Press.

Tang, D., Zhang, G., & Dai, S. (2009). Design as integration of axiomatic design and design structure matrix. *Robotics and Computer-integrated Manufacturing*, *25*(3), 610–619. doi:10.1016/j.rcim.2008.04.005

Technical Innovation Center. (2013). *40 Principles.* Retrieved June 25, 2017, from http://triz.org/index.php/triz/principles

Teymourian, K., Seneviratne, D., & Galar, D. (2016). Ergonomics Contribution in Maintainability. In D. Galar & D. Seneviratne (Eds.), *Proceedings of MPMM 2016: 6th International Conference on Maintenance Performance Measurement and Management* (pp. 180–186). Luleå, Sweden: Luleå Tekniska Universitet. Retrieved from http://www.diva-portal.org/smash/record.jsf?pid=diva2%3A1108271&dswid=-2491

Togay, C., Dogru, A. H., & Tanik, J. U. (2008). Systematic Component-Oriented development with Axiomatic Design. *Journal of Systems and Software*, *81*(11), 1803–1815. doi:10.1016/j.jss.2007.12.746

Tompa, E., Dolinschi, R., de Oliveira, C., Amick, B., & Irvin, E. (2010). A Systematic Review of Workplace Ergonomic Interventions with Economic Analyses. *Journal of Occupational Rehabilitation*, *20*(2), 220–234. doi:10.100710926-009-9210-3 PMID:19890618

Vandergrift, J., Gold, J., Hanlon, A., & Punnett, L. (2012). Physical and psychosocial ergonomic risk factors for low back pain in automobile manufacturing workers. *Occupational and Environmental Medicine*, *69*(1), 29–34. doi:10.1136/oem.2010.061770 PMID:21586759

Won, K., & Joon, Y. (2005, November). Mutual Compensation of TRIZ and Axiomatic Design. *Design*, 1–12.

Yang, K., & Zhang, H. (2000). A Comparison of TRIZ and Axiomatic Design. In *First International Conference on Axiomatic Design* (pp. 235–243). Cambridge. Retrieved from http://moodle.stoa.usp.br/file.php/1359/TRIZ_AD.pdf

Yi, J.-W., & Park, G.-J. (2005). Development of a design system for EPS cushioning package of a monitor using axiomatic design. *Advances in Engineering Software*, *36*(4), 273–284. doi:10.1016/j.advengsoft.2004.06.016

KEY TERMS AND DEFINITIONS

Adjustability: An object's capability to adjust or adapt itself to the requirements and need of different users.

Bench: A piece of furniture in which one or more people can sit to adopt a comfortable posture and decrease muscle fatigue.

Body Dimensions: Distances of body segments measured between two defined points. For instance, the height of a person is the distance measured from the ground up to the vertex of the person in standing position.

Ergonomic Design: Design that applies ergonomic tools, such as anthropometry, and allows objects to be adapted to human capabilities and limitations.

Forklift: Vehicle used in the industry to transport material on pallets, which are supported by two forks of the forklift.

Percentile: In statistics, a position value that indicates the percent of ordered data that are below that percentile. For instance, the 5th percentile is the value under which 5% of data are.

Postural Risk: Risk derived by postural load supported by different body segments when a person adopts an uncomfortable body posture. This risk is proportional to the time posture is adopted, its frequency of occurrence, and its deviation level of neutral posture.

Chapter 3

Modeling Simulation-Based Agents for Reducing Operative Deferral in Oncology in a Virtual Scenario using Ergonomics

Adriana García
Universidad Politecnica de Aguascalientes, Mexico

Alberto Ochoa-Zezzatti
Autonomous University of Ciudad Juárez, Mexico

Juan Luis Hernández Arellano
Autonomous University of Ciudad Juárez, Mexico

Aide Aracely Maldonado Macías
Autonomous University of Ciudad Juárez, Mexico

ABSTRACT

For public hospitals, the patients are generally placed on a waiting list prior to having their surgery; it means that they have an elective surgery. The elective surgery is what can safely be delayed for more than 24 hours, but this usually is moved when some external factors are present, one of these is attributable to medical skills. For this reason, it is necessary to analyze the behavior of the last one to determine its relationship with the elective surgery performance and how it increases the surgical deferral. For purposes of the desired analysis, it is necessary to use one of the analysis complex system techniques, in this case the modeling and simulation agent-based.

DOI: 10.4018/978-1-5225-5234-5.ch003

Copyright © 2018, IGI Global. Copying or distributing in print or electronic forms without written permission of IGI Global is prohibited.

INTRODUCTION

Elective surgery is one of the leading healthcare public hospitals that provide efficiency and quality of scheduled service directly influences the increase of the surgical deferral. "Surgical interventions produce at least 7 million disabling complications and one million deaths a year worldwide. Half of these complications could have been avoided if applied systematically and efficiently the basic standards of care in rich and poor countries" (Berrueta, Malagón, Alvarado, Fernández, & Sosa, 2008). One of the reasons attributable to the surgical deferral is the lack of personnel (Berrueta et al., 2008) and in this context is necessary to identify the influence of each surgeon in performing elective surgeries and the effect it has on the surgical deferral. Therefore, a previous analysis of medical competence is a complex system, i.e., the behavior of a social system or society of agents (Medina Ramírez, 2013). Simulation software environment that allows the same model of the skills and attitudes of surgeons and the result of integrating a simulated surgery department then seeks to develop a simulation model agent-based (Gilbert, 2004; Izquierdo, Galán, Santos, & Del Olmo, 2008).

General Objective

Determine the effect of heterogeneity in the surgeon skills that influence in surgical deferral by applying the technique of analyzing complex systems such as agent-based modeling.

Specific Objectives

- Characterize the surgeon skills in a surgical procedure.
- Build a conceptual model of the current allocation system using agent-based modeling.
- Develop a mathematical model that represents the relationship between the performance of elective surgical and medical skills in a surgical procedure.

BACKGROUND

Nowadays exist a quantitative assessment of the efficiency of the operating room, by identifying variables proposed by the working group such as start time of surgery (Secretaría de Salud Pública (SEP, 2002). Campaña Nacional por la Calidad. Lineamientos Operativos del Sistema de Monitoreo para la Gestión de la Calidad, 2002), the percentage of cancellations, time changes and long delays. Only collection and analysis of data are collected (Bejarano, 2011). Another research is what assesses the work performance of doctors in Cuba in which uses empirical methods such as questionnaires, interviews, and surveys applied to teachers and students. In each case, the universe and the corresponding sample, as well as the considerations for choosing a strategy to establish a specific assessment model, was established. (Salas Perea, 2010). Finally, it was found a research that presents a classification of competencies profile surgeon in Mexico, by the critical incident technique 998 positive and negative incidents, information was grouped by similarities for identifying functions and activities were collected (Abreu et al., 2008).

At the present time the surgical deferral, i.e., elective surgery is again scheduled, is considered a critical impact on hospitals, whether public or private, generating major problems initially presenting patients to the issuance of the order of the general practitioner; therefore, it is important to know the extent to which influence medical expertise in the surgical deferral (see Figure 1).

HYPOTHESIS

There is a relationship between the skills of surgeons performing elective surgery influencing surgical deferral. An important stage in the implementation of the analysis is to determine agents each agent related with Multiagent System involved, their environment and the interaction between them. As present the first version of the conceptual model of elective surgery in Figure 2.

PROCEDURE TO A CORRECT IMPLEMENTATION USING VIRTUAL REALITY AND ERGONOMICS

The relevance of this research lies in the development of a correct model of ergonomics to be able to work with the medical instruments associated with the simulation in virtual reality and to be able to help to give realism and specific criteria to be able to make correct decisions in the emulation of a hospital with several surgical rooms and as decided to use according to specific needs and requirements.

System Analysis

Using Agent-Based Modeling (MBA) is possible to determine the behavior of each of the stakeholders (all involved) and the interaction between them as shown in Figure 2. Subsequently, techniques R & R

Figure 1. Methodology for the application of modeling and simulation based on agents

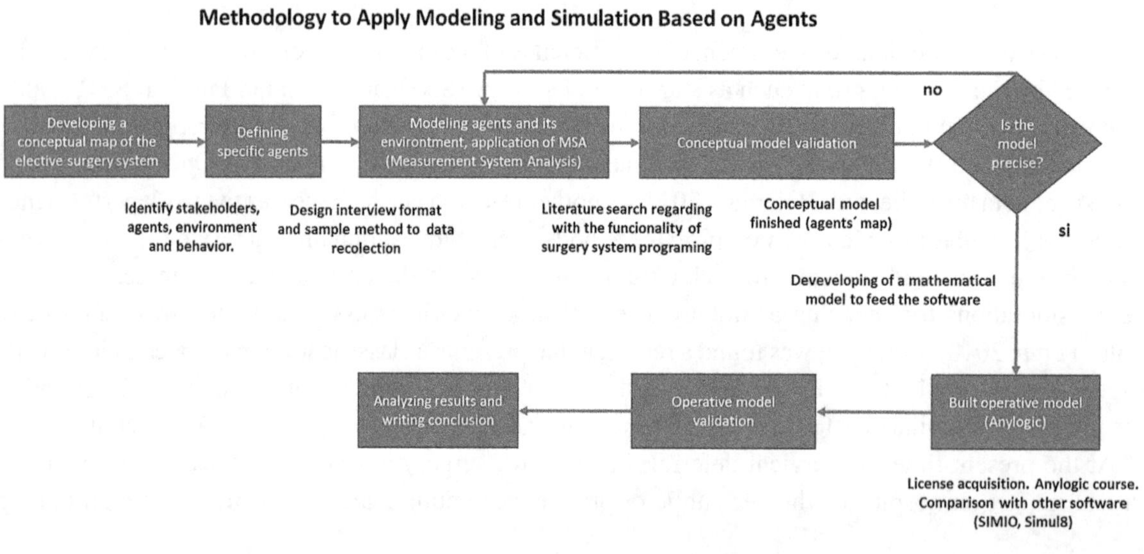

(Repeatability and Reproducibility) which are to detect those medical skills, which match each of the actors to define the types of medical skills, needed to avoid surgical deferral (Characterization of variables).

Mathematical Modeling

Characterized once the variables to be used are necessary to construct a mathematical model based on a deterministic optimization. The mathematical model will be developed using simple linear programming. So far, variables are defined only to be used:

Xi = type of competence of the surgeon. (1)

C = Number of times an elective surgery is rescheduled. (2)

T = Time that elective surgery is rescheduled. (3)

M: Factor influencing the deferral because of medical competence based on the following ranking factors: risk, frequency, severity, and frequency. x + y = z (4)

Figure 2. The first version of conceptual model of elective surgery, the stakeholders involved and the interaction between them, using the graphics that represent a probabilistic distribution

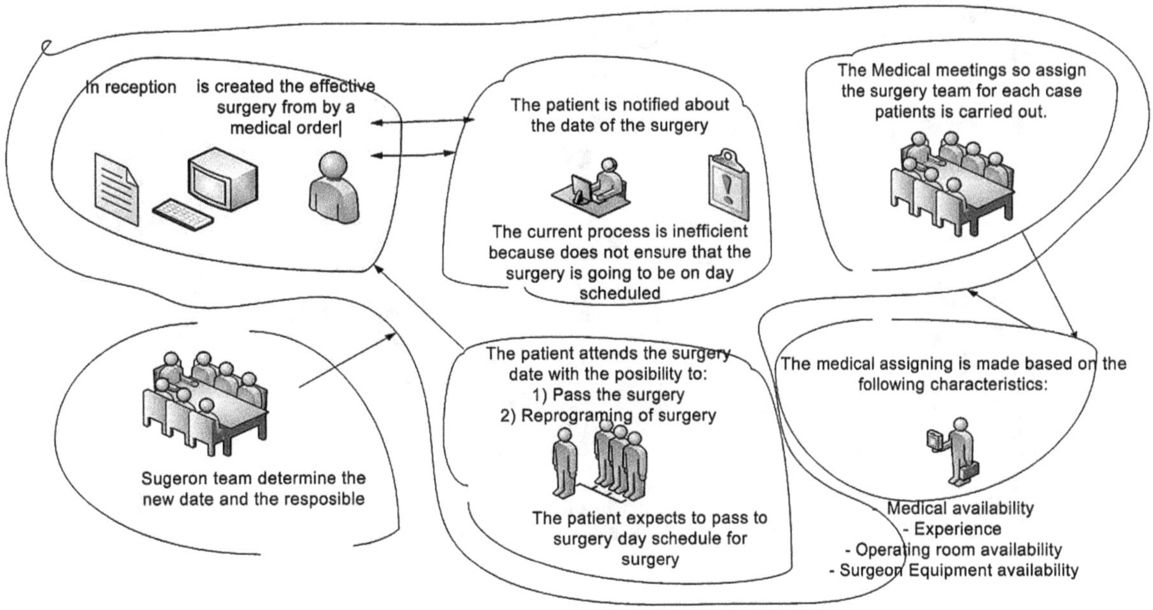

Simulation

To carry out the simulation of the analyzed complex system, it requires software that meets the basic requirements of the Agent-based modeling: allow social simulation, ie, complex systems and societies of agents to analyze their behavior, to solve a formal mathematical model, ie a model that is implemented and that can run on a computer. Other representative factors in choosing multi-agent software are based on an abstraction, design, and coding, software must perform logical inference and solution or multiple solutions for the modeler perform the corresponding analysis, interpretation, and application of the model finally if it meets the initial expectations of the study. On the market at least 36 software that can meet the above expectations, some of which are: (Arias, 2004) (Bernabeu, Sabater, & Maheut, 2010); (Pereda & Zamarreño, 2015), FIPA-OS (Commercial-Free), Grasshopper (Commercial-License) JACK (Commercial-Licence), among others (Borshchev & Filippov, 2004). However, for purposes of this study and the characteristics of Versatility, intuitive and simple to build and simulate models through causal diagrams, valuation decisions through interactive games and build apps DSS (Decision Support System), reporting EIS (Executive Information Systems), import and export data spreadsheets, is free for personal or academic version, business or government, among other functions covering the initial expectations of the investigation, With the use of Vensim PLE (Personal Learning Edition) software, the simulation accurately represent the performance of the current elective surgery will develop, which subsequently allow considering improvements on it, based on the establishment of standards in the protocol action or action of each of those involved in the system. In Figure 3, we show a virtual reality model associated with the correct ergonomic applied to a specialized hospital.

Figure 3. Virtual reality model of a specialized hospital with an adequate ergonomics to use this

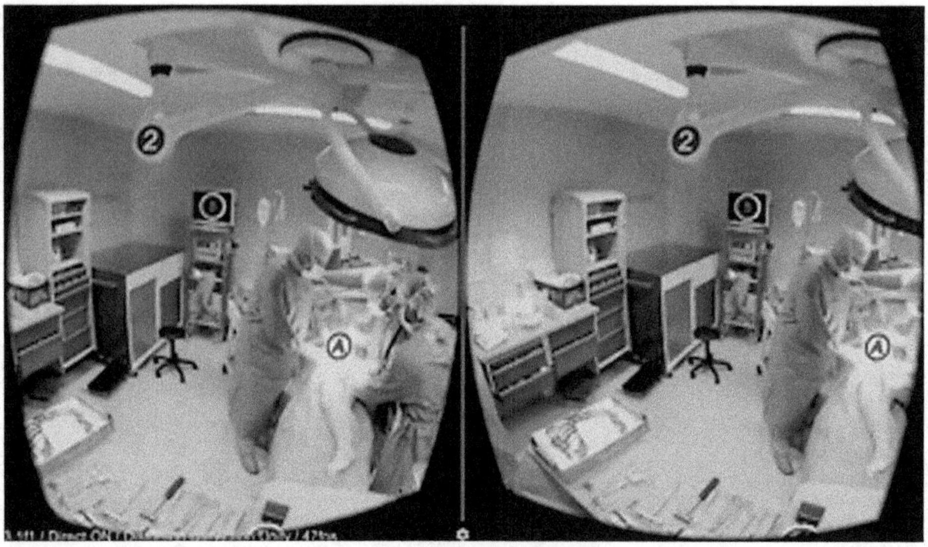

CONCLUSION

Analyze the behavioral factors influencing surgical deferral as the variability in applied medical skills, ie acting in the way each of the actors to schedule surgery, or performing elective surgery. It is necessary as in any complex system to perform an analysis using agent-based modeling to determine the actual behavior of the same based on both global perspectives as an individual and then develop a mathematical model and a simulation designed to show the current system, making changes without involving the corresponding cost.

Surgical deferral besides being one of the problems of elective surgery involves extending the health status of patients. For future research is important to analyze the rest of factors that influence in surgical deferral. The main contribution of this work is to develop a model of the powers of a surgeon through Agent-Based Modeling. There is no work at this level, other efforts to model performance of individuals but not in the area of the surgeon. There are no studies that analyze the behavior and skills of a surgeon by Modeling Technique Based on Agents. It will be possible to work on aspects that contribute to surgical deferral and thereby achieve decrease it.

REFERENCES

Abreu, L., Cid, A., Herrera, G., Lara, J., Laviada, R., Rodríguez, C., & Sánchez, J. (2008). *Perfil por competencias del médico general mexicano*. AMFEM.

Arias, J. (2004). *Propedéutica quirúrgica: preoperatorio, operatorio, postoperatorio*. Editorial Tebar.

Bejarano, M. (2011). Evaluación cuantitativa de la eficiencia en las salas de cirugía. *Revista Colombiana de Cirugía, 26*(4).

Bernabeu, X. P., Sabater, J. P. G., & Maheut, J. (2010). Arquitectura de un modelo de simulación basado en sistemas multi-agente de una red de suministro en el sector del automóvil. *4th International Conference On Industrial Engineering and Industrial Management*, 3–14.

Berrueta, É. A. R., Malagón, G. V., Alvarado, B. M., Fernández, M. del C. A., & Sosa, L. E. O. M. V. C. (2008). Tiempo perdido entre cirugías y su repercusión en el cumplimiento de la programación quirúrgica del Hospital General de Querétaro. *Cirujano General, 30*(4), 220–223.

Borshchev, A., & Filippov, A. (2004). From system dynamics and discrete event to practical agent based modeling: reasons, techniques, tools. *Proceedings of the 22nd international conference of the system dynamics society*, 22.

Gilbert, N. (2004). *Agent-based social simulation: dealing with complexity*. Guildford, UK: Centre for Research on Social Simulation, University of Surrey.

Izquierdo, L. R., Galán, J. M., Santos, J. I., & Del Olmo, R. (2008). Modelado de sistemas complejos mediante simulación basada en agentes y mediante dinámica de sistemas. *EMPIRIA. Revista de Metodología de Las Ciencias Sociales*, (16).

Medina Ramírez, R. (2013). *Main Factors of Surgical Deferment in Elective Surgery: A Systems Dynamics Analysis* (Unpublished Doctoral Dissertation). University of Guadalajara, Jalisco, México.

Pereda, M., & Zamarreño, J. M. (2015). Modelado Basado en Agentes: un Enfoque desde la Ingeniería de Sistemas. *Revista Iberoamericana de Automática E Informática Industrial RIAI, 12*(3), 304–312.

Salas Perea, R. S. (2010). Propuesta de estrategia para la evaluación del desempeño laboral de los médicos en Cuba. *Educación Médica Superior*, *24*(3), 387–417.

SEP. (2002). *Lineamientos Operativos del Sistema de Monitoreo para la Gestión de la Calidad*. Retrieved September 12, 2014, from http://www.salud.gob.mx/unidades/cdi/documentos/DOCSAL7494.pdf

Section 2
Ergonomic Product Design

Chapter 4
Aging Suit:
An Accessible and Low-Cost Design Tool for the Gerontodesign

Annika Maya Rivero
Universidad Autónoma del Estado de México, Mexico

ABSTRACT

The population pyramid is being inverted and the designers are facing challenges in the way they design products, spaces, and services for elder people. Although the design is composed of both cultural and social aspects, the functional aspect takes relevance for autonomy and empowerment of the person that gets older. To design products centered on the physical capacities of the elder population, the aging suit has being used as a tool for the gerontodesign. In countries like Mexico, its application is limited by the manufacturing cost. For this reason, as part of a design methodology for elder people, a low-cost suit was developed focus on the design students and professionals located in Latin America.

INTRODUCTION

The global aging of the population is a reality that has been glimpsed since some years ago. The main issue is not that the world has more elders than children, but the world is transforming, because as life expectancy increases, it is much more likely that grand grandfathers, grandfathers and parents live together, not necessarily in the same house. In fact, in some European countries, more than 40% women over 65 years old are currently living alone. In Denmark, is 55.1% (CSO, 2007:21) of women over 65. Global aging covers a diverse series of problems that are of the competence of every human being, regardless we are interested or not in aging processes. According to the global report about aging and health (Beard et al., 2015) aging and health needs to be addressed because several reasons: the first one belongs to human rights promotion, the second is to promote the sustainable development and the third one obeys the economic aspect.

Mexico is a country that counts with a population that exceeds 112 million people. Of them, a little more than 10 million has more than 60 years, this, based on the national population and housing census, carried out by the National Institute of Statistics and Geography (2010b:3). According with the recent

DOI: 10.4018/978-1-5225-5234-5.ch004

Copyright © 2018, IGI Global. Copying or distributing in print or electronic forms without written permission of IGI Global is prohibited.

inter-census survey of the same institution (INEGI, 2010b:35), is estimated that the population has increased to almost 120 million. Therefore, the number of elderly people has also increased. In Mexico 30% of older adults in Mexico are economically active (INEGI, 2010b: 132) but the reality observed in supermarkets where people over 60 can be seen as "volunteer packer", a work promoted by the National Institute of Older Adults (INAPAM, 2016) reflects that older people in Mexico perform tasks that require efforts that most of the adult population cannot perform efficiently. Voluntary packers require stand up for extended periods of time, in addition to having fine motor skills and hand-eye coordination. It is important to avoid that older people perform tasks that demand beyond their functional capacities. In order to promote older adults in tasks according to their functional capacities, it is necessary to have empathy towards the situation that many of them present.

To design a safe and inclusive environment that promotes a healthy performance of any activity, whether work or recreation, it is necessary to understand which capacities both functional and of various kinds an older adult person own. Just as the volunteer packers, we can observe in our day to day adults performing tasks that require physical skills such as gardening, blacksmithing, bakery, among many others. The environment is becoming a problem for people who remain economically active, but also for those who have a high degree of dependence, the older adults who require special care. The experience of aging may necessitate transitions in living environments, either through adaptations to current residences or through relocations to more supportive environments (Perry, Andersen & Kaplan, 2014: 75).

As a general objective of this chapter is to deepen the use that has been given to a design tool for older adults or gerontodesign, as well as to know how the aging suit has been applied in different places, a tropicalized proposal is made to the Mexican context, in order to reduce the costs of this tool as well as not only make it accessible to professionals and students of industrial design, architecture and other disciplines where the development of product and spaces is the objective, but to any individual who has to coexist, decide and propose some program, service or product for older adults.

BACKGROUND

Design for the older adult population, has been considered as part of design for minorities (Papanek, 1984) has become a hot topic, which is being addressed in various parts of the world. There are records indicating studies in 1800 to determine the relationship of man with environmental effects related to light, sound, weight and other variables (Kopec, 2012: 8), this is known as Environmental Psychology, which is related to the design for the aging according with The American Psychological Association (APA). In the other hand, the design for older adults has its beginning in the environmental gerontology initiated by E. Birren in the fifties (Hans-Werner Wahl & Weisman, 2003) Lawton and Nahemow defined it as the adaptation of the human to his environment and that environment presents alterations (1973). In Parmelee and Lawton referred to the environmental gerontology has been delayed by those who execute it in the field work, they specifically pointed to architects, interior designers, and planning specialists. Parmelee and Lawton said that those professionals did not generate theory (1990), thing that in our days is changing because the new perspective about the designer rol and for the implementation of new design approaches and new design tools as the aging suit. In recent years besides the design for aging, we talk about gerontodesign that emerged from the social anthropology of Portugal, which is defined as:

An agglutination of gerontology with design in the sense of having a design to project, conceive and adapt models to older adults and not maliciously conspire to sell models designed for people with Deficiencies (Parra-Marujo, 2006: 3).

To blend gerontology with design, scheming and adapting models for the elderly, it is necessary to apply design tools that provide valuable information and generate some empathy and understanding of the problems that the elderly person faces. Within these tools, is found the aging suit or old simulator. The aging suit has been known as a design tool for older adults since relatively little time, there is no record in the literature that indicates the first aging suit, but it is known that different academic entities have developed it, such as the AgeLab at the Massachusetts Institute of Technology, which has created the AGNES (Age Gain Now Empathy System) suit, which is designed to approximate the visual and motor skills as well as the flexibility and dexterity and strength of a person of About 75 years old (AgeLab, 2017). According to the "AGNES & Livable Communities" project report in 2013, when conducting a series of empathy exercises using the AGNES suit, researchers, students, developers, and vendors could realize that accessibility does not necessarily mean ease. This was exposed when wearing the suit to walk in the street heading to centers of morning activities. Participants became fatigued and concluded that distances should be shorter and rest points should exist (Coughlin, 2013).

In addition to AGNES, there is another age simulator suit named GERT, which offers young people the typical limitations of older people such as: Opacity of the eye lens, restriction of the visual field, high frequency deafness, restricted head mobility, joint stiffness, loss of strength, reduction of grip ability and reduction of coordination. This according to the German product website (GERT, 2017). In 2008, a study of the efficacy of the age simulation was carried out, as part of the doctoral thesis of Sascha Allan Filz, at the University of Würzburg in Germany. This study showed that of the 125 people who wore the GERT suit 95% of them improved their ability to understand the physical situation of an older person (Filz, 2008). GERT is now a product that is marketed online, its cost in Mexican pesos is just over 26,000 pesos (approximately 1,500 USD). This price includes the GERT age simulator that emulates the hearing, vision and motor skills of a person over 75 years of age. But if you want to add some kind of accessory to simulate hand tremor or back pain, among other features, you have to cover an extra cost, according to the GERT website (2017).

AGNES and GERT are not the only aging simulator suits, in fact, academic institutions such as the Design School of Loughborough University and the Engineering Design Center at the University of Cambridge have designed tools that simulate suffering in the hands of older people (Kullman, 2016: 75). The R70i suit is another suit produced by Applied Minds LLC, which according to H.L. Groza, S.B. Sebesi and D.S. Mandru (2017) is the suit that has the newest technology to generate a more real experience around feeling old, is a suit that is controlled wirelessly by means of a computer.

Companies from various sectors also have this design tool as an element that supports them in the projection of new products, such as Ford (CBS, 2014), Nissan (Reuters, 2008) and Blum, who has named its simulator: Age Explorer ® (Blum, 2017). It can be said that the aging suits or empathy generators have a recent history of less than 10 years of existence, although investigations like the one made by Patricia Moore more than 30 years ago (Moore & Paul, 1985), where the author crossed the United States disguised as an adult person, could be the direct antecedents of this tool that causes novelty today.

It is important to promote an environment which promotes the autonomy of elderly because as it is said by Dak Kopec this helps us to prevent passive and depended behaviors, which are preludes to an assortment of disabilities and perhaps institutionalization (Kopec, 2012:135).

DESIGNING A LOW-COST AGING SUIT

According to Kim Kullman, the old simulators "rather than being ready-made artifacts, simulations are open-ended prototypes that introduce both possibilities and limitations into the design (2016: 2)". The process of designing an age simulator is iterative, entails knowledge extracted from both documentary and field information. It involves getting to know older people in a deeper way through observation techniques, interviews, and talks. "When measured against ergonomic, medical and other existing data, then, wearable simulations rarely provide, 'a fully accurate match' (Hall et al., 2010:608)", viewing this from a cross-disciplinary perspective, the design of age simulators brings new insights into the visions and challenges when designing for older adults. "The purpose of his wearable simulations is rather to demonstrate that the body is more complex than designers might initially think (Kullman, 2016: 82)". The old simulators do not pretend to reduce the human being to a series of functions and movements but open the minds of the designers to understand that the human being is more complex than was believed and that each simulation can be redesigned as many times as necessary.

As part of a master's research (Maya-Rivero, 2012), supported by CONACYT, which sought to generate a series of gerontodesign strategies, it was decided to create a low-cost aging simulator based on the user-centered design method and taking as a guide suggested by the AgeLab suit (2017). At the beginning was unknown the existence of the old simulators, in fact, its creation came due to the need to be able to test a day in the life of the elderly person. A documentary research was carried out to know the process of functional aging in a person. This investigation began knowing the current state of health of the Mexican adult population based on studies such as ENASEM (National Study on Health and Aging), ENSANUT (National Health and Nutrition Survey of the State of Mexico) and statistical data on disability, old age and dependency by INEGI (National Institute of Statistics and Geography). In parallel, field research was carried out using ethnographic techniques such as non-participant observation and photography. The design of the empathy-generating costume had to be contextualized in the Mexican context as the subject of study focused on the problems older Mexican adults present when preparing food.

To contextualize the aging suit, it was known that the Mexican adult population in 2011 was close to ten million people (INEGI, 2010a), but to date, it is known that in Mexico there are 12.4 million people over 60 (INEGI, 2016). Of these, 36.1% have some functional limitation and 26% have some disability (INEGI, 2016: 4). Also, according to INEGI, the difficulties that older Mexican adults present are: Walk, up or down using their legs as well as having vision problems (2016: 5). The functional status of older adults does not has changed a lot, these limitations (sight and mobility in lower limbs) were considered when designing the low-cost age simulator. It is important to know that most of the older adults in Mexico are women (INEGI, 2016) because if we think about aspects of anthropometry and retrieve information from the topic we will know that the maximum height of Mexican adult women is 1.61 meters, but can measure less than 1.39 meters (Ávila-Chaurand et al., 2007). As we are aware of this, the design of the suit changes and seeks to bend the person who uses it to reach a height approaching the 95th percentile the older women in Mexico.

With the purpose of obtain the comfort angles for a population of older Mexican women, the linear ranges known and published in the book: "Anthropometric Dimensions of the Latin American Population: Mexico, Cuba, Colombia, Chile" (2007) was taken as a base and starting from the difference between the reach of the population under 65 and the range of the population over 65 years, it was obtained that the variation of comfort angles in the elderly is + - 20%. Here are some examples: 631 mm is the 100% of the frontal reach of the arm of the woman under 65 years and 508 mm is the reference to the frontal reach of the arm of a woman over 65 years. We obtain that 508 mm is 80% of the total reach, the reason why there is a variation of 20%. The lateral reach of the arm of a woman under 65 is 645 mm and the lateral reach of the arm of a woman over 65 is 499 mm, the total reach is reduced by 23%. This was corresponding with Yasmin Lavalle-Herrera (2014: 14) who states that the range of joint movement decreases approximately 25% in the third age. Together with the statistical information, a narrative review of the literature focused on the physiological changes of the person over 60 years old was carried out. The information was classified based on the senses (sight, taste, smell and ear) besides to known the diseases more common in the aged people. Below, the actions carried out around the design of the aging suit are described, considerations obtained as part of the documentary research carried out:

Mobility limitations were placed because osteoarthritis, along with diabetes and hypertension, are present in the majority of the adult Mexican adult population (Aguilar-Salinas, 2013). This condition affects the biomechanics of the joints resulting in pain, decreased mobility and deformity (Eduardo Wainstein, 2014). The simulation of joint stiffness, in addition, to imitating this condition was used to increase muscle fatigue and to obtain slowness in executed movements. In the case of the hands, the objective was also to decrease the force of pressure, as this decreases in the Mexican population (Beard et al., 2015). Limitations of mobility were placed on all possible joints such as neck (soft collar), fingers (latex gloves and masking tape), knees (elastic knee pads), elbows (elastic cloth elbows) as well as strings connecting the arm and waist of the person to reduce the mobility of the shoulder joints, since shoulder cuff tears are common in older people (Gumina et al., 2013). To shorten the pace and generate shorter and slower leg movements (slowing the walking speed), the movement of the hamstrings was limited by placing a tense ligature that joined the waist with the heel of each foot. Also with the aim of mimic, the stiffness of the aged body has added the use of a sash to further prevent rotation of the spine, in addition to providing heat to the body and promote physical wear due to pressure and temperature rise. With the application of a helmet used for protection in constructions and suspending cords to the shoulders, it was planned to simulate the compression and rotation of the vertebrae, giving the user a sensation of curvature of the spine that is produced with the aging.

The changes around the perception were realized making use of glasses of yellow color to hinder the vision of the colors, besides that with them it is difficult to see in low light. The glasses were adapted using grease on them to mimic visual problems such as cataracts and diabetic retinopathy since in Mexico most eye diseases are diabetic retinopathy, macular edema, retinal detachment and advanced cataract (Flores-González & Calonje, 2017). In fact, according to the World Report on Aging and Health (Beard et al., 2015), older adults have an opacity of the lens that results in blurred vision and decreased the ability to focus, while they can present hearing loss, so that earplugs were used to simulate difficulties with sharp sounds and soft tones. As for the loss of balance is emulated with footwear with plastic rubbers under the sole. Finally, the use of masks was implemented in order to imitate the lack of air, frequent in elderly people with respiratory problems

Elaborating and Testing the Suit

After having made the design brief of the elderly simulator suit, it was continued with its elaboration, the basic idea was to make it homemade way with materials that most people could obtain. The materials used were: Protective helmet for construction, elastic strings of different resistances designed for automobile towing, work overall, belt for abdomen, elbow pads, knee pads, soft collar, yellow safety goggles, masking tape, latex gloves, ankle and wrist gaiters, sandals, rubber to protect chair legs, earplugs, ear muffs, thread and needle. The cost of preparing the old age simulator suit was $ 1750 Mexican pesos (98 USD), reducing fifteen times the cost of purchasing the GERT suit.

The suit assembly was initially tested only with the latex gloves, which had the function of limiting joint movement in the hands as well as diminishing the ability of tactile perception. The gloves were placed and over them was placed masking-tape in order to tie up the fingers of the person. To finish were placed the gaiters of 500 gr on each wrist. The first experiment lasted 25 minutes and consisted of eating a breakfast a plate of fruit, coffee with milk, juice, and bread. The physical reactions that were had were: Cramps in both hands and pain in joints of the same.

Afterwards, the whole suit was tested (Figure 2). The initial test lasted a little more than 120 minutes, with activities such as: Going up and down stairs, walking, sitting and standing up from a chair, from a bed and from a toilet, besides performing the daily cleaning of a house such as picking clothes and placing them in a basket, opening and closing windows, and preparing food. The fatigue was increasing as time went on at par with frustration. The whole experiment was done in solitude, emulating the reality lived by many elderly people living alone, was an experience with high levels of stress. The activities in the kitchen were experienced with frustration when not being able to reach the objects, had to do things in a different way. The biggest obstacle was the vision as everything was done using imagination and memory, something that if not reflected all older adults can realize due to cognitive problems. Another obstacle was the reach: which was above the vision and below the knees, was impossible to take. The activities were done without much concern about the quality of what was produced. As time passed the body seemed to adapt gradually to the physical state in which it was, but the mind was difficult to control because the anxiety became present after the 20 minutes of the experiment. At the end of the test, the physical pain was notorious and it was concluded that the suit should not be used for long periods of time and that those who used it should be in good health (no heart problems, no osteoarthritis or arthritis, no sprains or problems of mobility).

Figure 1. Photograph of hand and joint limitations
Own authorship (2012).

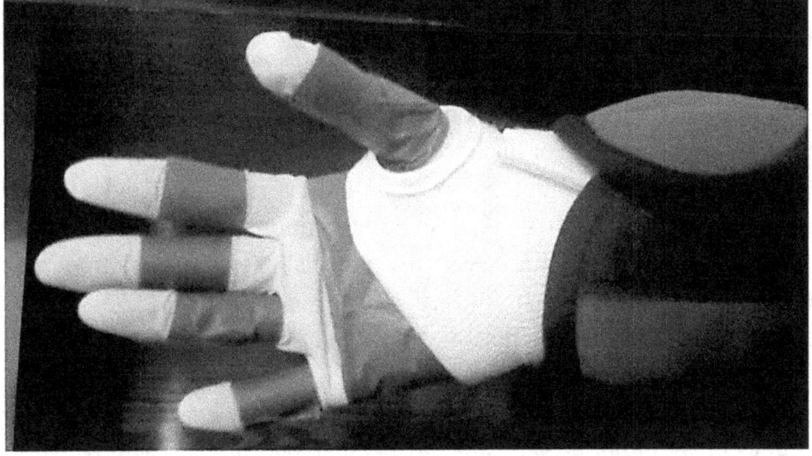

Figure 2. Photograph of the Empathy Generating Suit
Own elaboration. Own authorship (2012).

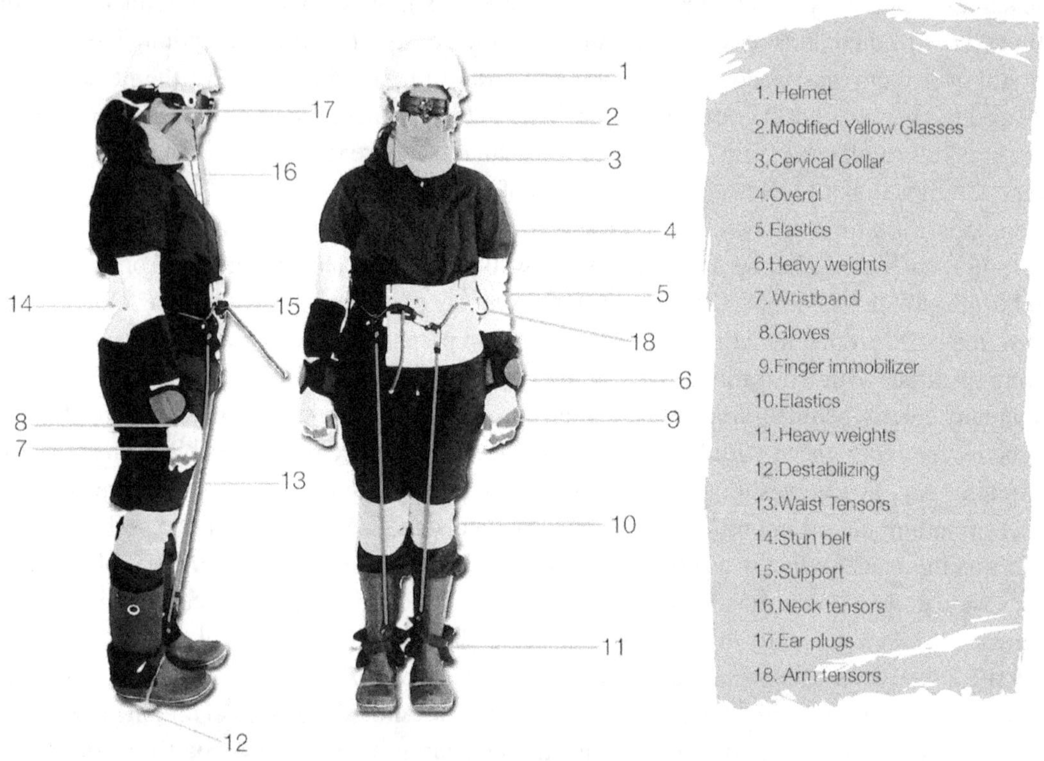

An Experimental Study

With the purpose of verify the efficiency and performance of the sympathy generator suit, as well as to continue with the design process of the same tool, an experimental study was carried out with ten subjects under sixty and over twenty-five years old, who were exposed to the use of the age simulator for a period of fifteen minutes, a period of time in which users of the suit performed a series of tasks focused on preparing food in a kitchen. They all performed the same tasks: Open a watertight container, eat using cutlery, open shelves, look for kitchen tools and open a can. We sought to relate their experience to some emotion emitted consciously or unconsciously, in order to find a causality between emotional state and physical difficulties. The volunteers were asked to perform a user-centered design technique called: Thinking out loud. Using this technique in collaboration with non-participant observation, the emotions of the ten subjects were obtained, which were related to the functional needs of the elderly person (represented in the aging suit).

The ten volunteers expressed their feelings in a written and verbal way:

Wearing the suit I never imagined what I would feel. I felt useless of not being able to realize things that are made simple to us being in full powers, now my perspective changed with respect to the old people and the difficulties that they face daily. It is encouraging that today designers are worried about this. Volunteer 5.

I wish there were no doors, drawers, shelves, Ah! And hopefully all the ovens were simple to use, unfortunately, most are more complicated. Volunteer 6.

Although it was a momentary experience, the anxiety, and importance that is felt or that causes, it made me think about how my life will be if it is that I arrive at least to the 65 years, it will be necessary utensils (equipment) that can give us an easier life. Volunteer 7.

Phrases issued at the time of the food preparation tasks indicated that the emotional state of the people and their level of confidence was related to the motor skills they had at that time. Some phrases were: *What if I electrocute myself? Is up? Ay do not tell me that!, Being this way no matter how you do things, but what you can do ... as you cannot be demanding as when are you okay!, Take it off because the water is going to fall on me!, This is easy opens for who?, It is easier to bend down than to climb, This easy open is pure story!, It is not difficult to open the freezer, the difficult thing is to identify if there are ice or not!* among many others. This was how a tool focused on the motor and functional aspects of older adults, in addition to providing information on the ergonomics and usability of products, also offered us a window on the sensations produced by physical limitations, promoting sympathy in whom used the elderly simulator.

Findings

Subsequent to the development of the experiment, each of the ten volunteers was asked to answer a questionnaire in which the objective was to reaffirm what was observed and to further specify the limitations that were most recurrent in the performance of the activities. Seven of the ten users of the aging suit said they presented greater problems with a spoon (compared to other activities), four of ten said that the greatest physical difficulty was to maintain a posture erected and three of ten thought that raising their arms was the most complicated. 100% of the volunteers said they had greater empathy with the elderly. The words most frequently used by wearers were: Uncertainty, Impotence, and limitation. These words are the same related to the moments in which the volunteers expressed: Having little visibility, lacking strength and not reaching the objects they wanted to get.

About the design of the aging suit it was noted that it was necessary to change the material of the torso band, because it raised the temperature of the users and they sweated too much (especially in the belt area) it was suggested to use a loader belt that decreases body rotation but has more ventilation than the previous one. During the use of the suit were added different motor obstacles, such as the overlapping of latex gloves (one over another) to generate greater limitation in the hands, in addition to the use of tape to limit the area of vision of the glasses with the purpose of diminishing peripheral vision. The use of elastic bands was a success since these could be adjusted according to the height of the person, after the first person who wore the suit it was decided to sew the knee pads and elbows to the overalls to expedite the placement of the age simulator. In addition, two sizes of one medium-sized and one medium-large suit were made, due to the diversity of heights and weights of the participants.

The empathy-generating suit collaborated in the generation of nine strategies of gerontodesign, among which are the promotion of accessibility and safety. Based on what was obtained, it was concluded that when designing for people over 60 years it is necessary to optimize the vision, thinking that everything you want to communicate should be contained within a range of 90° visibility, and lighting designed for older people requires giving importance to point lights, adjustable and movable. To help decrease the distortion of objects caused by a poor eye focus, we propose the use of high contrast and color plots, as well as contrasting shapes. It is necessary to think that when designing a space in which the aged person is, this, must have rest stations at strategic points. In regard to hearing it is essential to take into account the decibel ranges according to hearing loss. Also, it is of vital importance to contemplate that all the actions to be performed should be able to be performed with more than one joint, such as being able to write with the elbow instead of the fingers or use the upper limbs instead of the lower limbs (Maya-Rivero, 2012: 144). Focusing on the promotion of movements only at 90° or below the range, with regard to the reaches with upper limbs could be the ideal to avoid accidents in older people and to reduce the effort of them.

On the security aspect, it was proposed to promote lightweight products, also unbreakable, because it is known that many older adults have loss of muscle mass (sarcopenia). All product or system focused on the older adult must have a way to get feedback from an action made it by the user, doesn't matter if the action was correct or not, the important thing should be let him know, thanks to the aging suit it is proposed that feedback should be provided in more than one way, focusing on the 5 senses of the person: touch, taste, smell, sight, and hearing. Always remembering that the use of complementary senses will facilitate access to information, for example, the combination of sight and hear (Maya-Rivero, 2012: 152).

As it's been said, this research was part of a case of study, which focused on the cooking area for Mexican elders who were born in 1920, so many ethnographic research was made and this mixed with the aging suit could bring about a design proposal, which reflects the design strategies focus on the design for elderly people in the Mexican context. It turned out to a simple environmental design, but very thought-out, full of light and easy actions to follow, a stove with two buttons and two burners and a small refrigerator with doors that allow you to see inside. It was just a simple sketch result, which allows outlining how should be a kitchen area for most of the elderly in Mexico.

Controversies

Although this experimental study does not have statistical validity, it shows that the effectiveness of the GERT suit used by 125 people compared to the homemade costume used by 10 people is debatable. It is likely that in very specific aspects related to ranges of movement or visibility the GERT suit and especially the R70i, which is controlled by means of a computer, have better effectiveness and more precision. But if you talk about the current needs of universities and schools of design, it can be said that the homemade costume will give the necessary empathy to the students to be able to start a design thinking centered on the elderly person, since in fact one of the major problems facing the area of product development, spaces and services for older adults is the lack of professionals interested in the area. So as a first contribution, the low-cost aging simulator provides empathy for those who use it. It should be understood that the aging suit is a gerontodesign tool, but it is not the only way to design for older people, it should be seen as a complement to the design process, as a gateway to the functionality of the older person and a starting point for product design.

SOLUTIONS AND RECOMMENDATIONS

To design an aging suit more faithful to human biomechanics, it is recommended to carry out a previous anthropometric study with a group of older adults that meet the profile required in relation to the

Figure 3. Rendered image: The kitchen of the Mexican elder
Own authorship (2012).

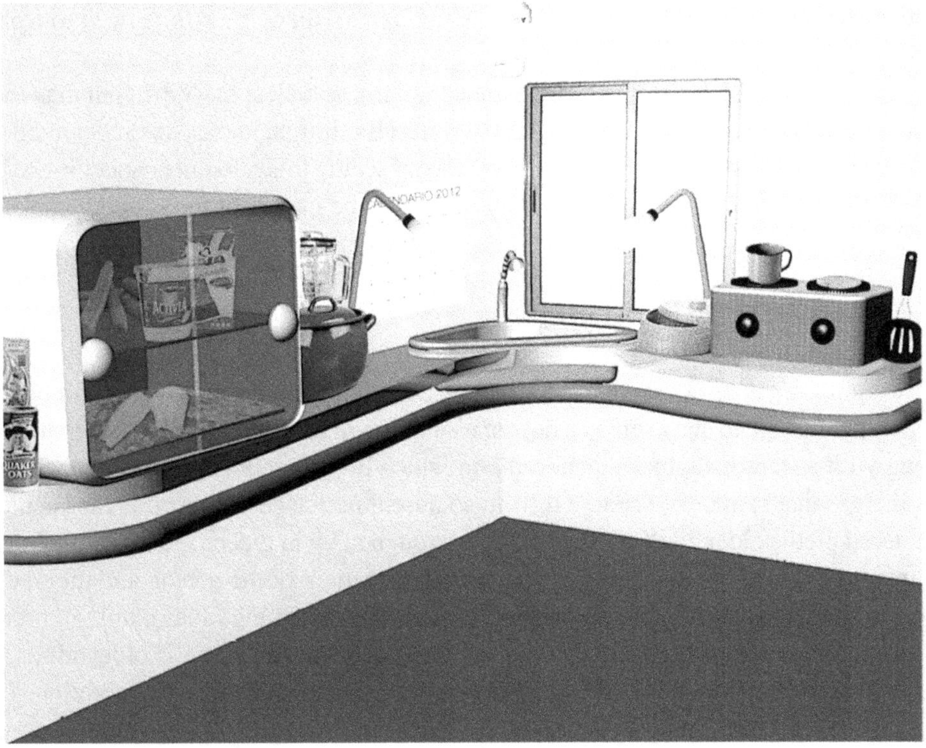

project to be worked on. And then, elaborate the custom aging simulator suit. Although it should be noted that the idea of age simulators is to act as mediators of a situation, outlining the capabilities of a population. In order to obtain reliable results, it is necessary to carry out an experimental study with a representative sample.

CONCLUSION

Designing for the elderly is not an easy thing because we have lived designing for youth and for children for decades. It is almost routine to carry out an architectural project or product design because it is assumed that the users are young people, this will no longer be so. To think that gerontodesign only applies for when you want to develop specific products for an older person is a mistake, gerontodesign is increasingly present in our daily lives, whether we like it or not, as older adults are present in our society for a long time, more than before. Having empathy with an aging person is the ideal for living together in this society, to promote spaces, programs, services, events, and objects that are not linked to eternal youth. Everything begins with empathy, from this are triggered the most innovative ideas, simple ideas that gave solutions. The society of youth often said that the children are the future, but we need to rethink this phrase, the future is full of elderly, elderly with wisdom, with problems and with experiences that could help to change the world that we as older humans will be experimenting. Countries like Mexico are not the young countries that they used to be, today there are more adults over sixty years old that children under five (Gutiérrez-Robledo & Arrieta-Cruz, 2015), there is a strong need to work in design for the aging population.

To know the aging human body is to change the paradigm, to look at the Vitruvian man in an unusual way, perhaps a little more stooped, perhaps with the arms less straight and close to the middle part of the body than to the line of the shoulders. Changing the way of seeing this referent is not going to be easy, but it is obligatory to collaborate with a better quality of life for the majors of today and those of tomorrow.

FUTURE RESEARCH DIRECTIONS

The use of the aging suit does not seem to be eliminated from future perspectives, it will be necessary to raise awareness, to design, to see itself as a design tool that will be imperative in universities and design offices, along with a transdisciplinary panorama in which both functionality and accessibility will be elementary themes that cannot be separated from socio-cultural aspects, because user-centered design and care centered on the older adult are flows that are toughen up in recent years. The aging suit should have different variants, each time more specific around the gender of the person and their percentage of muscle mass. It is important to encourage the co-design, by immersing older adults in product design processes, promoting innovations issued by the same older adult, as functionally older adults have undergone the gradual change of their body and this gives them certain possibilities of adaptation. The students of the design schools should generate their own tool based on this type of aging simulators and from the design process of the aging suit they will be deepened in the aging body knowledge, designing an aging simulator does not only works for the product design, but also for the awareness of how the people get old and what kind of strategies should be proposed by every each designer student and professional. Also, the aging suit fits perfectly inside social programs that boost empathy about the aging process.

REFERENCES

AgeLab. M. (2017). *Agnes (Age Gain Now Empathy System)*. Retrieved August 20, 2017, from: http://agelab.mit.edu/agnes-age-gain-now-empathy-system

Aguilar-Salinas, C. A. (2013). Comentarios a la ENSANUT 2012. *Salud Pública de México*, *55*(2), 347–350.

Ávila–Chaurand, R., Prado–León, L. R., & González–Muñoz, E. L. (2007). Dimensiones antropométricas de la población latinoamericana: México, Cuba, Colombia, Chile (2nd ed.). Guadalajara, México: Universidad de Guadalajara.

Beard, J., Officer, A., Cassels, A., Bustreo, F., Worning, A. M., & Asamoa-Baah, A. (2015). Informe mundial sobre el envejecimiento y la salud ed. WHO.

Blum. (2017). *Age explorer®: Preparing for the future by choosing quality hardware.* Retrieved August 20, 2017, from: https://www.blum.com/us/en/03/30/10/

CBS. (2014). *Ford car markers wear 'age suits' to design for older drivers.* Retrieved August 20, 2017, from: http://www.cbc.ca/news/canada/windsor/ford-car-makers-wear-age-suits-to-design-for-older-drivers-1.2762417

Coughlin, J. F. (2013). *Agnes & livable communities.* Retrieved August 3, 2017, from: http://agelab.mit.edu/files/finalreports/MITR22-3FP.pdf

CSO. (2007). *Ageing in Ireland.* Stationery Office, Dublin, Ireland. Retrieved July 31, 2017, from: http://www.cso.ie/en/media/csoie/releasespublications/documents/otherreleases/2007/ageinginireland.pdf

Eduardo Wainstein, G. (2014). Patogénesis de la artrosis. *Revista Médica Clínica Las Condes, 25*(5), 723–727. doi:10.1016/S0716-8640(14)70098-7

Filz, S. A. (2008). *"Instant aging"-selbsterfahrung des alterns (Doctoral dissertation).* Universität Würzburg, Medizinische Fakultät.

Flores-González, I., & Calonje, D. H. (2017). *Telemedicina para detección de enfermedades oculares con potencial de ceguera en México.* Revista Mexicana de Oftalmología; doi:10.1016/j.mexoft.2016.12.002

GERT. (2017). *Gert simulador de edad.* Retrieved August 20, 2017, from: http://www.simulador-de-edad.com

Groza, H. L., Sebesi, S. B., & Mandru, D. S. (2017). Age simulation suits for training, research and-development. In S. Vlad, & N. M. Roman (Eds.), *International conference on advancements of medicine and health care through technology* (pp. 77-80). Cham: Springer International Publishing.

Gumina, S., Carbone, S., Campagna, V., Candela, V., Sacchetti, F. M., & Giannicola, G. (2013). The impact of aging on rotator cuff tear size. *Musculoskeletal Surgery, 97*(1), 69–72. doi:10.100712306-013-0263-2 PMID:23588834

Gutiérrez-Robledo, L. M., & Arrieta-Cruz, I. (2015). Demencias en México: La necesidad de un plan de acción. *Gaceta Médica de México, 151*, 667–673. PMID:26526483

Hall, T. C., Nixon, M. F., Dias, J. J., Graham, T., & Cook, S. (2010). How accurately does a simulation glove reflect function compared to rheumatoid arthritis suffers? *Annals of the Royal College of Surgeons, 92*(7), 605–609. doi:10.1308/rcsann.2010.92.7.605 PMID:20529519

INAPAM. (2016). *Vinculación productiva para personas adultas mayores.* Retrieved August 20, 2017, from: http://www.inapam.gob.mx/es/INAPAM/Empleo_para_adultos_mayores

INEGI. (2010a). *Censo de población y vivienda.* INEGI.

INEGI. (2010b). *Perfil sociodemográfico de adultos mayores.* INEGI.

INEGI. (2016). *Estadísticas a propósito del día internacional de las personas de edad (1 de octubre)*. Aguascalientes. Retrieved July 31, 2017from http://www.inegi.org.mx/saladeprensa/aproposito/2016/edad2016_0.pdf

Kopec, D. (2012). *Environmental psychology for design* (O. T. Kontzias, Ed.). Fairchild Books, Inc.

Kullman, K. (2016). Prototyping bodies: A post-phenomenology of wearable simulations. *Design Studies*, *47*, 73–90. doi:10.1016/j.destud.2016.08.004

Lavalle-Herrera, Y. D. (2014). Diseño y ergonomía para la tercera edad. Aplicación al diseño de calzado. Coordinación de Estudios de Posgrado y el Programa de Maestría en Diseño Industrial de la Universidad Nacional Autónoma de México.

Lawton, M. P., & Nahemow, L. (1973). Ecology and the aging process. In C. Eisdorfer (Ed.), The psychology of adult development and aging (Vol. 6, pp. 619-674). Washington DC: American Psychological Association. doi:10.1037/10044-020

Maya-Rivero, A. (2012). *Gerontodiseño: Conceptualización y formulación de sus estrategias para una mejor calidad de vida del adulto mayor. Caso de estudio: Problemas del adulto mayor mexicano en la preparación de alimentos (Master Degree Dissertation)*. Toluca, México: Universidad Autónoma del Estado de México.

Moore, P., & Paul, C. (1985). *Disguised: A true story*. W Publishing Group.

Papanek, V. (1984). *Design for the real world. Human ecology and social change*. London: Thames and Hudson.

Parmelee, P. A., & Lawton, M. P. (1990). The design of special environment for the aged. In J. E. Birren & K. W. Schaire (Eds.), Handbook of the psychology of aging (pp. 465–489). New York: Academic Press. doi:10.1016/B978-0-12-101280-9.50033-4

Parra–Marujo, J. (2006). *Gerontodesign: A marca de água do design, do design ergoómico, da marca ou das marcas, branca*. Retrieved April 11, 2011, from: JMarujo.Artician website: http://jmarujo.artician.com/portfolio/gerontodesign/

Perry, T. E., Andersen, T. C., & Kaplan, D. B. (2014). Relocation remembered: Perspectives on senior transitions in the living environment. *The Gerontologist*, *54*(1), 75–81. doi:10.1093/geront/gnt070 PMID:23840021

Reuters. (2008). *Japan aging suit puts car makers in senior circuit*. Retrieved August 20, 2017, from: http://www.reuters.com/article/us-japan-nissan-suit-idUST9224920080417

Wahl, H.-W., & Weisman, G. D. (2003). Environmental gerontology at the beginning of the new millennium: Reflections on its historical, empirical, and theoretical development. *The Gerontologist*, *43*(5), 616–627. doi:10.1093/geront/43.5.616 PMID:14570958

KEY TERMS AND DEFINITIONS

Accessibility: The concept of accessible design, the ability to access and to use services, spaces, or products.

Elderly: Older adults are those who exceed the life expectancy of their own countries. In Mexico, the life expectancy is about 76 years old.

Empathy: Understand the needs, feelings, emotions, and thoughts about a special topic or situation.

Environmental Gerontology: Gerontology area whose goal is to analyze and modify the relationship between the elderly and their social and physical environment.

Environmental Psychology: Study the interrelationships between people and their physical surroundings (natural and built environments).

Gerontodesign: Conceptualized as the design for the elderly population. Is a term initiated by social anthropology. The union between gerontology and design.

Transdisciplinary: Cross disciplines. Cross the boundaries of two or more disciplines.

Chapter 5
Product and Space Analysis Using the Product Perception Method

Mario Osmar Moreno
Autonomous University of Ciudad Juárez, Mexico

Juan Luis Hernández Arellano
Autonomous University of Ciudad Juárez, Mexico

ABSTRACT

In this chapter, the data collection is presented, where the base is the study of a group of elderly people in eight nursery houses and two hospitals in Bogota, Colombia. The objective of this research is to look through industrial design and the analysis of existing products, in which way the designer could offer an improvement to the products and with that how to increase the life expectancy. The user perception method is applied, therefore, and the interview was performed as a sequence of questions, where the perception of an elder of existing personal hygiene products is analyzed through a collection of data. The necessity of designing an ergonomic chair that allows performing activities of personal hygiene in the shower was identified, and the requirements for the designing of the product were identified. Finally, the designer will dedicate to find products that best fit their daily needs.

INTRODUCTION

Old age is the last stage of the life cycle that begins just at 65 years old which has certain biological, psychological, and social characteristics (Duran, 2004) Currently, the design for elderly people is done in a limited and/or poor way, therefore, elderly people present difficulties to acquire products that allow them to struggle less and risk themselves less while they do their personal hygiene activities, and more emphasis is needed in that area to create products with right characteristics for the capacities and needs, that help them improving life quality (Martha, 2007). Wong & Díaz (2007) state that life quality of elderly people is a satisfactory life, subjective and psychological wellness, personal development and different

DOI: 10.4018/978-1-5225-5234-5.ch005

Copyright © 2018, IGI Global. Copying or distributing in print or electronic forms without written permission of IGI Global is prohibited.

representations of what a good life consists in, questioning the elderly, about how it gives meaning to their own life in a cultural context, the values they live in, in relation to their own life objectives.

In the same way, the concept of life quality, as pointed out by Sánchez-Sosa & González-Celis (2006), covers three global dimensions: the first one is what the person can do, functional state, the second is the access to resources and opportunities, and the third one is the feeling of wellness. The first two dimensions are designated as the objective of life quality while the last one is considered as subjective life quality to decrease the risk of accidents and to observe the interphase of the user. In this way, it is necessary to clear out what is understood by interphase, i.e. everything the user experiences, sees and does with the system. Some characteristics of the interphase are that it allows the intuitive catchment of the information of the system, product, and environment. Also, it makes visible the information and gives directions of how something can be used. Finally, it gives directions of the reason of a task and this becomes evident simply by watching, with or without instructions (Medina, 2006).

Problems that affect elderly people cannot deeply examined/analized without first evaluating the demographic development of this category. A development that even having very different dimensions of different regions or countries like Europe, North America, and southern countries, that have more than half the world population of 60 years old or more. In 2025, the portion of elderly people will reach the 70% of the entire population (OMS, 2009). Considering this market, the future will be more important because the demand and consumption of the products will be high due to most of these people will suffer typical age diseases and they must be taken care of in a better way. As a consequence, the purpose of this article is to determine the requirements of an specific product design: a chair to shower elderly people. Also, in this document, the development of the qualitative data collection is presented.

OBJECTIVE

To understand how furniture used for personal hygiene stimulates experiences on elderly people, regarding safety, need and appearance. In other words, to understand what the user thinks and perceives during their stay in the space and the interaction with the furniture characteristics.

Importance of the Project Development

The study is relevant due to the obtaining of veridic data related with the perception of the user regarding the product and with the interpretation what is liked in the market and not, finding different virtues and deficiencies. The interaction is a fundamental part to understand how the user interacts with the object.

Necessary information for the development of the project was collected, therefore, it was necessary to know that perception is a cognitive process of interpretation and process of the information. The cognitive system uses a series of resources simultaneously and depends on the complexity of the task, the learning level, or previous experience. The subject could use the product with efficiency. However, an excess of the mental load will occur causing deficiencies in the execution and perception of the product (Cañas & Waern, 2001). Several interview protocols were performed to achieve the goal of identifying the perception of the user and to determine the interphase product-user. Necessary techniques for the data collection, as role games, questionnaires, interview, etc., were applied.

The study considered two data collection methods: questionnaires and observation (taking into consideration descriptive, interpretative and inferential codes) which could be used by anyone who works in the same research topic (Fernández Núñez, 2006).

METHODOLOGY

The type of investigation used in this study was exploratory, due to what is looked for is to develop strategies and solutions to provide a better lifestyle for the elderly.

Participants

In this study, a total of 50 elderly people participated (25 men and 25 women), all of them above 65 years old without mental issues (they answered questions properly). The field work was performed in nursery homes, geriatric houses. Relatives or close people that live with them in the city of Bogota, Colombia.

Materials

The print questionnaire, 2 color images, an audio recorder, a cell phone, a notebook, and a pen. Also, a photographic camera, batteries, and a phone charger were used during the field work.

Method

1. Interest places (nursing homes, geriatric houses, etc.) were located through web pages like viejoportal.com, local address books, yellow pages, and internet adds. People who work as caretakers were contacted in advance in order to be ready for the study mentioned here or through recommendations of previously visited places.
2. The questionnaire (see Table 1) was applied in geriatric homes and/or where the interview was initiated. To avoid distractions, a quiet place free of noise was located. If the person is capable of standing, the interview was performed in the bathrooms, this way the people can get more contact with the objects and space, as a result, their answers will be more accurate not forgetting how important the comfort of our participants is. To avoid interruptions and opinions from the caretakers, the interview was performed without the presence of any caretaker. With this action, the answers reflect the real perception of the user.

To answer each question, a 5-point Likert scale was used. Table 2 shows the scale and scores used.

3. In a clear way, before starting the interview, corresponding indications were explained to the participants. The objective of the research was mentioned, giving the opportunity to feel a fundamental part of the research. It was emphasized that at any point, their privacy and their integrity was not aggravated, and if they're not comfortable with the interview, they can end it whenever they decide it. It was mentioned that all the answers were important for the analysis and everything was confidential. Also, permission to record the voice was asked.
4. Questionnaire 1 was performed.

Table 1. Questionnaire 1

Nº	Question
1	Are you concerned about your health?
2	Do you like to be assisted with your personal hygiene tasks?
3	Do you consider this place to make you feel like home?
4	Do you have physical limitations that prevent you or restrict you from performing your personal hygiene tasks?
5	Do you care about bathroom distribution?
6	Have you had an accident in the bathroom?
7	Have you been injured in a fall?
8	Do you need support to take a bath
9	Does the furniture in your bathroom help you performing your activities?
10	Does the staff that helps you, adapt the bathroom to your needs?
11	Do you feel safe when you take a shower?
12	Do you know safety products for bathing?
13	Do you have balance problems?
14	Do you have difficulties standing or sitting in your bath?
15	Do you get tired or exhausted when you bathe?
16	Do you find the staff cares appropriate?
17	Do you like the bathroom you use for your personal hygiene?
18	Do you have everything you need at your reach while you shower?

Table 2. Likert Scale

Value	Alternative
5	Always
4	Almost always
3	Sometimes
2	Almost never
1	Never

5. Two images were shown to the user (Figure 1). In this case, images of innovating bathrooms to identify which are the most important needs regarding space sizes, elements of safety, help devices, surfaces and aesthetic elements were considered. With these images, the user chooses the one he likes the most, i.e. the one that makes them feel safe, comfortable, doing a similar development to the semantic differential method created by Charles Osgood in 1957. This procedure allows identifying what emotions evocate the images in the user (González Cano, Aguayo González, Lama Ruiz, & Pérez Gutiérrez, 2009).
6. In order to get extra information, a second questionnaire was applied (see Table 3).

Figure 1. Sample images
http://www.asister.es/tienda/barras-de-bano-blancas/; http://www.decorablog.com/acondicionar-casa-personas-movilidad-reducida/

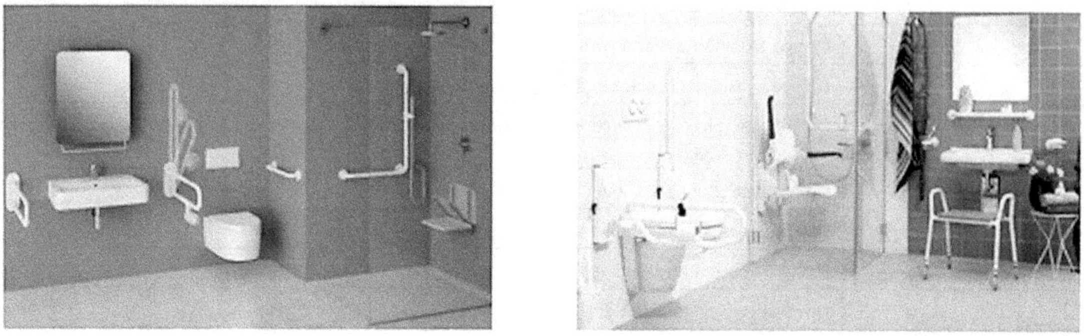

Table 3. Questionnaire 2 and objectives of methods

Question	Objective of the Question
Why did you choose this picture?	To know which of the two images presented, bring more attraction to the user
What is it that you like the most?	The user brings out the most important characteristics that will help get information about the project
Why do you like it?	To know the clear and explicit reasons that explain the election of the user eliminating ambiguities
What you dislike?	To identify the nonattractive elements to the user and doesn't get an added value for the product
How would you like it	The opinion of the user is considered a requirement of design through their subjective proposals
What do you think about the space?	The space perception of the user informs aspects that must be considered so personal hygiene activities can be performed comfortably
What do you think about the color?	Most predominating aesthetic tastes of the user are considered in the design of the product
You head to the bathroom: • Walking? • With help? • In a wheelchair?	Interior transport allows considering aspects of design

7. We thank the interviewee and given the chance to add extra details.
8. Before leaving the place, we asked permission to take pictures of the interest places. To avoid loss of data, all the information, images, and audios were checked.

Analysis of Information

With this, statistical analysis as average, standard deviation, and variance were calculated based on given information.

RESULTS

Results of Questionnaire 1

Before showing the results, it is important to mention that only the results of questions 1, 2, 4, 7, and 15 will be shown in this section.

Question 1: Are you concerned about your health?

Figure 2 (a) shows that the 82% of the patients said they were always concerned, while 16% mentioned that they do but sometimes they forget necessary cares, finally, 2% considered that sometimes they are concerned about their health. In comparison, there was a remarkable difference between man and women, for example, women usually are more cautious in the personal aspect.

Question 2: Do you like someone else to help you with your personal hygiene tasks?

Figure 2 (b) shows that, in consideration with question 2, it's observed that most of the patients don't like to be assisted in their personal hygiene tasks, although, there is a considerable percentage that feels affection by the staff and prefers to be assisted.

Question 4: Do you have physical limitations that prevent or restrict your performing personal hygiene tasks?

Figure 2 (c) shows that the most have physical limitations, some caused by accidents and other by natural aging process. It is important to mention that there are healthy patients that do not have any physical impediment.

Question 7: Have you been injured in a fall?

Figure 2 (d) shows that sometimes they have problems of injuries by accidents. Nevertheless, out of 100%, 58% has had some type of accident in the bathroom (falls, hits, slips, etc.)

Question 15: Do you get tired or exhausted when you take a shower?

Figure 2 (e) shows that they always struggle to perform movements from sitting to standing and vice versa, this by the high rate of ailments like arthrosis, osteoporosis, muscle dystrophy among others that difficult movements and strength of the patient when bathing. It is a moment of exhaustion for them because of previously mentioned limitations, emphasizing in fatigue and muscle pain.

Results of Questionnaire 2

While the information could be very subjective, the subjective sense and the subjective configurations are important not only because they allow to comprehend the individual action in their systemic character, but also it allows us to understand society in a new dimension, i.e. in their system of consequences about the

Figure 2. Results of questionnaire 1

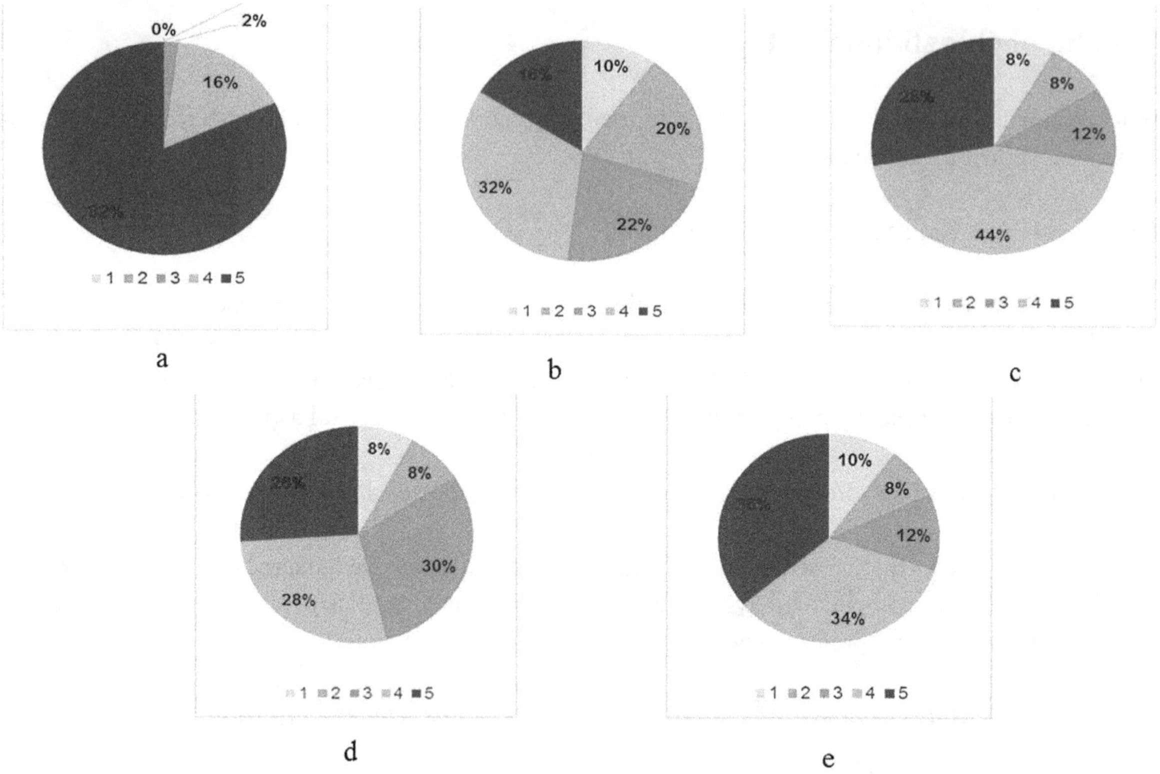

men and about the organization of its different spaces of social life (Domínguez-Sosa, Zavala-González, De la Cruz-Méndez, & Ramírez-Ramírez (2010); González-Rey, F. (2010). Thus, very important data was collected regarding perception and with this strengthen information of Table 1.

90% of the participants choose image number 1 and the rest the image number 2. The reasons were: image 1 is more elegant and formal, it looks neater, it called their attention easily, and most of them mentioned it evoked safety. Nonetheless, there were several comments stating more elements of safety were needed, due the floor of either picture was adequate to fit the need of safety in space. Regarding the question, “what was missing?” 80% mentioned a personal belonging organizer device and space in the showers for mobile chairs, 85% mentioned they would change one of these for theirs, and it would bring more comfort, and 25% expressed comfort with the space they have currently. About the images, 70% mentioned they liked the chair in the shower, 20% were indifferent (chair was fine), and the other 10% stated they do not need it. 60% said the space in the shower was not convenient. These results allowed to notice the elderly person is not used to innovation, due they think it is pretty, but they prefer to keep with same existing products and change not to be so drastic. Also, it is concluded that the most brought up elements and the ones we would need to focus on in our design are floor space and safety regarding falls considering elements of safety.

CONCLUSION

In conclusion, the use of products it is based on perception and the daily coexistence with objects, images, and spaces. Therefore, the design must be based on theoretical knowledge, information collection to find design requirements focused on the voice of the client, and the interpretation of their needs before design an object/product and defining functionality parameters as usability and aesthetics of an object. Hence, it was detected a need to design a duck-chair for the shower that allows assisting the elderly person in personal hygiene activities. Table 4 shows the requirements of design generated by the analysis of this article.

Table 4. Design requirements

Seat	• Ergonomic • Comfortable for the user • Anti-slippery texture to avoid accidents • Cushion for the comfort of the user • Hole for use in the toilet • Soft material • Inclination angles for users that need to be moved • Rotatory movement
Backrest	• Articulated movement for diverse angles • Ergonomic • Pushing system to move the user • Height adjustment • Cushion for comfort • Anti-slippery texture
Arm Support	• Cushion for comfort • Height adjustment • Side adjustment system • Option to be removed • Two kinds of holders for the transport of the chair • Side elements for subjections
Feet Support	• Option to be removed • Solid and firm structure • No stripped or pierced • Height adjustment
Tires	• Rubber material • Industrial sucker system to avoid non-desired movement • Breaking system at the reach of the user • Friendly and aesthetic structure
General	• Height adjustment • Spin system from a central axis • Máximum support of 130 kgs • Folding capacity • Aspect with light colors that represent cleanliness and health • Immobilization break system • Antitrust material (aluminum) lightweight for the legs • Adaptable to toilet • Light material, polymer (high-density polyethylene) for the seat and the back • Lifting system • Belt to avoid fetal position of the patient • Hygienic material to avoid bacteria • Considerable Price

With this background, the elements needed to fit the needs of design, the designer must always consider that the products help the elderly people to reach the adaptability to their environment and continue the experience of having a better life quality. Thanks to these methods, we could take into consideration the data collected for a better approach or concept of new products, and most of all, the requirements of the design of products for the developing of daily personal hygiene activities in nursing homes, giving them a plus in life quality to elderly people.

REFERENCES

Cañas, J., & Waern, Y. (2001). *Ergonomía cognitiva: aspectos psicológicos de la interacción de las personas con la tecnología de la información*. Granada.

Domínguez-Sosa, G., Zavala-González, M. A., De la Cruz-Méndez, D. del C., & Ramírez-Ramírez, M. O. (2010). Síndrome de sobrecarga en cuidadores primarios de adultos mayores de Cárdenas, Tabasco, México. Enero a mayo de 2008. *Revista Médicas UIS*, *23*(1), 28–37.

Duran, L. (2004). *La salud del adulto mayor en el Instituto Mexicano del Seguro Social*. DF, México.

Fernández Núñez, L. (2006). Fichas para investigadores ¿Cómo analizar datos cualitativos? *Bulletí LaRecerca*. : 1886-1946 / Depósito legal: B.20973-2006

González Cano, M., Aguayo González, F., Lama Ruiz, J. R., & Pérez Gutiérrez, J. R. (2009). Ingeniería Kansei para un diseño de productos centrado en los usuarios. *Técnica Industrial*, *280*(Marzo-Abril), 7.

González-Rey, F. (2010). (n.d.). Las categorías de sentido, -. *Google Académico*.

Martha, V. (2007). Significado de la calidad de vida del adulto mayor para sí mismo y para su familia. *Scielo*, *68*, 284–290.

Medina, A. (2006). Ergonomía cognitiva y usabilidad.

OMS. (2009). Programa Envejecimiento y Ciclo Vital. Envejecimiento activo: un marco político. *Organización Mundial de La Salud*, 74–105.

Sánchez-Sosa, J. J., & González-Celis, A. (2006). Evaluación de la calidad de vida desde la perspectiva psicológica. In *Manual para la evaluación clínica de los trastornos psicológicos: trastornos de la edad adulta e informes psicológicos* (pp. 473–492). Pirámide.

Wong, R., & Díaz, J. J. (2007). Health care utilization among older Mexicans: Health and socioeconomic inequalities. *Salud Pública de México*, *49*, s505–s514. doi:10.1590/S0036-36342007001000010 PMID:17724524

Chapter 6
Pedagogy as a Fundamental Part of School Furniture Design

Natalia Anaya Echeverría
Universidad de Guadalajara, Mexico

ABSTRACT

The teaching-learning processes are carried out through established objectives for each grade level and these are achieved with didactic activities based on principles, educational concepts, and pedagogical contents. Activities that are divided into specific tasks that are performed with physical aids such as furniture and classroom generate diverse forms of work. Through observations it was detected that the current furniture did not respond to the pedagogical and didactic practices and there is no data indicating that it is designed according to the anthropometric variability of the students of the metropolitan area of Guadalajara according to the school grade taken by the student. Therefore, the authors sought to know the physical and functional characteristics required by school furniture to be a facilitator of pedagogical and didactic practices to make a proposal of a furniture design that meets both the pedagogical demands and the anthropometric variability of schoolchildren.

INTRODUCTION

The traditional teaching was strict, the teacher imposed order from the front of the room where he made his authoritarian and obedient class. On the other hand, the pedagogical and didactic practices by competences, its objective is to create a teaching-learning environment to create a practical experience, which is linked to the student's learning and knowledge. Factors of time, resources, academic preparation and even the classroom change the pedagogical models among teachers, so that the fulfillment of the purposes, content and expected to learn of the programs require the distribution of time in class, omitting didactic projects or Permanent activities, as well as the organization of class in group, team or individual activities. The teaching-learning processes are carried out through established objectives for each grade level and are achieved through didactic activities based on principles, educational concepts and pedagogical contents. Activities that are divided into specific tasks that are carried out with physical aids such as furniture and classrooms; Generating diverse forms of work. From the general to the

DOI: 10.4018/978-1-5225-5234-5.ch006

Copyright © 2018, IGI Global. Copying or distributing in print or electronic forms without written permission of IGI Global is prohibited.

particular, we arrive at the school furniture since it is the physical object where the students and teachers have the greater interaction, reason why they must fulfill based on the pedagogical practices and with the ergonomic principles.

Because there is no data or research related to the subject in Mexico, this motive makes this research the first study carried out where the elements are determinants of furniture design, in addition to the anthropometric adjustments. The objective of this research is to propose an efficient, safe and comfortable school furniture that adapts to the pedagogical practices of our school system. For this, the functional and ergonomic aspects of the current school furniture were identified, according to the pedagogical and didactic practices carried out in the classroom. It was analyzed if the furniture adapts to the different pedagogies imposed by the teachers. And the dimensions were established based on the anthropometric data of the users to propose an ergonomic school furniture design.

BACKGROUND

Mexico is currently experiencing a crisis in Mexico's education system and the challenge for the authorities is to regulate the teaching and learning system to increase the academic level. Based on the program for the International Assessment of Students (PISA) students do not learn the minimum content and have a low reading level. Since the 1990s the National Union of Education Workers (SNTE) and the federal government have modified curricula to competency-based education. From 1992, an agreement was made between the SNTE and the Federal Government for the modernization of basic education; modifying the curriculum to competency-based education, establishing preschool, primary and secondary education as mandatory. The competency method has the capacity to respond to different situations and implies skills, knowledge, as well as values and attitudes. At the elementary level, the students coexist in the same space and are coordinated by a teacher, who works with all or most of the subjects, which allows them to locate meeting points between the subjects and establishing forms of work. The teacher aims to create a teaching-learning environment to create a practical experience, which is linked to the student's learning and knowledge.

Teachers have a "Teacher's Book" that is distributed by the Secretary of Public Education (SEP) according to the grade they teach. These contain the purposes of the subjects in progress, as well as the didactic proposals and permanent activities that students and teachers must carry out to achieve the objectives. Students also have their own free textbooks provided by the SEP with didactic activities and projects to be carried out by students during the school year.

Although the SEP provides teachers with books, materials, and training; these are the ones who generate the pedagogical activities within their classroom to try to capture the attention and interest of the students. Whether it is a teacher-centered pedagogy, student-centered pedagogy or a balance between both has an assignment of tasks and objectives, as well as a usable time. The time, resources, academic preparation and even the classroom (Martin, 2002) factors change pedagogical models among teachers, so the accomplishment of the purposes, content and expected learning program demands time distribution class in different moments of work omitting didactic projects or permanent activities, as well as the organization of the class in group, team or individual activities.

Current educational methods in Mexico, a competence-based education is proposed based on a constructivist approach to development is proposed. Although they are different currents at some point they connect. Constructivism postulates the need to provide the student with that allow him to create his own procedure to solve a problematic situation, which implies that their ideas are modified and continue to learn. It states that knowledge is developed individually and socially by students based on the interpretations of their experiences in the world.

Since knowledge cannot be transmitted, if not constructed, teaching should consist of experiences that facilitate the development of learning, Jonassen, (2000). The objective of the student is to interpret and solve the problem by helping to reinforce memory and increase cognitive flexibility. The fundamental idea revolves around the students' conversations about what is being learned and not about the interpretations of the teachers. In most constructivist learning environments, students need to explore, articulate what they have learned, speculate, manipulate the environment to elaborate and test their theories and models, and reflect on what they have done and what they have learned from these activities.

On the other hand, competences build in students the best performance in knowledge, concepts, skills, attitudes, values , and strategies in order to adequately address the various challenges of the present. Its purpose is related to productivity, in the sense of getting graduates to join the work situation and are better prepared to meet specific tasks such as the demands of human interaction in the work environment. The relationship between these educational methodologies and school furniture, which is the main object of the interaction of the student, manages to facilitate or deteriorates the didactic and pedagogical practices. Satisfactory performance can translate into satisfactory learning for the student, as well as the desire to return to class.

Rivlin and Rothenberg (1976) mentioned that a classroom is a system and that there is a complex relationship between the physical structure and the accommodation of the classroom, the teacher, the students and the distribution of space. This results in a physical space that influences the educational program and the behavior of the students. Studies report that there is a correlation between student achievement and infrastructure index where hygiene conditions, quality of construction and furniture (Cardemil et al., 1991; Trevino and Trevino, 2004) are included

From the general problem to the particular we landed in the design of school furniture since this implies the comfort of the students during the class sessions, the different accommodations that serve different purposes and that this could generate attitudes and advantage of the students. So far, there are few studies have focused on pedagogy in the design furniture, most are with an anthropometric approach only. In Mexico, no research has been carried out where the main objective is to design a school furniture that adapts to the pedagogical and didactic practices that it is provided by the "Book of the student" of the SEP. The activities and work suggested by this book to be carried out in the classroom can increase the students´ learning motivation as well as the organization and cooperation of teamwork. For such didactics, what surrounds and concentrates a classroom or in this case, a learning environment is an opportunity that both the student and the teacher have the appropriate the space to achieve more active dynamics. Encourage thinking and creativity of each student in their own time and space. To ensure that didactics and pedagogies that the teacher wants to implement flow, the furniture must be designed to achieve these objectives. Although the furniture must be designed based on pedagogical and didactic practices, it must also comply with ergonomic principles.

"The fact of using improper furniture can cause some problems with the musculoskeletal system, as well as loss of interest in learning, even during the most entertaining class" (Hira, 1980, pp. 213-221). The current furniture possessed by elementary schools in the urban area of Guadalajara Metropolitan Area does not meet the physical standards demanded by the National Institute of Physical Educational Infrastructure (INIFED) in the *Guide provision of furniture and equipment.* Despite data from the National Institute for Educational Evaluation (INEE) in 2007, 99% of schools in all denominations claim to have furniture; although they do not mention their condition. It is in the *Guide provision of furniture and equipment* where the dimensions suggested for the worktable for students of 1st to 3rd grade is 1.20 x 0.60 x 0.65 cm, while in classrooms dimensions are 1.00 x 0.40 x 0.67 cm of height. In spite of being 2 centimeters high on the worktable do not seem important, it is for students, who need to work with shoulder elevation causing more workload to the muscles and causing discomfort and pain in the shoulder region (García and Page, 1992)

Poor posture at an early age increases the risks of a musculoskeletal injury because the bones of young children are soft and vulnerable. It is, therefore, necessary to guide children to have a good sedentary stance by using school furniture that allows them to move freely in a range of good postures.

There is considerable variability in the physical characteristics of the students, despite being individuals of the same age, race and socioeconomic status, for this reason, the design of school furniture proposed here requires adjustments that must accommodate the highest population. These adjustments were based on recent anthropometric data held in Mexico (2007) children 6 to 8 years of age by Avila, Prado, and Gonzalez in Jalisco standardized positions.

Previously the school furniture was manufactured in such a way that the teacher could teach before a large group of students, they would sit in silence, listen and take notes. The classrooms were rectangular and the main attention was direct to the front, where the teacher would take control of the activities and contents of the program; the furniture was screwed to the floor so that there was no possibility of moving it (Cornell, 2002). Early furniture designs required that the backrests be upright to keep students disciplined and force them to look forward (Knight and Noyes, 1999).

The furniture has evolved by several factors such as furniture functionality, industrial manufacturing, ergonomics and teaching methods that have progressed from traditional education to other more active methods.

From the research of Herman and Van Gorp, the pioneer of folding school furniture Samuel Akbroit eliminated the screwed furniture with the aim of having a greater cleanliness and to avoid diseases of tuberculosis between the students and teachers. Then the designers of the time thought of some type of furniture that was quick to move when cleaning, so they joined the chair and the table, later to make the furniture lighter, removed the central table stand to place it sideways. From 1960-1970 the furniture became lighter and easier to accommodate compared to furniture nailed to the floor. This gave the teachers a greater option of appropriating space for didactic activities.

Much of the research analyzed turns to classrooms that act as learning spaces that must be planned according to the curricular objectives. We must think of spaces not as containers of things and more like theories of constructivist learning, designing through the activities that could be done by users in space.

Horne (2002) studied the relationship between the classroom environment and teacher practices, evaluating what happens inside the classrooms in a physical aspect and pedagogical practices of the teacher. The aim of this study was not to describe the "ideal" learning environment but to describe and analyze the existing environments. As well as the students can be coupled to space, the teacher must be able to create conditions to create learning stimulation in students. The ability to predict behaviors in certain types of accommodations may probably mean that teachers can distribute their accommodation to promote some particular action. As a result, noted that 79% of the class time has a "dominant teacher" class profile where pedagogy is centered on the teacher teaching in front of the class and there is no student-teacher interaction. Therefore, a large proportion of classrooms are not adjustable and their furniture is always in the same place. Also found that while the classroom is more flexible there is a trend that increases the mobility of the teacher, however, the more space the student has, the less the teacher can move around the class. On the pedagogical side, the classes with the highest proportion of "working students" have the most space. The rooms organized in a circle tend to a pedagogy centered on the teacher due to the control that maintains the same teacher. When teachers interact frequently with the whole group or individually, the classrooms tend to be arranged in columns. While pedagogy that is student-centered, classrooms tend to organize into teams and perform multiple activities.

According to Lippincott (2009), constructivists spaces should be learning spaces that support a flexible variety of learning and instruction taking into account ergonomic and easy to understand factors. To obtain the characteristics required to design an environment, Long and Holeton (2009) mention that verbs should be described in what students do and this helps to delimit the space that is required. Experience in the work environment is a very important component of the educational program, even though sometimes it is not practical, available, convenient or appropriate. To meet the objectives of competency-based methods, students need learning environments and assessments that simulate real-world environments. These are defined as spaces of experimentation, discovery, and manipulation, spaces where each student can learn at their own pace (Ferreiro Gravié, 2006). The student moves from being one of the classrooms to become a protagonist, has an active role in the teaching-learning process. And the teacher goes from being the person with the knowledge to be a guide and mediator to facilitate the learning process.

Analyzing the situation of furniture can be found on one side is the physical situation of students and on the other hand the pedagogical requirements. So far there are few investigations that have a reference design school furniture as an important part of education, most with an anthropometric approach where the school furniture is modified according to anthropometric data (Castellucci et al., 2009; Gastelum, H., 2004; Gouvali et al., 2006; Guzmán et al., 2009; Kayis et al., 1991; Mokdad et al., 2009; Parcells et al., 1999.), centered on the two-dimensional factor, leaving aside biomechanical factors, safety and of course the pedagogical practices that are carried out in the classroom. The furniture, as a facilitator of the learning process, has the purpose of promoting activities guided by the teacher.

It is worth mentioning that in Mexico only a study was realized by Gastélum, H. (2004), where he proposed a desk design, to improve the teaching and learning process in elementary education institute in the state of Sonora, Mexico. But he only performed an anthropometric study to third and fourth-grade students. With the obtained information realized adjustments of the dimensions for the studied population without modifying the design of the furniture. However, all previous authors concluded that only a size of available furniture does not accommodate anthropometric variability even among children of the same age so they recommend adjustable furniture or various sizes of chairs and tables.

METHOD

Study Design

1. Theoretical review of the pedagogic and didactic principles recommended for the educational practices of SEP programs
2. Direct observation of educational practices.
3. Teacher survey about their actual practices and functionality or shortcomings of the current school furniture.
4. Design furniture proposal based on the ergonomic principles applicable to furniture design and the theoretical review and direct observation.

Review of Didactic Activities Developed in the First and Second Grade of Elementary School

The concept of competencies not only implies the accumulation of knowledge but its implementation, that is why in the didactic projects are extended as planned activities involving coordinated and interrelated actions and reflections to achieve the expected learning. Students approach reality by working with problems that interest them. In this way, they investigate, propose and explain their hypotheses, discuss opinions and exchange comments with other colleagues.

To know the didactic activities carried out by first and second grade students with the help of textbooks for students, described each activity as mentioned in these books, dividing the activities for the corresponding block, the expected learning, topics, description of activities, suggestion of accommodation of furniture and posture of the teacher before the activity. The categories of individual work, work in pairs, work in trio, work of four people, teamwork of 5 or more students, work with the whole class and group divided in two were added. According to the descriptions of the activities of each subject, it was noted how the activity was suggested to do by the book. This in order to find out the types of activities that students perform with school furniture and the percentage of activities that they perform everyday life.

By having this information, the textbooks offer us we were able to know if in the classrooms the subjects are taught as they establish it in the books doing the activities as they are mentioned or instead they have better didactics in content. For the first elementary school, individual work (34.78%) was the activity with the highest percentage of subjects such as civic education and ethics (42.86%) and nature exploration (41.67%) as opposed to Spanish subjects (27.27%) where years back followed a traditional education and the student had to work individually.

In the second grade of elementary school, individual work is still the activity with the highest percentage (39.81%), nevertheless is not the only way to work the team of more than five students (27.83%) and work with the whole group (16.91%) complement half of the percentage. In this grade, the work in pairs (7.26%), as well as the work with the divided group (7.16%), decreased.

Table 1. Type of work according to activities of the book of the work of primary of primary

	Type of Job						
	Individual	**Couple**	**Triads**	**Four**	**More Than Five**	**Split the Group in Two**	**All the Group**
Mathematics	34.82%	26.79%	2.68%	0.89%	24.11%	10.71%	0.00%
Spanish	27.27%	13.64%	20.45%	2.27%	11.36%	25.00%	0.00%
Science	41.67%	16.67%	16.67%	0.00%	8.33%	16.67%	0.00%
Civics	42.86%	0.00%	0.00%	0.00%	14.29%	0.00%	42.86%
Arts	27.27%	9.09%	9.09%	0.00%	18.18%	18.18%	18.18%
TOTAL	34.78%	13.24%	9.78%	0.63%	15.25%	14.11%	12.21%

Source: First-grade workbooks 2011 (Secretary of Public Education, 2011).

Table 2. Type of work according to activities of the second-grade workbook

	Type of Job						
	Individual	**Couple**	**Triads**	**Four**	**More Than Five**	**Split the Group in Two**	**All the Group**
Mathematics	38.54%	16.67%	1.04%	1.04%	31.25%	10.42%	1.04%
Spanish	36.36%	6.06%	3.03%	0.00%	30.30%	0.00%	24.24%
Science	40.00%	10.00%	0.00%	0.00%	30.00%	0.00%	20.00%
Civics	55.56%	0.00%	0.00%	0.00%	33.33%	11.11%	0.00%
Arts	28.57%	3.57%	0.00%	0.00%	14.29%	14.29%	39.29%
TOTAL	39.81%	7.26%	0.81%	0.21%	27.83%	7.16%	16.91%

Source: Second-grade workbooks 2011 (Secretary of Public Education, 2011).

Direct Observation of Educational Practices

Observations were made at 11 classrooms with an average of 27.7 students per classroom with a total of 305 students and 11 teachers of the first and second grade of elementary, teachers were interviewed and observed the research was done at the Metropolitan Area of Guadalajara.

The type of sampling was convenient because few schools opened their doors to us to be part of this investigation. To make this possible, we talked with principals and teachers of the first and second grade of each elementary and explained what the research consisted of, as well as what would be observed, we also asked if it was possible to take video and photographs for further analysis.

Results of Direct Observation

As a result of the observations, it was found that all teachers work under a conventional class profile and their pedagogy is centered on the teacher. In each class, regardless of the subject or objective, there is an introduction, then the teacher teaches or explains the activity to be done, the student work on the activity and as a conclusion the activity, task or project is reviewed to continue with the next subject or topic. In activities that indicated the book of the student's work that should be done in groups or create a circle to the center of the room, were not made by four factors that were mentioned (See Table 3 and Figure 1).

The observations more recurrent in both first and second-grade classrooms were the posture that the students adopted when writing and reading where they hunched the back and neck and that somehow lay down on the work table to perform the activities. This factor is due to the great variability of the height of the children of the same age, it was possible to realize that the furniture is not suitable for a great majority of the students. The little ones do not have the possibility of placing their feet on the floor creating ischemia by compression in the legs and therefore discomfort that leads the student to move repeatedly and even stand up to work. It is impossible for them to write and read in a correct posture to carry out such activities they must stand from their seat or sit on their bent legs over the chair (see Figure 2).

Figure 1. Sitting position at the time of writing

Figure 2. Variability of height in girls of 7 years old in second grade of elementary

On the other hand, for the tall ones to sit correctly is a challenge since in some cases their legs met with the bottom of the table and/or the resting feet were uncomfortable to make sudden changes of posture. In the end, few decided to work with their legs open to get closer to the work table, however, the height of this is not suitable for equal way since its neck and column are curved at the moment of writing. The chairs and tables as mentioned have dimensions higher than the student needs. With the help of a digital scale, the chairs used at the elementary schools observed were weighed and their weight varied between 4 to 6 kg (see Figure 3). This weight surpasses the lifting standards that can be achieved by a child between 6 and 8 years old, therefore it is understood that it is difficult for the student to move the chair by itself. In order for the teacher to carry out activities where he has to modify the distribution, the students would have to drag the furniture, generating noise and wasted time in class.

Another important factor that was noticeable in the classroom observed is that the dimensions of the classroom do not correspond with the number of students that the classroom must have for the teacher to be able to work. The excess of students and the minimal dimensions of space do not help the teacher to do didactic activities where they can move around the classroom and create different learning environments (see Figure 4). This, in the end, generates in the pupil little importance in the subject and/or activity and for the teacher frustration because he or she cannot achieve its objectives.

Due to the lack of space to store books and teaching materials, students have their backpacks on the side of their chair or worktable. Today, thanks to the great diffusion that has been made so that children do not carry excess weight in their backpacks, the parents have invested in backpacks with wheels. While there are very useful for dragging are not so much to keep it them upright, so they create obstructions between the aisles of the work tables for both users; students who constantly stumble with these as for the teacher who cannot get close to the students who are at the back of the classroom. (see Figure 5)

At the end of classes in all schools observed, students have to place the chair on top of the table so that cleaning staff can enter the room to sweep and mop without moving the furniture around. This is done in both morning and evening shifts, even though this activity could be taken as a harmless movement is

Figure 3. School seat weight

Figure 4. Classroom with students

Figure 5. Messy rucksacks in hallways

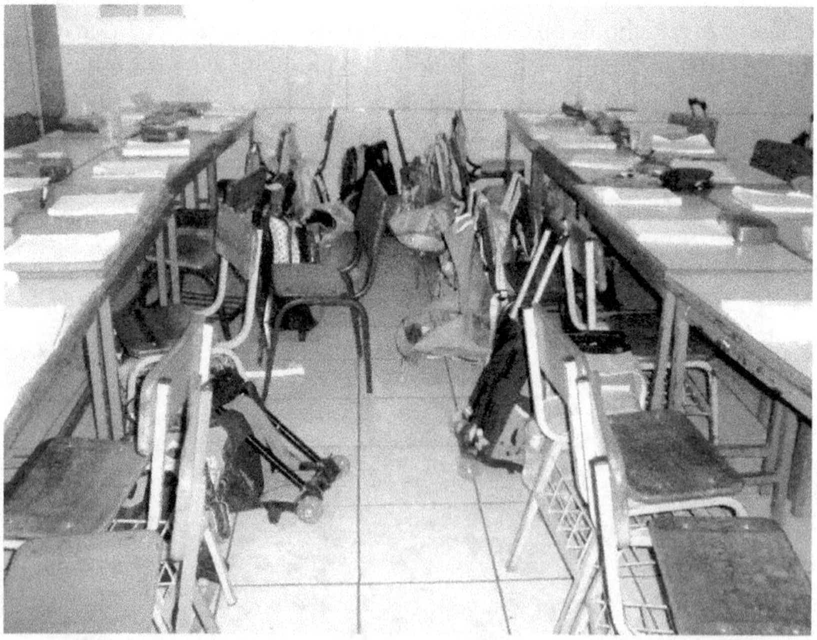

this the situation where more accidents occur. Due to 3 movements that students must make to achieve their goal; they have to lift, turn the chair upside down and load it on the table (see Figure 6). The most problematic moment is when they turn the chair since there is the possibility of an impact with the legs of the chairs towards their person or towards a companion. In the meantime, for the same weight of the chair (4 to 6 kg) there are students who have difficulty to place the chair on top of the table and takes longer to do these three movements if they do not get help from one of their peers. On the other hand, in an interview that was made to the cleaning staff mentioned that they have had some accidents because some chairs are not well placed on the tables and they fall on the person when they perform the cleaning.

Survey of Teaching Practice and Furniture Evaluation

A survey of 44 teachers (81.8% public schools and 18.2% private schools) was conducted through a physical interview (11) and an online survey (33). Because the survey was conducted online, teachers were asked to mention the current school grade where 18 preschool teachers, 22 elementary school teachers, and 4 high school teachers responded. Teachers indicated having a morning shift (77.3%) while few with the afternoon shift (22.7%). The majority (40.9%) reported having between 21 and 30 students in class, 27.3% had 31 to 40 students, while 13.6% of the teachers responded having more than 40 students in the classroom and 15.9% responded to have 21 to 30 students only 2.3% of teachers say they have 5 to 10 students.

Teachers Practices

60% of teachers responded that the characteristics of the classroom do not help the realization of pedagogical practices, even though the vast majority (56.8%) mentioned have complete freedom to appropriate the classroom and only 9% replied that had no freedom (see table 31). However, 20% of the teachers said that it is possible to do team activities without any difficulty while 52% responded to do activities in teams, but with difficulty. They were asked to select the three main distractions class student at the time, resulting companions invading each other's space (45%), objects on the table (34.1%) and backpack cluttering (31.8%).

Figure 6. Pupil lifting the chair on the worktable

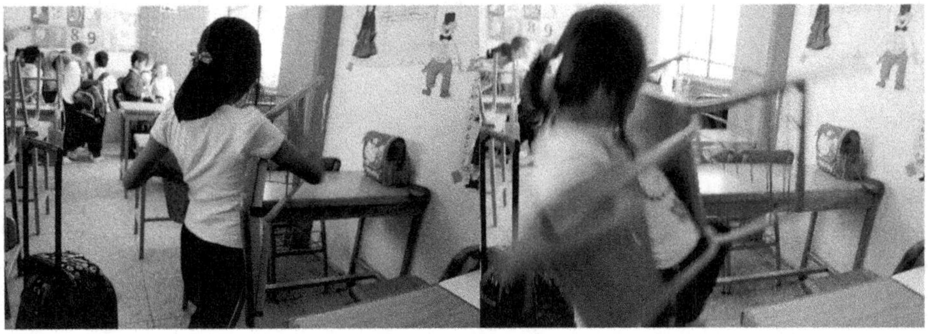

Table 3. Physical characteristics related to classroom teaching practices

		Frequency	Percentage
The characteristics physical help teaching practice	Do help	17	39.5
	Do not help	26	60.5
Freedom to appropriate classroom	Full freedom	25	56.8
	Little freedom	8	18.1
	No freedom	4	9.09
	Teachers did not respond	7	15.9
Conducting educational activities in teams	If they are done without difficulty	9	20.4
	If performed, but with difficulty	2. 3	52.2
	They are not performed	1	2.3
	Teachers did not respond	7	25.0
Three main distractions of students	Companions who invade the space of another	20	45.5
	Objects on the table	15	34.1
	Backpacks cluttering	14	31.8

Didactic Resources

A large percentage of teachers work in a general classroom (61.4%) where the worktable is individual (36.4%) and the work table for two (31.8%) are present. The 31.8% mentioned having a layout by columns, while the distribution by groups obtained 25% and only 2.3% mentioned having all kind of setups. To keep in mind the guidelines that should have the proposed school furniture teachers were asked to select three physical characteristics that they would like the school furniture to have, as a consequence the next keywords were selected. Light (38.6%), an accessory for hanging backpacks (34.09%) and storage space for school supplies (27.2%). A correlation was found with the answers given by teachers, between the main distractions that teachers perceive and the physical characteristics desired in the school furniture. (see Table 4)

Security

Of teachers surveyed 45% reported having witnessed an accident related to the school furniture and some responded tripping over the legs of the chairs, scratching with sticks arising from the deterioration of furniture, knocks with the corners of tables and when the student moves the chair up to the table they hit with the legs of the chair or hit their companion.

Preparation of Design Proposal

Sketches for Furniture Design

From the documentation and observations, this school furniture should be light, movable, safe among other categories to make it a facilitator of activities within the classroom. To be able to facilitate team-

Table 4. Type and equipping classrooms

		Frequency	Percentage
Types of classrooms	General classroom (work table and chairs)	27	61.4
	General classroom with reading corners and /or games	3	6.8
	Classroom with another type of furniture	3	6.8
Types of furniture	Individual desks for left or right handed	7	15.9
	Worktable for two people	14	31.8
	Individual work table	16	36.4
	Table for four or six people	6	13.6
Layout Furniture	Rows	14	31.8
	Groups	11	25.0
	Horseshoe or semicircle	2	4.5
	All	1	2.3
	Other	16	36.4
Physical characteristics desired in a school furniture	Lightweight	17	38.6
	Space to store school supplies	12	27.2
	Chair and table separately	3	6.81
	Scrollable	2	4.54
	Individual tables	7	15.9
	Stackable	5	11.3
	Easy to store	6	13.6
	Worktable for two students	1	2.27
	Accessory to hang the backpack	15	34.09
	Chair and table together	1	2.27
	Adaptable	eleven	25.0
	Another (less bulky)	1	2.27
	Other (tables for 4 or 6 people)	1	2.27
	Another (more comfortable chairs)	2	4.54

Table 5. Incidence relative to school furniture

	Frequency	Percentage
Existence of incident	15	45.5
None incident	18	54.5

work as couples or with more than 5 students. It began with a simple outline of what we were looking for, with simple lines expressed the worktable may have different angles to the time to work and that this together with the chair is achieved easily save.

To get a table that had the inclinations required to perform readings and avoid bad posture it was first envisaged to use a cover that would slope in entirely; drawing board style architects used for drawing. But thinking about children with books, notebooks and more teaching materials on the table, this wouldn´t be useful. Given this fact, it was proposed to divide the work table with a space to stay horizontally and the other could be tilted, but again thinking how the users will interact with the furniture this could cause it to fall suddenly and hurt the fingers of the students. From these unsuccessful attempts, the cover of a digital tablet was observed. We realized that it has the function that required malleability which offers the material would not serve us by its rude use. So, we opted to generate a small polycarbonate tablet which adheres to the cover by small holes; is according to the position it is placed can give us two angles that are needed.

It is so that the "reading tablet" is a piece with very lightweight, resistant and can be stored under the table cover when not occupied and avoid losing it. The inclinations (22 °, 45 ° and 90 °) of the table was an important factor since with these working degrees the student can work on the reading session and reinforcement work writing in a vertical position as it helps to work the child's wrist.

As part of a brainstorming we resorted to the image of a giraffe because they have short legs and a long neck so capturing it to the furniture part this translates into an organic design, which on both sides is balanced and representing the neck is the extension of the leg that can lengthened and contracted to settle at the necessary height. On the other hand, as mentioned above the table required to be move and store when needed without requiring physical effort that is why half of the rear leg is abducted by 10° so that a table could be stacked followed by another as observed in Figure 7. The wheels of the legs are an important factor; therefore, the table can be moved to all corners of the room to make sudden changes work. The first idea was to join the table with the chair but taking into account that activities are performed without the table, we confine ourselves to design a chair that adapts to the needs of the students. As it was perceived in the observations, the chair should be placed on the tables at the end of classes to give way to the cleaning staff to do their job. And because the movement to turn the chair and

Figure 7. Representation of the cover table legs

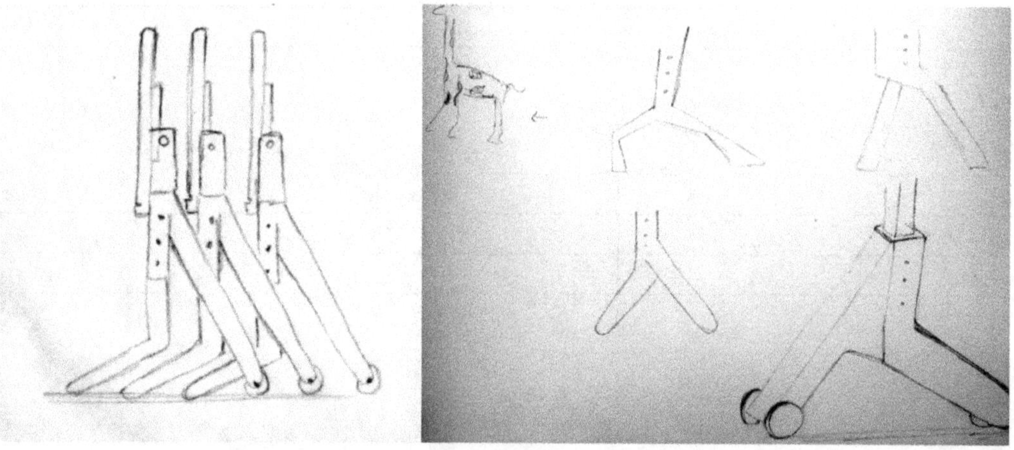

place it on the table was the most common problem of accidents. From this, it was aware that somehow the chair should be off the floor to make cleaning easier.

Proposed School Work Table

The work table proposal has a dimension of 0.60 x 0.43 cm with a thickness of 0.025 cm; it has space for a workbook, a notebook and a small box of pencils (see Figure 8). The table top has a concave shape on the front that was designed in such a way as to cause the user to stand in front of it and prevented from using it in the wrong way. It was established to use on the tabletop a matte white color so as not create distractions in the students at the time of work, it is proposed to use a table top made of polypropylene because it is a resistant material and easy to clean as well as a surface that will not lose their smooth shape. Unlike wood which over time and with the help of students begins to deteriorate and have subsidence due to misuse.

Yakimishyn & Magil-Evans (2002) recommend using vertical work areas for writing activities to promote stabilization of the wrist and posture control. On the other hand, a study of Burt and Benbow (2008) comment that academic delay of some children does not make them fit to work on horizontal surfaces up to seven years old. The worktable can be used both vertically and horizontally. Even though no classroom was seen working in this way, a recommendation could be made to teachers to implement this activity in their curriculum and help promote the beginning of writing in students.

At the time of writing children require shoulder stability to control movements of the elbow and wrist movements that allow firmness to control the fingers with those who write. This chain of action-reaction at the end is associated with the work surface as Bendix (1986) and Sciascia and Faenza (2003) an inclination of 22 ° to 45 ° improves reading posture but does not benefit writing. The table top has two holes with a female and male joining system for a small polycarbonate "reading tablet" which can be accommodated according to the desired inclination to read a book. Reading with the tablet, even when it is not the solution to correct bad habits, could promote good posture. "Reading tablet" has the option of being stored in the bottom of the working desk

Due to the large variability in size among first graders and second, the table is adjustable to meet the needs of each student. Adjustable surfaces help prevent higher inclinations, discomfort in the head, neck and trunk postures (Linton et al, 1994). The total height of the work table was defined with respect to the height of the elbow, authors like Garcia-Acosta and Lange-Morales, 2007; Molenbroek at al., 2003 mentioned that the elbow height sitting is related to the desk height. However, Parcells in 1999 defined the elbow height as the minimum for the desk height, this, in order to reduce the load of lumbar spine and considering the maximum height of the desk should not exceed 5 cm. Closely approximated this criterion Castellucci (2010, pg. 565) quotes Pheasant (1991) and Poulakakis and Marmaras (1998) who recommend that the desk height should be 3 to 5 cm greater than the dimension of the seated elbow. In the case of the children, if the desk height relative to the seat comes to have a greater height, children will need to work with a bent shoulder abduction and scapular elevation, causing more workload to the muscles and causing discomfort and pain shoulder region (Garcia-Molina, 1992).

To define the total dimensions, the percentiles 5 and 95 were taken without increasing more centimeters for the first and third adjustments. The second adjustment was defined by subtracting the height between the first and third adjustments. With this dimension, we obtain an inference of between + - 5 cm according to the adjustment matched by the user.

The work table was designed with the dynamics of work that could be achieved by the teacher if he had the freedom to move around and modify his work environments. That is why placing wheels to furniture generate the classroom is reconfigurable without forgetting the easy movement that would make elementary school students without the need to load furniture or drag them generating exhilarating noises and promoting collaboration among students. For safety to prevent involuntary movements on the work table. The front legs have two rollers that can only move in a forward and backward direction, the rear legs each have two wheels with locking that can rotate the 360° to move them to any corner of the classroom that is needed. The leg design was guided by the same need for displacement and distribution according to the teaching practice, the worktable has the ability to be stacked one next to another and in this way, expand the space and avoid the tables if they are not necessary. Green and blue colors used for both tablet reading and legs were considered because they are neutral colors that could be used for both girls and boys. The color palette was intended to be fresh and fun.

Proposed School Chair

The purpose of designing a simple chair was made thinking that it should be light, its weight shouldn´t exceed 10% of the weight users from 6 to 8 years could lift, so it can be easily manipulated by the main users. In the observations, the main security problem is that the chairs were heavy, have very prominent legs and students had to lift and turn the chairs on the tables so that cleaning staff could do their work, this action leads to possible accidents.

Figure 8. Worktable proposal
Source: Anaya Echeverría, N. (2017). Modelo Industrial de Escritorio Escolar Ergonómico. MX/f/2017/003341.

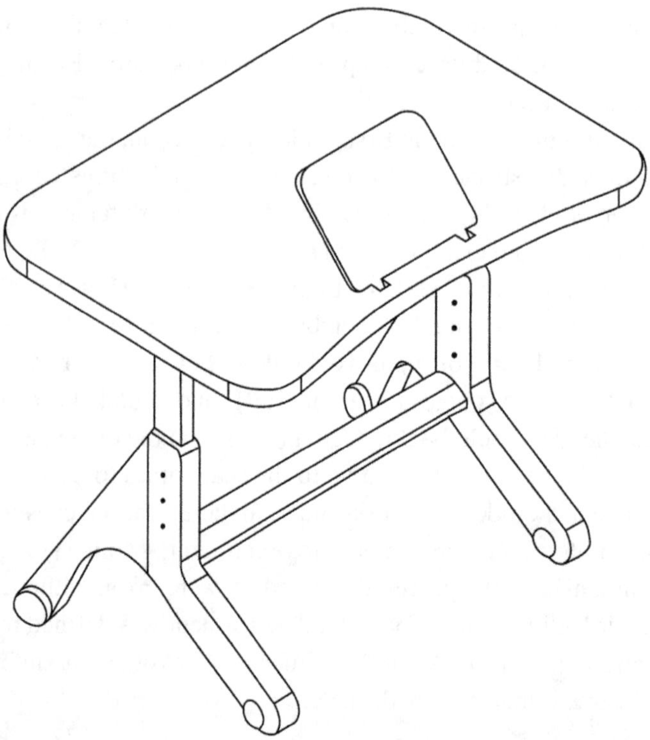

The proposed school chair has two projections under the seat, with the male and female joining system with the same size and system it was used on the "reading tablet" to avoid easily falls when the chair is on the deck. To place the chair over the table, the student would raise the chair from behind to insert the chair on the table top, with the projections inserted in the desk it will prevent the chair from falling or being knocked over. In order to avoid a chair with four straight legs, it was because it should be placed on top of the desk. The legs with extended triangular create a balance to prevent shaking, both legs are reinforced by various joints that work as a space to place the backpack or some school supplies.

As previously mentioned with respect to the work table due to variability in dimensions between students 6 and 8, the chair can be adjusted to three different heights. Popliteal seat height was used to obtain this dimension, some authors (Molenbroek and Ramaekers, 1996, Parcells et al, 1999) mention that the seat height should be no higher than 4 centimeters (Gutierrez and Morgado, 2001) or the 88% of popliteal height (Parcells et al., 1999) to avoid compression. If the seat is too low the thighs no longer rest on the seat and the pressure is concentrated on a very small area of the buttocks; in addition, decreasing the flexion angle at the knees and hips can be uncomfortable. (Garcia et al., 1992). The radius of the front edge of the seat must be adjusted to the recommended values (3° -5 °) so that the properly rounded edge prevents pressures and becomes uncomfortable (Garcia, Moraga, Page, Tortosa, and Green, 1992). In this way, the use of seat backing will support the lumbar spine without making compression on the popliteal surface, researchers say that to obtain the seat depth will have to take the 5th percentile buttock length popliteal dimension (BPL) (Gutiérrez and Morgado, 2001). In the case of a school, chair depth should be proportionally smaller to prevent users from collapsing on the seat and backrest (Garcia et al., 1992).

For proper adjustment of the seat width, it should be able to accommodate comfortably the buttocks and hips on the seat. For this purpose, researchers recommend using the 95th percentile (Evans et al., 1988; García et al., 1992; Gutierrez and Morgado, 2001) in order to avoid a lack of slack to enter and exit. In the sedentary position, most of the body weight is supported by the soft tissue of the buttocks and the proximal posterior of the thighs. If the pressure is excessive and prolonged it may appear numbness and pain.

The backrest with respect to the seat located at a minimum angle of 105 ° to achieve a minimum work of the trunk and thigh. In order to be able to move the trunk and arms properly, the subscapularis height should be higher than the top of the backrest and designed for the 5th percentile of the subscapular dimension (Garcia-Acosta and Lange-Morales, 2007). The solution to this conflict is an adequate support, a lumbar support that maintains the concavity of the column. According to the researcher, a contoured backrest adjusts the natural curves of the spine (Bendix et al., 1996) facilitates the lumbar lordosis (Aagaard-Hansen and Storr-Paulsen, 1995) and reduces kyphotic postures (Mandal, 1984; Bendix, 1987).

Thinking about the manipulation by secondary users as teachers and cleaning staff in the back of the chair was placed a handle with the dimensions of the 95th percentile size. To avoid stuck fingers while displacing the chair. Because the chair is the space where students will be sitting most of his time at school, the color green for the seat and back, the orange on the legs and the color blue in small pieces of subjections. The idea of designing it with bright colors is to create an aesthetic design that encourages students to want to use it continuously

Figure 9. School chair proposal
Source: Anaya Echeverría, N. (2017). Modelo Industrial de Silla Escolar Ergonómica. MX/f/2017/080312.

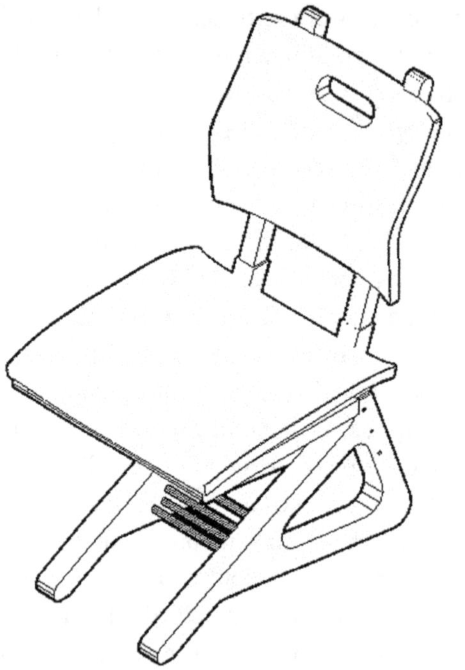

Considering that the rooms of the first and second grade of primary use the same furniture during these two years, it must be able to modified so as the growth of the boy or the girl. To avoid bad postures that lead to discomfort and later injury. This is why the adjustments of the chair and work table have the dimensions necessary to adapt 90% of the population between 6 and 8 years. This furniture can be used and manipulated by tall, low, thick and thin, light and heavy users. As an example, Figure 9, on the left-hand side, is the adjustment in 1st position of chair and table to adapt to a girl with a percentile 5 of 6 years old and on the other hand the chair and school table with the adjustments in the 3rd position may be used for the 95th percentile that corresponds to an 8-year-old boy (see Figure 10).

Figure 10. Figurines percentile 5th female aged 6 and percentile 95 male aged 8
Source: Anthropometric dimensions Latin American population, Ávila Chaurand, R., Prado Leon, L. and González Muñoz, E. 2007; Anaya Echeverría, N. (2017).

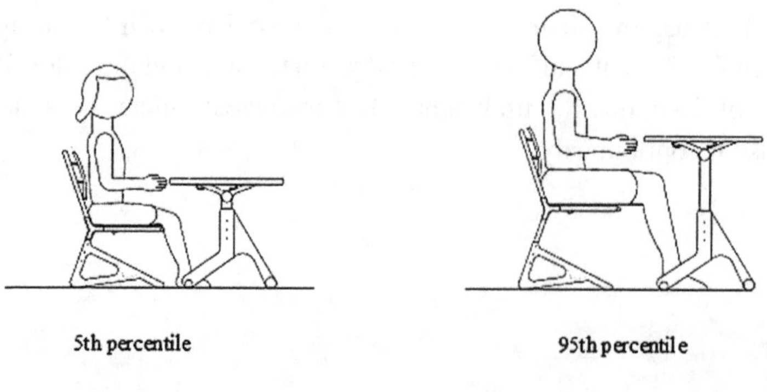

SOLUTIONS AND RECOMMENDATIONS

To carry out the gathering information for this project we had the support of teachers and principals, though not always have the same fate. Many doors were closed because teachers nor directors wanted an outsider to its school observe the process of the classes. However, talking to teachers and parents who agreed to observations agreed that a school furniture necessary to be safe for their children because they spend a large part of the day in the classroom. The furniture design presented here is based on the pedagogical and didactic practices SEP´s book and these characteristics could not be used in other contexts. To do so this would have to be evaluated to recognize the characteristics and determinants of the practices and if the furniture acts as a facilitator to these pedagogies.

FUTURE RESEARCH DIRECTIONS

Because it is the first project of its kind in Mexico, it is necessary to carry out more research on materials, design and manufacture, since it is necessary that, together with all the above-mentioned recommendations, the furniture can be assembled so that can to be safely distributed through all schools of the republic, which is easy to maintain as well as easy to repair if necessary. Likewise, an anthropometric research must be done in all states of the republic since the proposed school furniture is designed according to the data presented by Avila Chaurand et al. (2007) in the metropolitan area of Guadalajara. It should be noted that the data for children date from 1993-1999, and in the last 20 years, there have been important changes in the anthropometric dimensions of the population (obesity). Recall that there is great variability in the same classroom now if you compare a whole country where the context, location, and culture in each region are different. To evaluate in the future will have our school furniture manufacture and place in a classroom to determine whether it fulfills its role as a facilitator of teaching and helps teachers and students to generate more enriching educational practices.

Although the furniture design proposed in this research is clearly for students in the first and second grade as pedagogical and didactic practices will change as it increases in grade, as well as the social, historical and technological context. It would be necessary to investigate other grades to meet their educational and anthropometry of its users to determine the characteristics of furniture that is a facilitator of pedagogical practices and not a simple instrument in the classroom.

Ergonomics is a multidisciplinary science "new" in our country and has mistakenly misinformation that only influences anthropometrically, however, one should think about the features must have the furniture and/or space with respect to users. Unfortunately, designers have been told to be creative and generate something beautiful outpacing the characteristics, needs, and roles of users who are in direct contact with these spaces and furniture; however, the aesthetic is not fought with ergonomic. And it is something we should make it very clear because there is the misconception that a space or ergonomic product, for this reason, it is expensive or not aesthetic.

Because of this, it is necessary to continue to generate research in all educational fields since as mentioned above there is a lack of knowledge of elementary students and we as a designer will have to evaluate what is happening in the classroom and how design can influence positively or negatively on the user.

CONCLUSION

Despite efforts to modify the teaching-learning methodologies to a current trend, both teachers, spaces, and objects have remained in the past limiting teaching practices. Unfortunately, no teacher was observed who had a more active initiative towards the students to create learning environments with diverse activities and different ways of working that generate motivation for students to learn.

Even though there is no concrete data one could believe that the limiting factor towards a more active learning environment is the same space and furniture that teachers have.

But on the other hand, it could be the academic training that prevents them generating changes in their practices or may not want to experiment with new situations and get out of their comfort zone.

In order to know if the limitation to create educational and educational activities more active and motivating for the students is the same school furniture, it would be necessary to design based on the observations and the ergonomic requirements a school furniture that eliminates these limitations and becomes a facilitator of them.

The influence of the teacher on the furniture could be used properly to be a facilitator of pedagogical and didactic practices and achieve the goals or be used as a piece of furniture in the classroom. This factor may vary according to the teacher, their curriculum, age and even the same motivation. However, this study leads to the continuation of the research to do a pilot project where it can intervene and observe changes and see if they are favorable or not.

REFERENCES

Aagaard, J., & Storr-Paulsen, A. (1995). A comparative study of three different kinds of school furniture. *Ergonomics*, *38*(5), 1025–1035. doi:10.1080/00140139508925169 PMID:29105614

Anaya Echeverría, N. (2017). *Modelo Industrial de Escritorio Escolar Ergonómico. MX/f/2017/003341*. Zapopan, Jalisco: Instituto Mexicano de la Propiedad Industrial.

Anaya Echeverría, N. (2017). *Modelo Industrial de Silla Escolar Ergonómica. MX/f/2017/080312*. Zapopan, Jalisco: Instituto Mexicano de la Propiedad Industrial.

Ávila Chaurand, R., Prado León, L., & González Muñoz, E. (2007). *Anthropometric dimensions of Latin American population* (2nd ed.). Universidad de Guadalajara.

Bendix, T. (1986). Sitting postures-A review of biomechanics and ergonomic aspects. *Man and Medicine*, *2*, 77–81.

Bendix, T., & Bloch, I. (1986). How Should a seated workplace with a tiltable chair be adjusted? *Applied Ergonomics*, *17*(2), 127–135. doi:10.1016/0003-6870(86)90250-4 PMID:15676578

Burdek, B. (1994). History, theory and practice of industrial design. Gustavo Gili.

Burt, C., & Benbow, M. (2008). *Children and handwriting ergonomics. In Ergonomics for Children: Designing products and places for toddlers to teens* (pp. 689–720). London: Taylor & Francis.

Cardemil, C., Latorre, M., Filp, J., & Galvez, G. (1991). Factors affecting the improvement of learning in basic education. *Journal of Education*, 208.

Castellucci, H., Arezes, P., & Viviani, C. (2010). Mismatch Between classroom furniture and anthropometric Measures in Chilean schools. *Applied Ergonomics*, *41*(4), 563–568. doi:10.1016/j.apergo.2009.12.001 PMID:20031115

Castellucci, I., Gonçalves, M. A., & Arezes, P. (2009). Ergonomic considerations classrooms in school portuguesas first cycle. *Science. Job*, *11*(34), 184–187.

Chung, J., & Wong, T. (2007). Anthropometric evaluation for primary school furniture design. *Ergonomics*, *50*(3), 323–334. doi:10.1080/00140130600842328 PMID:17536771

Cornell, P. (2002). The impact of changes in teaching and learning on furniture and the learning environment. *New Directions for Teaching and Learning*, *2002*(92), 33–42. doi:10.1002/tl.77

Evans, W., Courtney, A., & Fok, K. (1988). The design of school furniture for Hong Kong schoolchildren: An anthropometric case study. *Applied Ergonomics*, *19*(2), 122–134. doi:10.1016/0003-6870(88)90005-1 PMID:15676654

Ferreiro Gravié, R. (2014). *New ways of learning and teaching, cooperative learning: theory of cooperative work, development of critical and creative thinking, strategies and teaching functions*. Academic Press.

Garcia, C., & Page, A. (1992). *Guide recommendations for designing ergonomic furniture* (1st ed.). Valencia Institute of Biomechanics of Valencia.

Garcia-Acosta, G., & Lange-Morales, K. (2007). Definition of sizes for the design of school furniture for schools Bogota based on anthropometric criteria. *Ergonomics*, *50*(10), 1626–1642. doi:10.1080/00140130701587541 PMID:17917903

Gastélum Gonzalez, H. (2004). Anthropometry in the classrooms of primary education in the State of Sonora. In *Sixth International Congress on Ergonomics*. Guanajuato: Society Ergonomists Mexico.

Gouvali, M., & Boudolos, K. (2006). Match Between school furniture dimensions and children's anthropometry. *Applied Ergonomics*, *37*(6), 765–773. doi:10.1016/j.apergo.2005.11.009 PMID:16442494

Gutierrez, M., & Morgado, P. (2001). Guide recommendations for the design of school furniture. Santiago de Chile: Ministry of Education and UNESCO.

Herman, F., Van Gorp, A., Simon, F. and Depaepe, M. (2011). The school desk: from concept to object. *History of Education, 40*(1), 97-117.

Hira, D. (1980). An ergonomic appraisal of educational desks. *Ergonomics*, *23*(3), 213–221. doi:10.1080/00140138008924735

Horne, S. (2002). The classroom environments and its effects on the practice of teachers. *Journal of Environmental Psychology*, *22*(1-2), 139–156. doi:10.1006/jevp.2001.0239

Jesús Hernández Guzmán, R., & Vásquez Quiroga, J. (2009). Analysis of anthropometry-dimensions Pupitres Relationship, likely to cause fatigue of Students of Industrial Engineering. *Research*, *4*(1), 10–16.

Jonassen, D. (2000). Designing constructivist learning environments. In C. Reigeluth (Ed.), *Instructional design theories and models. A new paradigm of instructional theory Part I* (1st ed., pp. 225–249). Madrid: Mc Graw Hill.

Knight, G., & Noyes, J. (1999). Children's Behavior and the design of school furniture. *Ergonomics, 42*(5), 747–760. doi:10.1080/001401399185423 PMID:10327894

Linton, S., Hellsing, A., Halme, T., & Åkerstedt, K. (1994). The effects of ergonomically designed furniture school on Pupils' attitudes, symptoms and Behavior. *Applied Ergonomics, 25*(5), 299–304. doi:10.1016/0003-6870(94)90044-2 PMID:15676981

Lippincott, J. (2009). Learning Spaces: Involving Faculty to Improve Pedagogy. *EDUCAUSE Review, 44*(2), 16.

Long, P., & Holeton, R. (2009). Signposts of the Revolution? What We Talk About When We Talk about Learning Spaces. *EDUCAUSE Review, 44*(2), 36.

Mandal, A. (1982). The correct height of school furniture. *Human Factors: The Journal of the Human Factors and Ergonomics Society, 24*(3), 257–269. doi:10.1177/001872088202400301

Martin, S. (2002). The classroom environment and Its Effects on the practice of teachers. *Journal of Environmental Psychology, 22*(1), 139–156. doi:10.1006/jevp.2001.0239

Mokdad, M., & Al-Ansari, M. (2009). Anthropometrics for the design of Bahraini school furniture. *International Journal of Industrial Ergonomics, 39*(5), 728–735. doi:10.1016/j.ergon.2009.02.006

Molenbroek, J., Kroon-Ramaekers, Y., & Snijders, C. (2003). Revision of the design of a standard for the dimensions of school furniture. *Ergonomics, 46*(7), 681–694. doi:10.1080/0014013031000085635 PMID:12745681

National Institute for Educational Infrastructure. (2014). Standards and specifications for studies, projects, construction and installation, Vol. 3 Livability and operation Volume III furniture. Author.

Parcells, C., Stommel, M., & Hubbard, R. (1999). Mismatch of classroom furniture and student body dimensions: Empirical findings and health implications. *The Journal of Adolescent Health, 24*(4), 265–273. doi:10.1016/S1054-139X(98)00113-X PMID:10227346

Quality Schools Program. (2010). Guide to facilitate the inclusion of students with disabilities in schools participating in the PEC. 1st ed. Secretary of Public Education. Curriculum 2011 Teacher's Guide Basic Primary Education First Degree.

Rivlin, L., & Weinstein, C. (1984). Educational issues, school settings, and environmental psychology. *Journal of Environmental Psychology, 4*(4), 347–364. doi:10.1016/S0272-4944(84)80005-5

Study Program 2011 Teacher's Guide Basic Education Elementary Second Grade. (2011). DF, Mexico: Ministry of Public Education.

Trevino E., & Trevino, G. (2004). Study on educational inequality in Mexico: the incidence of school in the academic performance of students and the role of teachers. *Research Papers Collection,* (5).

Vélez, E., Schiefelbein, E., & Valenzuela, J. (1993). *Factors affecting academic achievement in primary education*. Literature Review of Latin America and the Caribbean.

Yakimishyn, J. E., & Magill-Evans, J. (2002). Comparisons Among tools, surface orientation, and pencil grasp for children 23 months of age. *The American Journal of Occupational Therapy*, *56*(5), 564–572. doi:10.5014/ajot.56.5.564 PMID:12269511

Zorrilla, M. (2010)., Efficiency performance of pupils in primary schools between attendance and quality and access. In SITEAL (Ed.), Reflections from basic education in Mexico (8th ed.). Academic Press.

KEY TERMS AND DEFINITIONS

Competency-Based Learning: An approach to education that focuses on the student's demonstration of desired learning outcomes as central to the learning process.

Constructivist-Based Learning: The idea that learners construct knowledge for themselves—each learner individually (and socially) constructs meaning—as he or she learns.

Didactic Activities: The activities of educating or instructing; activities that impart knowledge or skill.

Educational Concept: The study of methods of training and teaching and effectiveness, and of problems experienced in learning formal material.

Learning Environment: The diverse physical locations, contexts, and cultures in which students learn.

Objective: Something that one´s effort or actions are intended to attain or accomplish; purpose; goal; target.

Pedagogy: The art or science of teaching; education; instructional methods.

Task: A definite piece of work assigned to, falling, to, or expected of a person; duty.

Chapter 7
Experimental Design to Analyze a Novel Stabilization Design of a Three-Wheel Vehicle

Luis Elias Dávila-Leyva
Autonomous University of Ciudad Juárez, Mexico

Ludovico Soto Nogueira
Autonomous University of Ciudad Juárez, Mexico

ABSTRACT

Given current trends in pedal vehicles, this chapter is intended to develop a product that is capable of driving without the use of hands, for people with disabilities or for simple recreation. An apparatus was developed to measure and record the necessary parameters to design the most adequate mechanism to achieve this objective. Twelve different positions were analyzed for the user, two vertical seat positions, two inverted angle configurations of the inverted pendulum, two free degree limits of the pendulum, two wheel rotation degrees regarding the body of the mechanism, and regarding the axis motion, two axial configurations. The conclusion resulted that the function of the mechanisms was optimal.

INTRODUCTION

Due to the increment of ecologic consciousness and personal care, more and more people around the world are using alternative means of transport to the internal combustion automobiles, as electric and pedal vehicles (Reiser 2nd, Peterson, & Broker, 2002). Focusing on these last ones, we found a growing and continue innovation market, which implies two main strands for this project: inclusion and diversification. Inclusion refers that all population segments to have the capacity to choose the mean of transport they prefer and suits them best. In this case, it pretends to include people with some dysfunction, injury or amputation of the upper limbs. Meaning, they cannot drive a bicycle or tricycle even if their lower motor skills are intact. Diversification is obtained by thinking in a different way of driving a vehicle; as long as you keep focus on the road, it will allow you to occupy your hands in various activities such as

DOI: 10.4018/978-1-5225-5234-5.ch007

Copyright © 2018, IGI Global. Copying or distributing in print or electronic forms without written permission of IGI Global is prohibited.

eating a fruit or an energetic shake on your way to college or work, take care of a fragile item, and even rest arms crossed while enjoying your drive.

Bicycles have been a means of transport widely used in all areas, both recreational and utilitarian because of its versatility, easy to use, and dynamic stability (Astrom, Klein, & Lennartsson, 2005). Even when its static stability is zero, dynamic stability resembles when a rod is balanced in one hand palm, which allows us to intuitively swing against the drop point (Schwab, 2012). In addition, the more speed it takes, the more it requires force to perform rotation, breaking the linear inertia, which paradoxically results in greater stability at a higher speed (Patterson & Leone, 2010). However, despite all the positive features of conventional bicycles, they do not meet the main requirement of diversification for the present project as they are not statically stable, that is, they do not stand on their own at rest. Static stability is essential for the application of a mechanical driving system without the use of hands. Not to mention that a recumbent bicycle is more efficient than the conventional bicycle in its biomechanics, ergonomics, kinematics, and power development (Ahmed, Qureshi, & Khan, 2015).

Even since 1896, when the first patent registration of a recumbent bicycle was made, just like the conventional bicycle, the general changes presented have been minimal; even when there are patents with some modifications in driving (Mighell, 2009), stability (James, 1941), and inclination (Roqueiro, Vieira, & Faria, 2010)

The main reason for searching a statically stable vehicle is because, if a person wants to drive without hands, the person should be getting on the vehicle should be done in the same way, which is not achieved with a bicycle. Taking into account that the most suitable position to get on and accommodate in some type of device is the sitting position (Mircheski, Kandikjan, & Sidorenko, 2014), the vehicle configuration should be one in which the user remains seated with his legs in front and his arms relaxed, starting this way from static to dynamic without altering his position. In addition, we take advantage of the fact that an ideal cardiovascular state is generated for cardiac rehabilitation, a highly recommended exercise (Kato, Tsutsumi, Yamaguchi, Kurakane, & Chang, 2011).

Starting from the premises described in the previous paragraphs, vehicles of three or more wheels will be taken into account because they have the characteristic of being statically stable, although these vehicles are not as dynamically stable as the two-wheeled vehicles. To overcome this disadvantage and introduce an innovative transport option, it is proposed to design and develop a completely mechanical system that generates a tilt in the vehicle by curving it, re-positioning the center of mass in relation to the centripetal and normal force vectors, at the time you reconfigure the position of the wheels, which would expect stability to increase.

All of this in order to be able to re-position along with the body a few centimeters of the frame and direct the vehicle without the use of hands or arms. The original design requires the investigation of all involved factors in the static and dynamic stability of the mechanism, with special attention in the immediate reaction and control derived from what will be a comprehensive analysis of the ergonomics and usability of it. In addition to the material resistance analysis and the selection of mechanical elements, submitting experimental tests to the models derived from the applied calculations.

OBJECTIVE

In the present study, it is intended to obtain all necessary data, both ergonomic and operational of the mechanism through experimentation in a mechanical system designed and manufactured to modify all positions in the seat and backrest as well as the position relative to the floor, the vertical, the inclination regarding the rotation, and the optimal graduation of the pendulum regarding the capacity of inclination of the body (see Figure 1)

METHODOLOGY

Study Design

A transversal, exploratory, experimental, and prospective study was developed to analyze a novel system of stabilization.

Sample

The criteria to participate in the experiment are shown in the following lines:

- Only men were accepted
- Age between 15 and 30 years
- No history of injuries and health problems
- Body weight between 70 and 100 kg

In the case of a volunteer presents injuries in back, neck, and legs their participation will be retired, and the data is not included in the analysis.

Figure 1. The different possible configurations of the device will be shown in detail

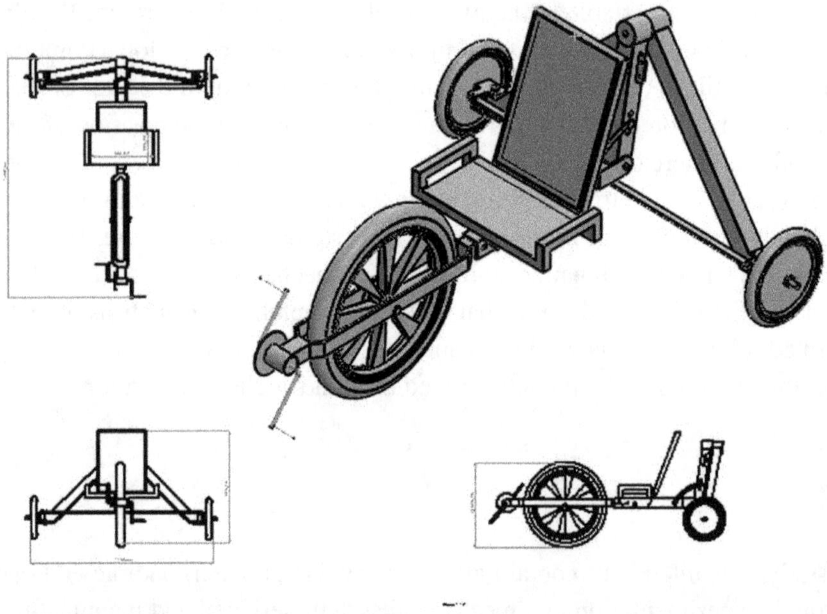

Materials

The following materials and devices were used to develop the experiment:

- The novel stabilization design cart
- Smartphone with video recorder, digital inclinometer, speedometer, and tracking distance apps.
- Orange security cones to delimitate the track.
- Metric tape
- Chronometer.

METHODS AND PROCEDURE

Taguchi Method

A qualitative and quantitative analysis about comfort perception regarding the inclination of the mechanism was made through the time spent to go over the circuit and the range movement measurements, inclination ranges, vehicle inclination, and position of the seat-backrest allowing the best performance in driving. Taguchi method (Taguchi, 1993) was applied to determine the importance of each factor to know the best performance of the device. Also, to know which is the most important factor. With the results, we will select the factors to build the final prototype.

Selection of the Factors

The selection of the factors derived from the dynamic analysis of three-wheeled vehicles to be analyzed, both static and mechanized, in the direction of inclination and adjustment of wheel configurations, taking as a key reference the inclination of the object when taking a curve, which is the repositioning of the mass center, counteracting the centripetal force generated by the friction in the steering wheels.

Taguchi methodology was applied (Taguchi, 1993) to analyze the device. This method consist is to determine all possible variables and determining which one is optimal. The factors and the test to be considered are shown in Table 1 and the tests to be performed in Table 2.

The inclination of the backrest will have two positions, which fluctuates in the optimum position for recumbent pedaling between 60 and 80 degrees. The degrees of freedom will measure how many degrees are recorded in the pendulum oscillatory movement when turning the vehicle; as well as the pendulum inclination that will start from 70 to 80 degrees regarding the vertical, changing the development of the

Table 1. Description and levels of the factors

Factor		Level 1 (degrees)	Level 2 (degrees)
A	Pendulum inclination (horizontal related)	70	80
B	Seat inclination (horizontal related)	80	60
C	Free pendulum degrees (vertical related)	30	40
D	Rear wheels turn (central axis related)	30	40

circumference described by it and by modifying the way in which the mechanism is rearranged to preserve the three support points on the ground. Finally, the relationship between the body of the mechanism, i.e. the central part, regarding the tires, parting from 0 degrees on the line to 45 degrees at the maximum curvature, or from 0 to 40, comparing the inclination achieved with the radius described in both cases and the vehicle performance.

Functionality ranges are given to each factor, then all the tests are done; each of them with all possible combinations, recording in each one the rating performance of the factors (see Table 2).

The tools to be used in the mechanism as a measuring device are: digital chronometer, a digital inclinometer, and a digital speedometer, with which you can determine with precision some of the necessary parameters for the present experimentation.

Before starting the experiment and using the function "Random. Among (lower, upper)" in an Excel spreadsheet, a random number between 1 and 8 was obtained. This number ensures that the tests order does not affect their result, that is, the volunteers will have different order when they perform the 8 different tests.

Measures

NASA-TLX

The NASA-TLX method (Hart & Staveland, 1988) was applied to assess the mental load generated by the vehicle through the different configuration planed. Participants were asked to rate six workload items – mental demands, physical demands, temporal demands, performance, effort, and frustration – each between 1 (Low) and 10 (High). An Overall Workload Index (OWI) score (ranging from 0 to 10) was calculated as the sum of the weighted ratings of the six sub-scales, divided by the number of factors as suggested by (DiDomenico, 2003). The method allows to divide the OWI into four levels depending on the final score: Level 1 = 1-25 points; no action; Level 2 = 26-50 points, medium, recommended actions; Level 3 = 51-75 points, high priority actions; and Level 4 = 76-100 points, very high, immediate actions needed (Garcia-Escutia, 2004).

Table 2. Rating performance of the factors

Test	Factor				Value			
	A	B	C	D	A	B	C	D
1	1	1	1	1	70°	80°	30°	30°
2	1	2	1	2	70°	60°	30°	40°
3	1	1	2	2	70°	80	40°	40°
4	1	2	2	1	70°	60°	40°	30°
5	2	1	1	1	80°	80°	30°	30°
6	2	2	1	2	80°	60°	30°	40°
7	2	1	2	1	80°	80°	40°	30°
8	2	2	2	2	80°	60°	40°	40°

1-10 Borg's Scale (Comfort)

Commonly, the RPE scale is used to measure the intensity of exercise. However, in this case the scale was used to rate the comfort of every test performed by the users. The RPE scale runs from 0 – 10. The numbers below relate to phrases used to rate how comfortable or not comfortable you find an activity. For example, 0 (nothing at all) would be how you feel at the end of the test if you found the test not comfortable; on the contrary, 10 (very, very comfortable) is how you feel at the end of the test if you found the test extremely comfortable.

Circuit Time Interpretation

For the unification question of comparative parameters, we proceed to make tabulation where ranges were divided from one to ten equal intervals since the other two factors that are being considered for the test analysis are based on one to ten. With which, an interaction between the three factors can be made based on the same quantification characteristic. Next, Table 4 shows the classification of the resulting time ranges, in which the longer the time of the route the less beneficial is for the performance of the test device. The lowest time (1.25 minutes) recorded circuit was taken as the minimum value and the longest time (2.12 minutes) taken as the maximum value.

Yi Calculation

The results of each test were reflected in three specific areas. The first was the general driving perception (determined by applying the NASA-TLX method). The second was the time it takes to finish the circuit. And the third was the comfort response of the mechanism to voluntary movements, the straight, and stability fidelity perceived in the curves. All of them were obtained from an interview at the end of each test. Figure 2 shows the formula to generate the objective function, that is the Yi.

Table 3. Comfort scale

Number	Interpretation
10	Very, very comfortable
9	
8	
7	Very comfortable
6	
5	
4	Somewhat comfortable
3	Moderate comfortable
2	Light
1	Very light
0	Nothing at all

Table 4. Classification of time ranges

Minimum	Minimum	Range
1.25	1.33	10
1.33	1.42	9
1.42	1.51	8
1.51	1.59	7
1.59	1.68	6
1.68	1.77	5
1.77	1.85	4
1.85	1.94	3
1.94	2.033	2
2.033	2.12	1

Values expressed in minutes

Figure 2. Objective function (Yi)

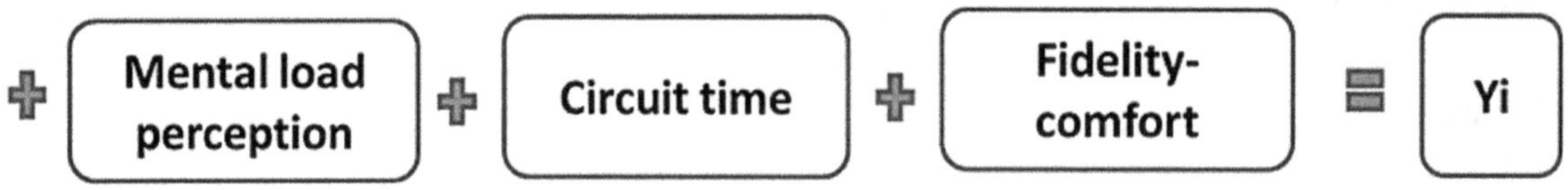

RESULTS

A total of 5 participants completed the study. The average stature was 1.76 mt, the average weight was 82 kg, and the average age was 22 years. All participants read and signed an informed consent to manifest their voluntary participation in the study.

Table 5 shows the results of the three variables measured at the end of the experiment. The test with the best results are shown in blue, that is, lower mental load, shorter circuit travel time, and greater comfort.

RESULTS EXPLANATION IN THE SUBJECTS' TABLES

The averages of both the time and the result of the sum of each arrangement were obtained, understanding that the shorter the circuit time, the best arrangement will be. As well as the addition, where the difference between comforts and the mental load sum plus the circuit time (Yi) is presented. Which, in turn, the higher the number to which it belongs the better the arrangement will be. This is shown in Table 6.

Then, Table 7 shows how the factors are divided at each level and all the tests are sum at that level. In this case, there were four at level 1 and four at level 2 in all cases. In addition, we calculate the average for each level and factors i.e. Yi is composed by the sum of the averages for every array, for example: at level Yi= *(20.8+20.2+22.6+19.8=83.4)(83.4/4=20.85).*

Lastly, a graph with the averages of each factor level is made as shown in Figure 3, which shows the difference between the levels of each factor using the Yi average for each level of every factor.

Table 5. Results for the three measures considered in the experiment

Subject 1	**Test**							
Measures	1	2	3	4	5	6	7	8
Mental load perception (NASA-TLX)	5	4	5	3	5	6	3	6
Circuit time (Track time)	8	9	9	9	8	8	9	8
Comfort-fidelity	8	8	9	8	5	6	7	5
Yi	21	21	23	20	18	20	19	19
Subject 2	**Test**							
Measures	1	2	3	4	5	6	7	8
Perception, mental load (NASA-TLX)	6	5	8	6	5	4	7	5
Circuit time	7	8	8	7	7	8	8	7
Comfort-fidelity	8	7	7	8	6	6	7	7
Yi	21	20	23	21	18	18	22	19
Subject 3	**Test**							
Measures	1	2	3	4	5	6	7	8
Perception, mental load (NASA-TLX)	7	6	8	5	4	6	9	6
Circuit time	9	9	9	8	9	9	9	9
Comfort-fidelity	8	7	7	8	7	6	6	8
Yi	24	22	24	21	20	21	24	23
Subject 4	**Test**							
Measures	1	2	3	4	5	6	7	8
Perception, mental load (NASA-TLX)	5	7	8	5	6	5	7	5
Circuit time	1	0	1	0	0	0	1	0
Comfort-fidelity	9	9	9	9	9	8	8	9
Yi	15	16	18	14	15	13	16	14
Subject 5	**Test**							
Measures	1	2	3	4	5	6	7	8
Perception, mental load (NASA-TLX)	6	5	7	7	5	4	5	5
Circuit time	9	8	9	8	8	8	9	8
Comfort-fidelity	8	9	9	8	8	7	8	8
Yi	23	22	25	23	21	19	22	21

Table 6. Average scores for Yi

	Yi Test Average							
Arrangement	1	2	3	4	5	6	7	8
Yi	20.8	20.2	22.6	19.8	18.4	18.2	20.6	19.2

Table 7. Factors average at each level

Factor	Level	Addition	Average
A. Pendulum inclination	1	83.4	20.85
	2	76.4	19.1
B. Seat inclination	1	82.4	20.6
	2	77.4	19.35
C. Free pendulum	1	77.6	19.4
	2	82.2	20.55
D. Rear wheels turn	1	79.6	19.9
	2	80.2	20.05

Figure 3. Average results for every factor

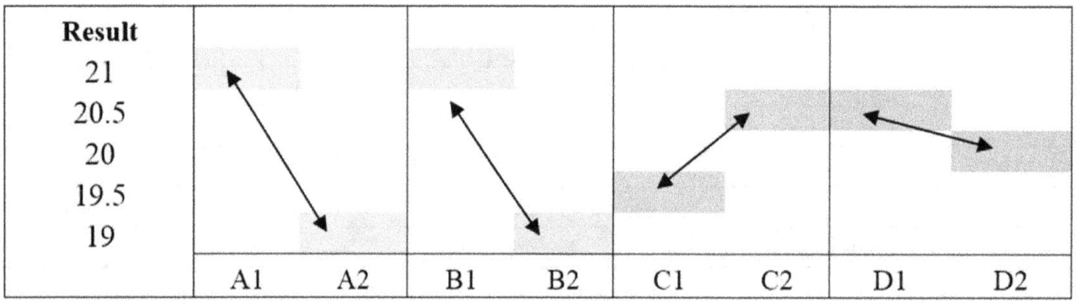

The table interpretation can be explained as follows: the more vertical the line is described by the union of the two levels of each factor, the greater its importance in the product design. In this specific case, as we mentioned before, while the higher the number, the device will perform better. The factors at the corresponding level that turned out to be the highest will be considered as the arrangement that could perform a better work. The result shows that the arrangement with the best result is 3, with the factors at their respective level as follows (see Table 8).

Considering the above, the more vertical the line is the greater the importance of said factor, we can notice that factor D is almost completely horizontal, which is why it is considered that the level at which it is used is not relevant.

Table 8. Best factors

Factor	Level	Result
A	1	80 degrees
B	1	80 degrees
C	2	30 degrees
D	1	40 degrees

That said, the arrangement that comes closest to the optimal result is arrangement number 3; which contains the following parameters in Table 8. Ones that will be used as a basis for the design and manufacture of a final prototype.

CONCLUSION

For the present article, an apparatus was designed and manufactured with the ability to change the configuration in order to carry out angle modifications and required restrictions, besides expanding and contracting to adjust to the height of each individual who performs the test. With this study, we can determine the best configuration to develop a final prototype considering the specifications to get an optimal performance.

REFERENCES

Ahmed, H., Qureshi, O. M., & Khan, A. A. (2015). Reviving a ghost in the history of technology: The social construction of the recumbent bicycle. *Social Studies of Science*, *45*(1), 130–136. doi:10.1177/0306312714560640 PMID:25803920

Astrom, K. J., Klein, R. E., & Lennartsson, A. (2005). Bicycle dynamics and control: Adapted bicycles for education and research. *IEEE Control Systems*, *25*(4), 26–47. doi:10.1109/MCS.2005.1499389

DiDomenico, A. T. (2003). *An investigation on subjective assessment of workload and postural stability under conditions of joint mental and physical demands*. Virginia Polytechnic Institute and State University.

Garcia-Escutia, M. (2004). *Manual de procedimientos: protocolos de prevención de riesgos laborales*. Valencia.

Hart, S. G., & Staveland, L. E. (1988). Development of NASA-TLX (Task Load Index): Results of Empirical and Theoretical Research. *Advances in Psychology*, *52*(C), 139–183. doi:10.1016/S0166-4115(08)62386-9

James, D. A. (1941, November 4). *Vehicle steering and stabilizing mechanism*. Google Patents.

Kato, M., Tsutsumi, T., Yamaguchi, T., Kurakane, S., & Chang, H. (2011). Characteristics of maximum performance of pedaling exercise in recumbent and supine positions. *Journal of Sports Science & Medicine*, *10*(3), 491. PMID:24150623

Mighell, R. (2009, February 10). *Tilting wheeled vehicle*. Google Patents.

Mircheski, I., Kandikjan, T., & Sidorenko, S. (2014). Comfort analysis of vehicle driver's seat through simulation of the sitting process. *Tehnicki Vjesnik (Strojarski Fakultet)*, *21*(2), 291–298.

Patterson, W. B., & Leone, G. L. (2010). The application of handling qualities to bicycle design. In *Proceedings, bicycle and motorcycle dynamics 2010 symposium on the dynamics and control of single track vehicles* (pp. 20–22). Academic Press.

Reiser, R. F. II, Peterson, M. L., & Broker, J. P. (2002). Understanding recumbent cycling: Instrumentation design and biomechanical analysis. *Biomedical Sciences Instrumentation*, *38*, 209–214. PMID:12085603

Roqueiro, N., Vieira, R. S., & Faria, M. G. (2010). Tilting control of a three-wheeled vehicle by steering. *CBA 2010—XVIII Cong. Brasileiro de Automatica*, 3464–3471.

Schwab, A. L. (2012). Bicycling safety and the lateral stability of the bicycle. *Proceedings, International Cycling Safety Conference*.

Taguchi, G. (1993). *Taguchi on robust technology development: bringing quality engineering upstream*. American Society of Mechanical Engineers. doi:10.1115/1.800288

Chapter 8
Design of Recreational Vehicles for Young and Adult People as an Alternative to Physical Activation in Open Spaces

Mayra Ivette Peña Ontiveros
Autonomous University of Ciudad Juárez, Mexico

Cesar Omar Balderrama Armendáriz
Autonomous University of Ciudad Juárez, Mexico

David Cortés Sáenz
Autonomous University of Ciudad Juárez, Mexico

ABSTRACT

Physical activity is an outstanding factor for daily life since it helps to maintain the physiological functions of the body, granting the possibility of increasing physical and mental health. Recreational vehicles of human propulsion are an alternative for skeletal muscle and cardiovascular system activation with the particularity of improving the system while enjoying the activity. This chapter presents an analysis of the problems of sedentary lifestyle to propose a user-focused alternative that meets the necessary requirements for healthy energy consumption. A prototype was designed and built to verify the metabolic expenditure through a telemetry equipment and another survey was applied to validate functional and user acceptance aspects. The results of the tests showed an average energy consumption of 50.2 kcal/min (Almost 6 METS) placing the prototype in a suitable range to perform physical activation. Finally, the resulting vehicle was cataloged by users as creative and attractive to be used in the future.

DOI: 10.4018/978-1-5225-5234-5.ch008

Copyright © 2018, IGI Global. Copying or distributing in print or electronic forms without written permission of IGI Global is prohibited.

INTRODUCTION

Physical activity plays an important role in the prevention of chronic diseases such as cardiovascular disease, musculoskeletal disorders, psychological disorders, lung diseases, cancer and metabolic disorders. There is a high percentage of 55% of adolescents and adults who do not perform any productive activity in their leisure time (Heyward, 2008). Besides helping to reduce the risk of these diseases, it also controls body weight, the formation and maintenance of bones, muscles and healthy joints, and creates a sense of psychological wellbeing, among others (Chomistek, Manson et al., 2013; Acero Mora, 2011).

According to the National Institute of Statistics and Geography (INEGI), poor diet, sedentary lifestyle, lack of physical activity are determinants of overweight and obesity, 72% of women and 66% of men over 19 years old suffer from overweight and obesity in Mexico (Olaiz-Fernandez Rivera-Dommarco et al., 2006).

Young and adult people have abandoned open-air activities, due to other activities such as work, school and home activities, video games, social networks, among others; this abandonment generates obesity, stress or depression (Muñoz & Salgado, 2011). This problem is increasingly growing, most of these people are not aware of this problem and do not perform any activity that causes a change (Patino & Marquez, 2009).

The variety of recreational mobility vehicles is scarce; traditional systems are commonly replicated in similar vehicles: bicycles, adapted and modified bicycles and tricycles, these create little expectation to attract the attention of young and adult people; the most attractive recreational systems are motorized, which reduce the possibility of physical activation (Superiores, 2006).

Recreational vehicles for fun and physical activation contribute to a better quality of healthy life, providing numerous advantages for people and society (Acero Mora, 2011). Changes and improvements in recreational vehicles increase the participation in recreational and physical activities, reducing stress levels and consequently preventing health risks (Hurlock, 2010).

The present work aims to demonstrate the importance of physical activation and proposes mobility systems that can be a viable alternative to help prevent diseases caused by sedentary lifestyle, and consequently improve the quality of health of individuals.

Sedentary Lifestyle

Sedentary lifestyle is called physical inactivity, which is a risk factor for cardiovascular diseases (Nosova, Yen et al, 2014); at least 60% of the world population do not perform a physical activity to obtain health benefits, this is due to the increase in the use of passive vehicles of transport and to the reduction of physical activity (WHO, 2010).

From the point of view of the time dedicated to physical activity, a sedentary individual is one who does not perform at least 30 minutes of moderate physical activity during most days of the week; from the point of view of energy expenditure, it can be defined as sedentary or inactive to a person who does not perform vigorous physical activity for at least 20 min, or who does not generate an energy expenditure of at least 3 METS (Metabolic Equivalent of Task) (WHO, 2015). Another definition of sedentary lifestyle is related to the time spent by a person sitting or lying down, also can be called sedentary to the individual who spends less than 1.5 METS per hour (Crespo et al., 2015).

The World Health Organization (WHO, 2010) estimates that sedentary lifestyle is the fourth leading cause of death worldwide, and influences in a major way in the global burden of disease (Lee, Shirona et al., 2012). Physical inactivity or sedentary lifestyle has been studied in recent years as predictable cause that determines a major number of diseases. It seems that the number of sitting hours has an association with an increased risk of mortality (Patel, Bernstein et al., 2010), the risk is higher in people who simultaneously is less physically active and adopt a sedentary behavior (Stamatakis, Hamer, & Dunstan, 2011).

Inactivity has become one of the worst enemies of physical well-being, 60% of the world's population does not perform the necessary activity to maintain their body in acceptable conditions (WHO, 2010). Figure 1 shows the results of men and women over 18 who are active and inactive, where 38.7% of men and 61.3% of women are inactive (INEGI, 2014).

Physical Activity

Physical activity is considered as the body movement produced by skeletal muscles that require energy expenditure (WHO, 2010). It is a complex and multidimensional behavior and a factor to consider in relation to a healthy lifestyle; the health benefits derived from physical activity are obtained after a regular practice of moderate intensity activity, which involves all the contexts of life, such as: leisure, transportation, work, and home activities (Casado et al., 2015). Studies have highlighted that physical activity reduces the risk of developing chronic, cardiovascular, and neurodegenerative diseases, breast or colon cancer, depression, diabetes and osteoporosis diseases; it also improves cardiorespiratory fitness and cognitive function, and reduces premature mortality (Heyward, 2008). Figure 2 shows a map that relates all these diseases.

In many countries, physical activity does not reach the recommended levels of at least 30 minutes, three times a day physical activity (Merí, 2005). The increase in chronic non-communicable diseases, such as hypertension and type 2 diabetes and others already mentioned, have become one of the main challenges of our health system, these conditions and some other conditions such as sedentary lifestyle, smoking, alcoholism and obesity have become the most important risk factors for the development of cardiovascular and cerebrovascular diseases (CEVECE, 2014).

In Table 1, the MET measure is used to express the intensity of physical activities; a MET is defined as the energy cost of sitting quietly and is equivalent to a consumption of 1 kcal/kg/h; the caloric con-

Figure 1. Physical activity condition
INEGI, 2014.

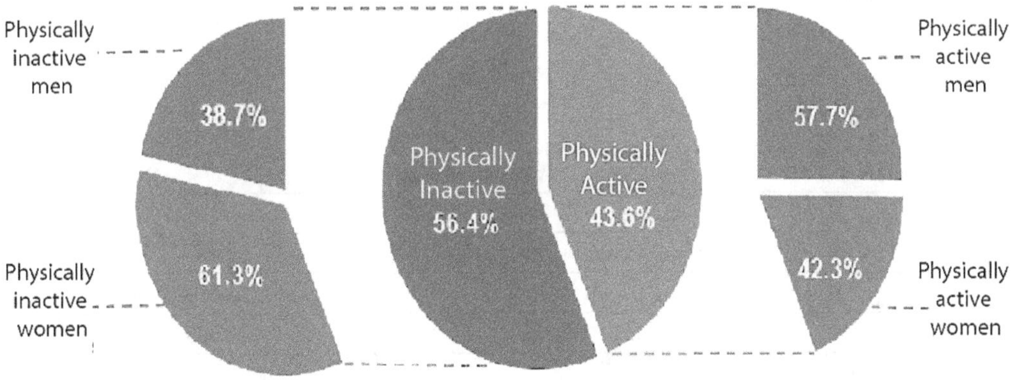

Figure 2. Map of physical activity and related diseases
Heyward, 2008.

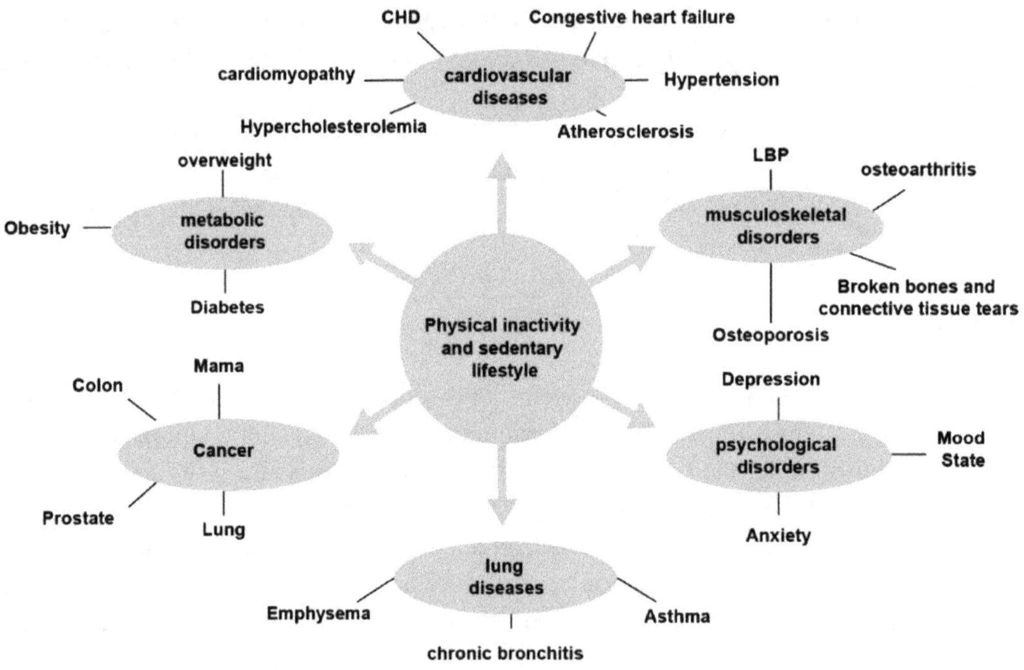

sumption is about 3 to 6 times higher (3-6 MET) when a moderate intensity activity is carried out, and more than 6 times higher (> 6 MET) when a vigorous activity is performed (WHO, 2015). The table evidences the need to carry out interactions and programs that promote activity (Casado et al., 2015).

The lack of physical activity or sedentary lifestyle threatens the health of young people considerably, which are less active than in the past, since they spend a lot of time watching television or playing video games. Although some of these activities can be educational and intellectually stimulating, it is important that this segment of the population be physically active to develop and grow healthy (Macarro, Romero and Torres, 2010). Table 2 shows the composition and description of factors of physical activity.

Table 2 shows the physical activity factors. Personal reasons in factor 1 are the dominant components formed by 10 of 15 items. Personal reasons related to time, the person and the interaction with others encourage the abandonment of the practice of physical-sport activity. Factor 2, environmental and social reasons, includes 3 items, which show that the pressure of physical activity, stress and the

Table 1. Level of physical activity

Intensity Level	Range of Activity and Time	Health Benefits
Inactivity.	No physical activity	None
Light	Activities of 1.1 MET to 2.9 MET.	Some
Moderate	Activities of 3.0 to 6 MET.	Substantial
Intense	Activities of 6 MET or more	Additional

WHO, 2010; Services, 2008.

Table 2. Composition and description of factors of physical activity

Factor	Designation	Variance Explained %	Items
1	Personal reasons	35,94	• Don't have time. • It is not fun. • I'm not clever enough • Others are better than me. • Problems with teacher, bosses, family. • Abandonment of classes. • Don't meet my expectation. • Injuries. • Not good at sports. • Unlikely to improve.
2	Environmental and social reasons	10,80	• Because I can't stand the pressure. • Competitive stress. • Work.
3	Attitudinal reasons	7,49	• I prefer doing other activities. • Laziness and unwillingness.

Macarro, Romero, & Torres, 2010.

influence of parents cause the abandonment, and factor 3, attitudinal reasons includes 2 items related to the preferences and the lack of interest (laziness and reluctance) to continue with the physical-sport activity (Macarro, Romero and Torres, 2010).

According to Marquez (2013), it is important to choose an active lifestyle from an early age, since sedentary lifestyle can cause serious diseases in adulthood, that is, encourage habits that allow:

- Stay active to feel good.
- Increase the daily routine in physical activities

The activities of young people in their spare time are not very productive for health. The habits during the spare time of men and women are different, and it is because men, unlike women, prefer to develop activities, such as attending sport events, playing video games, being on the computer; on the other hand, women prefer other activities such as reading, shopping, being with the family; Table 3 shows the preferences of activities of men and women in their spare time (Merlino & Roqué, 2004).

Recreational Transport Mobility

The concept of mobility refers to daily transportation and movement through public spaces (Hannam, Sheller, & Urry, 2006). The vehicle and the experience of visiting green areas, becomes a strong link since one can enjoy roads, landscapes, people; also transport is a form of recreation for people to continue going to parks and perform some activity outside of their daily lifestyle (Hallo, 2009; Veitch, 2013).

The use of recreational vehicle systems increases the motivation to perform healthy activities, involves physical activation and improves the quality of life; the vehicle must be efficient, attractive, interactive, non-polluting, and useful to move from a place to another without complications in smooth territories such as streets or sidewalks (Hurlock, 2010).

Table 3. Activities carried out by young people

Activities	Male	Female
Being with family	76.5%	94.9%
Being outside	70.6%	82.6%
Shopping	43.1%	72.3%
Watch videos	53.1%	68.4%
Eating out	39.8%	58.8%
Read magazines	37.6%	58.1%
Read books	43.7%	55.3%
Surfing the web	54.7%	46.6%
Go out to dance	58.9%	43.5%
Go to the gym	13.7%	26.9%
Go to the movies	35.9%	20.7%
Attending sporting events	37.1%	16%
Playing video games	27.6%	14.7%
Play an instrument	18.9%	4.1%

Merlino & G., 2004.

Some of the characteristics that distinguish this type of vehicles are: (Acero Mora, 2011; Hurlock, 2010):

- A silent, clean, healthy, affordable and sustainable vehicle.
- It is environmentally friendly.
- Improves social cohesion and allows greater independence and freedom.
- It is a very efficient mobility vehicle for medium and short distances.
- Increase leisure, sports and tourism opportunities.
- Improves physical and mental health.

The most outstanding and important example of vehicles for physical activity is the bicycle, which has become one of the favorite recreational sports.

Methods

In order to obtain information that leads to the design of a recreational vehicle that function as a means of physical activation, the characteristics presented in section 1.3 were used (Acero Mora, 2011; Hurlock, 2010). User needs were extracted by means of the survey to obtain necessary attributes of the new design. A non-probabilistic sampling of people aged 18 to 45 years was made to cover the function of the product. Using the image of an innovative recreational vehicle as a contrast, ten questions were asked with four possible answers. These were classified to obtain design requirements and propose innovative graphic alternatives that would provide a solution to the proposal. Subsequently, a design selection was made to meet the requirements.

To test the ergonomics of the device, a 3d simulation of the position and posture of the user regarding the product was made, based on 95, 50 and 5 percentiles simulating the conditions of the designed vehicle in relation to the distances of the pedals and the movements.

In the physical test the energy and the heart rate were measured with a Sunnto telemetric monitoring equipment using a data reception and calculation software called Team Manager (Suunto, Finland) through technology of computation developed by the company Firstbeat Tech., the evaluation was made to 5 people between the ages of 18 and 45, male and female, who were split it into two phases of the evaluation:

Phase One: The equipment to monitor the energy and heart rate was connected to the participant and was asked to remain seated at rest for 3 minutes.

Phase Two: Measurements of the energy expenditure performed on the vehicle during 10 minutes were made using the telemetric equipment. The differences between the two states (rest and activity) were compared and data was obtained on the relevance in the energy consumption when driving the vehicle.

Phase Three: To validate the appearance of the product, a survey was conducted containing the visual aspects of the proposal to 40 men and women between the ages of 18 and 45. It was applied in places with potential use of the product such as parks and recreational school areas. The results of the survey along with the physiological data were considered for the generation of the conclusions.

Results

According to the methodological proposal, data related to: the initial survey, the design proposal, the selection of the best design, ergonomic tests, physical activation tests and appearance tests were obtained. All these are presented below.

Survey Results

Considering the initial data obtained from the survey of potential users (Table 4), the proposed attributes and the design parameters were: 1) the vehicle will be intended for people who practice low physical activity, 2) it will have a design similar to the contrast proposal, 3) the vehicle will be designed for recreational areas, 4) the current design is for individual use, however it is proposed to obtain a version of the design for double use in the future, 5) the design must be different to the traditional vehicles, 6) the finish colors will be bright, 7) preferences on the type of exercise will not be considered, 8) must be comfortable to exercise 9) must be fun, and finally 10) the vehicle will be designed to be used 1 to 2 times a week. The aspects found by Acero Mora (2011) and Hurlock (2010) were also considered.

Design Proposals

In relation to the obtained design parameters, proposals were made to respond the user needs. Figure 3 shows a design proposal with circular shapes on the sides which work for displacement and protection, with side handles; in Figure 4 it is a vehicle with 3 wheels, two front and one rear, formed with a case for protection, the seat is at the rear with side handles; Figure 5 shows the third design proposal of two

Table 4. Survey results (2016)

<table>
<tr><td colspan="6">How often do you practice physical activities?</td></tr>
<tr><td>None</td><td>Low</td><td>Moderate</td><td colspan="2">Vigorous</td><td>Total</td></tr>
<tr><td>10</td><td>16</td><td>10</td><td colspan="2">4</td><td>40</td></tr>
<tr><td colspan="6">What is your intereston the object shown in the previous picture?</td></tr>
<tr><td>None</td><td>Low</td><td>Moderate</td><td colspan="2">High</td><td>Total</td></tr>
<tr><td>0</td><td>2</td><td>14</td><td colspan="2">24</td><td>40</td></tr>
<tr><td colspan="6">What is the interest in using a vehicle of this type in recreational areas?</td></tr>
<tr><td>None</td><td>Low</td><td>Moderate</td><td colspan="2">High</td><td>Total</td></tr>
<tr><td>0</td><td>6</td><td>22</td><td colspan="2">12</td><td>40</td></tr>
<tr><td colspan="6">How much would you like to do physical activity accompanied?</td></tr>
<tr><td>Nothing</td><td>Light</td><td>Moderate</td><td colspan="2">High</td><td>Total</td></tr>
<tr><td>2</td><td>9</td><td>12</td><td colspan="2">17</td><td>40</td></tr>
<tr><td colspan="6">What recreational vehicle would you like to use in parks or open areas?</td></tr>
<tr><td>Bicycles</td><td>Tricycles</td><td>Bicycles of load and improved</td><td colspan="2">New designs</td><td>Total</td></tr>
<tr><td>2</td><td>0</td><td>1</td><td colspan="2">37</td><td>40</td></tr>
<tr><td colspan="6">What kind of colors attract you in a recreational vehicle?</td></tr>
<tr><td colspan="2">Bright</td><td>Opaque</td><td colspan="2">Neutral</td><td>Total</td></tr>
<tr><td colspan="2">25</td><td>2</td><td colspan="2">13</td><td>40</td></tr>
<tr><td colspan="6">What means of exercise do you use?</td></tr>
<tr><td>Biciycle</td><td>Treadmills</td><td>Running</td><td>Fitness equipment</td><td>None</td><td>Total</td></tr>
<tr><td>7</td><td>3</td><td>10</td><td>10</td><td>11</td><td>40</td></tr>
<tr><td colspan="6">What interest would you have when doing physical activity with a recreational vehicle like the one in the picture?</td></tr>
<tr><td>None</td><td>Low</td><td>Moderate</td><td colspan="2">High</td><td>Total</td></tr>
<tr><td>0</td><td>2</td><td>20</td><td colspan="2">18</td><td>40</td></tr>
<tr><td colspan="6">What do you look for in a recreational vehicle?</td></tr>
<tr><td>Comfort</td><td>Exercise</td><td>Fun</td><td colspan="2">All</td><td>Total</td></tr>
<tr><td>2</td><td>1</td><td>8</td><td colspan="2">30</td><td>40</td></tr>
<tr><td colspan="6">How much would you use a recreational vehicle that gives you some physical activation?</td></tr>
<tr><td>Every day</td><td>1 or 2 times a week</td><td>3 or 5 times a week</td><td colspan="2">Less than 3 times a month</td><td>Total</td></tr>
<tr><td>5</td><td>18</td><td>14</td><td colspan="2">3</td><td>40</td></tr>
<tr><td colspan="6">Comments after regarding the picture</td></tr>
<tr><td colspan="6">Big</td></tr>
<tr><td colspan="6">Innovative</td></tr>
<tr><td colspan="6">Interesting</td></tr>
<tr><td colspan="6">Could be more colorful</td></tr>
<tr><td colspan="6">Not just a metal structure</td></tr>
<tr><td colspan="6">Like</td></tr>
<tr><td colspan="6">Different</td></tr>
</table>

continued on following page

Table 4. Continued

Attractive
Creative
It is practical?
Difficulty in the modo of transport
Comfortable and fun
Concern for safety
Curiosity of how it works
How mucho speed it reaches

Figure 3. Recreational vehicle Sketch 1

Figure 4. Recreational vehicle Sketch 2

Figure 5. Recreational vehicle Sketch 3

wheels, the rear is a cylinder, the upper part is mobile, the inner part is subject to the central structure and the front wheel the movement is generated for its displacement and its handlebars are lateral.

Design Selection

For the selection of the best proposal a comparative Pugh matrix was made using values associated with the proposal with greater compliance to the design parameters as shown in Table 5.

Subsequently, the results were analyzed, obtaining the proposal with the highest score. The selected design was modified based on color, mechanism and adjustments to make the product viable. Figure 6 shows the 3D model of the final proposal.

Ergonomic Test

A validation of ergonomic tests was carried out in the design, making the adjustments in the distances of 3 mannequins that meet the 95, 50, 5 percentiles, with heights of 1800mm, 1650mm and 1500mm (Figure 7).

Physical Activation

In the evaluation of physical activation, 5 people between the ages of 18 and 45 years were considered; the oxygen and the heart rate that was generated when using the vehicle were measured.

The evaluation was carried out in two phases, the first consisted of measuring the user in a resting state for 3 minutes and the second phase, using the prototype to examine the aerobic load for 10 minutes. We obtained data and graphic representations to show the heart rate and energy consumed in resting (short waves in this case the blue one, Figure 8) and activity (long waves in this case the green) states. Figure 8 shows a representation of the obtained data. In the top part there are two lines of the resting heart rate (blue color or the short one) and activity (green color or the long one). The lower part of figure 8 show, data of the energy consumed at rest (blue color or the short one) and activity (green color or the long one). Differences in energy consumption when using the recreational vehicle and in a resting state are observed.

Figure 9 shows an example of the Team Manager Software information extracted from the activity performed when using the device.

Table 5. Comparison of design proposals in pugh matrix (2016)

Puhg method. Selection of a recreational vehicle for physical activation.	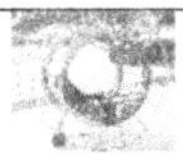	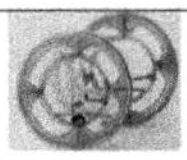		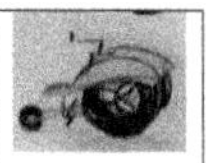
Assembly system		+	+	-
Indicators or instructions to arm		+	+	s
Minimum of 12 pieces to assemble		s	+	+
Screws in strategic points to assemble		s	+	+
Identification of functionality		+	+	+
Simple and understandable mechanisms for the user		+	+	+
Parts easy to replace if damaged		s	-	s
Easy maintenance		s	+	+
Adjustable height seat.		+	s	s
Adjustable angle seat.		s	+	s
Softness in the seat materials.		+	+	+
Cushioning		+	+	+
Anatomical design		+	+	+
Anthropometric design		+	+	+
4-point safety belt resistant.		s	s	s
Anti-slip materials in the handlebars.		+	+	+
Light structure		s	+	+
Reliable at a glance.		+	+	s
Resistance to the outside.		+	+	+
Resistance to the weight of a person.		+	+	+
Impact resistant		+	+	+
Physical activation.		+	s	+
Recreation device.		+	s	+
Measurement of energy expenditure		+	+	+
Attractive		s	+	+
Quality finishes		+	+	+
Differentiation of the competition		+	+	+
Long life cycle.		+	+	+
Σ+ (better than, cheaper then, easier than)		20	22	21
Σs (equally satisfies the criterion)		8	4	6
Σ- (worse than, more expensive that, harder to develop than)		0	1	1
	Sum	20	21	20
			strong	

Figure 6. 3D model of the final proposal

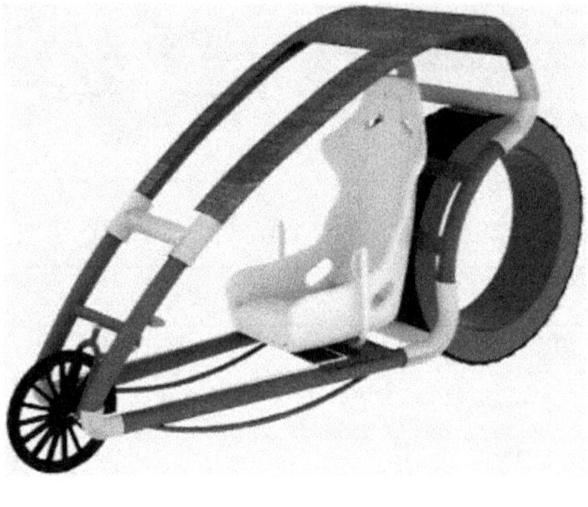

Figure 7. Dimensions of the product with respect to the 95, 50 and 5 percentiles (2017)

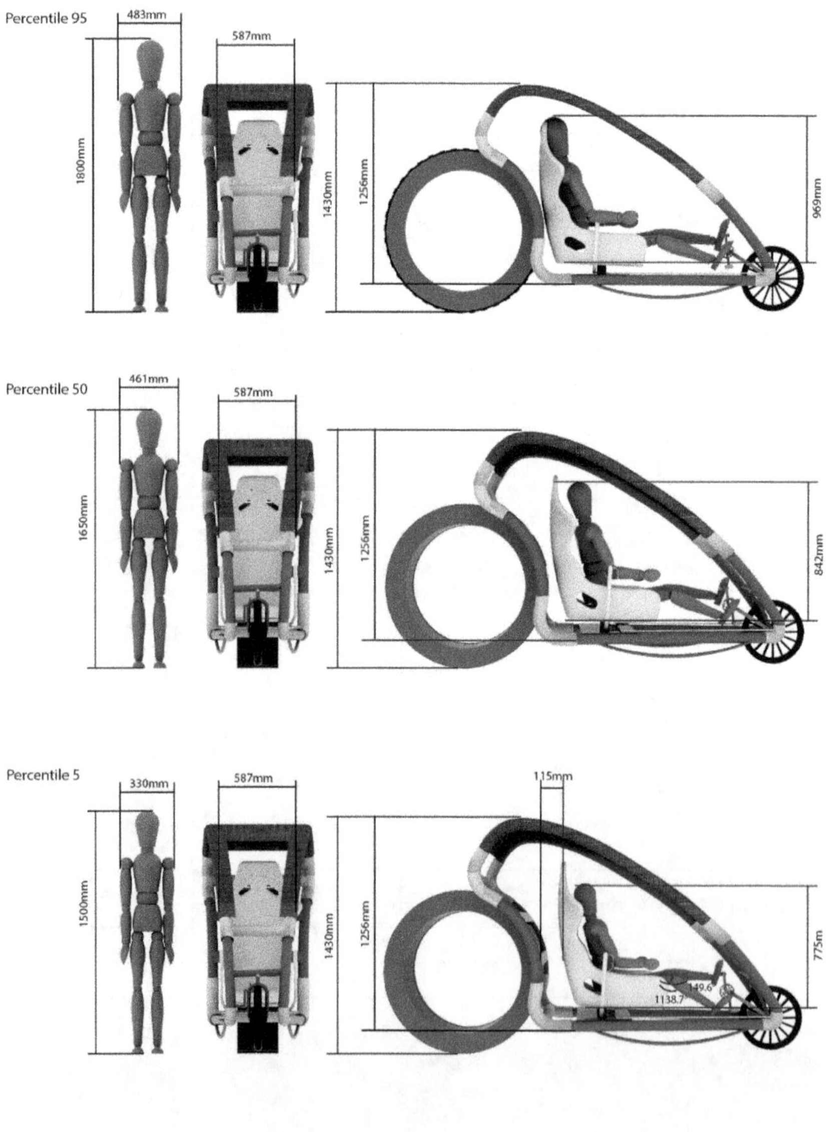

Figure 8. Heart rate and energy in resting and active states (2017)

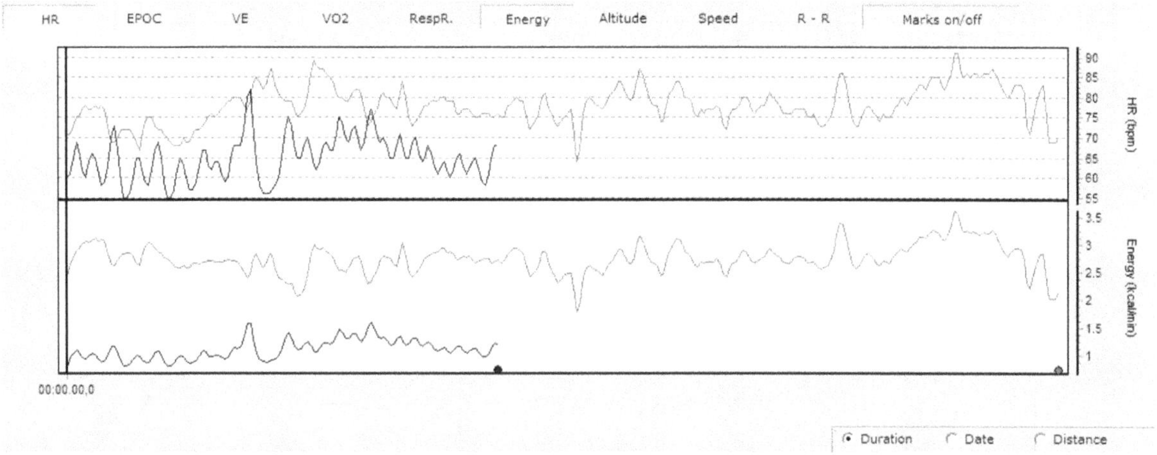

Figure 9. Results presented using the Team Manager software (2017)

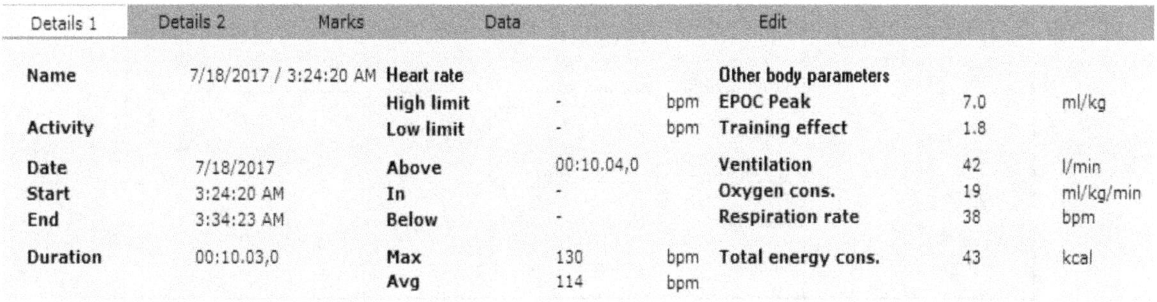

Details 1 | Details 2 | Marks | Data | Edit

Name	7/18/2017 / 3:24:20 AM	**Heart rate**			**Other body parameters**		
		High limit	-	bpm	**EPOC Peak**	7.0	ml/kg
Activity		**Low limit**	-	bpm	**Training effect**	1.8	
Date	7/18/2017	**Above**	00:10.04,0		**Ventilation**	42	l/min
Start	3:24:20 AM	**In**	-		**Oxygen cons.**	19	ml/kg/min
End	3:34:23 AM	**Below**	-		**Respiration rate**	38	bpm
Duration	00:10.03,0	**Max**	130	bpm	**Total energy cons.**	43	kcal
		Avg	114	bpm			

Table 6 shows the comparisons of oxygen, energy consumed, and heart rate of the people who performed the evaluations. A notorious difference can be observed in the energy consumption of the activity monitoring against the rest state (Graphics in Table 6).

Table 7 shows the comparison of average of energy consumption in kcal/min, the oxygen consumed in ml/kg/min and the heart rate in bmp; it is observed that the energy of 8.6 kcal/min was raised to 50.2 kcal/min (Almost 6 METS of difference), so it presents a physical activation when using the recreational vehicle.

Physical Appearance of the Vehicle

In the evaluation of the physical aspect, the survey results were analyzed. Figure 10 shows a graph of three questions based on: design, physical activity and recreational use. The results showed that most people were very attracted to the design; it seemed a vehicle which they would regularly use as a tool for physical activity. In addition, it was considered as a very recreational vehicle.

Figure 11 shows a graph of the results about the perception of the vehicle, users rated it as a fun product to use, they were also very attracted to the colors and shapes in the design.

Table 6. Comparison of oxygen, heart rate, and energy consumed of five users (2017)

Gender	Age	Rest	Active	Graphic
Male	24 years	Oxygen: 4 ml/kg/min Energy: 1 kcal/min Heart rate: 62 bpm	Oxygen: 11 ml/kg/min Energy: 3 kcal/min Heart rate: 80 bpm	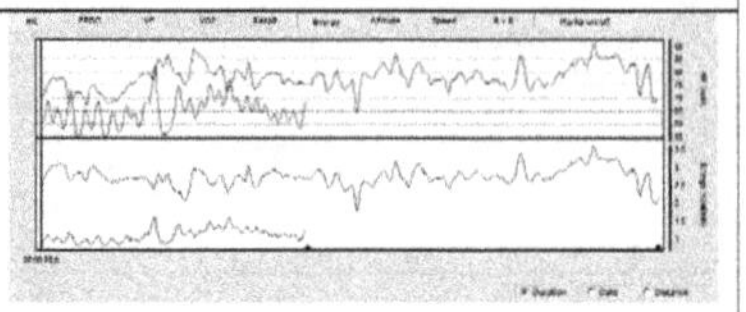
Female	28 years	Oxygen: 20 ml/kg/min Energy: 12 kcal/min Heart rate: 87 bpm	Oxygen: 27 ml/kg/min Energy: 65 kcal/min Heart rate: 119 bpm	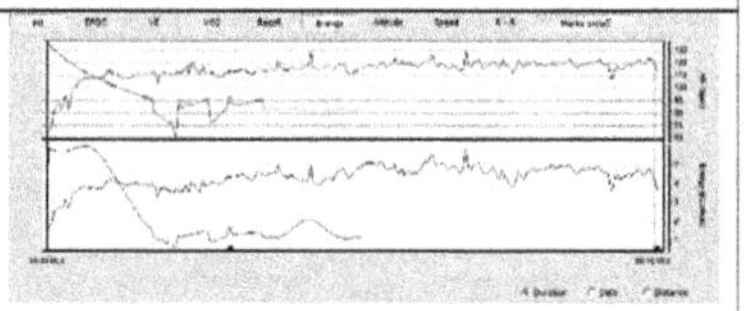
Male	31 years	Oxygen: 11 ml/kg/min Energy: 6 kcal/min Heart rate: 84 bpm	Oxygen: 28 ml/kg/min Energy: 52 kcal/min Heart rate: 142 bpm	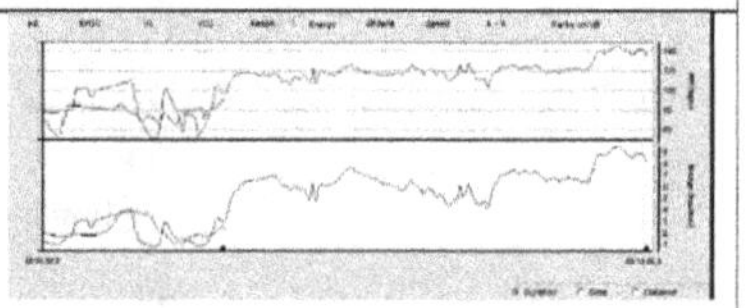
Male	19 years	Oxygen: 14 ml/kg/min Energy: 8 kcal/min Heart rate: 83 bpm	Oxygen: 24 ml/kg/min Energy: 50 kcal/min Heart rate: 125 bpm	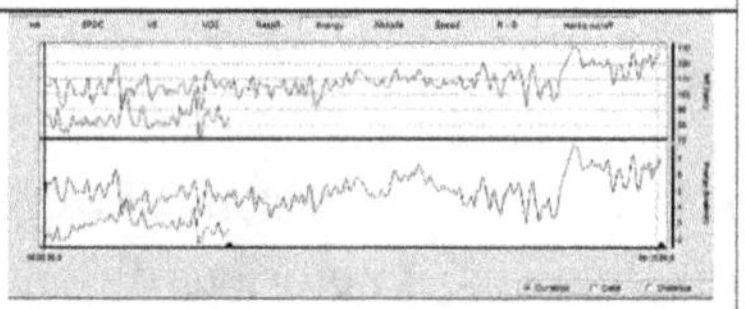
Male	44 years	Oxygen: 17 ml/kg/min Energy: 16 kcal/min Heart rate: 96 bpm	Oxygen: 26 ml/kg/min Energy: 81 kcal/min Heart rate: 135 bpm	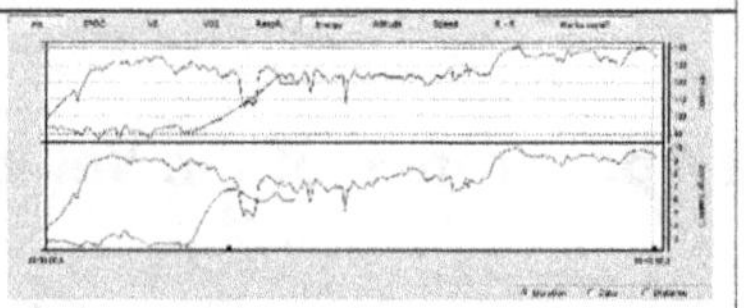

Table 7. Comparison table of average oxygen and energy consumption (2017)

	Average Oxygen Consumption	Average Heart Rate	Average Energy Consumption
Rest	13.2 ml/kg/min	82.4 bmp	8.6 kcal/min
Using the vehicle	23.2 ml/kg/min	120.0 bmp	50.2 kcal/min

Figure 10. Results of the evaluation of physical appearance of the vehicle 1 (2017)

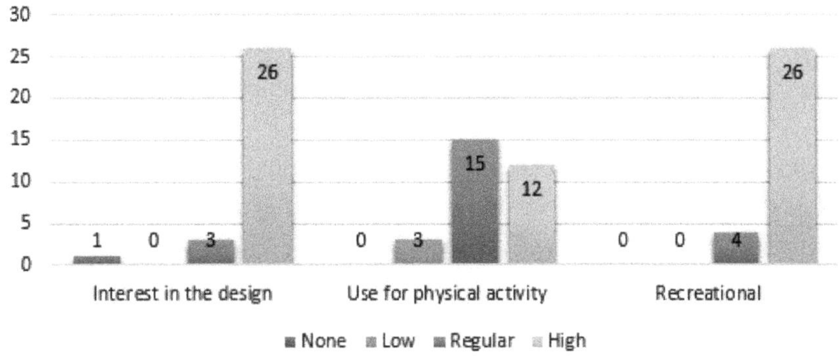

Figure 11. Results of the evaluation of physical appearance of the vehicle 2 (2017)

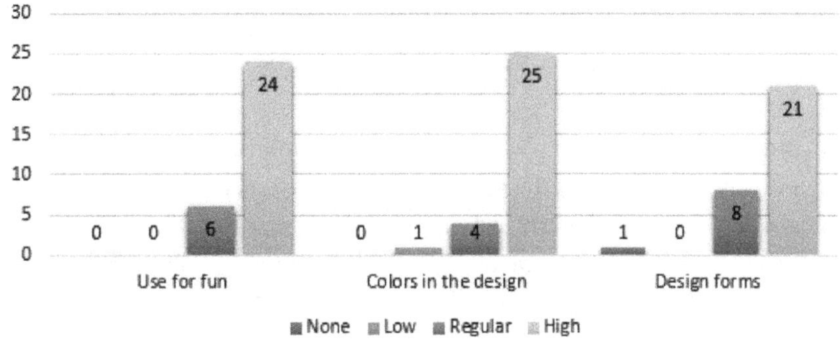

The results in Figure 12 show a graph that the vehicle was considered a very attractive, modern and creative design.

CONCLUSION

It is important to propose alternatives to reduce and prevent the consequences caused by the lack of physical activity. The result of the literature review and the surveys revealed the need to have attractive and different vehicles for practicing outdoor activities. The present work exposes an alternative to encourage people to carry out physical activity in a fun and healthy way.

Figure 12. Results of the evaluation of physical appearance of the vehicle 3 (2017)

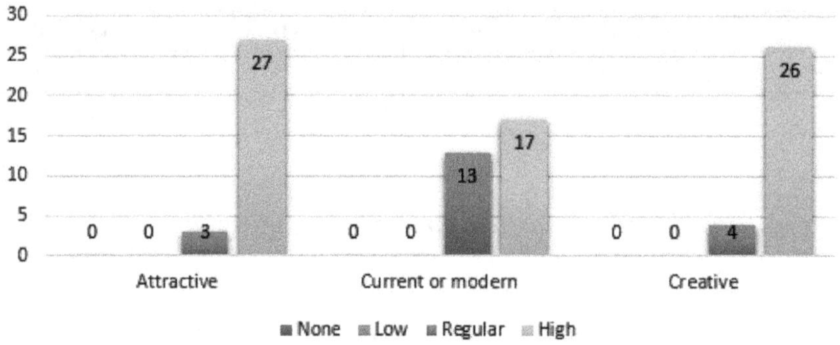

The proposal showed the potential to function as a recreational physical activation vehicle, demonstrating that new devices can be developed to have the necessary attributes to please users and at the same time serve as a means to preserve and promote health.

The use of design techniques oriented to ergonomic adaptation and preferences over an attractive appearance, are a means of putting in development of high level products of human value.

FUTURE RESEARCH DIRECTIONS

To obtain a more accurate research it is recommended to make some changes in the prototype and add research for greater accuracy in results; the changes and proposals are:

- Adjustments in the structure of the prototype for better mobility in its displacement, likewise perform a complete physical study on the properties and resistances of materials.
- Biomechanical study to avoid that the user applies the higher force to use the product
- Analysis of movement of all the interactions that the human body has when using the vehicle.
- Design a digital device that is incorporated in the vehicle, that it is possible to measure and observe the time, energy, heart rate, forces and METs that are generated in the use of the prototype.

REFERENCES

Acero Mora, J. (2011). *Los sistemas de bicicleta pública vistos desde la relación servicio - producto.* Bogota: UNC.

Carmichael, C., & Burke, E. (1997). *BICICLETA. Salud y ejercicio*. Ed Paidotribo.

Casado, C., Hernández, V., Jiménez, R., Fernández, C., Carrasco, P., & Palacios, D. (2015). Actividad física en población adulta trabajadora española: Resultados de la Encuesta Europea de Salud en España. *Atencion Primaria*, *1*(5), 10.

CEVECE. (2014). *Factores De Riesgo De Enfermedad Cardiovascular En Adultos Mayores*. Retrieved MARZO, 2015, from http://salud.edomex.gob.mx/cevece/doc/Documentos/Facriesgo_cardio_amayores.pdf

Chomistek, A., Manson, J., Stefanick, M., Lu, B., Sands-Lincoln, M., Going, S., & Eaton, C. (2013). Relationship of Sedentary Behavior and Physical Activity to Incident Cardiovascular Disease. *Journal of the American College of Cardiology*, *61*(23), 9. doi:10.1016/j.jacc.2013.03.031 PMID:23583242

Crespo, J. J., Delgado, J. L., Blanco, O., & Aldecoa, S. (2015). Guía básica de detección del sedentarismo y recomendaciones de actividad física en atención primaria. *Atencion Primaria*, *47*(3), 9.

González, J. M. (2003). *Actividad física, deporte y vida: beneficios, perjuicios y sentido de la actividad física y del deporte. Cuerpo y espíritu*. Basauri, España: Etor-Ostoa.

Hallo, J., & Manning, R. (2009). Transportation and recreation: A case study of visitors driving for pleasure at Acadia National Park. *Journal of Transport Geography*, *17*(6), 9. doi:10.1016/j.jtrangeo.2008.10.001

Hannam, K., Sheller, M., & Urry, J. (2006). *Mobilities*. Ed M.i.a. moorings.

Heyward, V. (2008). *Evaluación de la aptitud física y prescripción del ejercicio* (Vol. 5). Madrid: Médica Panamericana.

Hurlock, E. (2010). Interés y actividades recreativas. Cambios en los intereses. In Paidos (Ed.), Interés y actividades recreativas (pp. 51 - 55). Academic Press.

INEGI. (2014). *Módulo de Práctica Deportiva y Ejercicio Físico*. Instituto Nacional de Estadística y Geografía. Retrieved Abril, 2015, from http://www.inegi.org.mx/est/contenidos/proyectos/accesomicrodatos/encuestas/hogares/modulos/mopradef/?_file=/est/contenidos/proyectos/encuestas/hogares/modulos/mopradef/doc/resultados_mopradef_feb_2014.pdf

Lee, I., Shirona, E., Lobelo, F., Puska, P., Blair, S., & Katzmaryk, P. (2012). Effect of physical inactivity on major non-communicable diseases worldwide: An analysis of burden of disease and life expectancy. *Lancet*, *380*(9838), 10. doi:10.1016/S0140-6736(12)61031-9 PMID:22818936

Macarro, J., Romero, C., & Torres, J. (2010). Motivos de abandono de la práctica de actividad físico-deportiva en los estudiantes de Bachillerato de la provincia de Granada. *Review of Education*, *353*, 25.

Márquez, S. (2013). *Actividad física y salud* (E. D. d. Santos, Ed.). Ediciones Díaz de Santos.

Merí, Á. (2005). Fundamentos de fisiología de la actividad física y el deporte (E. M. Panamericana, Ed.). Academic Press.

Merlino, A., & Roqué, G. (2004). *Los nuevos jóvenes. Un estudio psicográfico de sus actitudes y estilos de vida*. Ed Brujas. Córdova.

Muñoz, I., & Salgado, P. (2011). Ocupaciones de tiempo libre: Una aproximación desde la perspectiva de los jóvenes. *Universidad de Chile*, *1*, 10.

Nosova, E., Yen, P., Chong, K., Alley, H., Stock, E., Quinn, A., & Grenon, M. (2014). Short-term physical inactivity impairs vascular function. *The Journal of Surgical Research, 190*(2), 11. doi:10.1016/j.jss.2014.02.001 PMID:24630521

Olaiz-Fernández, J., Rivera-Dommarco, J., Shamah-Levy, T., Rojas, R., Villalpando-Hernández, S., & Hernández-Ávila, M. (Eds.). (2006). INEG. *Instituto nacional de estadística y geografía*. Retrieved from http://www.inegi.org.mx/

OMS. (2010). *Inactividad física: un problema de salud pública mundial*. Retrieved MARZO, 2015, from http://www.who.int/dietphysicalactivity/factsheet_inactivity/es/

OMS. (2015). *Estrategia mundial sobre régimen alimentario, actividad física y salud*. Retrieved Abril, 2015, from http://www.who.int/dietphysicalactivity/physical_activity_intensity/es/

Patel, A., Bernstein, L., Deka, A., Feigelson, H., Campbell, P., Gapstur, S., & Thun, M. (2010). Leisure Time Spent Sitting in Relation to Total Mortality in a Prospective Cohort of US Adults. *American Journal of Epidemiology, 172*(4), 10. doi:10.1093/aje/kwq155 PMID:20650954

Patiño, A., & Márquez, J. (2009). *Actividad Física y Ejercicio Físico en salud: retos en un contexto globalizado*. Fúnambulos ed.

Sevices., D. o. H. H. (2008). *2008 Physical Activity Guidelines for Americans*. Retrieved 04, 20015, from http://www.health.gov/paguidelines/pdf/paguide.pdf

Stamatakis, E., Hamer, M., & Dunstan, D. (2011). Screen-Based Entertainment Time, All-Cause Mortality, and Cardiovascular Events: Population-Based Study with Ongoing Mortality and Hospital Events Follow-Up. *Journal of the American College of Cardiology, 57*(3), 8. doi:10.1016/j.jacc.2010.05.065 PMID:21232666

Superiores, I. T. (2006). Movilidad: Una visión estratégica (ITESO Ed. ITESO ed.). Guadalajara, México: ITESO.

Valero, A., Gómez, M., Gavala, J., & Ruiz, F. (2007). Porque no se realiza actividad fisco-deportivo en el tiempo libre. *FEADEF, 12*(5).

Veitch, J., Ball, K., Crawford, D., Abbott, G., & Salmon, J. (2013). Is park visitation associated with leisure-time and transportation physical activity? *Preventive Medicine, 57*(3). PMID:23933224

KEY TERMS AND DEFINITIONS

Anatomical Design: Which is built to suit the shape of the human body.

Design Parameters: Aspects of a component that are needed to make something. It decides cost, design, materials, and risk.

METS: Used to express physical intensity and are the ratio between the metabolism of the person during the performance of a job and their basal metabolism.

Non-Probabilistic: Sampling technique where the samples are gathered in a process that does not give all the individuals in the population equal chances of being selected.

Percentile: Each of the 100 equal groups into which a population can be divided according to the distribution of values of a particular variable.

Prototype: A first, typical, or preliminary model of something, especially a machine, from which other forms are developed.

Recreational: Relating to or denoting activity done for enjoyment when no one is working.

Chapter 9
Trials and Simulators Research in User Analysis and Sustainable Packaging Development

Berthana Ma Salas Dominguez
Universidad Autónoma Metropolitana – Xochimilco, Mexico

Silvia Ana María Oropeza Herrera
Universidad Autónoma Metropolitana – Xochimilco, Mexico

ABSTRACT

The packing sector is an area with a high consuming problem. The actual consequence is the elevated environmental impact that is developing a real problem. Everyday each home consumes a big variety of products and most of them come in a package that will be disposed of in a matter of minutes, hours, or days. They also require a massive set of structures, barriers, seals, and shock absorbers that assure the final product will be delivered in good condition to the user. Packaging esthetical design is remarkable and catch the attention of the buyer but if there is a design error the user will feel frustrated and the final purpose of a comfortable use will be lost. Incorporating ergonomics, user habits analysis, trials, and simulators in the design process and making users aware of the need of taking environmental actions can lead to a packaging design that promotes actions for a more sustainable world. The development of research methods in this area will lead to better design products.

INTRODUCTION

Current global conditions and environmental issues are themes that are part of our daily lives and a matter of concern. Climate change, biodiversity loss, and pollution are not isolated elements; they are closely linked with poverty, inequity and social problems. We need to raise awareness in industry and product developers about the risks involved in the constant use of processes without sustainable practices. With the objective of developing better and innovative solutions for packaging design, the academic field in an international scope has been working and researching in packaging development integrating ergonomic and usability concepts and not just from a sustainable point of view.

DOI: 10.4018/978-1-5225-5234-5.ch009

Copyright © 2018, IGI Global. Copying or distributing in print or electronic forms without written permission of IGI Global is prohibited.

The term "sustainability" commonly used today is more profound than the one used during the 80s and 90s called "green product". A product that addresses sustainability considers consumer needs and includes profits, people, and planet earth as a whole issue. In 1987 Bruntland's report says that "Sustainability is a concept that promotes a type of development whereby the needs of the present generation will not compromise the ability of future generations to meet their own needs, focusing on economic, environmental and social interdependence." (Martinho, 2015:58). Speaking of packaging this statement has not been applied, most of the packages are highly polluting and can´t be easily recycled.

The WWF (World Wildlife Fund) in 2012 established that global population uses more than 50% of the resources earth can provide, (as cited in Scott & Vigar, 2014:643), when a product is used it has a considerable quantity of hidden elements that the final user cannot see, materials, manufacturing processes, merchandising, workforce, etc., in such a way that consumers are not conscious of the real impact of products used on a daily basis. Considering that packaging is one of the products more commercialized in a global level, Pousette and Löfgren (2014) establish that "Consumer packaging could be described as an "everyday product" because its contents are used in everyday life" and they mention, "the average person has approximately 30 such experiences a day with approximately 30 packages". Therefore it is a must to make sustainability an inherent part of the design process among university students.

To fulfill this statement packaging characteristics have to be known, these have specific functions, one of the most important is preventing product loss by product protection and efficient transportation. Nevertheless to develop a sustainable packaging today has turned more into a commercial strategy for the majority of the companies than simply solving a real problem. Many times it is the consumers who demand solutions, this as part of the consumer's practices based on ethics, which has generated that some companies try to develop innovating clean packaging. For this Stern (2000) divided the consumers who are activists and environmentalists in different subtypes, based on the behavior, the generated impact and the actions that they perform as part of their daily activities. (as cited in Paco & Gouvei, 2016:467). These environmental activists are divided in different fields, starting with the most active and ending with the traditional activist, who is the one that never takes direct actions nor thinks in the short term but simply mentions that he is interested in the topic.

The development of a sustainable packaging requires the analysis of a specific group of factors: the first one is the consumer, it is important to know the commitment level he acquires of a sustainable behavior and from an ethical point of view. The second factor analyzes the consumer as a direct user, therefore it is important to know his physical characteristics and how the ergonomic principles will be applied, and these factors can give us the design parameters to achieve a more functional and safe object. The third factor is the environmental impact generated when the materials, the life cycle, the manufacturing process and the eco-friendly proposals are analyzed which will allow us to develop a better packaging. The fourth factor is the usability tests, the construction of simulators as part of the design method that allows to perform a more profound analysis with the user and in this way generate the redesigning elements before performing the final production. It can be said that a packaging that works is attractive and an attractive packaging becomes a product that will be bought.

PACKAGING FUNCTIONS IN PRODUCT MARKETING.

Celorio (1993) and Tarango (2016) say that the design of an adequate packaging should consider four basic functions that define the characteristics of each one and considers the external factors that will be part of the everyday handling of the packaging by the user, which are:

- **Containing:** It is important to consider the amount of product needed because each product has different characteristics; for example, it can be solid or liquid, poured or spread, and it is important how it is opened, used or kept. For example, sometimes the product must be poured and it can get the table or other elements dirty.
- **Protecting:** The lifespan of the product must be taken into consideration with the objective of preventing its alteration. Storage, distribution, selling and consuming conditions are part of the requirements of design, a packaging needs to become a barrier, this is why the selection of the material is fundamental, it has to cushion and prevent damage due to bumps or falls, it must decrease humidity and eliminate vibration.
- **Transporting:** It is necessary to consider the transference logistics and the marketing needed for this product.
- **Communicating:** All the products' characteristics have to be expressed in the packaging considering perception, retail graphic design concepts and at the same time, it must convince the buyer to acquire the product. The trademark and its prestige, where the product comes from and the market niche are elements that have to be transmitted through a rhetorical process. In the same way, it is necessary to manifest the ways it is going to be used and the possible misuse that this product can generate. Giovanetti (1999) and Celorio (1993) established that this type of communication is a silent seller and that its function is bigger because the packaging will end up being a representative of the company that could make the user to fall in love or to be disappointed, it can make itself present or be remembered, it can produce reflections or manipulate consuming, generate a simple use or a feeling of comfort and can achieve a change in the habits of other users.

Today, selling a product requires more research and analysis on the consuming habits and consumer behavior. "Merchandising" is an English term made up of the noun merchandise and an "ing" ending. Diez de Castro y Landa Bercebal (2006) mention that the analysis is made up for a series of techniques and tools whose purpose have better marketing results in the selling points, the consumer can be convinced if an analysis of how people perform and decide their purchases show their own selection strategies when they are shopping. As well as the lighting, auditory factors and ventilation, their perception of the hardness or softness of the flooring and the tiredness they feel at the moment of buying. All this is done to achieve the consumer's attention and to direct him to a final product so the purchasing process will be easier. The stores are based on the supply and demand and determine the position and distribution of a product on the shelves as well as the general logistics used for the distribution of all the products inside the entire store. Another element for the promotion of the product is those extra strategies like entertainment or demonstrations that are based on the taste and will make people buy certain products.

The graphic and industrial design combines these activities to generate an image that will be able to transmit a history that not only communicates the characteristics of the product and company data. It also transmits the ideal of the company and establishes the consumer´s stereotype. This way the consumer will be able to know if the product is expensive, only for women, for children, unisex, etc. Most of the time these products are backed by marketing campaigns directed to specific sectors of the market. An example could be the products considered "gay-friendly" as the Spanish Association of Gay Friendly Enterprises mention, these stores do not sell different products; they just want to make an inclusion or openness statement. Another example is the products focused on the DINK (Dual Income, No Kids) marriages sector, whose expenses per capita are bigger than for other families. (El Economista, 2013).

These studies combine the information in market niches that are based on established agreements performed among the suppliers and the stores, the merchandising studies have defined that the biggest affluence points are the heads of the aisles and this can determine if the buyer is going to walk down an aisle or not, the same happens with island displays or dumps as attraction elements that are positioned in the regular paths.

The packaging exposed in self-service stores become everyday objects to our eyes and lives, as Jean Braudillard (2007:14) mentions "The objects have a function, in first place the personification of human relationships, the settling of the shared space and to possess a soul" by this he means that they are part of our daily space and everyday nature, where a constant is a complete abundance; for example 40 years ago selecting a cereal in Mexico City was easy, there were only two brands, Kellogs and Maizoro, both had four different kinds of cereal: corn flakes, crisped rice, sugar corn flakes and chocolate crisped rice, today the options are endless, brand diversity and kinds of cereals, including subvariety of these products, so in each store visited the consumer has to be an expert in decision making, under a continuous advertising blitz and consuming pressure.

Today these elements have evolved with the e-commerce, therefore other perception facts have to be analyzed, this promotes any kind of buying, it can be done from any place and at any time. The information can include videos, audios, other buyers' opinions and an endless amount of elements developed to convince the consumer of buying. Buyers will receive more elements than the existing in a physical selling point and they are more effective than a packaging with an image. Finally, the main objective is client satisfaction and Pousette mentions, "Customer satisfaction reflects a customer´s evaluation of his or her purchase and consumption experiences with a product, service, brand or company" (Pousette, et.al. 2014:730). With this, the use of the web for evaluating products has allowed satisfaction levels to be considered by a bigger amount of consumers.

PURCHASE AND BUYING

As it was mentioned before, the way purchases are done has been changing with time, and it is important to know the consumer's habits; Magnier (2015) argued: "Consumer perceptions and behaviors are generally influenced by their prior attitudes and beliefs" (Magnier 2015,54). Today businesses want consumers to make impulse purchases and they support this practice with packaging design, Rook (1987) says "Impulse buying occurs when a consumer experiences a sudden, often powerful and persistent urge to buy something immediately" (as cited in Chen, Y., Wang, R., 2016:187). Fraj & Martinez (2007) mention that it is necessary to know the consumer's behavior perspectives from an ecologic point of view.

1. **Demographic and Socioeconomic Variables:** Analyzing contexts and economic conditions in which consumers manage.
2. **Amount of Information and Knowledge:** Their abilities to get to know the product completely, it can be through online reviews, by word of mouth information, etc.
3. **Psychographic Variables:** Analyze values, lifestyles, personality, characteristics, and attitudes with regard to the products they consume.

All this is usually based on two analyses: the first one is the rational and systematic use of the available information and the second one the consequences of these actions. These perspectives are related to three components:

1. **The Cognitive Component**: What the buyers believe, be it good or bad, and how they handle information.
2. **The Emotional Component:** The feeling generated by the buying or the use of the product.
3. **The Intentional Component:** The one that generates an intention to act in a certain way.

According to the National Association of Self-service and Department stores in Mexico (ANTAD), 75% of the purchases are done by impulse, and this is divided into three levels which are: remembrance, opportunity, and persuasion. The truth is that only a few people plan their purchases ahead and actually follow their plan. In malls, self-service and department stores especially, the factors of the buying decision are usually the same; the product has to be cheap, exclusive, nice looking, functional and it must generate personal satisfaction or as it was mentioned before, it needs an element that generates an impulsive buying which sometimes can be achieved with a simple packing design. The economic interest combined with a system of massive production focusing on designing consumer goods that are backed under a culture and a concept of "use and dispose", where the idea of comfort is translated into preventing the user to perform extra activities like washing, cleaning, storing and keeping, at the end when the users dispose of things, they do not even have to worry about what is going to happen with them.

Bauman (1999) establishes that self-service stores are strategically designed for the consumers to feel comfortable and the consuming takes place in a spontaneous and natural way, without having to reflect on the benefit of the purchase because "they are built in such a way that they keep people moving, looking around, constantly attracted and entertained [...] they are not encourage to stop, look, talk, think, analyze and debate something different from the objects exhibited [...]" (Bauman, 1999:37). The colors, smells, images, forms or textures of the packaging design must be worked on for a product to attract. Today universal design and sustainability concepts are included pretending to reach personal feelings. In this way emotions become a convincing factor, it can be done through a sober design, inclusive information, symbols or images that speak of a healthy product or less dangerous to the human being and the environment.

While walking the consumers see all the products on the shelves, therefore the design of the packaging in general, its graphic presentation and the structural form of packaging generate a concept that communicates and has the power of persuasion or dissuasion; the management of colors, sizes, and forms can convince the buyers of the acquisition of a product. The messages in the packages must be easily decoded because it is the packaging the one that sends all the information to the consumer. The messages that present a better effect are those that produce reasoning and reach the emotions and feelings of the consumers, as Grijelmo (2000) says, those that "attract", "Attraction comes out of an intellect, yes, but it is not directed to the rational zone of the one who receives the statement, but his emotions" (Grijelmo, 2000:37).

Nacional Financiera offers training for entrepreneurs online and in a classroom, in these courses, some guidelines are given, for example where to place the product in the store as a factor to lure the buying of the products. The logistics of placement in the self-service stores are based on the following[1]:

1. **Clear Conscience:** Are the products that are classified as healthy, basic or daily, their purchase does not generate guilt, because they are regularly consumed, for example, fruits and vegetables, this kind of products are positioned in an easy access point in the store, trying to be the first purchase. The consideration is that because it is a healthy product or an essential one it is the first thing that goes into the shopping cart, users consider their whole shopping as good. On the other hand, today, an option for many buyers is the store that displays the healthy, sustainable, natural, organic, fair trade, inclusive or under the concept of exchange products which want to break the paradigms of consumerism and want to keep a relation of win-win.
2. **Basic Food Basket:** It is made up of milk, tortillas or eggs placed in the far away corners so it is necessary to cross the whole store to reach them, this way other products are given the chance to be bought.
3. **Dangerous Products:** They are sales and bargains, usually placed in the busy spaces and based on the analysis of the most frequent circulation flows considering the visual impact.
4. **Last Minute Products:** Are placed at the cashier points to distract and generate spontaneous craving purchases, like sweets or chips. The display of magazines in this space is not profitable because the great majority that is waiting at the cashier read the magazine but a very few buy it, damaging them; nevertheless it is an excellent source of distraction for the shoppers, because when they are waiting in the line, they tend to reflect on what they have in the shopping cart.

The analysis of the consuming habits together with the packaging design and the activity in the shop allows mind control in the purchases and consumers persuasion through the unthinking purchasing, making the packaging a dangerous product of massive consumerism and useful life extremely short. A change of purchase paradigms and the use of the products through the elements of persuasion are necessary.

DECISION FACTORS IN THE PURCHASE OF A PACKAGING

As can be easily seen the sale of packaging is bound to a context that influences the senses and the unconscious, convincing the costumer of the purchase, bound to a global economy based on consumerism. The constant is the irrational making of decisions; impulse and fast thinking become important factors for the development of consumerism, food, clothing, and relationships are quick and momentary. When people walk in a shop they are bombed with images of products strategically placed because sight is one of the main convincing elements when talking about child population, the common resource is the use of pets or well-known characters with the objective that the child recognizes them and asks for the product.

Some packaging fulfills the function of protecting the aroma of the product because the smell is one of the senses that evoke strong feelings; for example, the packaging of coffee, because "aroma" is a factor that determines its flavor. Cretien Van Campen (2015) calls this ability the "Proust effect", the aromas have the ability to immediately retrieve the most ancient emotional and detailed memories beyond any sensory stimulation. Different studies confirm that old memories and experiences can be relieved through the path of recovery in such a way that if a product produces a happy feeling, the aroma will recreate the pleasant state, and with it the unconscious desire to acquire the product; this makes the aromas a powerful selling tool. When there is a pleasing memory, rash purchases are sparked off easily, this is the reason why some trademarks use inks with an aroma in the packaging or invite the consumer to open and enjoy its aroma in the advertising of the products.

Another important factor in buying decision is sampling, where smelling and taste are integrated to cover this need; the flavors are bound to the aromas because they are perceived through a combination of taste and smell. There are some discrepancies in the datum of some studies[2], these indicate that between 70 and 80% of people accept the tasting samples when they are offered to them and in a subconscious act of reciprocity there is an increase in the sales of the product during this tasting period. At the same time, the touch sensations are stimulated when the packages present attractive forms that not only call the attention at first sight but also invite the consumer to handle them, if it is easy to handle from the shelf, the consumer will be convinced of its easy use, because the handle is comfortable, the lid is easy to open, the portions are adequate for the handbag, the size is adequate for the refrigerator or the pantry, it has all the required elements needed for their use and can be handled by a child.

Easy opening and closing for adequate portion consuming is an important factor in the purchase decision, as comfort is a determining element in consuming, packaging practicality has inspired new designer generations, which consider physiology, perception, and commodity as an advantage over existent packaging. "Ergonomic packages facilitate user interaction with product and use. They are noteworthy and attractive in formal terms and language: colors, labeling, lids, etc. Additionally, they strengthen the product brand and convert into a guide for ranking. They are a competitive advantage" (Daniela Gómez, product manager of O-I Peldar)[3].

ENVIRONMENTAL IMPACT OF PACKAGING

The impact that generates product development and society consumerism is a common issue; nevertheless, the packaging sector is an area with a big consumption problem, the actual consequence has generated a high environmental impact and humans are facing it every day but they have not accepted this as a real problem. Every day a big variety of products like food, drinks, hygiene, cosmetics and entertaining are presented in a packaging that will be disposed of in a matter of minutes, hours or days, and in its production process logistic requires a massive structure set, that includes barriers, cushions and an endless amount of elements that assure the final product will be delivered in good conditions to the user. As a high demand product, the packaging industry requires elements in its production processes that cause atmosphere damages, land use, resource consumption like water, and waste disposal. As part of the packaging design process and from an ecologic point of view it is necessary to develop research that includes manufacturing processes, distribution, use, disposal and life cycle analysis.

Developing new materials, disposal management, recycling methods in packaging design lets consumer change its posture regarding packaging and its disposal. Even though there are no options for consumers and they have to acquire a product with an excess of elements in its packaging or that is highly polluting, they will keep the idea that something could be done for a change. "Sometimes consumers have a positive attitude and want to behave in an environmentally friendly manner, but they cannot act properly because they do not have sufficient means to do so." (Fraj & Martinez, 2007:31). But the lack of information in consumers generate inconsistency in their actions; when people cover their basic needs and reach satisfaction, they easily forget the problems that are generated because they are hidden. Only then the development of conscious moral reasons appears and they inspire the necessity of consuming recyclable and environmentally friendly products, reaching a positive attitude to go farther than obeying the social pressure generated in these days.

LIFE CYCLE

Analyzing the life cycle quantifies the environmental impact, and as part of the design process it is not just important to integrate the user from the ergonomic and universal design point of view, but to integrate environmental factors is also a significant element that improves the quality of life. Complete product life cycles in packaging have to be considered from the product conception. Packaging development requires a more profound analysis that visualizes possible interrelations between user-object in consumption and post-consumption; and life cycle has to lengthen its use with complex systems that include elements as returnable, refillable, reusable and interchangeable. Where implementation of life cycle principles will be considered in product development, production, distribution, commercialization, consumption, and post-consumption.

But working with the life cycle itself is not the only answer; developing analysis, evaluations and eco-friendly material development that generate new alternatives, together with clean energy use proposals and reduction in production costs have to be considered too. The ideal panorama is the development of packaging easy to assemble and disassemble, that can be cleaned, divided, reused and disposed of in an adequate way and that would not require a lot of effort from the users. The objective of connecting these principles with marketing is developing a product based in the "win-win", under the commercial principles where the businessman wins with material and energy reduction, it generates an environmentally friendly product, and is functional and more suited for the final user.

The biggest problem faced by packaging is when packages are mixed with general trash and they lose their recycling or reuse properties. When they are separated they become part of a new production cycle, raw material will be gained, energy expenses used in its production process will be diminished, less space in wastelands will be occupied and the indiscriminate use of raw materials, like wood, will be reduced and their extraction and process will allow a task based on environmental harmony. Knopman et.al. (1999) Suggest that the more citizens become aware of their own role in the problems interlinked with environmental degradation, the more willing they become to participate in solutions. (Paço 2016: 467). From the consumer side, there have been more favorable answers when an "environmental friendly" packaging design is consistent with its statement, its visual appearance, the materials used for manufacturing and its size.

The Sustainable Packaging Alliance (SPA) (2010) defines sustainable packaging as effective, when it blends the functional requirements of containing and protecting products with the minimum packaging; efficient, when it is manufactured with less raw material and energy resources; cyclic, when a minimum of resources and more recycled safe materials are used and safe, when all the materials used are harmless and do not cause any damage from their extraction to their disposal. (Scott & Vigar-Ellis, 2014:643). Part of this process includes changing factors like a reduction in ecotoxicity, acidification, land transformation, resources used, extraction damage, use of water resources, gas emissions, greenhouse effect and used energies.

DESIGN PARAMETERS

"Optimal package designs inspire an immediate understanding of use, opening (where and how), proper and accurate dispensing, re-closure and disposal" (De la Fuente, 2015, 157). The way to develop this concepts, whether for the industrial or familiar area, to be used immediately or to be divided into parts is the use the design process of trials and simulators, with the backing of design methods in the users analysis, universal design principles and the use of equipment that allows the researchers to observe the physical reaction of the materials used in the packaging.

Visualization of the packaging and based on the visibility principle is that action has to be considered, for example, how it is opened, clear instructions that allow a mental process, a precise visible message that communicates the function and the physical state of the product is the result. Consumers must classify the visual elements of the design, companies make a better and more efficient design and they do not necessarily want it to look ecological all around. "It is recommended that words like non-harmful, biodegradable and recyclable are used on packaging, as based on the findings, these properties are what respondents mostly identified within their understanding of environmentally friendly packaging" (Scott & Vigar-Ellis, 2014: 647). It is recommended that labels are easy to read.

Crul and Diehl (2006) suggest that the products must present the following list of options when a sustainable packaging is designed:

1. **Low Impact Material Selection:** Materials must be clean and renewable, they need to use low energy manufacturing processes and recycled or recyclable materials supplied by local producers.
2. **Materials' Reduction:** Materials must present volume and weight reduction for their transportation.
3. **Streamlining Manufacturing Techniques:** Alternative techniques should be used to reduce the production steps, cleaner energies, lessening raw material, less use of consumables and a clean and safe workplace.
4. **Streamlining of the Manufacturing Systems:** Lessening of packaging, which must be clean or reusable, more effective way of transportation, suitable logistics and the use of distributive economies.
5. **Impact Reduction:** Energy consumption must be reduced, clean energies used, less use of consumables and disposal reduction.
6. **Lifespan Streamlining:** Product design has to be reliable and lasting, allowing easy and local maintenance and repairs, modular structures must be used, a classic design that will not be out of fashion easily and achieve a product user relationship.
7. **End of Life System Streamlining:** The design of reusable products, remanufactured or renewable with recyclable materials or adequate materials for their incineration taking into consideration local recollection systems.

Alex Diener, mentions that Design for Disassembly (DfD) is based on the principles of disassembly: recycling, reuse of package, easy opening, without screws, glue or heat closed that allows an easier separation of the product so the disassembling times are reduced. Design teams have to be prepared to fulfill the design cycle, analyzing disposals and creating solutions that must be monitored for their design. When the phase of predesigning materials and production costs are reduced, Phil Berry argues, "Having designers understand how they can affect different groups in the organization is fundamental to designing more sustainable products" (as cited in Diener, 2010:37). This way the design phase is integrated to the previous characteristic materials analysis and their impact.

USER AND PACKAGING

How Packaging Functions Are Generated

Since humanity started using the same objects as part of their everyday life, the design of the packaging has been changing, and this has been done in stages, in the beginning the element or objects were used to contain recollected water or food, when they began to control harvests, these containers had to be modified in shape and size, when they had exceeding productions, storage areas required more space and products needed to be kept in good conditions. The function of protection during transportation had to be incorporated when they began traveling to near communities and later on to farther places for trading and selling and buying their products.

Dolores Vidales (1999) mentioned that during WWI and due to the lack of workforce, selling operations in the stores had to be changed, specially in the everyday products, employees had to cover different activities, the products were kept in bulk in the warehouses or the storage areas and were placed in bags, barrels, boxes or containers. When a consumer needed something the employees went to these areas to supply and pack the products and wait on the customer over the counter. At the same time, they had to convince the customer about the quality of certain goods so they could sell certain products. Clarence Saunders in 1916 proposed a new layout, he changed the movement flows on the stores, the customers had to enter one way and exit in another, he opened the warehouses and created aisles, placed similar products on certain shelves so the customer could go walking and choosing whatever he wanted to buy, every packaged product was tagged with a price, this method reduced considerably the amount of employees needed in a store, they only needed to work as cashiers. The predictions estimated that this new concept would reduce selling because there were no employees convincing costumers about the quality and characteristics of certain products. The truth is that sales grew considerably especially because people had the freedom to select what they wanted and at the same time they could come close to the products and choose among a bigger amount of products and new brands.

With this practice the concept of self-service was bond and packaging became more important for product selling. This new dynamic product packaging needed to fulfill four basic factors: contain, protect, transport and communicate. Even though all these functions have evolved based on technology development, and even though all of them are important for convincing the customer communication is decisive. It is important to consider that today the world works through images placed in ideal sceneries, for example, milk packaging, which shows a young slim woman for the low fat product in its image, knowing that the common buyer will be the overweighed woman that buys also the message that she will be like the model in the future, so this message that generates expectations in the consumer is ideal to convince people of buying the product.

Qi (2014) established that packaging has three levels: primary, it is the first wrap. Secondary, it manages a certain number of primary ones, and the tertiary is the adequate one for distribution and shipping. He also mentions that functions of packaging are divided into two basic areas: the first one is containment, the product needs to be protected and preserved, it has to be transported and at the same time has to give information and promotion of the characteristics and brand of the product. The second one is delivery, it has to be transported long distances, distributed, stored, retained, and used. Developing a better packaging considering these factors cannot be completed without doing a profound analysis of the consumer.

User and Packaging

Incorporating ergonomics, user habits analysis, trials and simulators in the design process and understanding the importance of making users aware of the need of taking environmental actions, can lead to a packaging design that promotes merging with and helping a more sustainable world. The development of research methods in this area will lead to better design products. Packaging design needs to address usability, efficiency, effectiveness, and satisfaction. Usability is the ability to use a product, analyzing the interaction between product and user can be done when the user characteristics are determined, like age, gender, body type, size, visual, auditory and biomechanical abilities. Pousette, et.al (2014) mentions that in the evaluation process effectiveness includes the ability to carry out the task and the quality of the result; Efficiency is the time taken and other resources used to reach the goal and satisfaction is the comfort, acceptance, and attitude to the product.

When analyzing usability, four components have to be taken into consideration: user, packaging, context, and task, which are engaged in a perceptional, cognitive and motor system. The interaction of all these factors leads to a better design, more functional and satisfactory. Poor consumer packaging design can lead to frustration and possible injury; packaging design shouldn't ignore functional limitations common to aging. All analyses done should consider movements and hand task strength dexterity also, like picking up, holding containers, removing caps and lids, opening bags and packets and dispensing contents. A good packaging will give independence to users.

Considering specific populations for the design of the packaging is important with some kind of products, one example is medicines, where openability is a predominant factor. Bottles with pills are designed with caps that are not easy to open by children, the problem is that most of the time they are difficult to be open by the elderly population, they commonly ask for help or use sharp things to open them, risking their own integrity in the process. On the other hand, children´s toys have to be open easily considering their size hand and ability. Also, a certain amount of kitchen or bath products have to be open with wet, oily or sticky hands, or by people with some hand or strength disability. So taking in consideration all the users that can use the packaging and product in the development process will lead to a more functional and universal design.

User Design

Packaging design needs to focus towards usability parameters, efficiency, effectiveness, and satisfaction, which can be achieved when incorporating ergonomic analysis, consumer's habits, user trials are performed with the help of simulators in the process of design and at the same time the consumer needs to understand the importance of looking after the environment, which is today´s objective. The term usability is defined as the ability to use the product, this can be done analyzing the interaction between the product and the consumer which can be done when the characteristics are determined, like age, gender, morphology, sizes and its visual, hearing and biomechanical abilities. Pousette et. Al. (2014) mention that in the process of evaluation, effectivity includes the ability to perform a determined action, and in the quality of the result, efficiency is the time that is taken and other resources that are used to reach the goal, and satisfaction means comfort, acceptance, and attitude towards the product.

When analyzing usability four components must be taken into consideration: consumer, packaging, context, and task, which are related to the perception, cognitive and kinetic systems. The interaction of these factors generates a more functional and satisfactory design because the wrong design of the packaging could lead to consumers' frustration and possible injury during handling. For example, a good design must consider movement, dexterity and hand strength, how to hold, carry, remove lids, opening bags and packages and the distribution of its contents. Good packaging will give the consumers independence and will take into consideration all the possible consumers and functional limitations due to age, for example, children and the elderly.

The esthetic design is very important in the handling of the packaging because it attracts the attention of the user, but when it presents a mistake in the packaging design the consumer at the end will be frustrated, for example, lids that are difficult to open or close, products that drip during pouring, seals that have leaks, illegible instructions, the loss of the product when it is applied or when due to the size of the packaging it is difficult to store. In these cases, the purpose to achieve a more comfortable use is lost.

Ergonomics

Ergonomics is applied directly to packaging in different ways, considering that the human body is a dynamic element and that every user has different habits, general ergonomic parameters have to be taken into consideration for the designing process. The first one is anthropometry, which applies dimensions and measurements of the final users; for example, objects for children, packaging and object dimensions have to be adequate to their body characteristics. The second parameter is biomechanics, as it was mentioned before the users' body is not static, so analyzing the possible movements and interactions of the user and the object are fundamental, this way anthropometry cannot be applied without a functional analysis. Openability, closing, and handling have to be analyzed during the process of designing and in the trial phases to be able to establish the redesign parameters (see Figure 1)

Another factor that must be considered is the one related to cognitive ergonomics, where perception aspects are analyzed from different points of view, packaging must be analyzed thinking about an everyday context and how it works in a shelf, the comparative colors, sizes and the relation contents-price considering competition, so the trademark can be remembered and it will be made more visible for the regular

Figure 1. Handling ability is an indispensable element in the design of packaging

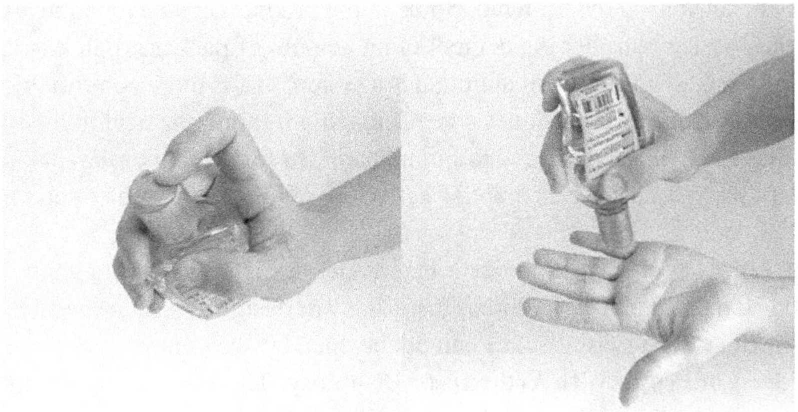

and new consumer. It is important to consider family traditions, because under this parameter price and competition are not an issue because many times consumers buy the brand they have used all their life.

For an ergonomic and usability analysis, in the design phase, using simulators and form models will let the possible users to perform tasks and evaluate pain and muscular strain. These trials can be done, for example, with anthropometry, electromyography or 2 axis electro goniometers. The anthropometric aspects have to be considered during the design and development of packaging, they have to be sustained under the parameters of the final users, for example, applying the 5th and 95th percentile principle in the different dimensions needed in the design. Celorio (1993) mentions that in packaging design for exportation the following elements have to be taken in consideration:

1. **Handle Diameter:** The correct diameter has to be applied. Therefore, the packaging could be handled easily, this is very common in containers bigger than a 2.5liter volume. The only exceptions are soda bottles that do not consider a useful handle diameter or even the possibility of having one.
2. **Body Diameter:** handling needs have to be considered in their easiest way, for example, a shampoo bottle which most of the time is used with soapy hands or with just one hand when it has to be poured.
3. **Opening Element or cap Diameter:** Depending on the type and form of the packaging, hand dimensions, force, ability, and dexterity has to be considered for the final design.
4. **Assembling or Filling Method:** Packaging production in general is massive but there are some specific cases like folding boxes that have to be done by hand, so steps and complexity needed for folding are productivity elements that will make a difference for the final users, this design needs to consider all the movements that have to be performed, this way productivity will increase and mistakes will be prevented.

Oropeza & Sanchez (2004) establish the points that have to be integrated into the handling of a container, not only dimensions have to be considered, also movements and functions:

1. **Moving or Handling:** Some of the containers are difficult to move or handle due to their weight or size. Incorporating handles that reduce the effort and allow a more comfortable loading of the object, so users can transport the object from the store to the house or to the designated place (see Figure 2).
2. **Multipackaging:** In some stores like Sam's Club or Costco, the size of the portions sold are very big, so they are not considered for retail. Some of the products can be found in a secondary collective packaging (which consider the content of an amount of packages), in this kind of stores this method shouldn't be seen as a retail option, it is a system that is more comfortable and lets the user decide how many individual portions he needs making it more practical in the final handling.
3. **Easy Opening Lids:** This concept is commonly applied to metallic or tetrapak packaging, where the design of this type of lids can simplify actions or eliminate the use of scissors or canopeners. (see Figure 3)
4. **Childproof Lids:** Especially in products like medicines is important to control the involuntary consumption of the product by children, these lids are designed with a mechanism that prevents its opening with just one twist, so they can not be open by accident.
5. **Lids for Elderly or People with Arthritis or Disability:** This kind of lids consider users that have a reduced force capability in their hands, most of them use the level principle with some other ele-

Figure 2. For heavy packaging is better than the final design has a handle that will let the object to be transported and manipulated when it is poured.

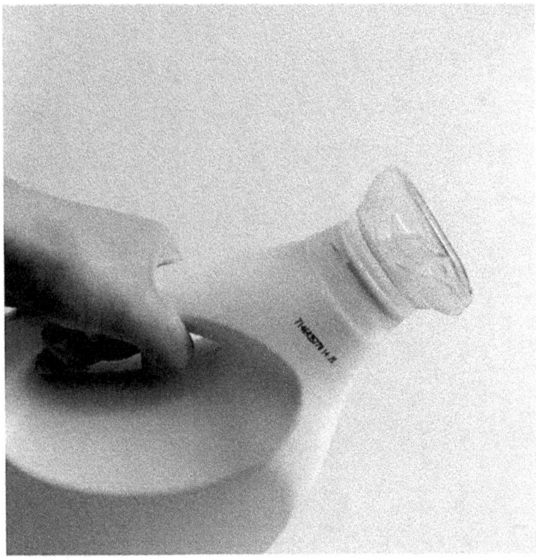

Figure 3. Opening lids easily has to combine a complete sealing and easy handling for openability.

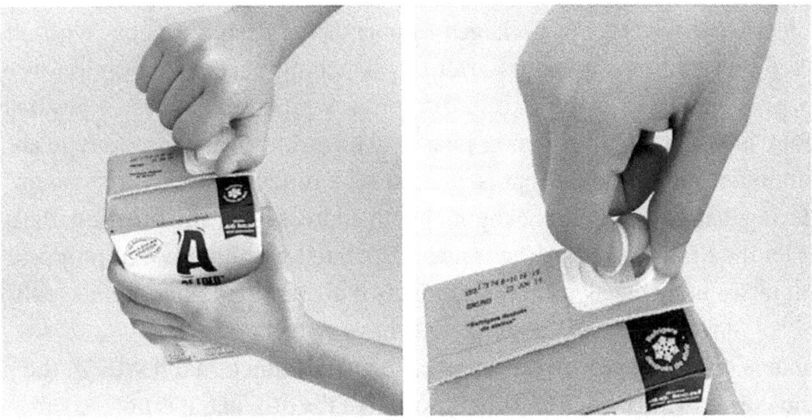

ment used frequently at the house like a pencil, spoon or knife, so it can be easily open or closed and a specific force can be applied so it will stay hermetic after use.

6. **Zip Fastener:** Used for bags that have to be constantly open and have to maintain a hermetical property to preserve the product
7. **Dispensers:** These are assorted considering material and the kind of service that is needed by the user. How the content will be applied, which through spraying, atomizing, dripping or the so-called "breaking flow" which lets the content in the bottle diminish the flow so the liquid will have a better application (see Figure 4).

Figure 4. Spraying caps are complex devices with a bigger amount of elements that will make this packaging more expensive but the comfort factor will compensate the price.

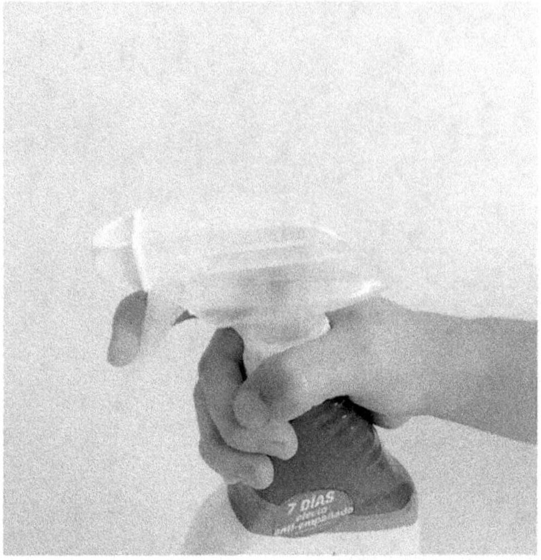

TRIALS AND SIMULATORS USE

To fulfill the objective of developing packages as part of the design process with all the characteristics mentioned before, building simulators after the sketching phase is an option to user trials, in the industrial design area and as a part of an interdisciplinary process, the use of physical prototypes is a way to make public a product in a beginning phase. “The terms physical prototype and model are used to describe a previous tridimensional representation of a product, service or system” (Hallgrimsson, 2013:6). Considering that a product or packaging has to address form and function, the use of a prototype for analysis will let the designer notice the issues associated with the design proposal. Using scales is also an important factor in this analysis, initial models will start being on a small scale to determine the form, but to analyze function real scales and details have to be built.

When fast models or prototypes are built, either formal or functional it will let the potential users or clients understand the real dimensions or form proposal of the product, it is not the same to see a proposal in a sketch, even though it is a perspective or a function graphic. With the model they can determine the sensorial conditions of handling, perceptions, griping or how easy it is to open or close the product. The main idea is determining the technical specifications in the concept, therefore, at the end, redesign parameters will be developed and a more comfortable and efficient product will be done. Using trials and simulators will let designers optimize material resistance with a physical structure analysis, where textures are not just a factor that will help the user to hold a product, they are also an element that favors structure and reduces the amount of material used, this will certainly impact the general cost of production.

Hallgrimson establishes the difference between models and prototypes, where prototype is a physical element that allows the analysis and the testing of use sequences of a product, also the manufacturing process can be checked and the final form, this reduces costs because it solves the problems that appear during the design process, which is evolving in form and function with user´s trials. When manufacturing is easy, the prototype is done with the final materials and processes. Prototypes must fulfill certain specifications like accuracy, this requires dimensional precision, and for this, the prototype has to be built with hard materials that will allow details. Finishing is important because the final user or client will have a better idea of what he is looking.

The model is the method used before producing a prototype, it can be divided between physical models and functional models, physical models just show dimensions and shape in general and functional models will present some of the basic functions, for example, opening and closing. Both can be done with cheap material like cardboard, plastics or wood, or soft like foam or dough. The objective is for the user to understand shape and dimension. Today the use of tools and technologies like CAD allows a quick creation of the first ideas and reduces the working time. In the first trial phases, tests can be done with coworkers but it is necessary to work with final users or consumers in posterior stages, in some cases it is needed to use specific populations to apply the universal design principles.

User Trials

User trials let the researchers analyze ergonomic factors, like dimensions, cognitive aspects, how they interact with the object and their opinion about the product. The use of ethnographic methods allows the designer to discover the problems that the proposal presents and how to solve them. The idea is to start with simple low-quality models that will be evolving along the process, It is needed to determine the kind of participants which will be making the user trials, is important to notice, that in this stage the functional tests can be done with models, but due to the materials characteristics, it is not recommended to perform resistance trials because it can create a risky situation. Trials can be performed in different spaces, in the beginning, they can be done in the design office or in the classroom, later as the redesigning develops, the proposal has to be done in a university or specialized laboratories where the ergonomic or resistance elements are analyzed. Nevertheless, in the final stage, it is necessary to perform the trial in the final context and with the final users, so the conditions of use can be tested under real and daily circumstances.

The design of the packaging also includes performing trials that can guarantee its effectiveness previous to the mass production, and this way the product can fulfill the requirements of transportation, storing and use. The trials can be performed in three ways. One is for the material, it has to be physically tested under the conditions the product will go through, for example, a trial can be performed on a greasy product covered with a plastic film to detect the UV rays impact, to prevent stalling. The second trial is done in the packaging, this can be done in the prototypes, and the third one is over the real material, which can be analyzing the technical data of its behavior, for example, permeability or barriers. This trials to be validated must follow a trial protocol with indicators that allow measuring the characteristics that guarantee the expected quality, the material, the structure, the dimensions, the printing system, text legibility and content, all this according to the government regulations.

Once the trials have been developed and the parameters determined accepting parameters are denominated "Technical specification", which characterize the material regarding color, width, barrier resistance, humidity or absorption, for example. At the end, the quality of the packaging will be tested under the regulations determined by the norm.

CONCLUSION

"Packaging can contribute to sustainability because it can promote product protection, reduce product deterioration and consequent waste and help economic development and create Jobs." (Martinho, et.al. 2015, 58), the design of products and packaging for everyday activities address comfort, easiness, quickness, and practicality. This has resulted in a world flooded with objects where the consumption is very fast and the production volume is high, packaging heads the list of products that in a few minutes will become waste.

The packagings I touch in a day

In the morning I use my toothpaste and my mouthwash to brush my teeth, shampoo and conditioner to wash my hair, when I finish showering I use deodorant, facial cream, perfume, body lotion when applying makeup I use more than five products. In my breakfast I have coffee, yogurt, and sliced bread; to teach I use a box of felt pens and at my coffee break, I have coffee and a cookie. To cook I use more than five bottled condiments and at dinner at least four more. Before going to bed I remove my makeup, I use again my toothbrush and mouthwash and this averages the use of more than 30 packaged products a day, some packaging only last an instant in my hands, the great majority will become waste in less than a month. Silvia Oropeza.

If we could develop a statistic analysis of the national or the world packaging consumption per day per person, most probably the numbers will be impressive, especially the speed in which they are thrown away. This is why we believe that packaging must be created with other characteristics. The first one that must be taken into consideration is to think of the universal design that will include everybody in a more sensitive way for the capacities of each group. The second characteristic is the design based on a concept of win-win, this means to create circular ideas that study cyclical, fair systems that generate the least negative impact on the environment. Reducing processes, materials, waste, and emissions, packaging should also be able to be accommodated taking advantage of the transportation systems, and it has to be useful, healthy, easy to use and inclusive this way the gain will be in the quality of life of the users.

Users must be conscious of what they consume, Kate Copper a marketing expert argues that convincing the customers to buy can be done with three keys, the first one is how information is written on the labels, it can give the idea of a cleaner or healthier product even though it is not completely true. The second one is the idea that using cutting-edge technology will automatically develop a better product but users would not think about the hidden processes used in the product. And the third one is the most dramatic, the power of voluntary ignorance because people prefer not to know when and how the products have been made and the damages they cause during production, marketing or consumption. This ignorance is what makes us repeat mistakes and only developing consciousness, knowledge, and understandings this paradigm will be broken.

The relationship between user and packaging has to be not only dimensional; it has to be more responsible in a sustainable and usability point of view. In the end product marketing is not just about consumption, "A business that wants to become and/or remain competitive will need to address sustainability issues" (Crul & Diehl, 2006:22), so research, innovation, and raising awareness by education is the next step that has to be addressed by universities and industry.

FUTURE RESEARCH DIRECTIONS

The Trials and Simulators Laboratory has proposed projects for short and middle term. The first one is the development of an international network denominated, Materials, Trials, and Simulators for Design; the objective is to collaborate in the development of new paradigms in design evaluation. This network will be integrated by researchers from the Milan Polytechnic, Italy, from Sao Paulo University, Brazil, from the Instituto Superior del Medio Ambiente, España, and IdEA Center, University at Buffalo, United States. Part of the project is to develop a book that allows us to spread the advances in research, and the position around the new design paradigms with a global vision, as a result of the 1st Materials and Simulators Congress done in the Universidad Autónoma Metropolitana, campus Xochimilco, Mexico. The idea is to establish a continuous collaboration network and having the basis for debating and giving a feedback of the advances in their own countries in future congresses. The second project is to establish a "Design Center" inside the Trials and Simulators Laboratory; this center will address specific population needs, from fast adjustments in their daily life technical aids to the development of design products under the parameters of universal design that will let them have a better quality of life.

The third project is to develop a "materials library" that lets students, researchers or industry to make the trials needed to verify the quality and conditions of the proposed object.

REFERENCES

Alonso, L. (2015). The Proust Effect. The Senses as Doorways to Lost Memories. *Revista investigación y Ciencia, 70*. Retrieved from http://www.investigacionyciencia.es/revistas/mente-y-cerebro/pensamiento-creativo-619/memoria-sensorial-12753

APR. (2013). "Dinks", la nueva clase de parejas mexicanas. *El economista*. Retrieved from http://eleconomista.com.mx/finanzas-personales/2013/07/16/dinks-nueva-clase-parejas-mexicanas

Bonnici, P. (1999). *Lenguaje Visual, el Medio Oculto de la Comunicación*. Luis Hernando Ramirez C.

Cawthray, R. & Denison, E. (1999). *Packaging*. McGraw Hill.

Celório, C. (1993). *Diseño del embalaje para exportación*. IMPEE.

Chen, Y., & Wang, R. (2016). Are Humans Rational? Exploring Factors Influencing Impulse Buying Intention and Continous Buying Intention. *Journal of Consumer Behaviour, 15*(2), 186–197. doi:10.1002/cb.1563

Conway, L. (1998). *Diseño de empaque*. Editorial Somohano.

Costa, J. (1999). *Diseñar para los ojos*. Barcelona: Gustavo Gilli.

Crul, M., & Diehl, J. (2006). *Design for Sustainability: A Practical Approach for Developing Economies*. Paris: UNEP.

Datschefki, E. (2002). *El Re-Diseño de Productos. Productos Sustentables, el regreso a los ciclos naturales*. McGraw-Hill.

De la Fuente, J., Gustafson, S., Twoney, C., & Bix, L. (2015). An affordance-based methodology for package design. *Packaging Technology & Science*, *28*(2), 157–17. doi:10.1002/pts.2087

De la Torre, G. (2000). *El lenguaje de los símbolos gráficos*. Limusa.

Diener, A. (2010). *Afterlife: An Essential Guide to Design for Disassembly*. Retrieved from http://www.core77.com/posts/15799/afterlife-an-essential-guide-to-design-for-disassembly-by-alex-diener-15799

Díez de Castro, E. C., Landa, F. J., & Navarro, A. (2006). *Merchandising: Teoría y práctica*. Pirámide.

Egaña, J. (2016). *¿Qué significa Gay Friendly? Empresas Gay Friendly y las lógicas susceptibilidades*. Retrieved from https://www.congresoegf.com/que-significa-gay-friendly-empresas-gay-friendly-y-las-logicas-suceptibilidades/

Fausset, C., Baranak, A., & Farmer, S. (2014). Developing a scoring matrix to evaluate the usability of consumer packaging: A pilot study. *Proceedings of the Human Factors and Ergonomics Society 58th Annual Meeting 2014*, 1802 – 1806. 10.1177/1541931214581377

Fraj, E., & Martínez, E. (2007). Ecological consumer behavior: An empirical analysis. *International Journal of Consumer Studies*, *31*(1), 26–33. doi:10.1111/j.1470-6431.2006.00565.x

García, B. (2008). *Ecodiseño, Nueva Herramienta para la Sustentabilidad*. Designo.

Giles, C. (2004). *Que es el packaging?* Barcelona: Editorial GG.

Gobe, M. (2008). *Branding Emocional*. Barcelona: Divine Egg Publicaciones.

Hallgrimsson, B. (2013). *Diseño de producto, maquetas y prototipos*. Promopress.

Magnier, L., & Schoormans, J. (2015). Consumer reactions to sustainable packaging: The interplay of visual appearance, verbal claim and environmental concern. *Journal of Environmental Psychology*, *44*, 53–62. doi:10.1016/j.jenvp.2015.09.005

Martinho, G., Pires, A., Portela, G., & Fonseca, M. (2015). Factors affecting consumer choices concerning sustainable packaging during product purchase and recycling. *Resources, Conservation and Recycling*, *103*, 58–68. doi:10.1016/j.resconrec.2015.07.012

Milton, A., & Rodgers, P. (2013). *Research methods for product design*. Laurence King Publishing Limited.

NNDB. (2014). Clarence Saunders. *NNDB, Tracking the entire world*. Retrieved from http://www.nndb.com/people/054/000166553/

Oropeza, S., & Sánchez, A. K. (2004). *Manual de Diseño de Envases*. IMPEE.

Paço, A., & Gouveia, R. (2016). Environmental activism and consumer's perceived responsibility. *International Journal of Consumer Studies*, *40*(4), 466–474. doi:10.1111/ijcs.12272

Pousette, S., Löfgren, M., Nilsson, B., & Gustafsson, A. (2014). An Extended Method to Measure Overall Consumer Satisfaction with Packaging. *Packaging Technology & Science*, *27*(9), 727–738. doi:10.1002/pts.2064

Qi, H. (2014). *Decision Making Framework For Sustainable Packaging Design Using Life Cycle Assessment* (Doctoral dissertation). Graduate School-New Brunswick Rutgers, The State University of New Jersey.

Rodríguez, J. A. (2006). *Manual de Ingeniería y Diseño en envase y embalaje.* IMPEE.

Rodriguez, J. A. (2008). *Enciclopedia IMPEE, Manual para elaboración de Especificaciones Técnicas para materiales de Envase y Embalaje.* IMPEE.

Scott, L., & Vigar-Ellis, D. (2014). Consumer understanding, perceptions and behaviors with regard to environmentally friendly packaging in a developing nation. *International Journal of Consumer Studies, 38*(6), 642–649. doi:10.1111/ijcs.12136

Urdiain, R. (2006). Las cuatro P. *Revista Entrepreneur.* Retrieved from https://www.entrepreneur.com/article/256375

Vidales, M. D. (1999). *El envase en el tiempo: Historia del envase.* Trillas – UAM.

KEY TERMS AND DEFINITIONS

Anthropometry: Measurement of size and proportions of the human body.
Biomechanics: The study of the action of external and internal forces on the living body.
Effectiveness: Ability to accomplish a purpose.
Efficiency: Ability to accomplish a job with a minimum expenditure of time and effort.
Ergonomic: A science that works with abilities, limitations, and requirements of users in the design of spaces or objects.
Goniometer: Instrument for measuring solid angles.
Life Cycle: Period of time where a product is developed, brought to the market, used, and disposed.
Sampling: The process of selecting a sample for testing.
Usability: Available or capable of being used.

ENDNOTES

1 For more information see: Online course – "Los retos de la empresa familiar", Desarrollo Empresarial http://www.nafintecapacita.com/suite/index.php?c=welcome&m=login,

2 For more information please see: Articles analized: http://www.dinero.com/internacional/articulo/regalar-comida-supermercados-buen-negocio/202456, http://www.bbc.com/mundo/noticias/2014/10/141010_degustaciones_gratis_supermercados_jgc, https://www.entrepreneur.com/article/264205

3 For more information please see: Interview: https://es.scribd.com/document/207413890/Unidad-1-de-Envase-Empaque-y-Embalaje, O-I Peldar is part of the Ardila Lülle and Owens Illinois group, Glass packaging bigger producer in Colombia.

Section 3
Anthropometrics

Chapter 10
Anthropometry Devices:
A Comparative Study

Julian Israel Aguilar-Duque
Universidad Autónoma de Baja California, Mexico

Guillermo Amaya
Universidad Autónoma de Baja California, Mexico

Victor M. Juarez
Universidad Autónoma de Baja California, Mexico

Diego Tlapa
Universidad Autónoma de Baja California, Mexico

Ulises J. Tamayo
Universidad Autónoma de Baja California, Mexico

Ana Y. Tovar-Hernández
Autonomous University of Ciudad Juárez, Mexico

ABSTRACT

This chapter exposes a comparative research of different anthropometric equipment that has a patent register, comparing analogic devices versus digital devices, with the aim to define the relevance and status of the anthropometric devices. The methodology used in this project is integrated in three stages. The first stage is a documental analysis research of the anthropometric patents devices registered in international offices with the aim to identify the anthropometric devices patented. The second stage is the description of the characteristics, features, and functionality for every anthropometric device founded during the first stage. Finally, the third stage is the generation of a matrix with similarities of the devices with the aim to have a comparison between the devices. As a result of this research, it was possible to describe the characteristics of 14 devices used to define anthropometric devices as well as their futures and utilities.

DOI: 10.4018/978-1-5225-5234-5.ch010

Copyright © 2018, IGI Global. Copying or distributing in print or electronic forms without written permission of IGI Global is prohibited.

INTRODUCTION

During the last decades, many international companies have been working in increase their capital and their business opportunities just to guarantee success in a globalized world (Groover, 2007). The efforts to achieve a successful organization has been registered an interesting evolution of different systems as a: commercial system, health systems, educational systems and government systems (Bitner & Sierak, 2017). As a consequence of this evolution, the systems have developed the economies creating a positive effect that improve their way to the success in international markets (Altinkemer, Ozcelik, & Ozdemir, 2011).

As a consequence of the race's companies to achieve business success and global market to survive, the manufacturer and service organizations have focused their efforts to develop new strategies, methodologies, proceedings, philosophies, etc., in order to streamline their process (Wheelen & Hunger, 2011). Obviously, the spread and adaptation of these concepts have allowed that companies as GM, Toyota, Motorola, among others, have been considered pioneers in the research of a productivity field (Corkery, Mikalsen, & Allan, 2017).

As some big companies have spent resources in research and improve their process, middle and small companies have spent their resources adopting the methodologies or strategies used by the big companies to increase their productivity. Those adaptations have resulted, in some cases, a successful result for some companies; however, others a complete failure for others (Kavanagh & Johnson, 2017; World Economic Forum, 2011).

Performance of human resources is one of the fields that have been covered interested by researchers in a manufacturing and services companies. The application of ergonomic analysis has gain importance as a strategy to determine different opportunities for improvement in manufacturing and service process. The use of anthropometric studies, biomechanics analysis, work-station design, are some of the most common strategies used to support the application of ergonomics in industries. As a consequence, all of them work as a complementary item to achieve successful results (Gatewood, Feild, & Barrick, 2015).

As it had been mentioned above? The use of different strategies has the aim to improve the condition of the systems. Under this restriction, the present chapter is focused on analyzing the different anthropometric devices that exist and are used as a part of the ergonomic strategy to improve productive systems. The devices considering during this research were restricted by the most common twenty-five measures determined by the device.

BACKGROUND

Ergonomics have been used as a strategy to improve the manufacture and service systems. This research has focused their efforts on determining the impact of many variables that affect the performance of humans during work activities. Cavassa (2006) mentions the effect of the importance of dimensions in work areas. In other words, from the ergonomists point of view, study the measure of the human body is the key to design the environment and the equipment to humans with the aim to assure the maximum benefit and capacity of the users. In order to determine or measure these performances exist different fields associated with ergonomics, one of them and the reason for this chapter is anthropometry.

Some concepts of anthropometry are "disciple that works with a physical measure of the human body and his proportions" (Cruz & Garcia, 2001), "is a branch of a human sciences that determine the measures of human body, the shape, the strong, the mobility, the flexibility and the work capacity" (Pheasant &

Haslegrave, 2005), etc. According to the concepts described above, the principal role of anthropometry is to determine body dimensions to be used in the design or re-design of equipment, workstations and also, environments used by humans.

Anthropometrics have an important role in the design of workstations because it helps to determine the conditions, restrictions, and characteristics of places or objects that will be used by individuals or groups. These conditions, restrictions, and characteristics determined by anthropometric studies are called anthropometric data. The importance of anthropometric data is relevant because they are used to design workstations, machines, tools, clothes, shoes, furniture, vehicles, etc., (Gatewood et al., 2015; Ramirez, 2006; Stanton, Hedge, Brookhuis, Salas, & Hendrick, 2006).

According to Shikdar (2003), the design of many components or work environments are created considering anthropometric information, with the aim of "satisfy the necessities of individuals or a particular population." It is here where the importance and vitality of the anthropometry gets relevancy due that the information generated by the anthropometry studies (generally human measures) representing the adequate dimensions of the entire population.

Cruz & Garnica (2007) mention that the shape of the work area needs to be dimensioned using the anthropometric dimensions of the users (arms reach, feet high, high, high in the seat position, etc.), the dimension of the equipment, and tools and equipment shapes. From the ergonomic point of view, study the dimensioning of the human body is the key to adapt the equipment, the machine and the environment to the human. It is here where the anthropometry is used by engineers, designers, and architects to assure the maximum benefit and capacity of the products and areas used by humans.

As it has been mentioned, anthropometry is considerate an adequate way to determine the input data when professionals are developing stuff for humans use. However, an anthropometry study is more complex than just measure. In this way, it is very important to considerate the final use of the areas or things, the tools or devices used to determine the measures, the protocol used during the study and, the final report supported by statistics.

Experts like Brown & Scurr (2012) y Calderoni et al. (2016) mention that the anthropometry protocols are the key of the procedures used to determine measures. In general, the protocol should consider at least three measures with devices certified and calibrated.It is important to use the same point of measure and the same person should develop the measure. Although the use or development of a standardized anthropometric protocol to evaluate human measures represent substantial challenges, the real challenge is to determine the measures that should be considered as the minimum measures required to reduce the measured error.

The anthropometric dimension lacks further application due that the measure of the individual is registered in a specific moment of his evolution. This means that the physic variability is constant in different stages of human development. Definitively, age, sex, diet, culture, labor activity, etc., are somatic particularities in continuous change (Salvendy, 2006). In fact, everybody has a different anthropometric composition. The only way to create an ergonomic environment and efficiently work area is to design the workplaces to adapting to individuals. As a consequence of the dimensional variability in populations, the measures should be statistically represented by percentiles.

Finally, anthropometry involves a customized method for every research. Nevertheless, this study should consider at least the principal aim of the study, the protocol and the results have been an essential part of the protocol, the correct selection and use of the measures advice. In this way, this chapter describes some of the anthropometry devices used to determine specific measures.

MAIN FOCUS OF THE CHAPTER

Issues, Controversies, Problems

Anthropometric has been covered importance since his creation. As researchers have described, this field of research has supported many productive areas as industrial, automotive, aerospace, etc. How, who, where, what, and why are the most important questions determined during the protocol of the study. They fit directly in the success of the study that is developed?

Across the time, the evolution of anthropometric field has registered important advances. One of the most important is the development of equipment. Since the big rules were used to determine the high of persons, until the use of electronic devices that determine muscular mass, the anthropometric devices have been in constant changes.

However, what instrument is the better? Is better for whom? Is better for what? Moreover, Why is it better? These questions are the frame of the present chapter. The present aims to expose comparative information of the most common anthropometric devices used to determine human measures. This research considered the restrictions that these devices should have a patent register and should determine at least five of the most common measures used by ergonomics in the design of work areas, tools, and machines.

According to the authors, this chapter brings important information to this book, due that Ergonomics and Product Design need human dimensions expressed as anthropometric data. These are used to identify, analyze and determine improves or creations in work areas, special or regular equipment and products used to them.

The methodology used is integrated by three stages.

Stage 1: A documental analysis research of the anthropometric patents devices registered in international offices. The documental analysis is restricted a period from 1950 to 2015. The analysis integrates devices that are used to determine the 25 most common measures and the use of the device that use electronic implements. The syntaxes used to search the devices considered were: anthropometric device and electronic component and data register. Under this restriction, it was necessary to identify at least one of the measures described below:

1. Height seat.
2. The height of eyes in seat position.
3. The height of shoulder in seat position.
4. The height of elbow in seat position.
5. Thigh high.
6. Height popliteus.
7. Functional scope.
8. Vertical functional scope.
9. Abdominal deep.
10. Knee height.
11. Length buttocks popliteus.
12. Length knee buttocks.
13. Inter elbow span.
14. Shoulder high stands up.
15. Waist width.

16. Height.
17. Crotch height,
18. Eyes height stands up.
19. Shoulder height stands up.
20. Bideltoid amplitude.
21. The width of the body of elbow height.
22. Height feet.
23. The width of the head.
24. The length of the head.
25. Weight.

Stage 2: Description of the characteristics, features, and functionality for every anthropometric device found during the first stage.

Stage 3: Generation of a matrix with similarities of the devices with the aim to develop a comparison among the devices.

RESULTS

As a result of the documental analysis research, it has been created a group of tables to sum up the organizations and patents of interest. Table 1 describes the information of patent offices where patents of anthropometric devices were found.

With the aim to consider as much as possible registers with this research, it was also considered the databases of designs of others offices of industrial property. Table 2 present results got it under this criteria.

Table 1. Information of offices that have registered anthropometric devices patented

Country/BD	Office	Web Page
México/SIGA	Mexican Institute of Intellectual Property.	http://siga.impi.gob.mx
United States of America / PatFT	United States Patent and Trademark Office (USPTO).	http://patft.uspto.gov/
Spain/ INVENES	Spain Patent and Trademark Office (OEPM) Spain (Sp).	https://www.oepm.es/es/index.html
Spain / LATIPAT	Spain Patent and Trademark Office (OEPM), database of Latin-American Inventions (LATIPAT)	http://lp.espacenet.com/
EPO / Espacenet	European Patent Office (EPO), Internet data base ESPACENET.	http://ep.espacenet.com/advancedSearch?locale=en_EP
OMPI / Patent Scope	World Intellectual Property Organization (WIPO),	http://patentscope.wipo.int/search/en/search.jsf
China / SIPO	China Patent Trademark Office (CPTO).	http://english.sipo.gov.cn/
Japan / IPDL Industrial Property Digital Library	Japan Patent Office (JPO).	http://www.ipdl.inpit.go.jp/homepg_e.ipdl

Table 2. Information of offices that have registered anthropometric devices patented, a database of the design

Country/BD	Office	Web Page
United States of America / PatFT	United States Patent and Trademark Office (USPTO).	http://patft.uspto.gov/
Spain / IVENES	Spain Patent and Trademark Office (OEPM).	https://www.oepm.es/es/disenos_industriales/index.html
OMPI / Hague Express	World Intellectual Property Organization (WIPO),	http://www.wipo.int/ipdl/en/hague/search-struct.jsp
Hungria / PIPACS	Hungarian Patent Office (HU)	http://www.hipo.gov.hu/en
United Kindom / -IPO	Intellectual Property Office, United Kingdom (UK),	http://www.ipo.gov.uk/types/design/d-os.htm
Australia / AU Designs Data Searching	Intellectual Property Australia (AU),	https://www.ipaustralia.gov.au/

As a consequence of the first stage of the research, there were found eleven anthropometric devices. The description and characteristics of them are described in the next stage.

Stage 2: Description of the characteristics, and functionality of every anthropometric device found. The descriptions of each device were a transcript from their respective site of reference.

X-Ray Photography Anthropometric Instruments, International Classification A61B6/00

The utility model relates to medical devices, is the body of a measuring instrument to do when the X-ray photography.

The body needs to do a lot of X-ray photography of the site, in practice with the corresponding box and angle measurement scale photography more. At present, the more widespread use of X-ray photography anthropometric instruments, because of its ruler and angle measurement instrument, such as measurement tools no designs on the same plane, and mounted on the same structure, so it has some deficiencies in use: base not solid, inconvenient to use, measuring ruler cannot be adjusted in the range of 360 degrees; straightedge and angle measuring instrument can be adjusted individually, making it difficult to achieve accurate measurement requirements, especially for children, it is difficult to measure accurately.

The utility model aims to provide an X-ray photography meter body, enabling it to overcome the shortcomings of the prior art, with a straightedge and angle measuring instrument can make individual adjustments can be achieved either portrait or landscape orientation adjustment measuring several directions, to complete the advantages of accurate quantitative measure of the human body photography correct position, and with the use of quick, easy operation.

The utility model to accomplish the above object, the technical programs is that it has a base mounted on the base of the first pillar, the second pillar and the third pillar, were installed on the pillars of the slider and the slider, the slider installs measuring ruler on the slider mounting angle instrument. The first pillar, the second pillar, and third pillar triangular distribution. Mounting rails on the slider, installation measuring ruler on track. Installation angle and rotation angle meter ruler on the slider.

The utility model obtained through clinical trials the following advantages: measuring ruler can be adjusted within the range of 360 degrees; at the same time can also be adjusted in either portrait or landscape; goniometer can also be individually adjusted angle measuring instrument and can use a ruler to measure several directions; can accurately measure the correct shooting position any part of the body; the base solid, easy to use and so on.

Brief Description

Figure 1 is a schematic diagram of the structure of the utility model.

Reference to the drawings, embodiments of the utility model for further explanation: Figure 1 is a base, a first pillar is installed on the base 2, the second pillar of 9 and 10 third pillar, the pillar installation slide 3 and 7, respectively, the slider, 3 Install measuring ruler on the slider 4, the slider mounting angle analyzer 6 and 7. Pillar 2, the second and third pillar 9 10 triangular distribution. Installed on the slider track 8, 3, and 4 installations measuring ruler on track 8. Mounting angle and rotation angle ruler ana-

Figure 1. X-ray photography anthropometric ("X-ray photograph anthropometer," 1999)
Source: Google patents, 2017.

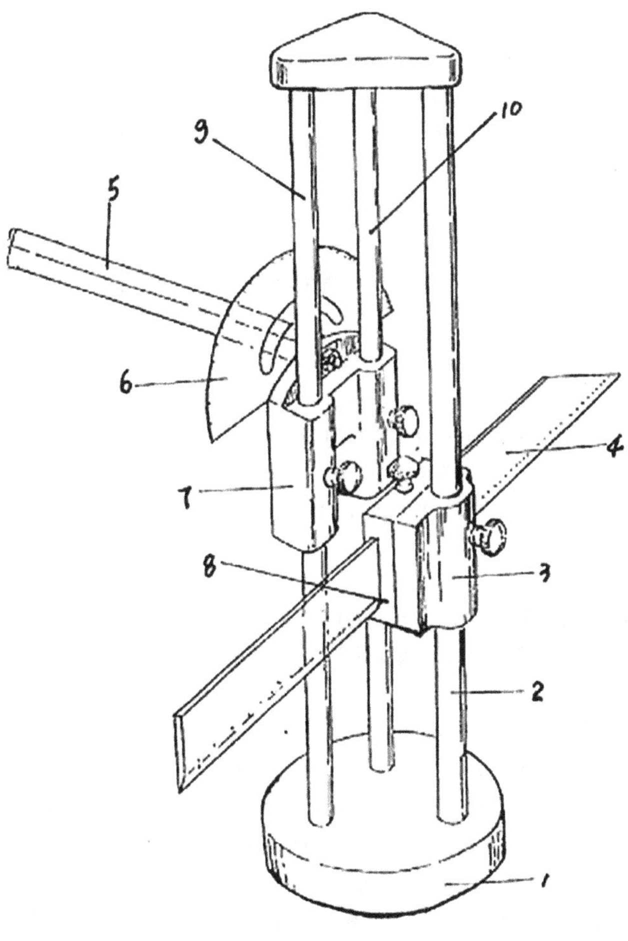

lyzer 6, 5 and 7 on the slider. Inclinometer and the rotation angle scale G 5 are fixed to the slider 7, the slider 7 is mounted on the second pillar 9 and the third pillar 10, the slider 7 may move up and down along the columns.The slider 3 can move up and down along the first leg 2, and may drive the measuring ruler is rotated in the range of 360 degrees, due to measurement ruler 4 is mounted on the inner rail 8, the measurement ruler 4 can be traversed along the inner rail 8, the track 3, 8 and slider combination enables measurement ruler 4 in portrait or landscape orientation be adjusted so that the utility model can accurately measure, especially in pediatric patients can be accurately measured.(Shaoqing, Lebin, & Tao, 1999; "X-ray photograph anthropometer," 1999).

Peniano Anthropometer Instruments, International Classification BR8506503A

This device is an instrument used by pediatric services to determine children measurements during the phase of?? grow and develop. It is used in ambulatory health offices, hospitals, clinics, and other health areas connected with surgery, genetic, and endocrinology with the aim to determine the final size of penne in baby's and boys in pre-scholar age diagnosed with early endocrines, genetic disorders, or tumor.

There is a variation in the size of the penis on children depending on the degree of development of the same, which increases with the installation of puberty. However, during periods of lactation and pre-school, the development of the penis is less pronounced, so the length of the penis has an enormous medical importance in the diagnosis of endocrine, genetic, and tumor pathologies.

The device consists of a millimeter ruler to measure the penis at rest with a ring of rigid material that allows, in use, the compression of the suprapubic fat completely exposing the body of the penis. The ring of rigid material that composes anthropometer has been stuck the front edge of the small millimeter ruler to remain upright and perpendicular to it. The penis is measured by the correct use of the device and the ring must be introduced in order to the penis rest in the rule (Helio de Oliveira, 1987; Oliveira, 1987). Figure 2 presents an image of the device.

Scholar Anthropometer Device, International Classification ES170025

Improvements in a scholar anthropometer apparatus characterized by having a base platform of rectangular, square shape polygonal, etc. In the center it will have mounted and fixed to its base of support and in vertical position a ruler graduated in either cylindrical or polygonal form any and made of metal wood or any other kind of material appropriate for it being graduated to its entire length in decimeters and centimeters with all accuracy for the measurement of the foot size *(Gutiérrez, 1945)*. In Figure 3 it shows an image of the device.

Portable Anthropometer for the Assignment of Linear Corporative Measures: International Classification BRMU 8403276

Equipment for measuring linear body measurements - length, height and height of the knee - made of aluminum, acrylic, and nylon, with a predominance of the first and approximate weight of 3.5kg. Features heel support, aluminum base with acrylic plates and nylon or aluminum bar with locking pressure for stem attachment; Two screws on the back for fixing the stabilizer bars; Aluminum base with stem support (nylon or aluminum bar with locking pressure); Calibration gauge made of aluminum with bar support to fit the rod and which guarantees the angle of the base 9018 and the gauge cursor with read-

Figure 2. Pinus anthropometer (Helio de Oliveira, 1987)
Source: Espacenet, 2017.

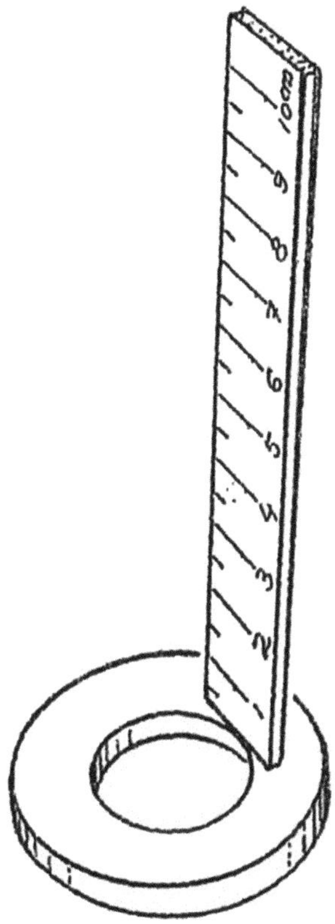

ing indicator. Two aluminum posture plates, Three containment sliders to define the positioning of the posture plates and the calibration slider; A 1.2m long rod, with a millimeter tape measure of fiberglass affixed and two bars for stabilization; A stretching rod measuring 1m, with tape measure throughout its length; With the possibility to measure height of up to 2.2m. "PORTABLE ANTHROMETER FOR THE ASSURANCE OF LINEAR CORPORATIVE MEASURES". Equipment for measuring linear body measurements - length, height and height of the knee - made of aluminum, acrylic, and nylon, with a predominance of the first and approximate weight of 3.5kg. Features heel support, aluminum base with acrylic plates and nylon or aluminum bar with locking pressure for stem attachment; Two screws on the back for fixing the stabilizer bars; Aluminum base with stem support (nylon or aluminum bar with locking pressure); Calibration gauge made of aluminum with bar support to fit the rod and which guarantees the angle of the base 9018 and the gauge cursor with reading indicator. Two aluminum posture plates, three containment sliders to define the positioning of the posture plates and the calibration slider; A 1.2m long rod, with a millimeter tape measure of fiberglass affixed and two bars for stabilization; A stretching rod measuring 1m, with tape measure throughout its length; With possibility to measure height of up to 2.2m. Aluminum base with stem support (nylon or aluminum bar with locking pressure); Calibration gauge made of aluminum with bar support to fit the rod and which guarantees the angle of the base 9018 and the gauge cursor with reading indicator. Two aluminum posture plates, three containment sliders to define the positioning of the posture plates and the calibration slider; A 1.2m long rod, with a millimeter

Figure 3.Scholar Anthropometer device (Gutierrez, 1945)
Source: Espacenet, 2017.

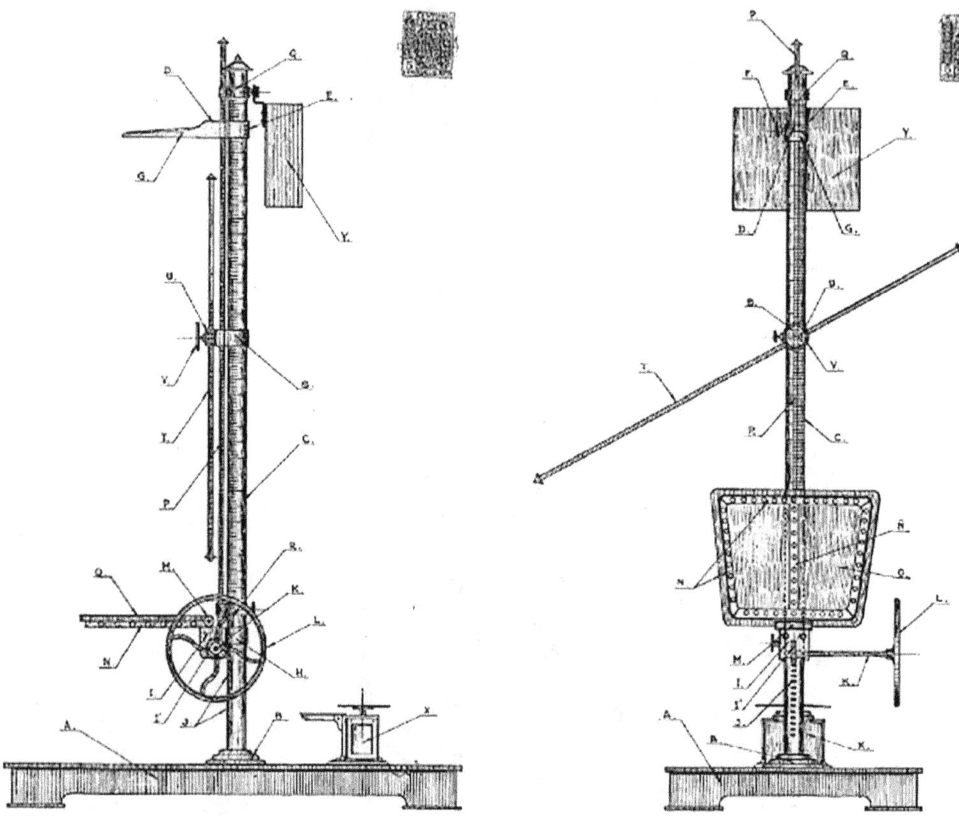

tape measure of fiberglass affixed and two bars for stabilization; A stretching rod measuring 1m, with tape measure throughout its length; With possibility to measure height of up to 2.2m. Aluminum base with stem support (nylon or aluminum bar with locking pressure); Calibration gauge made of aluminum with bar support to fit the rod and which guarantees the angle of the base 9018 and the gauge cursor with reading indicator. Two aluminum posture plates, three containment sliders to define the positioning of the posture plates and the calibration slider; A 1.2m long rod, with a millimeter tape measure of fiberglass affixed and two bars for stabilization; A stretching rod measuring 1m, with tape measure throughout its length; With possibility to measure height of up to 2.2m. Calibration gauge made of aluminum with bar support to fit the rod and which guarantees the angle of the base 9018 and the gauge cursor with reading indicator. Two aluminum posture plates, three containment sliders to define the positioning of the posture plates and the calibration slider; A 1.2m long rod, with a millimeter tape measure of fiberglass affixed and two bars for stabilization; A stretching rod measuring 1m, with tape measure throughout its length; With possibility to measure height of up to 2.2m. Calibration gauge made of aluminum with bar support to fit the rod and which guarantees the angle of the base 9018 and the gauge cursor with reading indicator. Two aluminum posture plates, three containment sliders to define the positioning of the posture plates and the calibration slider; A 1.2m long rod, with a millimeter tape measure of fiberglass affixed and two bars for stabilization; A stretching rod measuring 1m, with tape measure throughout its length; With possibility to measure height of up to 2.2m. With tape measure millimeter of fiberglass affixed and

Figure 4. Portable anthropometer for the assignment of linear corporative measures
Source: Espacenet, 2017.

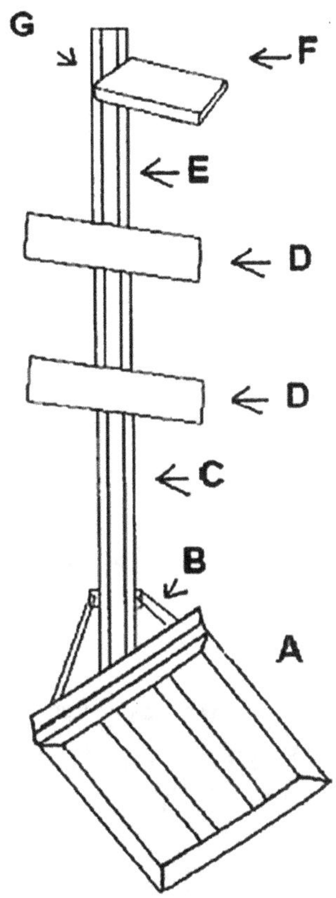

two bars for stabilization; A stretching rod measuring 1m, with tape measure throughout its length; With the possibility to measure the height of up to 2.2m. With tape measure millimeter of fiberglass affixed and two bars for stabilization; A stretching rod measuring 1m, with tape measure throughout its length; With the possibility to measure the height of up to 2.2m. Figure 4 represents an image of the device.

ANTHROPOMETER INTERNATIONAL CLASSIFICATION US 1946075 A

This invention relates to improvements in measuring instruments, particularly to an anthropometer, and it is the principal object of my invention to provide an instrument of this type enabling a clothing designer or tailor to furnish well-fitting clothes by enabling him to accurately measure abnormal or irregular body formations.

Another object of my invention is the provision of a measuring instrument of comparatively simple and inexpensive construction, yet durable and highly efficient in its operation.

A further object of my invention is the provision of a tailors measure including tape, strap, blade, rule measures, and gages allowing an accurate measuring by salesmen for instance for mail order houses etc., so that alterations in ready to wear clothes are avoided altogether or restricted to a minimum.

A still further object of my invention is the provision of a measuring instrument indicating the degree of shoulder slope and back and chest measures deviating from normal.

These and other objects and advantages of my invention will appear as the description thereof proceeds, and will then be specifically defined in the appended claim (Attorney, 1934). Figure 5 presents an image of the device.

AUTOMATIC ANTHROPOMETER INTERNATIONAL CLASSIFICATION US 8322043 B2

An automatic anthropometer for automatically measuring human height is provided. The automatic anthropometer includes a main post configured to stand vertically to a foot board and have a guiding rail formed integrally with an inner surface in a length direction, a sliding post configured to include a guide wheel that rotates along the guiding rail, and ascend and descend along the length direction of the main post under the guide of the guiding rail to measure height of a user, a head-touch bar configured to be coupled to the sliding post, and a lifting-driving device configured to raise and lower the sliding post (Fraval & Godfrey, 2013). Figure 5 presents an image of the device.

DIAGNOSTIC NETWORK PROCESSING SYSTEM, BILL COLLECTING SYSTEM, AND PHYSICAL DISTRIBUTION SYSTEM: INTERNATIONAL CLASSIFICATION 2002-222267

This diagnostic network processing system 10 comprises a server 3 provided in an information processing center 2, a computer 5 provided in a satellite 4, and an anthropometer 13 of the salon 4. The anthropometer 13 can measure a state of a diagnosis object member 6. The computer 5 sends measurement data of the anthropometer 13 and interview data, and receives and displays a diagnostic result to the diagnosis object member 6. The server 3 determines a diagnostic organization 8 based on each data from the computer 5, sends each data to a computer 9 of the diagnostic organization 8, and transmits diagnostic result data of the diagnostic organization 8 to the computer 5 (Kasuo & Kagekuni, 2002a, 2002b)

Figure 5. Anthropometer
Source: Google Patent, 2017.

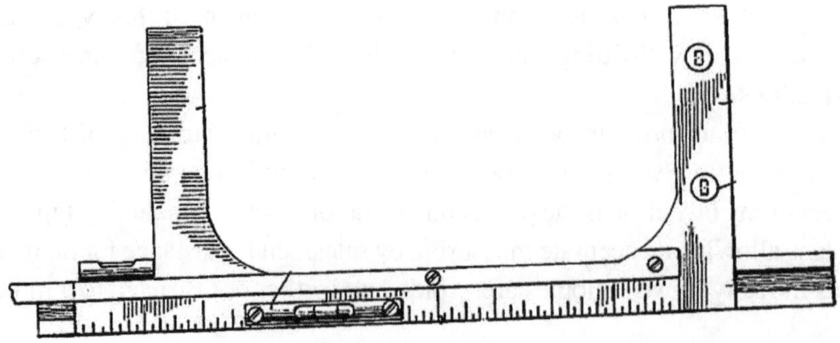

Figure 6. Automatic Anthropometer
Source: Google Patent, 2017.

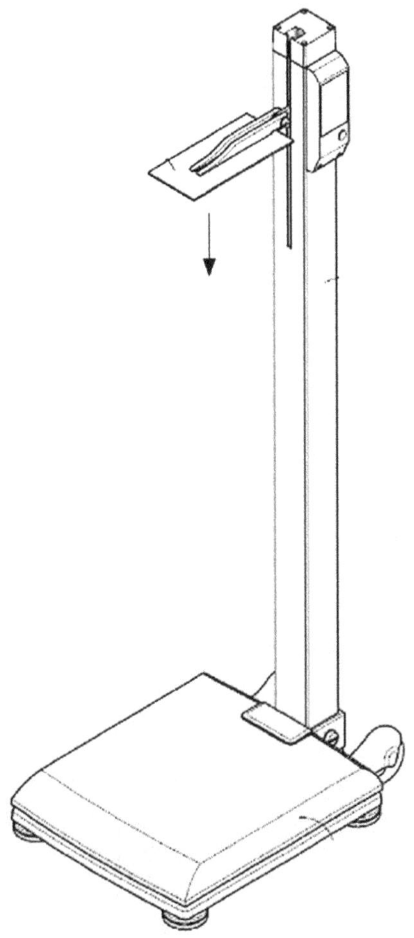

MOTOR DRIVEN ANTHROPOMETER INTERNATIONAL CLASSIFICATION G01B 11/02

To measure stature by one measuring a person and to perform processing corresponding to statistics by providing a measurement pole with a preset band, a measurement band, one preset position reading photosensor, and two displacement reading optical sensors.

CONSTITUTION: A cursor 14 is fitted through a motor-driven driving roller 16 to the measurement pole 13 stood on a foot plate 12. When the floating contact plate 20 of this cursor 14 abuts on the head of a person to be measured and moves against the energizing force of a spring 2, a limit switch 24 is turned on. Further, the cursor 14 is provided with the preset position reading optical sensor 30 corresponding to the preset belt 29 which has a reflection part 26 formed at a specific position on the foot plate 12 and two displacement reading optical sensors 34 and 35 which is formed of black painted parts 31 for reflecting no light and metallic ground parts 32 for reflecting light alternately and reads the quantity and direction of the displacement of the cursor 14 corresponding to the measurement belt 33.

Stature data is measured with the outputs of those sensors, and this data is read in a computer and also displayed (Takao, Yoshiyuki, Toshio, & Katsuhiko, 1986). Figure 7 presents an image of the device.

DEVICE TO MADE ANTHROPOMETRIC STUDIES INTERNATIONAL CLASSIFICATION A61B5/107

This invention only considers static measures. The device is integrated by 1.- Measure module in sitting position formed by a seat, feet support a moving wall with a set of rules for measuring heights and ranges. 2.- Measurement module in a standing position (Stright)formed by a mobile structure to take heights and by a section to take measurements horizontally (widths, lengths, and reaches). To take bodily measurements, the invention uses the basic principle of graduated rules and cursors, incorporating the novelty of including distance sensors. These sensors, placed in the rules and moving elements of the device send the data to display connected to control, which shows the value of the measurement (in

Figure 7. Motor Driven Anthropometer
Source: Japan Platform for Patent Information, 2017.

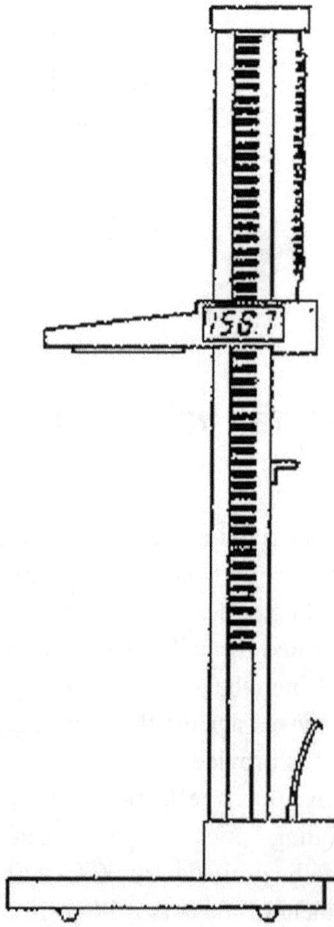

millimeters). The controller controls the movement (vertical and horizontal) of the rules and elements of the device (L. Uribe, n.d.; P. Uribe, 2008)

Stage 3: Generation of a matrix with similarities of the devices with the aim to develop a comparison among the devices. Table 3 describes similitudes between the devices.

Whit the aim to determine advantages, it was necessary to create a table to describe the invention cause and their advantages (see Table 4).

FUTURE RESEARCH DIRECTIONS

During the development of the present research, results have been expressed under specific restrictions. These restrictions were the number minimum of positions considering to be measured by the advice, the equipment with digital or analog components, with software or without software and automatized. Due that the syntaxes were too specific, the results involved just some of many devices that are used in anthropometry studies. In fact, the results just considered the devices with the patent register and not all the existents devices on the internet without a patent.

For future research, it is recommended to increase the field increasing the period of time that the syntaxes considered, the use of the device, the set of mechanism that is used as a complement of the device, the specific measure that it could determine, and the automatization of the device.

Table 3. Comparative of the interesting characteristics among the devices

Device	Characteristics of the Device				
	Determine Anthropometric Measures	Need Supports for the Components	Use Image Processor Images	Create a Database of the Measures	Have Tutorial Software
X-ray Photograph Anthropometer	Yes	Yes	No	No	No
Pennies Anthropometer	No	No	No	No	No
Portable Anthropometer	Yes	Yes	No	No	No
Anthropometer	Yes	Yes	No	No	No
Anthropometer	Yes	Yes	No	No	No
Scholar Anthropometer	Yes	Yes	No	No	No
Automatic Anthropometer	Yes	Yes	No	No	No
Diagnostic Network Processing System, Bill Collecting System, and Physical Distribution System	Yes	No	No	No	No
Moto-Driven Anthropometer	Yes	Yes	No	No	No
Device to made anthropometric studies	Yes	Yes	No	Yes	No
Anthropometer	Yes	Yes	No	No	No

Table 4. Comparative of functions and advantages in anthropometer

Device	Type	Invention Cause	Advantages
X-Ray Photography Anthropometric Instrument	Medical	Difficulty in gets precise measures, especially in children.	• Measures of different parts of the body. • Fast • Easy Operation. • Range of 360º
Pennies Anthropometer	Medical	Determine the anthropometry of the pennies in Brazilian children and teenagers	Straight Rule graduated in millimeters.
Portable Anthropometer	General Application	Determine the High of person.	• Wireless measure. • Deployment of high in the display. • Operation keyboard.
Anthropometer (1)	Industrial	Increase precision measures.	• Reduce register mistakes. • Deployment of measures in the display.
Anthropometer (2)	Medical	*Unknown*	• Measure adequate for medical areas. • Seating height measurement • Measurement of weight
Scholar Anthropometer	Scholar	Improves on scholar measures	• Tangential head stop. • Measurement of standing height-seating, chest, and fathoms. • Greater accuracy, • Measurement of weight.
Automatic Anthropometer	General Application	Measure automatically human height	• Automatic measurement. • Deployment of high in the display.
Diagnostic Network Processing System, Bill Collecting System and Physical Distribution System	General Application	Streamline measurement through a diagnostic processing system.	• TIC's adaptation. • Fast.
Device to made anthropometric studies	General Application	Static anthropometric studies.	• Measure straight-seat. • Distance sensors. • Control Knob • Data base • Variable Armable
Anthropometer	Bicycles	Measures for bike selection.	Horizontal measurement for the crotch of the subject.

With the advance of technology, the possibility of developing new anthropometric devices probably will include the analysis internal and external using invasive and not invasive methods. These concepts should open a new field of research. Finally, is important to emphasize that this chapter only describes a search focus in specific anthropometric devices with the aim to compare their characteristics.

CONCLUSION

At the beginning of this project, there was insufficient information on the different types of anthropometric measurement instruments that exist around the world. One of the questions formulated at the beginning of this research was focused on the similarity of these instruments to determine specific anthropometric measures. Based on the results obtained by the Patent Technical Information request and according to the patent registers regarding anthropometric measurement instruments, it could be summarized that none of the patents found obtains all the anthropometric measures data under the patent application and conditions required. Although all anthropometers listed in Table 2 fulfill their main function of obtaining anthropometric dimensions, most of them do so through simple mechanisms (current standard) installed or tools, which together with the instrument itself, allow sizing the subject. Some of the instruments are usually suitable for a short range of measurements (usually standing-seated measurement), while others focus on specific measurements, such as the Penean or Lodorsimeter anthropometers.

Among the instruments found, highlight those whose measurement capabilities go beyond tools and mechanical supports using the technology factor. Instruments such as the Automatic Anthropometer - with automatic body measurements -, the Portable Anthropometer - with wireless measurements - DNPS - adapting different TICs - in which the Lodorsimeter can be included again, offer variety and agility in anthropometric data collection. In addition, most of these display the results on a screen, making the work even easier for the user. The Device for Conducting Anthropometric Studies could be the only anthropometer that resembles the characteristics of the general anthropometer since it also creates databases and has distance sensors to obtain measurements; however, it lacks software for the determination of anthropometric data and the use of computer equipment for image capture.

Finally, after the comparison of the devices found, it is important to highlight that any device includes the principal characteristics of anthropometer device for general proposes. These should be at least the most common measures of a general position in groups of stand up and seat, the way to register the data in electronic databases, the way to be portable and easy to use and the reliability of the data with minimum measures.

ACKNOWLEDGMENT

This research was supported by Universidad Autónoma de Baja California.

REFERENCES

Altinkemer, K., Ozcelik, Y., & Ozdemir, Z. (2011). Productivity and performance effects of business process reengineering; a firm level analysis. *Journal of Management Information Systems*, *27*(4), 129–162. doi:10.2753/MIS0742-1222270405

Attorney, B. Y. (1934). *Automatic Anthropometer*. Retrieved from https://www.google.com/patents/US1946075

Bitner, M., & Sierak, J. (2017). Institute of Economic Research Working Papers Kristóf Gyódi. In *9th International Conference on Applied Economics Contemporary Issues in Economy, Institute of Economic Research, Polish Economic Society Branch in Torun, Faculty of Economic Sciences and Management, Nicolaus Copernicus University* (pp. 22–23). Torun, Poland: Institute of Economic Research.

Brown, N., & Scurr, J. (2012). The need for a standardized anthropometric protocol for objective assessment of pre-and postoperative breast surgery. *Gland Surgery*, *1*(27), 1–4. doi:10.3978/j.issn.2227-684X.2012.10.01 PMID:25083437

Calderoni, D. R., Motta, M. M., & Kharmandayan, P. (2016). Development and implementation of an anthropometric protocol to evaluate results of otoplasty. *Revista Brasileira de Cirurgia Plástica (RBCP) –. British Journal of Plastic Surgery, 31*(1), 66–73. doi:10.5935/2177-1235.2016RBCP0010

Cavassa, D. (2006). *Ergonomía y Productividad*. LIMUSA.

Corkery, J., Mikalsen, M., & Allan, K. (2017). *Corporate social responsibility: The good corporation*. Centre for Commercial Law.

Cruz, A., & Garnica, A. (2007). Ergonomía. *Evaluación E Investigación, 1*(2), 96–97. Retrieved from http://www.saber.ula.ve/bitstream/123456789/24086/2/resena2.pdf

Fraval, J. T., & Godfrey, M. T. (2013). United States Patent, 1(12), 5. 10.1126/science.Liquids

Gatewood, R., Feild, H. S., & Barrick, M. (2015). *Human resource selection. Nelson Education*. Nelson Education.

Groover, M. (2007). Fundamentos de Manufactura moderna: materiales, procesos y sistemas. Pearson Prentice-Hall.

Gutierrez, A. (1945). *Un aparato antropometro escolar*. Retrieved from http://lp.espacenet.com/publicationDetails/originalDocument?CC=ES&NR=170025A2&KC=A2&FT=D&ND=4&date=19450701&DB=lp.espacenet.com&locale=es_LP

Gutiérrez, R. (1945). *Un Aparato Antropometro Escolar*. Academic Press.

Helio de Oliveira, S. (1987). *Antropometro Peniano*. Retrieved from http://lp.espacenet.com/publicationDetails/originalDocument?locale=en_LP&CC=BR&NR=8506503A&ND=4&KC=A&DB=lp.espacenet.com&FT=D&date=19870630

Kasuo, S., & Kagekuni, A. (2002a). Diagnostic Network Processing System, Bill Collecting System and Physical Distribution System. *Japan Platform for Patent Information*. Retrieved from https://www19.j-platpat.inpit.go.jp/PA1/cgi-bin/PA1DETAIL?MaxCount=1000&PageCount=1000&SearchType=0&TempName=wuZalqa&MaxPage=1&DispPage=1+1000&HitCount=1&ResultId=I00727005602&CookieId=2&DetailPage=1&Language=ENG&Reserve1=DetailPaging&Reserve2=do8yRaE58_c

Kasuo, S., & Kagekuni, A. (2002b). *Diagnostic Network Processing System, Bill Collecting System and Physical Distribution System*. Japon.

Kavanagh, M., & Johnson, R. (2017). *Human resource information systems: Basics, applications, and future directions*. S. Publications.

Oliveira, S. (1987). *Antropómetro Peniano*. Brasil.

Pheasant, S., & Haslegrave, C. (2005). *Bodyspace: Anthropometry*. Ergonomics and the Design of Work.

Ramirez, C. (2006). Ergonomía y Productividad. LIMUSA.

Salvendy, G. (2006). Handbook of human Factors and Ergonomics. Academic Press. doi:10.1002/0470048204

Shaoqing, L. (1999). *Tao*. China: X-Ray Photograph Anthropometer.

Shikdar, A., & Sawaqed, N. (2003). Worker productivity, and occupational health and safety issues in selected industries. *Computers & Industrial Engineering*, *45*(4), 563–572. doi:10.1016/S0360-8352(03)00074-3

Stanton, N., Hedge, A., Brookhuis, K., Salas, E., & Hendrick, H. (2006). *Handbook of Human Factors and Ergonomics Methods*. Washington, DC: CRC.

Takao, I., Yoshiyuki, K., Toshio, M., & Katsuhiko, H. (1986). *Motor-Driven Anthropometer*. Japón.

Uribe, L. (n.d.). *Dispositivo para realizar estudios antropometricos*. Retrieved from http://siga.impi.gob.mx/newSIGA/content/common/busquedaSimple.jsf

Uribe, P. (2008). Dispositivo para realizar estudios antropométricos.

Wheelen, T., & Hunger, J. (2011). *Concepts in strategic management and business policy*. Pearson Education India.

World Economic Forum. (2011). *The global competitiveness report 2011-2012*. 10.4102ajip.v40i1.1117

X-ray photograph anthropometer. (1999). Retrieved from https://www.google.com/patents/CN2347543Y?cl=en

Chapter 11

An Ergonomic Analysis on Working Postures of Construction Site Workers:

A Framework for Construction Site Workers

Suchismita Satapathy
KIIT University, India

ABSTRACT

Risk factors related to work activity and ergonomics can enhance the probability that some persons may develop a MSD (musculoskeletal disorder). Usually the MSD develops due to high task repetition, forceful exertion, and repetitive/sustained awkward postures. MSD is also found in some cases where workers are engaged with working in awkward postures, cold temperatures, contact stress, heavy load, static postures, and vibration, etc. Many studies explain the problems for MSD, but in this chapter an effort is taken to rank the maximum body movement and body parts as per the different types of work flow system such that ergonomics design can be planned. Physical disorders can be avoided by finding and ranking the difficult task and the affected body part due to that kind of work. The prioritization of task will help the organization to think of sustainable designs of working procedure or instruments or machines to provide maximum comfort to humankind. It may also help to frame policies for occupational safety and hazards in workplace.

INTRODUCTION

Rapid development in industrialization methods and technology demands higher production rates. Modern technology increases the demand for products and makes industries competitive. To survive in the business human race has to deal with sophisticated machines and processes. This interaction between human and machine interaction or ergonomics uses the principle of natural laws of work to solve anatomical & physiological problems between man & machine. Many companies like construction, manufacturing,

DOI: 10.4018/978-1-5225-5234-5.ch011

Copyright © 2018, IGI Global. Copying or distributing in print or electronic forms without written permission of IGI Global is prohibited.

mining demand more human interactions. Basically, construction companies play a vital role and act as the building blocks for the development of a country. Improper management in the workplace will reduce the productivity of construction company and ergonomic performance of various workers. To study the ergonomic performance of workers in heavy working conditions like construction company a questionnaire-based survey is conducted. The send questionnaires are classified under the following three primary classifications or factors: (a) human/labor related factors; (b) task-related factors; and (c) equipment/tools-related factors based on various age groups and occupations of the construction workers by literature review. Industry practitioners and researchers can use the primary outcomes of this study in developing systems to enhance and improve health and safety of the construction workers for effective management of construction labor workforce and to achieve a competitive level of quality and a cost-effective project. Again, the body parts and various body movements of Carpenter, Plumber, Welder & Masons anthropometric analysis is done to improve worker protection and avoid ergonomic risk factor. Some important physical motions like stress, strain, and overexertion, including vibration, awkward postures, forceful exertions, repetitive motion, and heavy lifting etc are conducted and workers are ranked by studying their body parts and body movement with multicriteria decision-making techniques like fuzzy AHP and a framework is designed by QFD.

LITERATURE REVIEW

Ergonomic design can be achieved but its evaluation can be varied. This varied needs and results can be further experimented and scrutinized in a very controlled environment. But the very big problem is also the relationship between eco-friendly effectiveness and restriction of autonomous values. To carry out a lot of experiments is also difficult. But with the use of new supercomputers and by running direct diagnosis one can find results by running computational experiments with a virtual set up. The basic objective of ergonomics is to fit man and machine together to improve the worker's performance, reduce stresses and fatigue at work. Application of ergonomics is very significant in an area where manual activities directly affect the physical and mental health of the employee (Parkes et al., 2005). Kushwaha and Kane (2016) have discussed that intervention of ergonomics in the workplace reduces the mismatch between man and machine and makes workplace comfortable for work. Ramos et.al.(2016) have developed a model to perform a financial and economic cost-benefit analysis related to OSH projects was developed and it was applied in the case of the prevention of WMSDs in a Portuguese hospital. An analysis of the accidents and corresponding costs has been made in six of the services of the hospital. Financial and an economic cost-benefit analysis have been made and the benefit-cost ratio (B/C) has been calculated. Charles et.al. (2012) have suggested that a brief hazard perception test could potentially be used in the assessment of drivers across the adult lifespan. Ergonomics is not only analyzed for construction, industries but also it is useful for all types of sectors.(i.e.education,health sector etc).Many modifications are also done in the workplace, work technique/procedures, equipment for giving the maximum comfort of humankind. Assunção et.al. (2013) have examined the effect on the prevalence of back pain of the mismatch between school furniture dimensions, backpack weight and the anthropometric characteristics of 138 adolescents (1215 years old) with differing maturity levels. These results highlight the importance to study the school environment to establish preventive programmes for back pain in youths.

Habidin et.al. have purposed to explore lean healthcare practice adopted by Malaysian healthcare industry and the relationship between lean healthcare practices and supply chain innovation performance.

Mishra et. al. (2015) have developed a framework and provide empirical evidence to support the conceptual and descriptive statements in the literature regarding the impact of human dimensions (HDs) to improve supply chain management (SCM) performance. Lyonnet et. Al. (2016) have developed a new method based on a multi-criteria approach to identify critical resources. Eight prioritization criteria involved in the company's global performance are proposed. The originality of our method is to take into account both material and human resources of the company.

Sylla et al. (2014) have surveyed PSA Peugeot Citroen factories and figured out that workstation requires high precision which results in manual stress. Liu and Li (2014) have studied the difference which is in between conventional and digital main control rooms in nuclear power plants. Factors were listed as per the task components and complexity dimensions. The study was done on three levels, namely task components, complexity dimensions and individual factors. This study can be used in improving the working standard of people working in nuclear power plant. Mohanty and Mahapatra (2014) have developed ergonomically designed office chair which will improve the comfort level of customer. Ray et al. (2015) have studied about Manual Material Handling tasks like lifting and carrying materials like bricks and heavy jack pipes at a construction site in India. They identified different risk factors and measures to prevent, eliminate and minimize their effects. Dul and Neumann (2005) have come up with a new strategy that ergonomics in addition to occupational health & safety, can also be used to evaluate business performance. Narayanan et al. (2013) have stated that Ergonomics is an important tool for a textile industry. Parida and Ray (2015) have discovered various problems in construction industry sectors like poor management of workforce, working methods, tools, and improper working conditions. Openshaw and Taylor (2006) have written ergonomics is the main factor considered while designing product. Lindroos (2009) has explained that when people deviate from their respective course of work, it results in accidents. Also, sometimes deviation leads to less productivity. Det et al. (2009) have analyzed the ergonomics challenges in minimally invasive surgery (MIS) which produced both mental and physical challenges for operating team.

However, there will be always domains and fields where people will have to perform a certain level of dirty, unpleasant work. Chedi (2015) has discussed the implementation of Ergonomics in Technical and Vocational Education and Training (TVET). Rundcrantz et al. (1991) have compared dentists with and without occupational cervicobrachial disorders with respect to mobility of shoulder & neck and posture of shoulder muscles. Ayhan (2013) has used FUZZY AHP in a Gear motor company for selecting the best supplier. Rouyendegh and Erkan (2012) have used Fuzzy Analytic Hierarchy Process for selecting academic staff by evaluating the suitability of the candidate. Garnaik et al. (2016) have implemented QFD & AHP for optimizing product design of furniture for Pankaj Enterprises, Odisha, India. Only human factor sciences can improve the work environment for such individuals. All these basic processes though driven by factors from science and technology will always have an impact on peoples' lives. Its impact on culture and civilization will come at par with the internet's and mobile telecommunication's impact. Through development ergonomics always has to consider the human side of the communication with machines and other technological artifacts.

RESEARCH METHODOLOGY

To avoid MSDs for workers like carpenters, masons, welders & plumbers in construction sites and for providing occupational safety in the workplace, many theories found in literatures,but still this problem remains un resolved. To find most adverse repetitive work and awkward postures which creates MSD in workers/labors, a survey conducted for construction site works categorically carpenters, masons, welders & plumbers. Their body movements studied and noted and different body parts and various body movements are compared with each other by expert analysis. Then, eventually with the help of FUZZY AHP different body parts and various body movements ranked to find which movement is maximum and the part of the body suffered due to frequent movement. Such that some technology must be developed to help workers in completion of heavy work and occupational hazards can be avoided. Body movements can be categorized to Abduction, Adduction, Circumduction, Dorsiflexion, Eversion, Flexion, Hyperextension, Inversion, Plantarflexion, Pronation, Rotation, Supination as per literature review. To study and analyze body movements of construction site workers a standard questionnaire is designed and distributed among 282 workers(i.e., Carpenter, Mason, Plumber, Welder). By personal contact, the questionnaire is explained to them and their views are collected. It is observed that Body parts involved in construction work are the ankle, arm, eye, finger, foot, knee, knuckle, leg, palm, shoulder, spine, toe, and wrist. The survey conducted particularly for 282 workers (i.e, Carpenter, Mason, Plumber, Welder) in different building construction sites of Bhubaneswar and nearby apartment construction sites among them 207 data are collected. The survey found for workers of age 20 to 40, including women workers. They worked for 8 hrs and some of them doing overtime also. To improve the ergonomic performance of workers, the ergonomics value measured for 30 factors and total performance is divided into three major categories type of workers, as per body movement and body parts. As different types of tasks like repetitive work, continuous work like brick, laying, bending, squatting, standing, stretching, Twisting, Kneeling, Stooping, Suddenly Changing Position even loading and unloading etc. occured in work place So, their body movement is categorized by an anatomical term as follows.

- **Abduction:** It is a type of movement which draws a limb away from the median plane of the body like Swinging the hands from the side of the body up to the shoulder
- **Adduction:** It is the movement of a body part toward the body's midline. A person has their arms straight out at the shoulders and brings them down to their sides, it is adduction.
- **Circumduction:** This is the Joint Movements section about the types of movements between bones at joints in the body. Circumduction is a conical movement of a limb extending from the join.
- **Dorsiflexion:** It is the Backward bending of hand or foot.
- **Eversion:** Motion denoting the movement of the sole of the foot away from the median plane.
- **Flexion:** It describes a bending movement that decreases the angle between a segment and its proximal segment. For example, bending the elbow, or clenching a hand into a fist, are examples of flexion.
- **Hyperextension:** The extension of a limb or other body part beyond its normal range of movement.
- **Inversion:** Inversion is a movement in which the plantar surface (sole) of the foot rotates towards the mid-line of the body.
- **Planer Flexion:** It refers to extension or flexion of the foot at the ankle.
- **Pronation:** It is the way the foot rolls inward when you walk and run.

- **Rotation:** Rotation of body parts is referred to as internal or external, referring to rotation towards or away from the center of the body.
- **Supination:** It is movement in which the forearm occurs when the forearm or palm are rotated outwards. Supination of the foot refers to turning of the sole of the foot inwards.

Then Fuzzy AHP is implemented to Prioritized the workers as per their body parts movement.

The Saaty Scale is shown in Table 1. According to this method, the criteria were written in columns and in rows too. Then, the comparison was started between each criterion among other criteria. If the column criterion had greater importance than the corresponding row criterion, then the reciprocal of the Saaty Scale was assigned to it. And if the row criterion was of greater importance than the column criterion, then the normal scale was used. For example, in our case, in Table 1, Carpenter had lower importance than Mason, therefore, Mason was given 1/7, 1/8, 1/9 with respect to Carpenter as Mason was in the column and Carpenter was in a row. Similarly, when Mason was in a row and Carpenter was in column, Mason was given, 7, 8, 9 with respect to Carpenter. Let these fuzzy triangular values be known as d_{ij} (where i = no. of rows and j = no. of columns). Table 1 shows the Saaty Scale - Linguistic terms and the corresponding triangular fuzzy numbers.

The comparison was done between the criteria or alternatives via linguistic terms and Saaty Scale by Decision Maker as shown in Table 2-10. Table 2 Shows the fuzzy input for different workers.

Table 3 shows carpenter body movement with the motions of their body in construction site

Table 4 shows carpenter body movement of their body in construction sites.

Table 5 shows Mason body movement with the motions of their body in construction sites.

Table 1. Saaty Scale-Linguistic terms and the corresponding triangular fuzzy numbers

Saaty Scale	Definition	Fuzzy Triangular Scale
1	Equally Important	(1, 1, 1)
3	Weakly Important	(2, 3, 4)
5	Fairly Important	(4, 5, 6)
7	Strongly Important	(6, 7, 8)
9	Absolutely Important	(9, 9, 9)
2	The Intermittent values between adjacent scales	(1, 2, 3)
4		(3, 4, 5)
6		(5, 6, 7)
8		(7, 8, 9)

Table 2. Input of construction site workers

	Carpenter	Mason	Plumber	Welder
Carpenter	1, 1, 1	1/7, 1/8, 1/9	1/4, 1/5, 1/6	9, 9, 9
Mason	7, 8, 9	1, 1, 1	9, 9, 9	6, 7, 8
Plumber	4, 5, 6	1/9, 1/9, 1/9	1, 1, 1	1, 1/2, 1/3
Welder	1/9, 1/9, 1/9	1/6, 1/7, 1/8	1, 2, 3	1, 1, 1

Table 3. Carpenter body movements

Movements	Abduction	Adduction	Circumduction	Dorsiflexion	Eversion	Extension	Flexion	Hyperextension	Inversion	Plantar Flexion	Pronation	Rotation	Supination
Abduction	1, 1, 1	7, 8, 9	9, 9, 9	4, 5, 6	5, 6, 7	1/9, 1/9, 1/9	1, 1, 1	5, 6, 7	7, 8, 9	6, 7, 8	1, 1, 1	5, 6, 7	1, 1, 1
Adduction	1/7, 1/8, 1/9	1, 1, 1	6, 7, 8	7, 8, 9	5, 6, 7	1/7, 1/8, 1/9	1, 1, 1	1/4, 1/5, 1/6	1, 2, 3	5, 6, 7	1, 1, 1	7, 8, 9	1, 1, 1
Circumduction	1/9, 1/9, 1/9	1/6, 1/7, 1/8	1, 1, 1	1/4, 1/5, 1/6	2, 3, 4	1/5, 1/6, 1/7	1/5, 1/6, 1/7	1/9, 1/9, 1/9	3, 4, 5	1/4, 1/5, 1/6	1/6, 1/7, 1/8	1/6, 1/7, 1/8	1/7, 1/8, 1/9
Dorsiflexion	1/4, 1/5, 1/6	1/7, 1/8, 1/9	4, 5, 6	1, 1, 1	1, 1, 1	1/6, 1/7, 1/8	1/7, 1/8, 1/9	1/7, 1/8, 1/9	1, 1, 1	1, 1, 1	1/4, 1/5, 1/6	2, 3, 4	1/5, 1/6, 1/7
Eversion	1/5, 1/6, 1/7	1/5, 1/6, 1/7	1/2, 1/3, 1/4	1, 1, 1	1, 1, 1	1, 2, 3	2, 3, 4	1, 1/2, 1/3	1, 1, 1	1, 2, 3	1/5, 1/6, 1/7	1, 1/2, 1/3	1/6, 1/7, 1/8
Extension	9, 9, 9	7, 8, 9	5, 6, 7	6, 7, 8	1, 1/2, 1/3	1, 1, 1	1, 2, 3	1, 1/2, 1/3	1/2, 1/3, 1/4	7, 8, 9	1/6, 1/7, 1/8	4, 5, 6	1/6, 1/7, 1/8
Flexion	1, 1, 1	1, 1, 1	5, 6, 7	7, 8, 9	1/2, 1/3, 1/4	1, 1/2, 1/3	1, 1, 1	1, 1, 1	1/3, 1/4, 1/5	6, 7, 8	3, 4, 5	7, 8, 9	4, 5, 6
Hyperextension	1/5, 1/6, 1/7	4, 5, 6	9, 9, 9	7, 8, 9	1, 2, 3	1, 2, 3	1, 1, 1	1, 1, 1	6, 7, 8	7, 8, 9	1/5, 1/6, 1/7	2, 3, 4	1/5, 1/6, 1/7
Inversion	1/7, 1/8, 1/9	1, 1/2, 1/3	1/3, 1/4, 1/5	1, 1, 1	1, 1, 1	2, 3, 4	3, 4, 5	1/6, 1/7, 1/8	1, 1, 1	2, 3, 4	1/6, 1/7, 1/8	5, 6, 7	1/7, 1/8, 1/9
Plantar Flexion	1/6, 1/7, 1/8	1/5, 1/6, 1/7	4, 5, 6	1, 1, 1	1, 1/2, 1/3	1/7, 1/8, 1/9	1/6, 1/7, 1/8	1/7, 1/8, 1/9	1/2, 1/3, 1/4	1, 1, 1	1/7, 1/8, 1/9	1/2, 1/3, 1/4	1/7, 1/8, 1/9
Pronation	1, 1, 1	1, 1, 1	6, 7, 8	4, 5, 6	5, 6, 7	6, 7, 8	1/3, 1/4, 1/5	5, 6, 7	6, 7, 8	7, 8, 9	1, 1, 1	7, 8, 9	1, 1, 1
Rotation	1/5, 1/6, 1/7	1/7, 1/8, 1/9	6, 7, 8	1/2, 1/3, 1/4	1, 2, 3	1/4, 1/5, 1/6	1/7, 1/8, 1/9	1/2, 1/3, 1/4	1/5, 1/6, 1/7	2, 3, 4	1, 1, 1	1, 1, 1	1/9, 1/9, 1/9
Supination	1, 1, 1	1, 1, 1	7, 8, 9	5, 6, 7	6, 7, 8	6, 7, 8	1/4, 1/5, 1/6	5, 6, 7	7, 8, 9	7, 8, 9	1, 1, 1	9, 9, 9	1, 1, 1

Table 4. Carpenter body parts

Body Part	Ankle	Arm	Eye	Finger	Foot	Knee	Knuckle	Leg	Palm	Shoulder	Spine	Toe	Wrist
Ankle	1, 1, 1	1/7, 1/8, 1/9	5, 6, 7	1/6, 1/7, 1/8	4, 5, 6	1, 1, 1	1/9, 1/9, 1/9	1/6, 1/7, 1/8	1/7, 1/8, 1/9	1/9, 1/9, 1/9	1/7, 1/8, 1/9	9, 9, 9	1/5, 1/6, 1/7
Arm	7, 8, 9	1, 1, 1	6, 7, 8	1, 1, 1	4, 5, 6	5, 6, 7	2, 3, 4	3, 4, 5	1, 1, 1	4, 5, 6	1/6, 1/7, 1/8	7, 8, 9	1/5, 1/6, 1/7
Eye	1/5, 1/6, 1/7	1/6, 1/7, 1/8	1, 1, 1	1/4, 1/5, 1/6	1/2, 1/3, 1/4	1/4, 1/5, 1/6	1/3, 1/4, 1/5	1/3, 1/4, 1/5	1/5, 1/6, 1/7	1/9, 1/9, 1/9	1/7, 1/8, 1/9	1/2, 1/3, 1/4	1, 2, 3
Finger	6, 7, 8	1, 1, 1	4, 5, 6	1, 1, 1	6, 7, 8	7, 8, 9	1, 1, 1	4, 5, 6	2, 3, 4	1/5, 1/6, 1/7	1, 1, 1	9, 9, 9	1/5, 1/6, 1/7
Foot	1/4, 1/5, 1/6	1/4, 1/5, 1/6	2, 3, 4	1/6, 1/7, 1/8	1, 1, 1	7, 8, 9	1/5, 1/6, 1/7	1, 1, 1	1/4, 1/5, 1/6	1/6, 1/7, 1/8	1/6, 1/7, 1/8	7, 8, 9	1/2, 1/3, 1/4
Knee	1, 1, 1	1/5, 1/6, 1/7	4, 5, 6	1/7, 1/8, 1/9	1/7, 1/8, 1/9	1, 1, 1	1/6, 1/7, 1/8	1/9, 1/9, 1/9	1/7, 1/8, 1/9	1/4, 1/5, 1/6	1/6, 1/7, 1/8	5, 6, 7	1/3, 1/4, 1/5
Knuckle	9, 9, 9	1/2, 1/3, 1/4	3, 4, 5	1, 1, 1	5, 6, 7	6, 7, 8	1, 1, 1	1/4, 1/5, 1/6	1/5, 1/6, 1/7	1/6, 1/7, 1/8	1/9, 1/9, 1/9	6, 7, 8	1, 1, 1
Leg	6, 7, 8	1/3, 1/4, 1/5	3, 4, 5	1/2, 1/3, 1/4	1, 1, 1	9, 9, 9	4, 5, 6	1, 1, 1	1/6, 1/7, 1/8	1/5, 1/6, 1/7	1, 1, 1	7, 8, 9	3, 4, 5
Palm	7, 8, 9	1, 1, 1	5, 6, 7	5, 6, 7	4, 5, 6	7, 8, 9	5, 6, 7	6, 7, 8	1, 1, 1	1/4, 1/5, 1/6	1, 1, 1	7, 8, 9	1, 1, 1
Shoulder	9, 9, 9	1/4, 1/5, 1/6	9, 9, 9	1, 1, 1	6, 7, 8	4, 5, 6	6, 7, 8	5, 6, 7	4, 5, 6	1, 1, 1	1/9, 1/9, 1/9	6, 7, 8	1, 1, 1
Spine	7,8, 9	7, 8	8, 9	1/9	6,8	6,8	9,9	1,1	1,1	9,9	1, 1	7,9	6, 7,8
Toe	1/9, 1/9, 1/9	1/7, 1/8, 1/9	2, 3, 4	5, 6, 7	1/7, 1/8, 1/9	1/5, 1/6, 1/7	1/6, 1/7, 1/8	1/7, 1/8, 1/9	1/7, 1/8, 1/9	1/6, 1/7, 1/8	1/7, 1/8, 1/9	1, 1, 1	1/6, 1/7, 1/8
Wrist	5, 6, 7	5, 6, 7	1, 1/2, 1/3		2, 3, 4	3, 4, 5	1, 1, 1	1/3, 1/4, 1/5	1, 1, 1	1, 1, 1	1/6, 1/7, 1/8	6, 7, 8	1, 1, 1

Table 5. Mason body movements

Movements	Abduction	Adduction	Circumduction	Dorsiflexion	Eversion	Extension	Flexion	Hyperextension	Inversion	Plantar Flexion	Pronation	Rotation	Supination
Abduction	1, 1, 1	5, 6, 7	1, 1/2, 1/3	1/5, 1/6, 1/7	1/6, 1/7, 1/8	1, 1, 1	1, 1, 1	1/7, 1/8, 1/9	1/6, 1/7, 1/8	1/7, 1/8, 1/9	1/4, 1/5, 1/6	1, 1, 1	1/3, 1/4, 1/5
Adduction	1/5, 1/6, 1/7	1, 1, 1	1/5, 1/6, 1/7	1/7, 1/8, 1/9	1/2, 1/3, 1/4	1, 1, 1	1, 1, 1	1/3, 1/4, 1/5	1/2, 1/3, 1/4	1/5, 1/6, 1/7	1/3, 1/4, 1/5	1/6, 1/7, 1/8	1/5, 1/6, 1/7
Circumduction	1, 2, 3	5, 6, 7	1, 1, 1	1, 1, 1	3, 4, 5	6, 7, 8	6, 7, 8	1, 1, 1	4, 5, 6	5, 6, 7	6, 7, 8	1, 1, 1	1, 1, 1
Dorsiflexion	5, 6, 7	7, 8, 9	1, 1, 1	1, 1, 1	1, 1, 1	5, 6, 7	5, 6, 7	3, 4, 5	1, 1, 1	1, 1, 1	1, 1, 1	1, 1/2, 1/3	1, 1/2, 1/3
Eversion	6, 7, 8	2, 3, 4	1/3, 1/4, 1/5	1, 1, 1	1, 1, 1	6, 7, 8	6, 7, 8	1/6, 1/7, 1/8	1, 1, 1	1, 1, 1	1, 1, 1	1/2, 1/3, 1/4	1/2, 1/3, 1/4
Extension	1, 1, 1	1, 1, 1	1/6, 1/7, 1/8	1/5, 1/6, 1/7	1/6, 1/7, 1/8	1, 1, 1	1, 1, 1	1, 1, 1	1/5, 1/6, 1/7	1, 1, 1	1, 1, 1	1/6, 1/7, 1/8	1/6, 1/7, 1/8
Flexion	1, 1, 1	1, 1, 1	1/6, 1/7, 1/8	1/5, 1/6, 1/7	1/6, 1/7, 1/8	1, 1, 1	1, 1, 1	1/4, 1/5, 1/6	1/5, 1/6, 1/7	1, 1, 1	1, 1, 1	1/5, 1/6, 1/7	1/5, 1/6, 1/7
Hyperextension	7, 8, 9	3, 4, 5	1, 1, 1	1/3, 1/4, 1/5	6, 7, 8	1, 1, 1	4, 5, 6	1, 1, 1	1, 1, 1	5, 6, 7	1, 1, 1	6, 7, 8	6, 7, 8
Inversion	6, 7, 8	2, 3, 4	1/4, 1/5, 1/6	1, 1, 1	1, 1, 1	5, 6, 7	5, 6, 7	1, 1, 1	1, 1, 1	1, 1, 1	1, 1, 1	1/3, 1/4, 1/5	1/3, 1/4, 1/5
Plantar Flexion	7, 8, 9	5, 6, 7	1/5, 1/6, 1/7	1, 1, 1	1, 1, 1	1, 1, 1	1, 1, 1	1/5, 1/6, 1/7	1, 1, 1	1, 1, 1	1, 1, 1	1/5, 1/6, 1/7	1/5, 1/6, 1/7
Pronation	4, 5, 6	3, 4, 5	1/6, 1/7, 1/8	1, 1, 1	1, 1, 1	1, 1, 1	1, 1, 1	1, 1, 1	1, 1, 1	1, 1, 1	1, 1, 1	2, 3, 4	2, 3, 4
Rotation	1, 1, 1	6, 7, 8	1, 1, 1	1, 2, 3	2, 3, 4	6, 7, 8	5, 6, 7	1/6, 1/7, 1/8	3, 4, 5	5, 6, 7	1/2, 1/3, 1/4	1, 1, 1	1/2, 1/3, 1/4
Supination	3, 4, 5	5, 6, 7	1, 1, 1	1, 2, 3	2, 3, 4	6, 7, 8	5, 6, 7	1/6, 1/7, 1/8	3, 4, 5	5, 6, 7	1/2, 1/3, 1/4	2, 3, 4	1, 1, 1

Table 6 shows masons body movement with the motions of their body in construction site.
Table 7 shows plumber's body movement with the motions of their body in construction sites.
Table 8 shows carpenter body movement with the motions of their body in construction sites.
Table 9 shows welder body movement with the motions of their body in construction site
Table 10 shows welder body movement in construction sites.
Then, the fuzzy weights were found out using the following 3 steps,

1. Vector Summation of r_i was done.
2. The Reciprocal of the summation vector was done and then was arranged in increasing order.
3. The fuzzy weights, w_i were found out

Table 11 shows the geometric means, fuzzy weights and averaged and normalized relative weight for construction workers.

Table 12 shows the geometric means, fuzzy weights and averaged and normalized relative weight for carpenter's body parts.

Table 13 shows the geometric means, fuzzy weights and averaged and normalized relative weight for carpenter's body movement.

Table 14 shows the geometric means, fuzzy weights and averaged and normalized relative weight for mason's body parts.

Table 15 shows the geometric means, fuzzy weights and averaged and normalized relative weight for mason's body movement.

Table 16 shows the geometric means, fuzzy weights and averaged and normalized relative weight for plumber's body parts

Table 17 shows the geometric means, fuzzy weights and averaged and normalized relative weight for plumber's body parts.

Table 18 shows the geometric means, fuzzy weights and averaged and normalized relative weight for welder's body parts.

Table 19 shows the geometric means, fuzzy weights and averaged and normalized relative weight for welder's body movement.

Since we were still stuck with fuzzy triangular values, they were then de-fuzzified by Centre of Area method which was proposed by Chou and Chang. These values were called as M_i (where i = 1, 2, 3 … n). This calculation was done using the following equation:

$$M_i = (lw_i, mw_i, nw_i)/3$$

Then, M_i was normalized using the following function as it was a non-fuzzy number,

$$N_i = \frac{Mi}{\sum_{i=1}^{n} Mi} . M\Sigma_{i=1}^{n} M_i \frac{M_i}{\sum_{i=1}^{n} M_i} .$$

Table 6. Mason body parts

Body Part	Ankle	Arm	Eye	Finger	Foot	Knee	Knuckle	Leg	Palm	Shoulder	Spine	Toe	Wrist
Ankle	1, 1, 1	5, 6, 7	6, 7, 8	6, 7, 8	1, 1, 1	1, 1, 1	7, 8, 9	1, 1, 1	5, 6, 7	1/4, 1/5, 1/6	1/9, 1/9, 1/9	1, 1, 1	1, 1, 1
Arm	1/5, 1/6, 1/7	1, 1, 1	1, 1, 1	1, 1, 1	1, 1, 1	1, 1, 1	4, 5, 6	1, 1, 1	6, 7, 8	1/3, 1/4, 1/5	1/9, 1/9, 1/9	4, 5, 6	2, 3, 4
Eye	1/6, 1/7, 1/8	1, 1, 1	1, 1, 1	1/5, 1/6, 1/7	1/4, 1/5, 1/6	1/4, 1/5, 1/6	1, 1, 1	1/5, 1/6, 1/7	1, 1, 1	1/5, 1/6, 1/7	1/9, 1/9, 1/9	1/4, 1/5, 1/6	1/2, 1/3, 1/4
Finger	1/6, 1/7, 1/8	1, 1, 1	5, 6, 7	1, 1, 1	1/6, 1/7, 1/8	1/6, 1/7, 1/8	1, 1, 1	1/7, 1/8, 1/9	1, 1, 1	1/7, 1/8, 1/9	1/9, 1/9, 1/9	1/6, 1/7, 1/8	1, 1, 1
Foot	1, 1, 1	1, 1, 1	4, 5, 6	6, 7, 8	1, 1, 1	1, 1, 1	6, 7, 8	1, 1, 1	5, 6, 7	1/3, 1/4, 1/5	1/9, 1/9, 1/9	1, 1, 1	4, 5, 6
Knee	1, 1, 1	1, 1, 1	4, 5, 6	6, 7, 8	1, 1, 1	1, 1, 1	9, 9, 9	1, 1, 1	6, 7, 8	3, 4, 5	1/9, 1/9, 1/9	1, 1, 1	5, 6, 7
Knuckle	1/7, 1/8, 1/9	1/4, 1/5, 1/6	1, 1, 1	1, 1, 1	1/6, 1/7, 1/8	1/9, 1/9, 1/9	1, 1, 1	1/9, 1/9, 1/9	1, 1, 1	1/6, 1/7, 1/8	1/9, 1/9, 1/9	1/5, 1/6, 1/7	1/2, 1/3, 1/4
Leg	1, 1, 1	1, 1, 1	5, 6, 7	7, 8, 9	1, 1, 1	1, 1, 1	9, 9, 9	1, 1, 1	6, 7, 8	1, 1, 1	1/9, 1/9, 1/9	1, 1, 1	6, 7, 8
Palm	1/5, 1/6, 1/7	1/6, 1/7, 1/8	1, 1, 1	1, 1, 1	1/5, 1/6, 1/7	1/6, 1/7, 1/8	1, 1, 1	1/6, 1/7, 1/8	1, 1, 1	1/9, 1/9, 1/9	1/9, 1/9, 1/9	1/7, 1/8, 1/9	1, 1, 1
Shoulder	4, 5, 6	3, 4, 5	5, 6, 7	7, 8, 9	3, 4, 5	1/3, 1/4, 1/5	6, 7, 8	1, 1, 1	9, 9, 9	1, 1, 1	1/9, 1/9, 1/9	6, 7, 8	7, 8, 9
Spine	9,9, 9	9,9	9,9	9, 9,9	9, 9,9	9, 9,9	9, 9,9	9, 9,9	9, 9,9	9, 9,9	1, 1,1	9, 9,9	9, 9,9
Toe	1, 1, 1	1/4, 1/5, 1/6	4, 5, 6	6, 7, 8	1, 1, 1	1, 1, 1	5, 6, 7	1, 1, 1	7, 8, 9	1/6, 1/7, 1/8	1/9, 1/9, 1/9	1, 1, 1	1, 1, 1
Wrist	1, 1, 1	1/2, 1/3, 1/4	2, 3, 4	1, 1, 1	1/4, 1/5, 1/6	1/5, 1/6, 1/7	2, 3, 4	1/6, 1/7, 1/8	1, 1, 1	1/7, 1/8, 1/9	1/9, 1/9, 1/9	1, 1, 1	1, 1, 1

The values of r_i, w_i, M_i, N_i are shown in Tables 11 - 19. In the last and final step, each alternative was multiplied with respectively related criteria and scores for each case were calculated and the one with the highest score was found to be of greatest importance. These values and the ranks are shown in Tables 20 - 21.

Table 7. Plumber body movements

Movements	Abduction	Adduction	Circumduction	Dorsiflexion	Eversion	Extension	Flexion	Hyperextension	Inversion	Plantar Flexion	Pronation	Rotation	Supination
Abduction	1, 1, 1	3, 4, 5	1/2, 1/3, 1/4	5, 6, 7	6, 7, 8	1/2, 1/3, 1/4	1/2, 1/3, 1/4	1, 1, 1	6, 7, 8	5, 6, 7	1/7, 1/8, 1/9	1/2, 1/3, 1/4	1/7, 1/8, 1/9
Adduction	1/3, 1/4, 1/5	1, 1, 1	1/3, 1/4, 1/5	4, 5, 6	5, 6, 7	1/3, 1/4, 1/5	1/3, 1/4, 1/5	1, 1, 1	5, 6, 7	4, 5, 6	1/7, 1/8, 1/9	1/3, 1/4, 1/5	1/7, 1/8, 1/9
Circumduction	2, 3, 4	3, 4, 5	1, 1, 1	5, 6, 7	6, 7, 8	1/2, 1/3, 1/4	1/2, 1/3, 1/4	1, 1, 1	6, 7, 8	5, 6, 7	1/7, 1/8, 1/9	1/2, 1/3, 1/4	1/7, 1/8, 1/9
Dorsiflexion	1/5, 1/6, 1/7	1/4, 1/5, 1/6	1/5, 1/6, 1/7	1, 1, 1	1, 1/2, 1/3	1/9, 1/9, 1/9	1/9, 1/9, 1/9	1/4, 1/5, 1/6	1, 1/2, 1/3	1, 1, 1	1/9, 1/9, 1/9	1/4, 1/5, 1/6	1/9, 1/9, 1/9
Eversion	1/6, 1/7, 1/8	1/5, 1/6, 1/7	1/6, 1/7, 1/8	1, 2, 3	1, 1, 1	1/9, 1/9, 1/9	1/9, 1/9, 1/9	1/4, 1/5, 1/6	1, 1, 1	1, 2, 3	1/9, 1/9, 1/9	1/4, 1/5, 1/6	1/9, 1/9, 1/9
Extension	2, 3, 4	3, 4, 5	2, 3, 4	9, 9, 9	9, 9, 9	1, 1, 1	1, 1, 1	4, 5, 6	6, 7, 8	1/9, 1/9, 1/9	1/3, 1/4, 1/5	3, 4, 5	1/3, 1/4, 1/5
Flexion	2, 3, 4	3, 4, 5	2, 3, 4	9, 9, 9	9, 9, 9	1, 1, 1	1, 1, 1	4, 5, 6	6, 7, 8	1/9, 1/9, 1/9	1/3, 1/4, 1/5	3, 4, 5	1/3, 1/4, 1/5
Hyperextension	1, 1, 1	1, 1, 1	1, 1, 1	4, 5, 6	4, 5, 6	1/4, 1/5, 1/6	1/4, 1/5, 1/6	1, 1, 1	2, 3, 4	2, 3, 4	1/5, 1/6, 1/7	1, 1/2, 1/3	1/5, 1/6, 1/7
Inversion	1/6, 1/7, 1/8	1/5, 1/6, 1/7	1/6, 1/7, 1/8	1, 2, 3	1, 1, 1	1/6, 1/7, 1/8	1/6, 1/7, 1/8	1/2, 1/3, 1/4	1, 1, 1	1, 2, 3	1/9, 1/9, 1/9	1/4, 1/5, 1/6	1/9, 1/9, 1/9
Plantar Flexion	1/5, 1/6, 1/7	1/4, 1/5, 1/6	1/5, 1/6, 1/7	1, 1, 1	1, 1/2, 1/3	9, 9, 9	9, 9, 9	1/2, 1/3, 1/4	1, 1/2, 1/3	1, 1, 1	1/9, 1/9, 1/9	1/4, 1/5, 1/6	1/9, 1/9, 1/9
Pronation	8, 9	78,9	7, 89	9, 99	9, 99	3, 45	3, 45	5, 67	9, 99	9, 99	1, 11	4, 56	1, 11
Rotation	2, 3, 4	3, 4, 5	2, 3, 4	4, 5, 6	4, 5, 6	1/3, 1/4, 1/5	1/3, 1/4, 1/5	1, 2, 3	4, 5, 6	4, 5, 6	1/4, 1/5, 1/6	1, 1, 1	4, 5, 6
Supination	7, 8, 9	7, 8, 9	7, 8, 9	9, 9, 9	9, 9, 9	3, 4, 5	3, 4, 5	5, 6, 7	9, 9, 9	9, 9, 9	1, 1, 1	1/4, 1/5, 1/6	1, 1, 1

Table 8. Plumber body parts

Body Part	Ankle	Arm	Eye	Finger	Foot	Knee	Knuckle	Leg	Palm	Shoulder	Spine	Toe	Wrist
Ankle	1, 1, 1	1/9, 1/9, 1/9	1/2, 1/3, 1/4	1/9, 1/9, 1/9	1, 1, 1	1/2, 1/3, 1/4	1/9, 1/9, 1/9	1/2, 1/3, 1/4	1/4, 1/5, 1/6	1/6, 1/7, 1/8	1/4, 1/5, 1/6	1, 1, 1	1/6, 1/7, 1/8
Arm	9, 9, 9	1, 1, 1	6, 7, 8	1, 1, 1	9, 9, 9	6, 7, 8	1, 1, 1	6, 7, 8	1, 1, 1	2, 3, 4	2, 3, 4	9, 9, 9	1, 1, 1
Eye	2, 3, 4	1/6, 1/7, 1/8	1, 1, 1	1/6, 1/7, 1/8	3, 4, 5	1, 2, 3	1/5, 1/6, 1/7	3, 4, 5	1/6, 1/7, 1/8	1/6, 1/7, 1/8	1/6, 1/7, 1/8	3, 4, 5	1/7, 1/8, 1/9
Finger	9, 9, 9	1, 1, 1	6, 7, 8	1, 1, 1	9, 9, 9	9, 9, 9	1, 1, 1	5, 6, 7	1, 1, 1	1, 1, 1	4, 5, 6	9, 9, 9	1, 1, 1
Foot	1, 1, 1	1/9, 1/9, 1/9	1/3, 1/4, 1/5	1/9, 1/9, 1/9	1, 1, 1	1, 1/2, 1/3	1/9, 1/9, 1/9	1/2, 1/3, 1/4	1/9, 1/9, 1/9	1/7, 1/8, 1/9	1/5, 1/6, 1/7	1, 1, 1	1/9, 1/9, 1/9
Knee	2, 3, 4	1/6, 1/7, 1/8	1, 1/2, 1/3	1/9, 1/9, 1/9	1, 2, 3	1, 1, 1	1/7, 1/8, 1/9	1, 1, 1	1/7, 1/8, 1/9	1/5, 1/6, 1/7	1/3, 1/4, 1/5	2, 3, 4	1/9, 1/9, 1/9
Knuckle	9, 9, 9	1, 1, 1	5, 6, 7	1, 1, 1	9, 9, 9	7, 8, 9	1, 1, 1	7, 8, 9	1, 1, 1	3, 4, 5	4, 5, 6	9, 9, 9	1, 1, 1
Leg	2, 3, 4	1/6, 1/7, 1/8	1/3, 1/4, 1/5	1/5, 1/6, 1/7	2, 3, 4	1, 1, 1	1/7, 1/8, 1/9	1, 1, 1	1/9, 1/9, 1/9	1/6, 1/7, 1/8	1/4, 1/5, 1/6	1, 2, 3	1/7, 1/8, 1/9
Palm	4, 5, 6	1, 1, 1	6, 7, 8	1, 1, 1	9, 9, 9	7, 8, 9	1, 1, 1	9, 9, 9	1, 1, 1	2, 3, 4	4, 5, 6	9, 9, 9	1, 1/2, 1/3
Shoulder	6, 7, 8	1/2, 1/3, 1/4	6, 7, 8	1, 1, 1	7, 8, 9	5, 6, 7	1/3, 1/4, 1/5	6, 7, 8	1/2, 1/3, 1/4	1, 1, 1	2, 3, 4	5, 6, 7	1/4, 1/5, 1/6
Spine	4, 5, 6	1/2, 1/3, 1/4	6, 7, 8	1/4, 1/5, 1/6	5, 6, 7	3, 4, 5	1/4, 1/5, 1/6	4, 5, 6	1/4, 1/5, 1/6	1/2, 1/3, 1/4	1, 1, 1	3, 4, 5	1/3, 1/4, 1/5
Toe	1, 1, 1	1/9, 1/9, 1/9	1/3, 1/4, 1/5	1/9, 1/9, 1/9	1, 1, 1	1/2, 1/3, 1/4	1/9, 1/9, 1/9	1, 1/2, 1/3	1/9, 1/9, 1/9	1/5, 1/6, 1/7	1/3, 1/4, 1/5	1, 1, 1	9, 9, 9
Wrist	6, 7, 8	1, 1, 1	7, 8, 9	1, 1, 1	9, 9, 9	9, 9, 9	1, 1, 1	7, 8, 9	1, 2, 3	4, 5, 6	3, 4, 5	1/9, 1/9, 1/9	1, 1, 1

RESULT AND DISCUSSION

Though over the years, technology has given many comfort to human being in the workplace due to mechanized or automated technology. Still, the majority of the tasks are still performed manually in different industries.As labors are the backbone of developing industries like India, so their occupational safety is the most important issue. In this research, it is observed from the Table.20 that body parts like spine, shoulder, leg, and knee are most strained body part for carpenters, masons, plumbers, and welders. But among them from Table 20, it was found that for carpenter spine is the more affected parts for Mason's shoulder, for plumber was the shoulder, and for welder was the leg. Table 21 shows that the difficult

Table 9. Welder body movements

Movements	Abduction	Adduction	Circumduction	Dorsiflexion	Eversion	Extension	Flexion	Hyperextension	Inversion	Plantar Flexion	Pronation	Rotation	Supination
Abduction	1,1, 1	1, 1,1	2, 3,4	2, 3,4	4, 5,6	4, 5,6	2, 3,4	9, 9,9	6, 7,8	4, 5,6	6, 7,8	6, 7,8	4, 5,6
Adduction	1,1, 1	1,1, 1	2,3, 4	2,3, 4	4,5, 6	4,5, 6	2,3, 4	9,9, 9	6,7, 8	4,5, 6	6,7, 8	6,7, 8	4, 5,6
Circumduction	1/2, 1/3, 1/4	1/2, 1/3, 1/4	1, 1, 1	6, 7, 8	2, 3, 4	4, 5, 6	6, 7, 8	9, 9, 9	6, 7, 8	6, 7, 8	4, 5, 6	2, 3, 4	4, 5, 6
Dorsiflexion	1/2, 1/3, 1/4	1/2, 1/3, 1/4	1/6, 1/7, 1/8	1, 1, 1	1, 1, 1	1/2, 1/3, 1/4	1/4, 1/5, 1/6	9, 9, 9	1/2, 1/3, 1/4	1/6, 1/7, 1/8	1/4, 1/5, 1/6	1/6, 1/7, 1/8	1/2, 1/3, 1/4
Eversion	1/4, 1/5, 1/6	1/4, 1/5, 1/6	1/2, 1/3, 1/4	1, 1, 1	1, 1, 1	1/4, 1/5, 1/6	1/4, 1/5, 1/6	9, 9, 9	1/6, 1/7, 1/8	1/2, 1/3, 1/4	1/4, 1/5, 1/6	1/2, 1/3, 1/4	1/6, 1/7, 1/8
Extension	1/4, 1/5, 1/6	1/4, 1/5, 1/6	1/4, 1/5, 1/6	2, 3, 4	4, 5, 6	1, 1, 1	6, 7, 8	9, 9, 9	4, 5, 6	2, 3, 4	6, 7, 8	4, 5, 6	2, 3, 4
Flexion	1/2, 1/3, 1/4	1/2, 1/3, 1/4	1/6, 1/7, 1/8	4, 5, 6	4, 5, 6	1/6, 1/7, 1/8	1, 1, 1	9, 9, 9	2, 3, 4	1, 1, 1	4, 5, 6	6, 7, 8	2, 3, 4
Hyperextension	1/9, 1/9, 1/9	1/9, 1/9, 1/9	1/9, 1/9, 1/9	1/9, 1/9, 1/9	1/9, 1/9, 1/9	1/9, 1/9, 1/9	1/9, 1/9, 1/9	1, 1, 1	1/6, 1/7, 1/8	1/4, 1/5, 1/6	1/9, 1/9, 1/9	1/9, 1/9, 1/9	1/9, 1/9, 1/9
Inversion	1/6, 1/7, 1/8	1/6, 1/7, 1/8	1/6, 1/7, 1/8	2, 3, 4	6, 7, 8	1/4, 1/5, 1/6	1/2, 1/3, 1/4	6, 7, 8	1, 1, 1	2, 3, 4	6, 7, 8	4, 5, 6	4, 5, 6
Plantar Flexion	1/4, 1/5, 1/6	1/4, 1/5, 1/6	1/6, 1/7, 1/8	6, 7, 8	2, 3, 4	1/2, 1/3, 1/4	1, 1, 1	4, 5, 6	1/2, 1/3, 1/4	1, 1, 1	4, 5, 6	6, 7, 8	6, 7, 8
Pronation	1/6, 1/7, 1/8	1/6, 1/7, 1/8	1/4, 1/5, 1/6	4, 5, 6	4, 5, 6	1/6, 1/7, 1/8	1/4, 1/5, 1/6	9, 9, 9	1/6, 1/7, 1/8	1/4, 1/5, 1/6	1, 1, 1	1/2, 1/3, 1/4	1/6, 1/7, 1/8
Rotation	1/6, 1/7, 1/8	1/6, 1/7, 1/8	1/2, 1/3, 1/4	6, 7, 8	2, 3, 4	1/4, 1/5, 1/6	1/6, 1/7, 1/8	9, 9, 9	1/4, 1/5, 1/6	1/6, 1/7, 1/8	2, 3, 4	1, 1, 1	6, 7, 8
Supination	1/4, 1/5, 1/6	1/4, 1/5, 1/6	1/4, 1/5, 1/6	2, 3, 4	6, 7, 8	1/2, 1/3, 1/4	1/2, 1/3, 1/4	9, 9, 9	1/4, 1/5, 1/6	1/6, 1/7, 1/8	6, 7, 8	1/6, 1/7, 1/8	1, 1, 1

Table 10. Welder body parts

Body Part	Ankle	Arm	Eye	Finger	Foot	Knee	Knuckle	Leg	Palm	Shoulder	Spine	Toe	Wrist
Ankle	1, 1, 1	9, 9, 9	1/9, 1/9, 1/9	9, 9, 9	1, 1, 1	1, 1, 1	9, 9, 9	1, 1, 1	9, 9, 9	2, 3, 4	4, 5, 6	1, 1, 1	4, 5, 6
Arm	1/9, 1/9, 1/9	1, 1, 1	2, 3, 4	4, 5, 6	1/9, 1/9, 1/9	1/9, 1/9, 1/9	1, 1, 1	1/9, 1/9, 1/9	2, 3, 4	4, 5, 6	4, 5, 6	4, 5, 6	1, 1, 1
Eye	9, 9, 9	1/2, 1/3, 1/4	1, 1, 1	9, 9, 9	9, 9, 9	9, 9, 9	9, 9, 9	9, 9, 9	9, 9, 9	9, 9, 9	9, 9, 9	9, 9, 9	9, 9, 9
Finger	1/9, 1/9, 1/9	1/4, 1/5, 1/6	1/9, 1/9, 1/9	1, 1, 1	1/4, 1/5, 1/6	1/6, 1/7, 1/8	2, 3, 4	1/9, 1/9, 1/9	1, 1, 1	4, 5, 6	1/2, 1/3, 1/4	1, 1, 1	1, 1, 1
Foot	1, 1, 1	9, 9, 9	1/9, 1/9, 1/9	4, 5, 6	1, 1, 1	1, 1, 1	6, 7, 8	1, 1, 1	6, 7, 8	6, 7, 8	4, 5, 6	6, 7, 8	6, 7, 8
Knee	1, 1, 1	9, 9, 9	1/9, 1/9, 1/9	6, 7, 8	1, 1, 1	1, 1, 1	6, 7, 8	1, 1, 1	6, 7, 8	4, 5, 6	4, 5, 6	1, 1, 1	6, 7, 8
Knuckle	1/9, 1/9, 1/9	1, 1, 1	1/9, 1/9, 1/9	1/2, 1/3, 1/4	1/6, 1/7, 1/8	1/6, 1/7, 1/8	1, 1, 1	1/9, 1/9, 1/9	1, 1, 1	2, 3, 4	1/9, 1/9, 1/9	1/6, 1/7, 1/8	1/2, 1/3, 1/4
Leg	1, 1, 1	9, 9, 9	1/9, 1/9, 1/9	9, 9, 9	1, 1, 1	1, 1, 1	9, 9, 9	1, 1, 1	9, 9, 9	9, 9, 9	9, 9, 9	1, 1, 1	9, 9, 9
Palm	1/9, 1/9, 1/9	1/2, 1/3, 1/4	1/9, 1/9, 1/9	1, 1, 1	1/6, 1/7, 1/8	1/6, 1/7, 1/8	1, 1, 1	1/9, 1/9, 1/9	1, 1, 1	1/6, 1/7, 1/8	1/6, 1/7, 1/8	1/2, 1/3, 1/4	1, 1, 1
Shoulder	1/2, 1/3, 1/4	1/4, 1/5, 1/6	1/9, 1/9, 1/9	1/4, 1/5, 1/6	1/6, 1/7, 1/8	1/4, 1/5, 1/6	1/2, 1/3, 1/4	1/9, 1/9, 1/9	6, 7, 8	1, 1, 1	1/6, 1/7, 1/8	2, 3, 4	2, 3, 4
Spine	1/4, 1/5, 1/6	1/4, 1/5, 1/6	1/9, 1/9, 1/9	2, 3, 4	1/4, 1/5, 1/6	1/4, 1/5, 1/6	9, 9, 9	1/9, 1/9, 1/9	6, 7, 8	6, 7, 8	1, 1, 1	2, 3, 4	4, 5, 6
Toe	1, 1, 1	1/4, 1/5, 1/6	1/9, 1/9, 1/9	1, 1, 1	1/6, 1/7, 1/8	1, 1, 1	6, 7, 8	1, 1, 1	2, 3, 4	1/2, 1/3, 1/4	1/2, 1/3, 1/4	1, 1, 1	2, 3, 4
Wrist	1/4, 1/5, 1/6	1, 1, 1	1/9, 1/9, 1/9	1, 1, 1	1/6, 1/7, 1/8	1/6, 1/7, 1/8	2, 3, 4	1/9, 1/9, 1/9	1, 1, 1	1/2, 1/3, 1/4	1/4, 1/5, 1/6	1/2, 1/3, 1/4	1, 1, 1

Table 11. Geometric means, fuzzy weights and averaged and normalized relative weight for people

Type of Work	r_i			w_i			M	N
Carpenter	0.753015	0.688725	0.638959	0.110369	0.10475	0.100659	0.10526	0.105077
Mason	4.409334	4.738137	5.045379	0.646276	0.720635	0.794832	0.720581	0.719332
Plumber	0.816476	0.725961	0.686555	0.119671	0.110413	0.108158	0.112747	0.112552
Welder	0.05407	0.064202	0.071173	0.05407	0.064202	0.071173	0.063149	0.063039
Total	6.347728	6.574951	6.822682					
Reciprocal of Totals	0.157537	0.152092	0.14657					
Increasing order of Reciprocals	0.14657	0.152092	0.157537					

Table 12. Geometric means, fuzzy weights and averaged and normalized relative weight for carpenter's body parts

Body Part	r_i	w_i			M	N		
Ankle	0.455206	0.438494	0.423937	0.022952	0.023067	0.023367	0.023129	0.023067
Arm	1.916479	2.135509	2.326009	0.096632	0.112337	0.128207	0.112392	0.112095
Eye	0.303294	0.264249	0.234728	0.015293	0.013901	0.012938	0.014044	0.014007
Finger	1.846602	1.983445	2.097859	0.093108	0.104337	0.115632	0.104359	0.104084
Foot	0.572477	0.528189	0.49414	0.028865	0.027785	0.027237	0.027962	0.027888
Knee	0.376005	0.348247	0.326352	0.018959	0.018319	0.017988	0.018422	0.018373
Knuckle	1.064374	1.049207	1.040109	0.053667	0.055193	0.05733	0.055397	0.05525
Leg	1.322357	1.346924	1.366415	0.066675	0.070854	0.075315	0.070948	0.070761
Palm	2.184028	2.27208	2.362587	0.110122	0.119521	0.130223	0.119955	0.119639
Shoulder	2.55164	2.766337	2.962681	0.128657	0.145521	0.1633	0.145826	0.145441
Spine	3.812943	4.123273	4.414588	0.192254	0.216901	0.243328	0.217494	0.21692
Toe	0.207986	0.193956	0.181578	0.010487	0.010203	0.010008	0.010233	0.010206
Wrist	1.529186	1.560015	1.60188	0.077104	0.082063	0.088294	0.082487	0.082269
Total	18.142578	19.009923	19.832862					
Reciprocal of Totals	0.055119	0.052604	0.050421					
Increasing order of Reciprocals	0.050421	0.052604	0.055119					

Table 13. Geometric means, fuzzy weights and averaged and normalized relative weight for carpenter's body movement

Movements	r_i	w_i	M	N				
Abduction	2.497368	2.737042	2.959048	0.134051	0.152618	0.171893	0.152854	0.152536
Adduction	1.321598	1.425976	1.500816	0.070939	0.079513	0.087184	0.079212	0.079047
Circumduction	0.293992	0.278294	0.264093	0.015781	0.015518	0.015341	0.015547	0.015514
Dorsiflexion	0.465836	0.446355	0.428334	0.025005	0.024889	0.024882	0.024925	0.024873
Eversion	0.600964	0.569437	0.543917	0.032258	0.031752	0.031597	0.031869	0.031803
Extension	1.661487	1.588972	1.558067	0.089183	0.088602	0.090509	0.089431	0.089246
Flexion	1.848386	1.810123	1.811853	0.099215	0.100933	0.105252	0.1018	0.101589
Hyperextension	1.1.484034	1.716229	1.880438	0.079658	0.095697	0.109236	0.094864	0.094667
Inversion	0.708563	0.693702	0.68419	0.038033	0.038681	0.039745	0.03882	0.038739
Plantar flexion	0.368588	0.30866	0.270999	0.019785	0.017211	0.015743	0.017579	0.017543
Pronation	2.671343	2.891544	3.099687	0.143389	0.161233	0.180063	0.161562	0.161226
Rotation	0.478029	0.463066	0.44647	0.025659	0.025821	0.025936	0.025805	0.025752
Supination	2.814252	3.004501	3.182101	0.15106	0.167532	0.184851	0.167814	0.167465
Total	17.214439	17.933901	18.630013					
Reciprocal of Totals	0.058091	0.05576	0.053677					
Increasing order of Recipro	0.053677	0.05576	0.058091					

Table 14. Geometric means, fuzzy weights and averaged and normalized relative weight for mason's body parts

Body Part	r_i	w_i	M		N			
Ankle	1.487778	1.556057	1.618444	0.068826	0.073762	0.078827	0.073805	0.07374
Arm	1.027448	1.070713	1.104906	0.047531	0.050755	0.053815	0.0507	0.050655
Eye	0.349421	0.304815	0.273065	0.016164	0.014449	0.0133	0.014638	0.014625
Finger	0.408283	0.386834	0.368904	0.018887	0.018337	0.017968	0.018397	0.018381
Foot	1.432173	1.505524	1.572123	0.066253	0.071367	0.076571	0.071397	0.071334
Knee	1.805066	1.949615	2.077619	0.083504	0.092418	0.101191	0.092371	0.092289
Knuckle	0.296377	0.26915	0.249044	0.013711	0.012759	0.01213	0.012866	0.012855
Leg	1.731756	1.816929	1.893845	0.080112	0.086128	0.09224	0.08616	0.086084
Palm	0.317031	0.294458	0.276299	0.014666	0.013958	0.013457	0.014027	0.014015
Shoulder	2.435033	2.684638	2.913214	0.112646	0.12726	0.141889	0.127265	0.127153
Spine	7.600449	7.600449	7.600449	0.351603	0.360285	0.370183	0.36069	0.360371
Toe	1.110132	1.137443	1.161478	0.051356	0.053918	0.05657	0.053948	0.0539
Wrist	0.530667	0.519048	0.507203	0.024549	0.024604	0.024704	0.024619	0.024597
Total	20.531613	21.095673	21.616592					
Reciprocal of Totals	0.048705	0.047403	0.046261					
Increasing order of Reciprocals	0.046261	0.047403	0.048705					

Table 15. Geometric means, fuzzy weights and averaged and normalized relative weight for mason's body movements

Movements	r_i			w_i			M	N
Abduction	0.464821	0.405359	0.366384	0.026776	0.024488	0.023347	0.02487	0.024786
Adduction	0.347026	0.288466	0.249446	0.01999	0.017426	0.015895	0.017771	0.01771
Circumduction	2.344883	2.741407	3.081391	0.135076	0.165608	0.196351	0.165678	0.165117
Dorsiflexion	1.832356	1.77436	1.773303	0.105552	0.107189	0.112998	0.10858	0.108212
Eversion	1.147787	1.114306	1.093598	0.066118	0.067315	0.069686	0.067706	0.067477
Extension	0.449844	0.417165	0.390959	0.025913	0.025201	0.024913	0.025342	0.025256
Flexion	0.415832	0.377428	0.347714	0.023954	0.0228	0.022157	0.02297	0.022893
Hyperextension	2.211346	2.388448	2.550786	0.127384	0.144286	0.16254	0.144737	0.144247
Inversion	1.177133	1.188593	1.197873	0.067808	0.071803	0.07633	0.071981	0.071737
Plantar flexion	0.801137	0.776107	0.755824	0.046149	0.046885	0.048162	0.047065	0.046906
Pronation	1.173479	1.283763	1.37019	0.067598	0.077552	0.087311	0.077487	0.077224
Rotation	1.516854	1.650035	1.752426	0.087378	0.099678	0.111667	0.099575	0.099237
Supination	1.810785	2.148139	2.429799	0.10431	0.129769	0.154831	0.129636	0.129197
Total	15.693282	16.553578	17.359694					
Reciprocal of Totals	0.063722	0.06041	0.057605					
Increasing order of Reciprocals	0.057605	0.06041	0.063722					

Table 16. Geometric means, fuzzy weights and averaged and normalized relative weight for plumber's body parts

Body Part	r_i			w_i			M	N
Ankle	0.314764	0.270481	0.241104	0.01512	0.013536	0.012647	0.013768	0.013732
Arm	2.793158	3.080603	3.320778	0.134168	0.154162	0.174196	0.154175	0.153778
Eye	0.519088	0.555249	0.573546	0.024934	0.027786	0.030086	0.027602	0.027531
Finger	2.841516	2.966507	3.075721	0.136491	0.148452	0.161341	0.148761	0.148378
Foot	0.28467	0.249732	0.227918	0.013674	0.012497	0.011956	0.012709	0.012676
Knee	0.416084	0.413516	0.408096	0.019986	0.020693	0.021407	0.020696	0.020642
Knuckle	3.069063	3.304333	3.512761	0.147421	0.165358	0.184267	0.165682	0.165254
Leg	0.385855	0.392908	0.39044	0.018534	0.019662	0.020481	0.019559	0.019509
Palm	2.889721	2.990647	3.107452	0.138806	0.14966	0.163006	0.150491	0.150103
Shoulder	1.761738	1.767743	1.786049	0.084624	0.088463	0.09369	0.088926	0.088696
Spine	1.142085	1.106626	1.085877	0.05486	0.055379	0.056961	0.055733	0.055589
Toe	0.426034	0.369352	0.334371	0.020464	0.018483	0.01754	0.018829	0.018781
Wrist	2.219661	2.515242	2.754237	0.10662	0.125869	0.144477	0.125656	0.125332
Total	19.06344	19.98294	20.81835					
Reciprocal of Totals	0.05246	0.05004	0.04804					
Increasing order of Reciprocals	0.04804	0.05004	0.05246					

Table 17. Geometric means, fuzzy weights and averaged and normalized relative weight for plumber's body movement

Movements	r_i			w_i			M	N
Abduction	1.099845	1.023967	0.978716	0.05298	0.051834	0.052426	0.052413	0.052217
Adduction	0.770262	0.719037	0.682469	0.037104	0.036398	0.036557	0.036686	0.036549
Circumduction	1.223612	1.212531	1.211382	0.058942	0.061379	0.064889	0.061737	0.061506
Dorsiflexion	0.288339	0.239364	0.210579	0.01389	0.012117	0.01128	0.012429	0.012382
Eversion	0.2756	0.285303	0.285851	0.013276	0.014442	0.015312	0.014343	0.01429
Extension	1.682192	1.843222	1.974017	0.081033	0.093305	0.105741	0.093359	0.09301
Flexion	1.682192	1.843222	1.974017	0.081033	0.093305	0.105741	0.093359	0.09301
Hyperextension	0.868517	0.852207	0.843214	0.041837	0.043139	0.045168	0.043381	0.043219
Inversion	0.309419	0.308451	0.3003	0.014905	0.015614	0.016086	0.015535	0.015477
Plantar flexion	0.597969	0.489486	0.427144	0.028805	0.024778	0.022881	0.025488	0.025392
Pronation	4.59323	5.108199	5.574707	0.22126	0.258579	0.298617	0.259485	0.258513
Rotation	1.566222	1.842107	2.065367	0.075446	0.093248	0.110634	0.09311	0.092761
Supination	3.711023	3.987805	4.231683	0.178763	0.201864	0.226676	0.202434	0.201676
Total	18.668422	19.754899	20.759445					
Reciprocal of Totals	0.053566	0.05062	0.048171					
Increasing order of Reciprocals	0.048171	0.05062	0.053566					

Table 18. Geometric means, fuzzy weights and averaged and normalized relative weight for welder's body parts

Movements	r_i			w_i			M	N
Ankle	2.16764	2.314419	2.433517	0.10366	0.112833	0.120971	0.112488	0.112432
Arm	0.866832	0.988199	1.09252	0.041453	0.048177	0.05431	0.04798	0.047956
Eye	6.085269	5.898356	5.769307	0.291006	0.287558	0.286794	0.288453	0.288309
Finger	0.471647	0.458155	0.447164	0.022555	0.022336	0.022229	0.022373	0.022362
Foot	2.46553	2.707498	2.931261	0.117905	0.131997	0.145714	0.131872	0.131806
Knee	2.14809	2.331099	2.497963	0.102725	0.113646	0.124175	0.113515	0.113458
Knuckle	0.318907	0.298313	0.282915	0.015251	0.014543	0.014064	0.014619	0.014612
Leg	2.75687	2.75687	2.75687	0.131837	0.134404	0.137045	0.134429	0.134361
Palm	0.311935	0.279499	0.256615	0.014917	0.013626	0.012756	0.013767	0.01376
Shoulder	0.451238	0.423537	0.401943	0.021579	0.020648	0.019981	0.020736	0.020726
Spine	0.898837	0.930481	0.95193	0.042984	0.045363	0.047321	0.045222	0.0452
Toe	0.75908	0.746156	0.73577	0.0363	0.036377	0.036575	0.036417	0.036399
Wrist	0.414671	0.379271	0.353375	0.01983	0.01849	0.017566	0.018629	0.01862
Total	20.116546	20.511855	20.911151					
Reciprocal of Totals	0.04971	0.048752	0.047821					
Increasing order of Reciprocals	0.047821	0.048752	0.04971					

Table 19. Geometric means, fuzzy weights and averaged and normalized relative weight for welder's body movement

Movements	r_i			w_i			M	N
Abduction	3.218802	3.922821	4.572742	0.144029	0.190131	0.242951	0.19237	0.190493
Adduction	3.218802	3.922821	4.572742	0.144029	0.190131	0.242951	0.19237	0.190493
Circumduction	2.829902	3.124119	3.395084	0.126627	0.15142	0.180381	0.152809	0.151318
Dorsiflexion	0.484668	0.38668	0.326388	0.021687	0.018742	0.017341	0.019257	0.019069
Eversion	0.449439	0.366825	0.313526	0.020111	0.017779	0.016658	0.018183	0.018005
Extension	1.830574	2.058369	2.245418	0.081911	0.099765	0.119299	0.100325	0.099346
Flexion	1.420678	1.478121	1.525837	0.06357	0.071642	0.081068	0.072093	0.07139
Hyperextension	0.144468	0.140336	0.136966	0.006464	0.006802	0.007277	0.006848	0.006781
Inversion	1.173515	1.231673	1.276958	0.05251	0.059697	0.067845	0.060017	0.059432
Plantar flexion	1.249011	1.239701	1.237764	0.055888	0.060086	0.065763	0.060579	0.059988
Pronation	0.506616	0.454929	0.416784	0.022669	0.02205	0.022144	0.022287	0.02207
Rotation	0.76603	0.745749	0.726234	0.034277	0.036145	0.038585	0.036336	0.035981
Supination	1.529186	1.560015	1.60188	0.068425	0.075611	0.085108	0.076381	0.075636
Total	18.821692	20.632158	22.348324					
Reciprocal of Totals	0.05313	0.048468	0.044746					
Increasing order of Reciprocals	0.044746	0.048468	0.05313					

Table 20. Body parts ranking

Body Part	Carpenter	Mason	Plumber	Welder	SUM	RANKS
Ankle	0.00242381	0.05304354	0.00154556	0.0070876	0.06410052	6
Arm	0.01177861	0.03643776	0.001730802	0.0030231	0.06854749	5
Eye	0.00147181	0.01052023	0.00309867	0.01817471	0.03326542	13
Finger	0.01093683	0.01322204	0.01670024	0.00140968	0.04226879	9
Foot	0.00293039	0.05131283	0.00142671	0.00830892	0.06397884	7
Knee	**0.00193058**	**0.06638643**	**0.0023233**	**0.00715228**	**0.07779259**	4
Knuckle	0.0058055	0.00924701	0.01859967	0.00092113	0.03457331	12
Leg	**0.00743535**	**0.06192298**	**0.00219578**	**0.00846998**	**0.08002409**	**3**
Palm	0.01257131	0.01008144	0.01689439	0.00086742	0.04041455	11
Shoulder	**0.0152825**	**0.09146522**	**0.00998291**	**0.00130655**	**0.11803718**	**2**
Spine	**0.0227933**	**0.25922639**	**0.00625665**	**0.00284936**	**0.29112571**	1
Toe	0.00107242	0.03877199	0.00211384	0.00229456	0.04425281	8
Wrist	0.00864458	0.01769341	0.001410637	0.00117379	0.04161814	10

body movements like Supination, Circumduction, Hyperextension are the dangerous movements for all types of workers. For Carpenter Supination is more, for mason circumduction, for plumber supination, and for welder circumduction are maximum body movements. This paper shows the most harmful body movements and mostly affected body part due to these movements. So proper ergonomic design must be suggested by modifying the workplace and work system. So, the diseases can be avoided and safety in the workplace can be maintained. To avoid or less these body movements new advanced equipment can be designed. Table 20 shows the body parts ranking.

Table 21 shows the body movements ranking.

Table 21. Body movements ranking

Movements	Carpenter	Mason	Plumber	Welder	SUM	RANKS
Abduction	0.01602803	0.01782936	0.00587713	0.01200849	0.051743	9
Adduction	0.00830602	0.01273937	0.00411366	0.01200849	0.03716754	13
Circumduction	**0.00163016**	**0.11877394**	**0.00692262**	**0.00953894**	**0.13686567**	**2**
Dorsiflexion	0.00261358	0.07784035	0.00139362	0.00120209	0.08304964	6
Eversion	0.00334176	0.04853837	0.00160837	0.00113502	0.05462351	8
Extension	0.0093777	0.01816745	0.01046846	0.00626267	0.04427628	10
Flexion	0.01067467	0.01646767	0.01046846	0.00450035	0.04211115	12
Hyperextension	**0.00994732**	**0.10376148**	**0.00486438**	**0.00042747**	**0.11900066**	**3**
Inversion	0.00407058	0.05160272	0.00174197	0.00374653	0.0611618	7
Plantar flexion	0.00184337	0.03374099	0.00285792	0.00378158	0.04222386	11
Pronation	0.01694114	0.05554969	0.02909616	0.00139127	0.10297826	4
Rotation	0.00270594	0.07138435	0.01044044	0.00226821	0.08679893	5
Supination	**0.01759672**	**0.09293554**	**0.02269904**	**0.00476802**	**0.13799931**	**1**

Designing a Framework to Avoid Health Problems for Construction Site Workers

Fifty useful responses were collected and tested to examine the validity and reliability of the scale so as to obtain a quantitative and statistically proven identification of the responses. The test for the quantitative variable was conducted by factor analysis on 50 responses using the principal component method followed by varimax rotation to ensure that the variables are important and suitable for the model using Minitab 15.0. Twenty-two items were categorized under nine social factors constituting various variables for the proposed instrument to measure the relation between occupational safety management and firm performance of construction site workers as perceived by safety officers and workers. These nine social factors are defined as a Preventive procedure, OHS training, Risk management, Work practices, Periodic plans, Continuous improvement, Injury avoidance, Emergency plans and Healthcare. The items that failed to get loaded more than 0.5 were not considered for further analysis. This was true for refer to items 5, 6, 8, 10, 21, and 22. Percentage of total variance explained was found to be 75% which is an acceptable value for the principal component varimax rotated factor loading procedure. The internal consistency of the actual survey data was tested by computing the Cronbach's Alpha. Table 22 shows the factor's loading.

After getting seven important parameters/items for the improvement of policies of occupational health and safety by interview with experts, QFD was applied to identify system design requirements for occupational health and safety is studied

- **Customer Needs (What's):** A structured list of requirements derived from expert's feedback.
- **Design Requirements (How's):** A structured set of relevant and measurable services/characteristics which are required for fulfilling what's.
- **Planning Matrix (Left Matrix):** Gives customer/expert perceptions observed in surveys. It includes the relative importance of requirements
- **Interrelationship Matrix (Center Matrix):** Gives the expert's perceptions of interrelationships between design requirements and customer needs. An appropriate scale is applied and illustrated using symbols or figures. Filling this portion of the matrix involves discussions.
- **Design Correlation (Top) Matrix:** Used to identify where design requirements support or impede each other in the system or product design.

Customer rating $=Z_i+ (1/n-1(\Sigma^{i}_{j=1}B_{ij}Z_j))$

B_{ij} is i^{th} initial customer rating and Z_j denote the relationship between customer need and customer rating respectively

n number of customer need

$B_{ij}\ [B_{1ij}, B_{2ij}]$.

The individual rating for each design requirement is obtained from the center matrix using the following relation:

Table 22. Factor loading score

Factor	Item n°	F1	F2	F3	F4	F5	F6	F7	F8	F9	Cronbach's Alpha
Preventive procedures	1	0.877									0.825
	2	0.557									
	3	0.593									
	4	0.892									
	13	0.892									
OHS training	9		0.626								0.782
	10		0.683								
	11		0.787								
	12		0.697								
Risk management	5			0.614							0.516
	6			0.655							
	7			0.578							
	18			0.627							
Work practices	8				0.754						0.714
	21				0.648						
Periodic plans	19					0.748					0.760
	28					0.774					
Continuous improvement	7						0.517				0.656
	15						0.853				
Injury avoidance	16							0.613			0.747
	17							0.820			
Emergency plan	14								0.843		0.824
Healthcare	20									0.888	0.788

$$\text{Design requirement} = \Sigma^{n}_{j} \sum_{j}^{n} \cdot ({}_{i}X_{j})$$

where A_{ij} and X_j denote the relative importance of the ith characteristics with respect to the Jth customer need in the relationship matrix and the importance of Jth customer needs (customer ratings) and n is the number of customer needs. Table 23 shows the design requirements and Table 24 shows the employee requirements.

A QFD matrix was prepared and the respondents were asked to assign numeric relative importance scores to the various customer requirements, from the lowest to the highest, on a scale from 1 to 9. They were also asked to express, in numeric values, the strong-moderate-weak relationship between what's and the How's as 0.2, 0.4, 0.6, and 0.8 with symbols. After the responses were collected, scores for each of the columns was computed. The design requirements are prioritized as per their importance of design requirement. Figure 1 shows HOQ for Construction site workers.

Table 23. Design requirement

N°	System Design Requirement for Workers in Construction Site
1	Preventive measure
2	Emergencies plan
3	Injury avoidance
4	Health plan
5	OHS training
6	Periodic plan
7	Risk Management

Table 24. Employee requirements table

No	Employee Requirement/Needs for Construction Site
1	Safety training programs are conducted in the organization on regular basis
2	Instruction manuals being followed before executing any hazardous job activity
3	Immediate medical facilities are provided for personal injuries
4	Alarms/beeping sound/alarming light and emergency lights work effectively when a crisis situation arises
5	A Safety equipments are supplied by the organization to avoid accidents
6	The company readily provides health subsidies when accidents occur in the workplace
7	The employees work in health and safety condition
8	Regular auditing is done by OHS
9	Company undertakes a stress risk assessment in the workplace

CONCLUSION

Although technology is changing day by day, man-machine interaction is still the main rule of modernization. Human accidents and health hazards are important points of discussion among many researchers. From the whole analysis, the inference found was that considering all 4 kinds of people, the body movement which did hurt and injured the most and which brought out the maximum discomfort out of all 13 movements was Supination (wrist curling upwards). And the body part which was damaged the most out of all 13 body parts was Spine. The proper ergonomic design can change or modify the adverse effect generating from man-machine interaction. The main objective of this paper is the application of FUZZY-AHP in understanding role of ergonomic development in the construction sites and the reason for the origin of discomfort of people working there. This will eventually help in rising of the Construction Companies. The above analysis will help them in understanding the problems faced by plumber, mason, welder, and carpenter. It may help in reducing accidents and injuries that are very common on their construction sites. The data collected in this study were self-reported; hence, there is a chance of bias and measurement errors. The sample size for certain groups of employees is small; hence this might affect the results obtained. In future, the studies can be done for different workers. A model can be generated and tested to validate the body movement and body parts rank.

Figure 1. House of quality

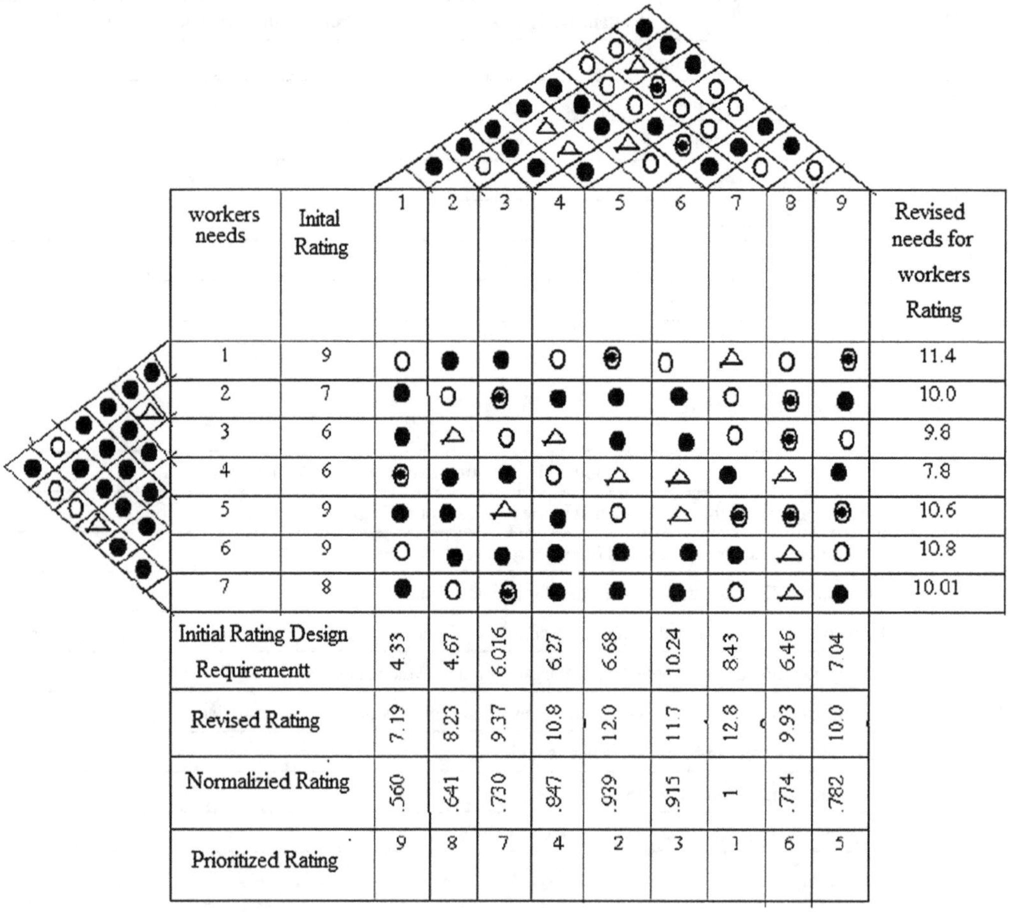

FUTURE RESEARCH DIRECTIONS

In future, more data can be collected on construction site workers of industries also. The analysis can be done for skilled and unskilled workers in heavy working conditions. Other sectors like automobiles, mining sectors can be analyzed by studying their workers. Interpretive structural modelling and SEM can be useful for providing relations among design requirements. More surveys can be done to get a better result.

REFERENCES

Assunção, A., Carnide, F., Vieira, F., Silva, S., & Araújo, J. (2013). Mismatch of school furniture and back pain in adolescents with different maturation levels. *Int. J. of Human Factors and Ergonomics*, *2*(1), 66–81.

Ayhan, M. B. (2013). A fuzzy ahp approach for supplier selection problem: A case study in a gearmotor company. *International Journal of Managing Value and Supply Chains*, *4*(3), 11–23. doi:10.5121/ijmvsc.2013.4302

Charles, T. S., Micheline, C. D., & Jennifer, D. (2012). Hazard perception in older drivers. *Int. J. Human Factors and Ergonomics*, *1*(3), 221–233. doi:10.1504/IJHFE.2012.050869

Chedi, J. M. (2015). The Need for the Incorporation of Ergonomics into Curriculum of the Technical and Vocational Education and Training for Teacher/Trainer Professionalization. *3rd UPI International Conference on Technical and Vocational Education and Training*, 105-109.

Det, M. J., Meijerink, W. J. H. J., Hoff, C., Totte, E. R., & Pierie, J. P. E. N. (2009). Optimal ergonomics for laparoscopic surgery in minimally invasive surgery suites: A review and guidelines. *Surgical Endoscopy*, *23*(6), 1279–1285. doi:10.100700464-008-0148-x PMID:18830751

Dun, J., & Neumann, W. P. (2005). *Ergonomics Contribution to Company Strategies. 10th International Conference on Human Aspects of Advanced Manufacturing: Agility and Hybrid Automation (HAAMAHA)*, San Diego, CA.

Garnaik, A., Kumar, S., & Satapathy, S. (2016). A Case Study of Furniture: Optimizing Product Design by implying Analytic Hierarchy Process and Quality Function Deployment. *International Journal of Engineering Science and Management, 2*(1), 1-7.

Iqbal, N. J., Vivek, K., & Amithabha, D. (2015). Manual material handling and occupational low back disorders: A narrative literature review emphasising maximum acceptable weight of load. *Int. J. of Human Factors and Ergonomics*, *3*(3/4), 376–392. doi:10.1504/IJHFE.2015.073010

Kushwaha, D. K., & Prasad, V. K. (2016). Ergonomic assessment and workstation design of shipping crane cabin in steel industry. *International Journal of Industrial Ergonomics*, *52*, 29–39. doi:10.1016/j.ergon.2015.08.003

Lindroos, O. (2009). Relationships between observed and perceived deviations from normative work procedures. *Ergonomics*, *52*(12), 1487–1500. doi:10.1080/00140130903197461 PMID:19941182

Liu, P., & Li, Z. (2014). Comparison between conventional and digital nuclear power plant main control rooms: A task complexity perspective. *International Journal of Industrial Ergonomics*, *51*, 10–20. doi:10.1016/j.ergon.2014.06.011

Mohanty, P. P., & Mahapatra, S. S. (2014). A Compromise Solution by VIKOR Method for Ergonomically Designed Product with Optimal Set of Design Characteristics. *Procedia Materials Science, 6*, 633-640.

Narayanan, A., Mathew, C., & Baby, V.Y. (2013). Improvement of Ergonomic Factors That Affects Employees in a Textile Industry. *International Journal of Engineering Science and Innovative Technology, 2*(1), 276-283.

Openshaw, S., & Taylor, E. (2006). *Ergonomics and Design: A Reference Guide*. Allsteel.

Parida, R., & Ray, P. K. (2015). Factors influencing construction ergonomic performance in India. 6th International Conference on Applied Human Factors and Ergonomics. *Procedia Manufacturing, 3*, 6587-6592.

Parkes, K. R. P., & Carnell, S. E. (2005). *Farmer Research Report on Musculoskeletal Disorder, Mental Health and Work Environment, Health and safety executive*. University of oxford.

Pavlovic-Veselinovic, S., Hedge, A., & Veselinovic, M. (2016). An ergonomice xpert system for risk assessment of work-related musculo-skeletal disorders. *International Journal of Industrial Ergonomics*, *53*, 130–139. doi:10.1016/j.ergon.2015.11.008

Ramos, Pedro M. A.& Paulo, A. (2016). Analysis of the return on preventive measures in musculoskeletal disorders through the benefit–cost ratio: A case study in a hospital. *International Journal of Industrial Ergonomics*. doi:10.1016/j.ergon.2015.11.003

Ray, P. K., Parida, R., & Sarkar, S. (2015). Ergonomic analysis of construction jobs in India: A biomechanical modelling approach. 6th International Conference on Applied Human Factors and Ergonomics. *Procedia Manufacturing*, *3*, 4606-4612.

Rouyendegh, B. D., & Erkan, T. E. (2012). Selection of academic staff using the fuzzy analytic hierarchy process (fahp): A pilot study. *Technical Gazette*, *19*(4), 923-929.

Rundcrantz, B. L., Johnsson, B., & Mortiz, U. (1991). Occupational cervico-brachial disorders among dentists: Analysis of ergonomics and locomotor functions. *Swedish Dental Journal*, *15*(3), 105–115. PMID:1876977

Sylla, N., Bonnet, V., Colledani, F., & Fraisse, P. (2014). Ergonomic contribution of ABLE exoskeleton in automotive industry. *International Journal of Industrial Ergonomics*, *44*(4), 475–481. doi:10.1016/j.ergon.2014.03.008

KEY TERMS AND DEFINITIONS

AHP: Analytical hierarchy process is a structured technique for organizing and analyzing complex decisions, based on mathematics and psychology.

ANN: Artificial neural network is a computational model based on the structure and functions of biological neural networks. Information that flows through the network affects the structure of the ANN because a neural network changes—or learns, in a sense—based on that input and output.

HOQ: House of quality is a voice of customer analysis tool and a key component of the quality functional deployment technique.

ISM: Interpretive structural equation modeling is a well-established methodology for identifying relationships among specific items, which define a problem or an issue. This approach has been increasingly used by various researchers to represent the interrelationships among various elements related to the issue.

QFD: Quality function development is a structured approach to defining customer needs or requirements and translating them into specific plans to produce products to meet those needs. The "voice of the customer" is the term to describe these stated and unstated customer needs or requirements.

SEM: Structural equation modeling is a multivariate statistical analysis technique that is used to analyze structural relationships. This technique is the combination of factor analysis and multiple regression analysis, and it is used to analyze the structural relationship between measured variables and latent constructs.

Chapter 12
Body Scanner Measurements for Apparel Design in Mexican Women

Lilia Roselia Prado-León
Universidad de Guadalajara, Mexico

Carlos Aceves-González
Universidad de Guadalajara, Mexico

ABSTRACT

This chapter aims to report data on body dimensions and proportions for apparel design in Mexican women and to analyze changes in dimensions and proportions among groups according to the age. Measurements of 1131 female subjects aged from 15 to 65 years from the metropolitan zone of Guadalajara, Mexico were obtained using the TC2 3D body scanning system. Mean, standard deviation, and percentiles 5, 50, and 95 were calculated for 28 dimensions of six age groups. Body proportion was calculated for 26 dimensions with respect to stature, using mean data of those groups. Results show changes in body dimensions and proportions with age. Weight increases gradually from 15 to 65 years of age, with gains in all groups as age advances. Stature was very similar between the younger groups, but it began to decrease the ages of 35-44. The authors emphasize the relevance of addressing such differences for optimal anthropometric garment fitting.

INTRODUCTION

In the sphere of Ergonomics, there is a clear need for anthropometric data in designing space, equipment and furnishings, with a large number of research projects establishing ergonomic guidelines for product or workspace design (Castellucci, Arezes, & Viviani, 2010; Dianat, Karimi, Asl Hashemi, & Bahrampour, 2013; Laios & Giannatsis, 2010; Panagiotopoulou, Christoulas, Papanckolaou, & Mandroukas, 2004; Yokota, 2005). However, apparel design should also include knowledge of the user population's anthropometric data (Roebuck, Jr., 1995). Apparel fitting is important because contributes to the confidence and comfort of the wearer, but also because consumers tend to blame themselves when the clothes do

DOI: 10.4018/978-1-5225-5234-5.ch012

Copyright © 2018, IGI Global. Copying or distributing in print or electronic forms without written permission of IGI Global is prohibited.

not fit their bodies (Laitala, Klepp, & Benedicte, 2011). The close relationship between clothing, body measurement and shape, and proportion is still a challenge for ergonomic engineering (Konz & Johnson, 2000). Hsu (2008) points out that the application of ergonomics in the apparel should considers anthropometric data, as well as body shape and proportions.

In line, Loker, Ashdown, & Schoenfelder (2005) mention that one way of improving the sizing system is to change basic specifications for each size so they correspond more closely to proportions for the majority of the user population. Also, Pheasant (1996) has suggested that variations in body dimensions for different groups should be observed not only according to body size but body proportion. Body proportions are the relation calculated between one body dimension and a specific referent dimension. Proportions for various corporal dimensions change over the course of a life (Roebuck, Kromer, & Thomson, 1975).

Currently, there are many consumers who state frustration about the sizing systems and the incorrect use of the system (Laitala et al., 2011). For instance, the Kurt Salmon Associates (2000) study reported that 50% of women in the United States complain of not being able to find clothing that fits well. According to Laitala et al. (2011), there are disparities within clothing sizes used today, which could cause confusion and fit problems among customers. These authors also mention that the customer groups that have most problems are mainly women. For instance, in a survey of 106 young college women in Mexico, Corona (2014) found that 41.5% of participants did not know with certainty their size due to the variability of measurements for each size between brands.

In Mexico, there are some anthropometric studies known to have been undertaken to date. Among these are Ramos-Galvan (1975) and Faulhaber (1976), who carried out studies of anthropometry for child growth and development in Mexico City, measuring 5,333 children and 523 adolescents and reporting upon a total of 11 to 14 measurements. Liu, Sanchez-Monroy, & Parga (1999) undertook a study on the Mexico-U.S. border in which they collected data on 12 measurements from 110 assembly plant workers, between 17 and 40 years of age. Sanchez-Monroy (2007) conducted a study in which 974 bus drivers, between 18 and 68 years old, were measured; and reported on 60 measurements. The most extensive study to have been reported is that conducted by Avila-Chaurand, Prado-León, & González, Muñoz (2007) which measured 8228 subjects from the Guadalajara metropolitan area, aged between 2 and 90; reporting on 56 measurements. Another study has been recently reported by Hernandez-Arellano, Talavera-Aguirre, Serratos-Perez, Maldonado-Macias, & Garcia-Alcaraz (2016) where 36 body dimensions were measured of 140 bachelor students in the Northern of Mexico. Despite the existence of these efforts, there are no reported anthropometric studies which focus on apparel design. This situation might cause inefficiency in the clothing industry, economic loss, and frustration among consumers as a result of clothing designs whose inadequacy is due to a lack of current data that would otherwise permit up-to-date sizing. Moreover, Laitala et al. (2011) point out that more than ever before, the apparel industry is faced with customers occupying a larger spectrum of sizes due to both migrations from different cultures. In line with this, Mexican population is in a large proportion in the United States (US) and other countries due to this migratory movement. Newcomb & Istook (2011) state that the US Hispanic market (among them Mexicans) has created a consumer group that many industries cannot afford to ignore.

It is worth noting that, the method that is commonly used by Mexican manufacturers in creating sizes is based on the measurements of an ideal model, from which regular proportional increases and decreases are determined. On the other hand, the measurement of lengths and girths is usually done with a measuring tape across the body's surface. However, technological development allows anthropometric studies to be undertaken using 3D scanner imagery. This modern anthropometric measurement system offers an interesting alternative to traditional measurements. These scanning systems are attractive for the large quantity of measurements that may be taken in a short time, for their precision and low level of invasiveness. The anthropometric data of Mexican women presented in this work are part of the first reported anthropometric study which gathered 3D measurements in Mexico (Prado-León, 2014). Such study aimed to provide a survey of sizes for apparel design in Mexico and was undertaken by the Ergonomics Research Center of the University of Guadalajara. Thus, there are anthropometric measurements of a representative sample for 2103 subjects of both sexes, aged between 15 and 65 from the Metropolitan Zone of Guadalajara.

Historically, anthropometric measurements in the United States were conducted for the purposes of sizing U.S. military equipment and apparel. During the mid-1990s, a bold new project was proposed to provide civilian measurement information in addition to military information, using a full body 3D scanner. This project, known as CAESAR (Civilian American and European Surface Anthropometry Resource), was put together jointly by the Society of Automotive Engineers and the U.S. Air Force and was designed around a new 3D laser body scanning technology. That study was designed to generate information that could serve multiple industries including automotive, aerospace, furniture, apparel, and others (Textile Clothing Technology Corporation, $[TC]^2$ (TC^2, 2003). The CAESAR study preceded $[TC]^2$'s SizeUSA study, which was completed in 2004. The objective of the SizeUSA National Sizing Survey was to measure the body dimensions of a representative sample of the U.S. population, so the variety of age, ethnicity, gender, and geographic area was considered when determining sampling methodology. The study reported in this chapter has been conducted with similar methodology and equipment than SizeUSA (TC^2, 2017).

Given the lack of data of body dimensions and proportions of Mexican women for apparel design, the aim of this chapter is twofold: 1) to report the body dimensions of a representative sample of women from the Metropolitan Zone of Guadalajara, Mexico, and 2) to calculate body proportions with respect to stature. Additionally, the authors analyze the change of dimensions and proportions of Mexican women aged from 17 to 65 years.

MATERIAL AND METHODS

Participants

Data were collected from a sample of 1131 women in the Guadalajara Metropolitan Area, Mexico. The data are part of an anthropometric study (Prado-León, 2014), where the sample was one of convenience, with measuring posts set up in such various locations as schools, commercial expositions, hospitals, and workplaces; places where numbers of people might wish to participate in a field study. Participants were invited verbally by the researchers, but also visual information was used in form of flyers and posters. Participants were invited to take part in the study and they were offered to receive a copy with their data and the clothing they were provided with as part of the study. Women's ages were between 15 and 65 years, and they were divided into the following age groups: 15-17, 18-24, 25-34, 35-44, 45-54, and 55-65. Body measurements were obtained via the TC^2 3D Body Scanning System.

Instruments

Measurements were obtained using two procedures: 1) manually measuring weight and height, and 2) automatically taking the remaining measurements with the [TC]^2 3D Body Scanning System. The manual measurements used one anthropometer built following the Martin type specifications, with an accuracy of 1.0 mm. The accuracy was tested prior to the study. Weight was recorded by means of a Secca Alpha scale, model 770 which was constantly calibrated for accuracy. Body measurements were obtained via the TC2 3D Body Scanning System, a Body Measurement System designed by the Textile and Clothing Technology Corporation [TC]^2 with volume scanning capacity for 1.1 m width, 1.1 m depth and 2.0 m height. It uses a principle of white-light phase profilometry to obtain a complete 3D representation of the body. Figure 1 shows the image of a body in frontal and side views which is an example obtained from the scanner.

Procedure

The procedure began with explaining the measurement process to prospective participants, assuring them of anonymity as well as privacy during scanning. A consent form was offered for them to sign before participating in the study. They were then asked to complete an anonymous questionnaire which elicited demographic data and sizes, sports-related and shopping habits. Participants then proceeded to the measuring area. Measurement included two stages, one of manual measurement where height and weight were taken, and another where automatic measurement by scanner collected the remaining data. Each participant was provided with scanwear prior to entering the change room/scanning area: a form-fitting pair of shorts and a bra top. In the scanning area, there was always a technical assistant who instructed and assisted the participants during the process. Once each subject entered the scanning chamber, she was scanned a minimum of two times, to assure that all measurements were obtained.

Figure 1. Frontal and side-view images, taken by the 3D body scanner

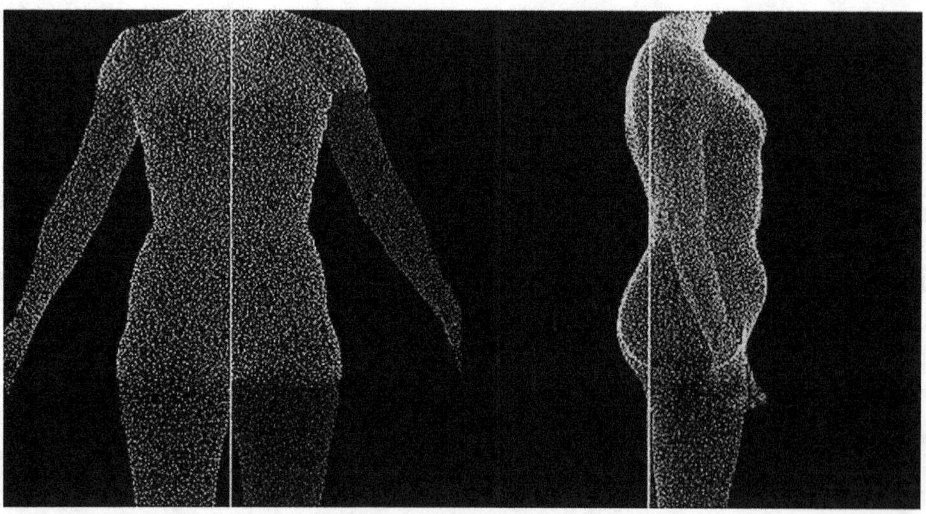

Definition of Dimensions

While the Body Measurement System permits extraction of over 100 measurements, the 28 dimensions presented here are the principal ones required by apparel manufacturers. Table 1 presents the definition of each of dimension considered in this study. Definitions of these dimensions were taken from the Internacional Standard ISO 8559 (Garment Construction and Anthropometric Surveys Body Dimensions)*, with the rest defined from a previous anthropometric study (Avila-Chaurand et al., 2007)** and from definitions included with the scanner itself ***.

Data Analysis

The mean, standard deviation and 5, 50 and 95 percentiles were calculated. The aggregate data for each dimension was tested to assure that they represented a normal distribution using Symmetry indices (Skinessis and Kurtosis). Body proportions relating to 26 dimensions were obtained, using measurement data from 6 age groups. Percentages of change in proportions were arrived at by using the difference in the value of the first group's proportions as related to the last age group's value. Statistical analyses were done with the SPSS V20 software.

RESULTS

Anthropometric Data

The mean, standard deviation, and percentiles 5, 50 and 95 for weight and 27 body dimensions of 1131 Mexican women living at the Metropolitan Zone of Guadalajara, divided into 6 age groups are shown in Table 2 to Table 7 respectively.

Table 8 displays the mean of Anthropometric data by age group and differences between the first and last groups. It may be observed that weight increases gradually from 15 to 65 years of age, with gains in all groups as age advances but the increase is more noticeable with the younger groups. For instance, the 18-24 group when compared with the 25-34 year olds the gain (around 7 kg) was much higher than other comparisons. Regarding stature, it was very similar between the younger groups 15-17, 18-24 and 25-34, but from the ages of 35-44 it began to decrease; Such that the difference in stature between 15-17-year-olds and the 55-65 group was 62 mm.

Body Proportions

Body proportion was obtained for 26 dimensions with respect to stature, in six age groups within the Mexican participants (see Table 9). The last column of this Table presents the percentage increase between the first and last age groups. It may be observed that all proportions change with age, if those for the 15-17 year-old group are compared with the 55-65 age group. However, in comparing one age group with the next closest, no remarkable change is observed. Although while changes are not homogeneous, we

Table 1. Definitions of anthropometric dimensions considered in this study

Dimension	Definition
1. Weight [kg]	Subject's total mass, measured in kilograms on a clinical scale, precise to 100 grams..**
2. Stature [mm]	The vertical distance between the crown of head and ground, with standing subject's head oriented in the Frankfort plane. **
3. Front neck base height [mm]	Distance between center front neck (through suprasternal point) and ground. ***
4. Long Shoulder height [mm]	Distance between the level of right armhole at highest shoulder point, and ground. ***
5. Armscye height [mm]	Distance between right underarm level and ground. ***
6. Armscye Girth [mm]	Circumference of right armscye measured from underarm midpoint, around arm, over shoulder point and back to midpoint.. ***
7. Waist height [mm]	Vertical distance from natural waist level at center back and ground, with subject standing upright.*
8. Upper hip height [mm]	Distance from a point located midway between top hip and waist levels (highest point on hip bone), and ground.***
9. Top hip height [mm]	Distance from a midway point, located between upper and lower hip heights, to the ground.***
10. Hip height [mm]	Vertical distance between the fullest part of buttocks (trochanteric projections) and ground, with the subject standing upright. *
11. Crotch height [mm]	Distance in a straight line from crotch to ground, with standing subject's legs slightly apart and body mass distributed equally.*
12. Center back neck point to waist-Contoured [mm]	Distance from the 7th cervical vertebra, following center back contour, to the waist. *
13. Side waist to hip height [mm]	Distance along the right side of the body from the small of back at waist level to the hip level (fullest part of buttocks), following the contour of the hip.***
14. Underarm length [mm]	Distance from the right underarm level at back, to the inside of wrist at the level of its most prominent bone (ulna); measured with the arm hanging naturally. *
15. Arm Length Straight [mm]	Distance from the right armscye shoulder line intersection, straight to the crease of hand immediately below most prominent wrist bone (ulna).***
16. Total crotch length [mm]	Distance from the center front at natural waist level, through the crotch, to center back at waist level.*
17. Neck column Girth [mm]	Horizontal circumference of the neck measured below Adam's apple, at level of center back neck point (7th cervical vertebra).*
18. Across Back width [mm]	Horizontal distance between lower ends of two back armscye folds, following the curvature of the back.*
19. Upper arm Girth [mm]	Maximum girth of the right upper arm at armpit level, measured with subject standing upright, arm hanging naturally straight.*
20. Chest Girth [mm]	Maximum circumference of chest measured from the bust level at center back, under armpits and above breasts. ***
21. Bust Girth [mm]	Girth measured over/under the shoulder blades, under armpits and across bust points, with standing subject breathing normally.*
22. Underbust Girth [mm]	Horizontal circumference of the body immediately below the breasts. *
23. Waist Girth [mm]	Circumference of the waist, measured on level derived from small of back (lumbar zone, where back curves closest to the center of the body). ***
24. Upper hip Girth [mm]	Circumference of upper hip measured at upper hip level. (Midway between waist and top hip level/iliac crest).***
25. Top hip Girth [mm]	Circumference of top hip measured midway between upper and lower hip levels.***
26. Thigh Girth [mm]	Maximum horizontal circumference of right thigh (measurement is taken at the upper right thigh, no pressure exerted, with the subject standing upright). *
27. Knee Girth [mm]	Horizontal circumference of right knee measured with leg straight.***
28. Calf Girth [mm]	Maximum circumference of the right calf, measured with subject standing legs slightly apart, body mass equally distributed.*

Source: * ISO 8559.
** Prado-León, Avila-Chaurand and González-Muñoz, 2007.
*** Body Scanning System.

Table 2. Anthropometric data of female aged from 15 to 17 (n= 184)

Dimensions	Mean	SD	Percentiles		
			5	50	95
1. Weight [kg]	55.2	10.6	42.7	52.7	76.1
2. Stature [mm]	1588	63	1482	1586	1700
3. Front neck base height [mm]	1324	56	1236	1320	1418
4. Long Shoulder height [mm]	1305	57	1213	1300	1403
5. Armscye height [mm]	1212	51	1134	1209	1300
6. Armscye Girth [mm]	380	30	336	378	439
7. Waist height [mm]	973	47	894	973	1055
8. Upper hip height [mm]	933	43	859	932	1014
9. Top hip height [mm]	892	41	817	889	969
10. Hip height [mm]	809	37	756	805	875
11. Crotch height [mm]	724	37	670	720	790
12. Center back neck point to waist- Contoured [mm]	414	28	372	410	462
13. Side waist to hip height [mm]	250	42	168	252	312
14. Underarm length [mm]	448	24	406	448	484
15. Arm Length Straight [mm]	530	31	482	532	586
16. Total crotch length [mm]	659	55	573	656	758
17. Neck column Girth [mm]	309	17	284	309	345
18. Across Back width [mm]	319	29	265	319	366
19. Upper arm Girth [mm]	253	32	206	246	318
20. Chest Girth [mm]	870	59	779	870	985
21. Bust Girth [mm]	905	73	792	894	1039
22. Underbust Girth [mm]	763	67	663	760	887
23. Waist Girth [mm]	752	76	641	735	909
24. Upper hip Girth [mm]	824	76	713	810	981
25. Top hip Girth [mm]	858	72	759	848	999
26. Thigh Girth [mm]	545	50	461	535	638
27. Knee Girth [mm]	350	26	312	346	405
28. Calf Girth [mm]	339	30	292	332	399

may yet observe two principal tendencies: stature and some other heights and lengths, such as armscye and crotch heights and underarm length, decrease slightly in the upper age groups. Conversely, weight and perimetral measurements increase to a greater degree. Two extreme cases of decreased or increased proportions are side waist-hip height, and top hip girth.

Figure 2 illustrates 11 proportions pertaining to the 55-65 age group. These proportions belong to the variables of front neck base height, long shoulder height, armscye height, waist height, crotch height, arm length straight, neck column girth, upper arm girth, waist girth, thigh girth and calf girth.

Table 3. Anthropometric data of female aged from 18 to 24 (n= 209)

Dimensions	Mean	SD	Percentiles		
			5	50	95
1. Weight [kg]	56.9	9.8	43.5	55.5	74.8
2. Stature [mm]	1589	57	1493	1586	1687
3. Front neck base height [mm]	1326	52	1240	1322	1404
4. Long Shoulder height [mm]	1304	54	1218	1304	1394
5. Armscye height [mm]	1209	51	1125	1209	1299
6. Armscye Girth [mm]	393	29	347	391	449
7. Waist height [mm]	979	53	889	986	1061
8. Upper hip height [mm]	936	51	846	942	1017
9. Top hip height [mm]	895	47	818	899	974
10. Hip height [mm]	811	41	743	811	886
11. Crotch height [mm]	726	42	661	726	791
12. Center back neck point to waist- Contoured [mm]	415	27	375	411	460
13. Side waist to hip height [mm]	222	30	167	225	266
14. Underarm length [mm]	457	30	410	45	505
15. Arm Length Straight [mm]	530	27	486	529	576
16. Total crotch length [mm]	676	71	536	684	777
17. Neck column Girth [mm]	327	17	299	325	360
18. Across Back width [mm]	331	26	289	329	378
19. Upper arm Girth [mm]	267	32	223	265	329
20. Chest Girth [mm]	898	66	802	891	1016
21. Bust Girth [mm]	928	80	815	917	1065
22. Underbust Girth [mm]	782	71	678	777	906
23. Waist Girth [mm]	782	79	679	766	926
24. Upper hip Girth [mm]	854	82	738	838	1009
25. Top hip Girth [mm]	974	75	869	965	1102
26. Thigh Girth [mm]	570	51	488	562	657
27. Knee Girth [mm]	360	26	323	359	403
28. Calf Girth [mm]	347	28	304	344	401

DISCUSSION

Differences in dimensions and bodily proportions suggest that morphological characteristics between the different age groups are dissimilar, so it is important to address such characteristics for optimal anthropometric garment fitting. Weight gain across the different age groups corresponds to advances in age, and shows a definite tendency in the Mexican population, rising slightly in the earlier age groups and more markedly in older groups. This coincides with Liu et al. (1999), who presented data for a Mexican female working population.

Table 4. Anthropometric data of female aged from 25 to 34 (n= 195)

Dimensions	Mean	SD	Percentiles		
			5	50	95
1. Weight [kg]	63.6	13.6	44.8	61.5	88.7
2. Stature [mm]	1587	61	1480	1587	1689
3. Front neck base height [mm]	1316	57	1222	1317	1403
4. Long Shoulder height [mm]	1300	54	1213	1297	1388
5. Armscye height [mm]	1202	50	1127	1200	1283
6. Armscye Girth [mm]	409	43	348	405	487
7. Waist height [mm]	972	53	886	970	1055
8. Upper hip height [mm]	931	49	856	927	1008
9. Top hip height [mm]	890	45	820	885	963
10. Hip height [mm]	808	40	745	804	875
11. Crotch height [mm]	713	43	643	710	785
12. Center back neck point to waist- Contoured [mm]	417	28	374	417	464
13. Side waist to hip height [mm]	232	54	130	238	309
14. Underarm length [mm]	438	29	393	437	484
15. Arm Length Straight [mm]	525	33	475	525	580
16. Total crotch length [mm]	700	71	597	699	823
17. Neck column Girth [mm]	334	41	281	324	410
18. Across Back width [mm]	341	33	289	339	394
19. Upper arm Girth [mm]	284	38	228	281	357
20. Chest Girth [mm]	159	12	139	159	179
21. Bust Girth [mm]	974	99	828	963	1167
22. Underbust Girth [mm]	831	96	705	812	1020
23. Waist Girth [mm]	836	117	683	827	1085
24. Upper hip Girth [mm]	906	118	753	892	1171
25. Top hip Girth [mm]	943	104	784	927	1146
26. Thigh Girth [mm]	583	49	498	583	672
27. Knee Girth [mm]	366	25	327	366	413
28. Calf Girth [mm]	355	30	310	353	408

As for changes in stature, in analyzing the Mexican women it is observed decreased stature starting with the 35 to 44 age group. Liu et al. (1999) did not find stature decreasing with age in Mexican women, but the fact that their age group was from 17 to 39 may explain the difference between these two studies. Various factors have been identified which explain the decrease in stature shown in the cross-sectional studies. One is the secular tendency referring to generational changes, those which occur over time. It has been demonstrated that younger generations are taller than older ones, which may be explained by certain changes in other factors such as a population's nutrition, hygiene, and health status (Peebles & Norris, 1998). Another factor is the biological change that occurs in the normal aging process. The

Table 5. Anthropometric data of female aged from 35 to 44 (n= 202)

Dimensions	Mean	SD	Percentiles		
			5	50	95
1. Weight [kg]	68.0	13.1	50.2	65.0	93.7
2. Stature [mm]	1563	61	1459	1568	1667
3. Front neck base height [mm]	1300	53	1211	1302	1395
4. Long Shoulder height [mm]	1281	57	1189	1281	1377
5. Armscye height [mm]	1178	53	1100	1181	1261
6. Armscye Girth [mm]	432	44	361	429	511
7. Waist height [mm]	948	52	862	949	1035
8. Upper hip height [mm]	909	49	828	910	991
9. Top hip height [mm]	870	46	794	871	948
10. Hip height [mm]	793	43	723	793	866
11. Crotch height [mm]	692	45	622	694	764
12. Center back neck point to waist- Contoured [mm]	427	28	384	425	483
13. Side waist to hip height [mm]	183	39	120	179	254
14. Underarm length [mm]	426	29	381	426	473
15. Arm Length Straight [mm]	518	33	464	517	572
16. Total crotch length [mm]	703	67	598	703	821
17. Neck column Girth [mm]	352	34	300	351	409
18. Across Back width [mm]	356	32	304	356	407
19. Upper arm Girth [mm]	306	38	248	299	373
20. Chest Girth [mm]	983	80	862	982	1113
21. Bust Girth [mm]	1031	100	876	1028	1208
22. Underbust Girth [mm]	882	94	737	872	1053
23. Waist Girth [mm]	910	120	736	904	1132
24. Upper hip Girth [mm]	963	107	794	959	1154
25. Top hip Girth [mm]	1016	113	868	1006	1235
26. Thigh Girth [mm]	592	47	519	590	673
27. Knee Girth [mm]	379	31	333	378	436
28. Calf Girth [mm]	365	31	318	361	411

most important change is thinning of the intervertebral discs and vertebrae, and consequent increase in curvature of the spine (Stoudt, 1981).

In referring to changes in perimeter with advancing age, Liu et al. (1999) found that the dimensions of bi-acromial and bi-trochanteric breadth increased with age. This coincides with our analysis of proportions in which we found that breadth and perimeters increase with age. These authors present data on breadth but not perimeters, but it is logical that if the breadths increase the perimeters will as well.

The normal aging process also offers some explanations for changes in the human body's size and shape. For example, breadths, primarily at the level of shoulders, chest, abdomen, and hips, are related

Table 6. Anthropometric data of female aged from 45 to 54 (n= 260)

Dimensions	Mean	SD	Percentiles		
			5	50	95
1. Weight [kg]	69.6	11.8	53.5	67.2	93.4
2. Stature [mm]	1542	58	1450	1544	1636
3. Front neck base height [mm]	1282	50	1193	1284	1370
4. Long Shoulder height [mm]	1263	53	1175	1265	1358
5. Armscye height [mm]	1163	46	1088	1165	1239
6. Armscye Girth [mm]	432	42	370	429	500
7. Waist height [mm]	933	44	860	932	1004
8. Upper hip height [mm]	896	42	826	895	964
9. Top hip height [mm]	858	40	792	858	924
10. Hip height [mm]	784	38	719	783	844
11. Crotch height [mm]	677	40	607	678	740
12. Center back neck point to waist- Contoured [mm]	432	30	380	431	476
13. Side waist to hip height [mm]	191	43	125	188	265
14. Underarm length [mm]	418	25	372	420	455
15. Arm Length Straight [mm]	512	27	469	510	556
16. Total crotch length [mm]	714	71	600	715	841
17. Neck column Girth [mm]	365	33	313	361	420
18. Across Back width [mm]	362	31	308	361	416
19. Upper arm Girth [mm]	312	32	263	309	375
20. Chest Girth [mm]	1010	70	899	1004	1138
21. Bust Girth [mm]	1064	86	928	1052	1215
22. Underbust Girth [mm]	923	88	798	913	1094
23. Waist Girth [mm]	950	101	795	940	1139
24. Upper hip Girth [mm]	1004	99	847	997	1180
25. Top hip Girth [mm]	1049	95	906	1036	1228
26. Thigh Girth [mm]	593	52	509	586	698
27. Knee Girth [mm]	381	30	333	380	439
28. Calf Girth [mm]	366	30	323	365	421

to a tendency toward the accumulation of fat and lessening of muscular tissue with the passage of years, generally starting in the 30s and 40s (Stoudt, 1981). Alternatively, it should be mentioned that changes in age groups must be viewed with a certain amount of caution, since there has been research suggesting that cross-sectional and longitudinal studies do not yield the same results (Borkan, Hults, & Glynn, 1983; Clement, 1974; Friedlaender et al., 1977).

Results of this study suggest that, in fitting apparel, it is necessary to consider proportions between age groups since there appears to exist a considerable difference between young and mature women. Given the above, the problem of anthropometric apparel fitting becomes more complex, requiring gar-

Table 7. Anthropometric data of female aged from 55 to 65 (n= 81)

Dimensions	Mean	SD	Percentiles		
			5	50	95
1. Weight [kg]	70.6	15.1	51.5	68.5	100.3
2. Stature [mm]	1526	59	1436	1527	1642
3. Front neck base height [mm]	1268	55	1179	1273	1364
4. Long Shoulder height [mm]	1254	53	1151	1254	1349
5. Armscye height [mm]	1154	45	1076	1159	1239
6. Armscye Girth [mm]	435	52	352	434	534
7. Waist height [mm]	930	44	842	936	1003
8. Upper hip height [mm]	894	42	811	900	967
9. Top hip height [mm]	858	40	782	862	928
10. Hip height [mm]	785	40	706	787	862
11. Crotch height [mm]	666	43	580	669	741
12. Center back neck point to waist- Contoured [mm]	424	29	374	420	476
13. Side waist to hip height [mm]	173	44	107	165	242
14. Underarm length [mm]	413	31	352	416	466
15. Arm Length Straight [mm]	517	28	464	520	563
16. Total crotch length [mm]	744	80	612	739	912
17. Neck column Girth [mm]	374	39	311	374	451
18. Across Back width [mm]	353	41	285	347	431
19. Upper arm Girth [mm]	317	44	251	318	382
20. Chest Girth [mm]	1016	95	882	998	1196
21. Bust Girth [mm]	1074	108	920	1069	1285
22. Underbust Girth [mm]	947	116	783	944	1179
23. Waist Girth [mm]	979	135	789	955	1282
24. Upper hip Girth [mm]	1043	128	870	1015	1322
25. Top hip Girth [mm]	1091	130	919	1065	1343
26. Thigh Girth [mm]	581	65	489	572	689
27. Knee Girth [mm]	384	39	333	376	468
28. Calf Girth [mm]	363	38	300	358	433

ments to be specially fitted for the changing proportions of older women. This is facilitated by using 3D data on body shape, contour and perimetral dimensions. Robinette & Daanen (2006) have reported that precision when measuring anthropometric dimensions with a 3D scanner is as good as or better than manual measurements.

Although data reported in this study are from a much-localized population, results concur with other authors (J. Y. Lee, Istook, Nam, & Park, 2007; Y.-S. Lee, 2000) regarding the need to conduct further anthropometric research nationally. This would be aimed at determining variability in the user population's morphology, in order to satisfy necessary fitting of clothing items and permit us to generalize

Table 8. Mean of each anthropometric dimension by age group

Dimensions	Age Groups						Differences Between First and Last Group
	Mean 15-17	Mean 18-24	Mean 25-34	Mean 35-44	Mean 45-54	Mean 55-65	
1. Weight [kg]	55.2	56.9	63.6	68.0	69.6	70.6	15.4
2. Stature [mm]	1588	1589	1587	1563	1542	1526	-62
3. Front neck base height [mm]	1324	1326	1316	1300	1282	1268	-56
4. Long Shoulder height [mm]	1305	1304	1300	1281	1263	1254	-51
5. Armscye height [mm]	1212	1209	1202	1178	1163	1154	-58
6. Armscye Girth [mm]	380	393	409	432	432	435	55
7. Waist height [mm]	973	979	972	948	933	930	-43
8. Upper hip height [mm]	933	936	931	909	896	894	-39
9. Top hip height [mm]	892	895	890	870	858	858	-34
10. Hip height [mm]	809	811	808	793	784	785	-24
11. Crotch height [mm]	724	726	713	692	677	666	-58
12. Center back neck point to waist-Contoured [mm]	414	415	417	427	432	424	10
13. Side waist to hip height [mm]	250	222	232	183	191	173	-77
14. Underarm length [mm]	448	457	438	426	418	413	-35
15. Arm Length Straight [mm]	530	530	525	518	512	517	-13
16. Total crotch length [mm]	659	676	700	703	714	744	85
17. Neck column Girth [mm]	309	327	334	352	365	374	65
18. Across Back width [mm]	319	331	341	356	362	353	34
19. Upper arm Girth [mm]	253	267	284	306	312	317	64
20. Chest Girth [mm]	870	898	159	983	1010	1016	146
21. Bust Girth [mm]	905	928	974	1031	1064	1074	169
22. Underbust Girth [mm]	763	782	831	882	923	947	184
23. Waist Girth [mm]	752	782	836	910	950	979	227
24. Upper hip Girth [mm]	824	854	906	963	1004	1043	219
25. Top hip Girth [mm]	858	974	943	1016	1049	1091	233
26. Thigh Girth [mm]	545	570	583	592	593	581	36
27. Knee Girth [mm]	350	360	366	379	381	384	34
28. Calf Girth [mm]	339	347	355	365	366	363	24

more from the analysis. As is also noted by Y.-S. Lee (2000), the complexity of many new requirements and system modifications means that anthropometric studies which include 3D information, along with comparative data between countries, have now become a necessity for apparel manufacturers in addressing consumer satisfaction and well-being.

From an ergonomic point of view, it is imperative to note that along with anthropometric fit in apparel design, it must be considered such additional aspects as fabric composition for different activities and climates, garment design that will contribute ease and comfort to those activities for which they are

Table 9. Comparison of body proportions between age groups

Dimensions	Age Groups						% Change by Age
	15-17	18-24	25-34	35-44	45-54	55-65	
1. Front neck base height	0.834	0.834	0.829	0.832	0.831	0.830	-0.4
2. Long Shoulder height	0.822	0.821	0.819	0.820	0.819	0.821	-0.1
3. Armscye height	0.763	0.761	0.757	0.754	0.754	0.756	-0.7
4. Armscye Girth	0.239	0.247	0.258	0.276	0.280	0.285	4.6
5. Waist height	0.613	0.616	0.612	0.607	0.605	0.609	-0.4
6. Upper hip height	0.588	0.589	0.587	0.582	0.581	0.585	-0.3
7. Top hip height	0.562	0.563	0.561	0.557	0.556	0.562	0.0
8. Hip height	0.509	0.51	0.509	0.507	0.508	0.514	0.5
9. Crotch height	0.456	0.457	0.449	0.443	0.439	0.436	-2.0
10. Center back neck point to waist-Contoured	0.261	0.261	0.263	0.273	0.280	0.277	1.6
11. Side waist to hip height	0.157	0.14	0.146	0.117	0.124	0.113	-4.4
12. Underarm length	0.282	0.288	0.276	0.273	0.271	0.270	-1.2
13. Arm Length Straight	0.334	0.334	0.331	0.331	0.332	0.338	0.4
14. Total crotch length	0.415	0.425	0.441	0.450	0.463	0.487	7.2
15. Neck column Girth	0.195	0.206	0.21	0.225	0.237	0.245	5.0
16. Across Back width	0.201	0.208	0.215	0.228	0.235	0.231	3.0
17. Upper arm Girth	0.159	0.168	0.179	0.196	0.202	0.207	4.8
18. Chest Girth	0.548	0.565	0.58	0.629	0.655	0.665	11.7
19. Bust Girth	0.570	0.584	0.614	0.66	0.69	0.703	13.3
20. Underbust Girth	0.480	0.492	0.524	0.564	0.599	0.620	14.0
21. Waist Girth	0.474	0.492	0.527	0.582	0.616	0.641	16.7
22. Upper hip Girth	0.519	0.537	0.571	0.616	0.651	0.683	16.4
23. Top hip Girth	0.540	0.608	0.589	0.650	0.68	0.714	17.4
24. Thigh Girth	0.343	0.359	0.367	0.379	0.385	0.380	3.7
25. Knee Girth	0.220	0.227	0.231	0.242	0.247	0.251	3.1
26. Calf Girth	0.213	0.218	0.224	0.234	0.237	0.237	2.4

worn (Entwistle, 2000). Results from this study suggest that special attention is needed in clothing design for older people. Designing without considering the age groups characteristics can create difficulties in using, putting or removing a garment (Civitci, 2004; Gupta et al., 2013).

A potential limitation of this study might be that it was restricted to a single metropolitan area, more research needs to be conducted to gather anthropometric data nationally (as also mentioned by J. Y. Lee et al., 2007; Y.-S. Lee, 2000). This would be aimed at gathering countrywide anthropometric data to better inform garment design and permit generalizations from the analysis.

Figure 2. Illustration of eleven proportions for Mexican women 55-65 years of age

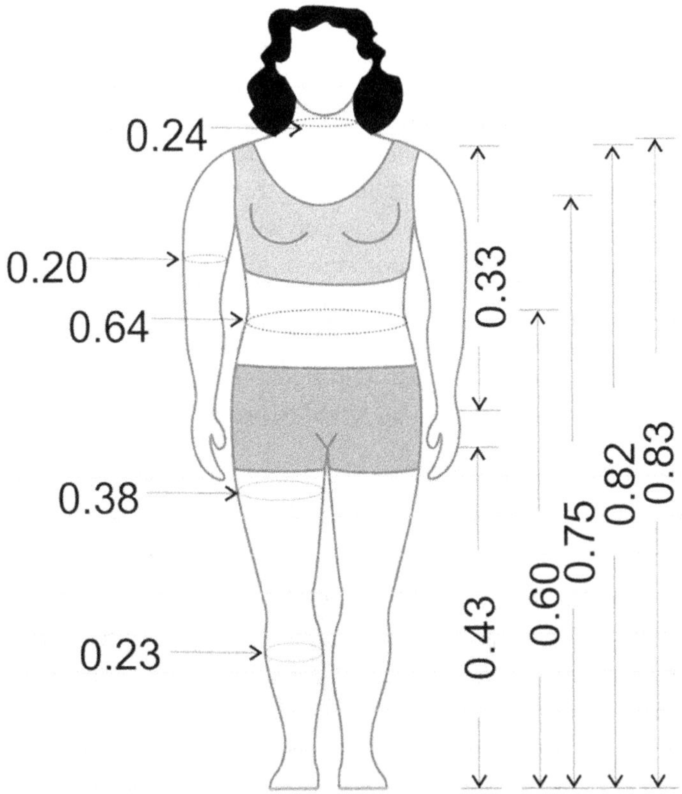

CONCLUSION

The research detailed in this chapter was undertaken in response to the lack of anthropometric data for apparel design in Mexican women. Accordingly, the study aimed to report the body dimensions of a representative sample of women from the Metropolitan Zone of Guadalajara, Mexico, and to provide data on body proportions calculated with respect to stature. Thus, this chapter summarizes data of 28 body dimensions and 26 body proportions that are essential for apparel design. In addition, it has been analyzed differences and changes of dimensions and proportions through the different age groups who took part in the study. Knowledge on anthropometric measurements and body proportions of this groups of users could help apparel companies to produce and meet demands for comfortable and well-fitting clothes.

Results of this study suggest that, in fitting apparel, it is necessary to consider proportions between age groups since there appears to exist a considerable difference among age groups, especially between young and mature women. The authors emphasize the need of paying attention on these variances among these different aged groups when designing clothing, otherwise, it might create difficulties for users and discouraging them in buying certain types of clothing. It can be concluded that this anthropometric survey provides a set of data that can help in designing more ergonomic clothing.

FUTURE RESEARCH DIRECTIONS

Based on the results of this study, several research directions could be considered. Firstly, the authors suggest that further research is needed which should consider more locations in the country for a complete measure of the Mexican population, as this might be an important factor in how data behaves.

Since main changes were found in the elder age groups, more research is needed in populations aged 65 years old and over, which were not included in this study. It would be expected to find deeper differences on anthropometric dimensions and proportions among these populations due to the aging process. Furthermore, as the population is aging, future research in fitting apparel should also include measurements of other characteristics of these populations as reach, force, and dexterity capabilities.

More and more, 3D Anthropometry studies have been reported in the literature. It is assumed that using these measurements helps to make better apparel fitting, however, research is needed in order to see how these data are being or could be used in the real world to produce clothing within an industrial process.

ACKNOWLEDGMENT

This research was supported by The National Council of Science and Technology, [CONACyT -39996]. Special thanks to the following researchers of the Ergonomics Research Center who participated in the data collection process: Rosalío Ávila-Chaurand, Enrique Herrera-Lugo, Rosa A. Rosales-Cinco and Mónica P. López-Alvarado.

REFERENCES

Avila-Chaurand, R., Prado-León, L. R., & González, Muñoz, E. L. (2007). *Dimensiones Antropométricas de Población Latinomericana* (2nd ed.). Guadalajara, Mexico: Universidad de Guadalajara.

Borkan, G. A., Hults, D. E., & Glynn, R. J. (1983). Role of Longitudinal Change and Secular Trend in Age Differences in Male Body Dimensions. *Human Biology*, *55*, 629–641. doi:10.2307/41464710 PMID:6642483

Castellucci, H. I., Arezes, P. M., & Viviani, C. A. (2010). Mismatch between classroom furniture and anthropometric measures in Chilean schools. *Applied Ergonomics*, *41*(4), 563–568. doi:10.1016/j.apergo.2009.12.001 PMID:20031115

Civitci, S. (2004). An ergonomic garment design for elderly Turkish men. *Applied Ergonomics*, *35*(3), 243–251. doi:10.1016/j.apergo.2004.02.001 PMID:15145287

Clement, F. J. (1974). Longitudinal and Cross-Sectional Assessments of Age Changes in Physical Strength as Related to Sex, Social Class, and Mental Ability. *Journal of Gerontology*, *29*(4), 423–429. doi:10.1093/geronj/29.4.423 PMID:4833758

Corona, G. P. (2014). *Propuesta de mejora en el ajuste de prendas de vestir para dama por análisis de elementos antropométricos de la Zona Metropolitana de Guadalajara*. University of Guadalajara.

Dianat, I., Karimi, M. A., Asl Hashemi, A., & Bahrampour, S. (2013). Classroom furniture and anthropometric characteristics of Iranian high school students: Proposed dimensions based on anthropometric data. *Applied Ergonomics*, *44*(1), 101–108. doi:10.1016/j.apergo.2012.05.004 PMID:22695080

Entwistle, J. (2000). *The Fashioned Body: fashion, dress and modern social theory*. Cambridge, UK: Polity Press.

Faulhaber, J. (1976). *Investigación Longitudinal del Crecimiento*. Instituto Nacional de Antropología e Historia. Departamento de Antropología Física.

Friedlaender, J. S., Costa, P. T. Jr, Bosse, R., Ellis, E., Rhoads, J. G., & Stoudt, H. W. (1977). Longitudinal Physique Changes Among Healthy White Veterans at Boston. *Human Biology*, *49*, 541–558. doi:10.2307/41464475 PMID:590953

Gupta, D., Das, A., Jain, V., Soneja, T., Dang, P., & Srivastava, P. (2013). Design of Ergonomic Garment for Elderly Indian Women For Authors Design of Ergonomic Garment for Elderly Indian Women. *Research Journal of Textile and Apparel*, *17*(3), 118–12743. 10.1108/RJTA-17-03-2013-B012

Hernandez-Arellano, J. L., Talavera-Aguirre, G., Serratos-Perez, J. N., Maldonado-Macias, A. A., & Garcia-Alcaraz, J. L. (2016). Anthropometrics of University Students in Northern Mexico. *Open Journal of Safety Science and Technology*, *6*(4), 143–155. doi:10.4236/ojsst.2016.64011

Hsu, C.-H. (2008). Applying a Bust-To-Waist Girth Ratio Approach To Develop Body Measurement Charts for Improving Female Clothing Manufacture. *J Chin Inst Industrial Eng*, *25*(3), 215–222. doi:10.1080/10170660809509085

Konz, S., & Johnson, S. (2000). *Work Design: Industrial Ergonomics* (5th ed.). Holcomb Hathaway Publishers.

Laios, L., & Giannatsis, J. (2010). Ergonomic evaluation and redesign of children bicycles based on anthropometric data. *Applied Ergonomics*, *41*(3), 428–435. doi:10.1016/j.apergo.2009.09.006 PMID:19837401

Laitala, K., Klepp, G., & Benedicte, H. (2011). Materialised Ideals: Sizes and Beauty. *Culture Unbound*, *3*(3), 19–41. doi:10.3384/cu.2000.1525.11319

Lee, J. Y., Istook, C. L., Nam, Y. J., & Park, S. M. (2007). International Journal of Clothing Science and Technology. *International Journal of Clothing Science and Technology International Journal of Clothing Science and Technology Journal of Fashion Marketing and Management: An International Journal*, *19*(3), 374–391. doi:10.1108/09556220710819555

Lee, Y.-S. (2000). 3-d anthropometric information for designing the wearable. *Proceedings of the Human Factors and Ergonomics Society Annual Meeting*, *6*(38), 751–755. doi:10.1177/154193120004403818

Liu, W. C. V., Sanchez-Monroy, D., & Parga, G. (1999). Anthropometry of female maquiladora workers. *International Journal of Industrial Ergonomics*, *24*(3), 273–280. doi:10.1016/S0169-8141(98)00036-5

Loker, S., Ashdown, S., & Schoenfelder, K. (2005). Size-specific analysis of body scan data to improve apparel fit. Journal of Textile and Apparel. *Journal of Textile and Apparel, Technology and Management*, *4*(3), 1–15.

Newcomb, E., & Istook, C. (2011). Confronting stereotypes: Apparel fit preferences of Mexican-American women. *Journal of Fashion Marketing and Management: An International Journal*, *15*(4), 389–411. doi:10.1108/13612021111169915

Panagiotopoulou, G., Christoulas, K., Papanckolaou, A., & Mandroukas, K. (2004). Classroom furniture dimensions and anthropometric measures in primary school. *Applied Ergonomics*, *35*(2), 121–128. doi:10.1016/j.apergo.2003.11.002 PMID:15105073

Peebles, L., & Norris, B. (1998). *Adultdata: The handbook of adult anthropometric and strength measurements : data for design safety*. Dept. of Trade and Industry. Retrieved from https://books.google.com.mx/books/about/Adultdata.html?id=ZcdetwAACAAJ&redir_esc=y

Pheasant, S. (1996). *Body Space: Anthropometry, ergonomics and the design of work*. London: Taylor and Francis.

Prado-León, L. R. (2014). *Antropometría para el diseño industrial, de espacios y de vestuario: datos y aplicaciones*. Guadalajara, Mexico: Universidad de Guadalajara.

Ramos-Galvan, R. (1975). Somatometría pediátrica : Estudio semilongitudinal en ninõs de la ciudad de México. *Archivos de Investigacion Medica*, *6*(1).

Robinette, K. M., & Daanen, H. A. M. (2006). Precision of the CAESAR scan-extracted measurements. *Applied Ergonomics*, *37*(3), 259–265. doi:10.1016/j.apergo.2005.07.009 PMID:16202970

Roebuck, J., Kromer, K., & Thomson, W. (1975). *Engineering Anthropometry Method* (2nd ed.). New York: Wiley.

Roebuck, J. A. Jr. (1995). *Anthropometric Methods: Designing to Fit the Human Body*. Santa Monica, CA: Human Factors & Ergonomics Society.

Sanchez-Monroy, D. (2007). Operadores de autotransporte de 18 a 68 años. In R. Avila-Chaurand, L. R. Prado-León, & E. L. González-Muñoz (Eds.), Dimensiones Antropométricas de Población Latinoamericana (2nd ed.). Guadalajara, Mexico: Academic Press.

Stoudt, H. W. (1981). The Anthropometry of the Elderly. *Human Factors: The Journal of the Human Factors and Ergonomics Society*, *23*(1), 29–37. doi:10.1177/001872088102300104 PMID:7228046

TC^2. (2003). *SizeUSA project notebook*. Author.

TC^2. (2017). *SizeUSA*. Retrieved from http://www.tc2.com/size-usa.html

Yokota, M. (2005). Head and facial anthropometry of mixed-race US Army male soldiers for military design and sizing: A pilot study. *Applied Ergonomics*, *36*(3), 379–383. doi:10.1016/j.apergo.2005.01.009 PMID:15854582

KEY TERMS AND DEFINITIONS

3D Body Dimensions: Measurements of body dimensions made with a scanner that provides three-dimensional data.

Anthropometry: It can be defined, in general terms, as the anthropological technique that measures the human body.

Body Proportions: Relationship of correspondence between different parts of the body and a selected body dimension as constant.

Ergonomic Design: Design of products, tools, clothes, systems, or environments that considers the characteristics of people who will use them.

Fitting Apparel: Adjustment of clothing to body shape and dimensions.

Older People: People aged 60 and more.

Younger People: People aged 59 and less.

Section 4
Usability and User Experience

Chapter 13
Usability on Standard Work Visuals in Manufacturing

Guadalupe Hernández-Escobedo
Instituto Tecnológico de Tijuana, Mexico

Karina Cecilia Arredondo-Soto
Universidad Autónoma de Baja California, Mexico

Carlos Aceves-González
Universidad de Guadalajara, Mexico

Arturo Realyvásquez
Institituto Tecnológico de Tijuana, Mexico

ABSTRACT

This chapter presents a proposed checklist to determine the usability levels of standard work visuals (SWVs) used in manufacturing processes. The checklist considers attributes in terms of easy to use and methods facilitating their design. It is because there were not enough studies that address both attributes of use and methods to design visuals. Specifically, flow diagrams, procedures lesson, and one-point lessons as visuals are studied in this chapter. The theoretical departure covered diverse areas, and activity theory served as conceptual framework. So, the overall purpose of this chapter is to (1) explore the defined specifications for design visuals, (2) propose a checklist to determine the levels of usability, and (3) evaluate the visuals utilized in two manufacturing companies serving as cases of study. Finally, the results have permitted the convenience of using the proposed checklist for checking the design and utilization of the visuals in manufacturing processes to be proven.

INTRODUCTION

The visuals are widely used in the manufacturing industry. Therefore, signs, labels, cards, posters, graphics, quick lessons, banners, displays or other means of information are commonly placed in the physical work areas (Brady, 2012). With their use, engineers and managers intend to create safe and efficient workplaces seeking the reduction and at best, the elimination of constant training and supervision. Apart

DOI: 10.4018/978-1-5225-5234-5.ch013

Copyright © 2018, IGI Global. Copying or distributing in print or electronic forms without written permission of IGI Global is prohibited.

from the expected benefits, the use of these pursues other assets in terms of productivity and safety. They assume that if employees have available the information required or needed at a given time in their work areas, they can perform your activities efficiently, effectively and safely (Greif, 1991). So, the types of visuals used in manufacturing are named visuals, work visuals, or standard work visuals (visuals, WVs or SWVs -used interchangeably-).

Particularly, the visuals have been developed by lean practitioners as communication aids and are used to help on driving operations and processes in real time (Parry& Turner, 2006). Those visuals serve to support the consolidation of standards and to point out abnormalities in the production areas. It is common that personnel or individuals deal with visuals aimed to reinforce and guarantee the operational standard work by implicitly exhibiting efficient work methods. The practice on designing visuals seeks to eliminate or reduce variations resulting in defects, which increase quality problems. This is stressed by some studies that have discussed the importance of visualization and visual means in the manufacturing sector (Bevilacqua, et al., 2013; Steenkamp, et al., 2017; Tjell, & Bosch-Sijtsema, 2015). For example, work visuals have been used in manufacturing processes in the industrial sectors considered of low propensity to error, such as the medical, aeronautical, among others sectors. These visuals are primarily thought to exhibit the efficient work process underscoring expected activities from individuals.

Then, this opened new areas of study giving importance to use of visuals in context. According to Tjell, & Bosch-Sijtsema (2015), Visual Management (VM), as area of interest, has emerged during the past decades within manufacturing and service organizations, as a system that through visualization enables the employees to better understand their role and contribution in relation to both their own organizational values and customer needs. For instance, 5S, Standard Work (SW), and Total Productive Maintenance (TPM) are three concepts extracted from the Lean Manufacturing (LM) that are commonly applied to capture the relation between the organizations and their customers into the form of work visuals. The origin of the link formed between the uses of visuals in LM is named Visual Factory (VF). In order to remind this, it is usual that flow diagrams, procedures, and lessons of a point are collocated into the work areas to achieve it. So, these visuals are seen as a product to reinforce that relationship and should be designed with this aim.

In the current literature, there were found some suggestions to design those visuals (Niebel, Freivalds & Osuna, 2004; Niebel & Benjamin, 1996). Nevertheless, there are not enough studies that enlist specifications for designing SWVs and how to address those specifications in a holistic approach considering the expected activities from individuals. So, in order to fill this gap of knowledge, it is important to capture the relationship between the organization and customers through the development of work efficiently, effectively and safely. Additionally, the consideration of the individuals, who are the organizational elements, that creates value in this relationship.

So, the overall purpose of this chapter is to (a) explore the defined specifications for designing visuals, (b) propose a checklist to determine the level of usability of the visuals; and, (c) evaluate the visuals of two manufacturing companies serving as cases of study. Initially, it presents a theoretical departure covering LM and VF, usability and AT areas. After, some methodologies are explored continuing with the methodology that serves to put the basis to develop the checklist. Then, the proposed checklist is presented. It is followed by the cases of study showing the evaluation of the level of usability of work visuals used in two manufacturing organizations. Finally, conclusion, recommendations and the future direction of research are displayed.

THEORETICAL DEPARTURE

Diverse SWVs are collocated within the organization's companies to support their processes and reduce the predisposition of errors within them. In order to decrease them, it is important to focus on specific processes, production configurations and their intrinsic elements (Graif, 1991). Particularly, those visuals serve to support the consolidation of standards and to point out abnormalities in the production areas. It is common that personnel deal with visuals aimed to reinforce and guarantee the standard work of operations by implicitly exhibiting an efficient working method (Brady, 2012). The practice seeks to eliminate or reduce variations, which should result in defects increasing quality problems (Melton, 2005).

For example, visuals have been used in manufacturing processes in the industrial sectors considered of low propensity to error, such as the medical, aeronautical, automotive among others sectors (Dickson et al. 2009). Specifically, companies in those sectors implement LM initiatives to reduce that propensity of error. The incorrect execution of an operation endangers the life of the end user of the product. A clear illustration in the manufacturing of medical devices is the quote "*if I touch the product, I touch the patient*". Consequently, this industrial sector exhibits its needs of strategies that facilitate the work of its employees helping permanently to diminish the defective products. This is based on the fact that visuals are intended to support employees to perform operations thus reducing the error, remembering that an error involves the possibility of losing a life and costly demands (Dickson et al., 2005). As a result, the existing regulations for the elaboration of these products are rigorous and demanding (Durivage, 2016).

Then, the SWVs are those visuals that aid to assure that works are generally executed in the most efficient method where alterations are eradicated to decrease errors or defects. Their objective is to boost consistency and accuracy so that many types of visual area used to achieve it (Brady, 2012). In this chapter, three types of visuals are studied: dance charts or flow diagrams, procedures, and lessons of a point or one-point lessons. Dance charts or flow diagrams (DC or FD -used interchangeably-) are those visuals deployed in the work area indicating and remembering to individuals the sequence of activities to make products and the personnel needed to act those activities (Brady, 2012). They also involve the production configurations as production cells or lines used in manufacturing. Procedures are the list of phases needed and described to accomplish certain activity in the manufacture of products. They are also generally confined to a single physical area within the production area that suggests specifically to a workstation. Although organizations have usually employed visuals, they linked procedures with training videos, which can be observed or demanded in real time (Brady, 2012). Lessons of a point or one-point lessons (LP or OPL –used interchangeably-) are delineated explanations of critical operations or points that should be partially blown by individuals. These lessons regularly refer to an approval or rejection criteria (quality, cosmetics, or technical) and workstations could necessitate several of them (Brady, 2012). An important issue is the information needs of the individuals in the workstations because critical operations are linked with the activities that change within the production processes and configurations (Garcia Criollo, 2000).

Lean Manufacturing and Visual Factory

LM is a philosophy for production management that aims to favor the continuous flow in the assembly line. Visuals are the most effective way to enforce standards that are implemented during LM activities. So, one initiative linked to the use of visuals is the concept called Visual Factory (Brady, 2012). This initiative has as primary objective to allocate critical information in those work areas where it is required or needed within the application of other concepts of LM. Diverse visuals are situated in the organizations in order to support their activities, which are closely associated with the required information by individuals performing activities. For instance, productivity and safety are essential elements within VF as well as socio-cultural aspects (Graif, 1991). In line, Steenkampa et al., (2017) developed a visual management system for managing manufacturing resources for the Stellenbosch Technology Centre's Laboratory for Advanced Manufacturing (STC-LAM). This system gathers shop floor data and displays it in a dashboard reducing the process of seeking information. Additionally, Jiun et al., (2014) proposed a visual board to facilitate production flow regulation at sequential processes. However, their approach consisted in a work-in-progress board to avoid significant inventory build-up, shorten average flow time, and ensure maximum production leveling. Furthermore, Bevilacqua et al., (2013) presented some benefits of implementing a visual workplace. It enlisted some advances that could be verified on the effects of the new operative approach on the mental worker's workload.

On the other hand and providing some guidelines on utilizing the initiative, Bilalis et al. (2002) presented the visual characteristics principles used in the development of safety systems, visual factory design, and advanced training practices. Principally, they focused their efforts on that information required in the processes and activities exhibiting what information is mandatory and what information could be excluded in the visuals. Table 1 provides the standard and variable information that is generally monitored and shows the interest of a VF.

Similarly and considering human aspects, Lindlöf & Soderberg (2011) investigated the pros and cons of lean visual planning at four technology-based companies. They argued that there are few industrial cases investigated outside Toyota. Later, Lindlöf (2014) studied VM as a way of improving an organization's information processing capability. His research was focused on task communication, i.e., how the process, the tasks, and the deliverables are communicated. As a result, he found that visualization plays an important role within cognitive functions and knowledge transfer. Similarly, visualization is seen an attractive communication strategy (Lindlöf, 2014; p. 6).

However, none of the above-mentioned investigations addressed specific considerations for the design of flow diagrams, procedures and one-point lessons, which are relevant in LM and specifically within VF. Particularly, these studies opened the opportunity to research and consider companies in diverse

Table 1. Information must be included in visuals

Standard Information	Variable Information
• Safety signs color-coded and correct sized • Motivational signs (slogans, mission, vision) • Basic safety and job-related signs • Mimic diagrams of special installations • Colour-coded boundaries of all areas • Information signs • Directional signs • Hygiene procedures • Cleaning procedures • Start-up and shut-off procedures • Lockout and tag out procedures • Floor layout posters • Evacuation and emergency procedures • Standard settings • Equipment limitations and restrictions	• Operating procedures depending on product • Palletization procedures depending on product • Quality control procedures depending product • Product specifications information • Production, customer satisfaction, incident, waste, and energy consumption shift rates against target and best-achieved rates • Variable settings (product related) • Production • Equipment limitation • Restrictions depending on the product • Staff performance

Source: Adapted from Bilalis et al., 2002.

industrial sectors as manufacturing of medical devices, automotive and aerospace, that usually implement LM initiatives to correct deviations in their processes. Besides, the results suggest to putting attention to cognitive processes of individuals, who finally design and use those visuals performing activities in the manufacturing of products. Also, the proposed information requirements are factors that should be included into the guidelines to design and elaborate visuals. In other words, the guidelines should include cognitive processes of individuals, information requirements and socio-cultural aspects in the visuals. In other words, these suggest the attention of them as product attending user requirements in context.

Usability

Usability is a term introduced in 1984 in the context of computing and there are several definitions in the current literature. One of them suggests attributes of quality of products in terms of easy to use (Nielsen, 1993). Another is related to those methods used to facilitate the use during the design process (Avila-Chaurand, 2013). In other words, one of them is related to the definition of usability and the other, guidance for designing usability (Jokela et al., 2003). In this chapter, both definitions are used to capture the expected result and the process to obtain it. Five elements conform usability: learnability, efficiency of use, ability of infrequent users, frequency, seriousness of errors, and subjective user satisfaction (Nielsen, 1993). In order to extend the use of these concepts in any product, both definitions capture effectiveness, efficiency, and satisfaction of products permitting to achieve specific objectives of particular users within a peculiar context of unique use (Jokela et al., 2003). This suggests that products should be designed for particular objectives, uses, users, and contexts at the same time in order to satisfy the users or customers (used interchangeably).

Here, the satisfaction of customers is the departure point in the design process considering relevant that SWVs mediate the work as an individual or collective activity. At this point, visuals are seen as a product resulting of an easy process. This product should be designated to achieve particular objectives and uses of determined users in certain contexts that must satisfy them at the same time. For these reasons, visuals are seen as mediators of diverse associated activities: perception, involuntary and voluntary attention, discovering meaning on them and motor actions (Avila-Chaurand, 2013), Figure 1 exhibits these chained activities. In other words, activities begin with the perception of visuals at work. Then, users move from their involuntary attention to their voluntary attention, discovering the meaning of visuals within the work area. Finally, they continue with the required motor actions and shown in the visuals. Particularly, this shows that visuals can mediate the work required in the achievement of the operations in the production processes initiating with the perception activity. This activity is seen as a cognitive process in which individuals use principally their senses. In the case of visuals, the sense of sight is the principal sense utilized in the manufacturing processes. It serves to pick up the information contained in the visuals displayed in the work areas.

Consequently, the effects on the design come from the usability concept in which perception seen as the activity is a relevant element in consideration. This concept also contemplates aspects of form and background (a type of product and process) and its objective is to understand and facilitate their use, risk perception and mental stress (Rossi et al. 2013). In this line, there are diverse effects of visualization seen here as visual perception, which is a recognized studied area from the cognitive perspective. Its results have shown that vital cognitive processes are correlated with the visualization effects. Similarly, these processes aid with the functions of communication and thinking (Lindlöf, 2014), among others. For example, some researchers have highlighted the use of visualization to reinforce different cognitive

Figure 1. Chained activities

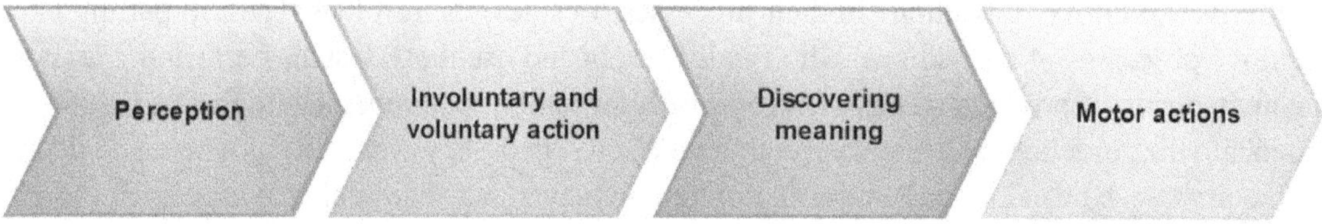

functions, as is shown in Table 2. It presents different cognitive functions focusing primarily on the information processing capacity and identifying patterns. Primarily, these functions served to put the basis implicitly and/or explicitly in the norms and regulations in the visual aids presented in the following paragraphs related to those regulations.

It is important to underscore that there are regulations which norm the design of visual specifically aids with the strong relationship with security and safety. Some examples are the safety signs, electrical labeling, pipe marking, hazardous chemical labels, lockout/tag out procedures, permit required confined space placards, physical hazard marking, among others regulations. Particularly, three organizations (American National Standard Institute –ANSI-, International Organization for Standardization –ISO- and Occupational Safety and Health Administration –OSHA-) are recognized as specialized bodies that establish these norms. In the case of Mexico, there are some regulations emitted by governmental offices, which are aligned to those issued by ANSI and ISO (Aceves-Gonzalez; Prado-Leon and Avila-Chaurand, 2010). This suggests the global use of standards related to safety and security issues.

Continuing with the term usability, it primarily suggests the consideration of users. Here it is proposed to ponder and recognize equally aspects, which impose companies in design and elaboration of visuals as artifacts or tools. Similarly, it is important to make systemic considerations that allow us to explore into details that are generally left out (Mwanza, 2002). As a manner of detail, the attention to sociocultural aspects is one of them. These refer to the values, customs, habits, tastes, and preferences of employees in the use of visuals. For these reasons, it is relevant to seek and find a framework that permits to include those contemplations. Consequently, one conceptual framework, which has been used in different areas of knowledge and permitted to explore details on objects of study, is AT. For example, in past research in the User-Computer Interaction, it has been used to develop a methodology used to evaluate that interaction. It is because this framework works well incorporating sociocultural aspects and putting atten-

Table 2. Cognitive functions reinforced by visualization

Cognitive Function	Authors
Information processing capacity	Miller, 1956; Larkin and Simon, 1987
Identifying patterns (Gestalt psychology)	Koffka, 1935; Ellis, 1938; Glenberg and Langston, 1992
Memory	Kosslyn, 1980; Shepard and Cooper, 1982
Learning	Mandil and Levin, 1989; Weidenmann, 1989
Problem solving	Rieber, 1985; Finke, 1990
Reasoning	Bauer and Johnson-Laird, 1993; Novick, 2001
Comprehend verbal information	Kosslyn and Koenig, 1992

Source: Adapted from Lindlöf, 2014.

tion to experiences and knowledge as mediators (Engeström, 2000). This idea refers to past and actual interactions between users and visuals at work. Hence, AT is proposed to be the conceptual framework that should be utilized to develop the basis of a checklist taking into account systemic considerations. Under this frame, visuals are particularly seen as mediators of the work activities mentioned before, specifically in the perception activity. Visuals are seen as graphic instruments that should mediate the mental and motor activities of the individuals in a short period of time (see Figure 1).

Methodologies in Literature

In the current literature, diverse methodologies referring to usability were found. Some of them were focused on evaluating the work as a product in terms of usability. The methodologies come from diverse perspectives and approaches in order to evaluate specific tasks. For instance, the task performance is commonly explained and analyzed through the use of the task analysis method. This helps to describe how users interact with complicated systems putting major attention on cognitive tasks (Gómez-Bull, Hernández-Arellano, and Ibarra-Mejía, 2015). These tasks and its components are important in the design of production processes to minimize the costs of operation (Lehto and Landry, 2013), referred as accidents and occupational diseases. The idea is to design a place in which productivity and quality of products are increased. Here, the main objective is to increment the level of usability in the workplace as a product. In these cases, different methods to gather and analyze data from the places are used. For example, video and mathematical approaches are utilized (Gomez-Bull et al, 2015; Weisner and Deuse, 2014; Rossi et al., 2013). They have provided methodological richness to the evaluation of designs in terms of usability and increased the number of methods to evaluate a similar number of contexts in which methodologies could be applied.

On the other hand, other methodologies are been focused on taking care of details of the procedures of design in terms of usability. These practices imply effective forms of work designing usability in work and taking care of the steps included into those methodologies (Mwanza, 2002). Diverse methods are also developed to assist usability in context. One example is a methodology that was used to detect and mitigate potential usability problems in the design and use of medical equipment (Bligärd and Osvalder, 2013). The principal objective was to evade use errors when that equipment is handled. Another illustration is the design of Internet web pages (Prado-Leon and Cardenas-Vazquez, 2005). The research put on manifest diverse factors that should be considered in the design of the web pages. Principally, the methodology should reflect the definition and analysis of the user requirements; the system design and convenient software; the implementation and tests, and the integration and system test. Additionally, other individuals should participate in the design of the web pages covering other approaches designing under usability conditions.

Similarly, the usability assessment of both product evaluation and methodologies utilizes diverse forms to measure it. The methods are employed to gather information related to that assessment. Particularly, the measurements are connected with the product or methodology in use. For example, in the case of products, usability is aided with protocols and basic requirements to evaluate its performance (IEA, 2008). In order to offer credibility in the results, subjective and objective information should be gathered and an evaluation report should be delivered exhibiting the usability level of the product or products in the study. The result could give an idea of the success of products in the market (Nael, 2011). In the case of methodologies, the procedure itself provides subjective and objective information concerning the usability levels of the methods. Interviews are a common method to gather information, which are similarly occupied to assess usability in products (Young and Staton, 2005). It provides a high grade of environmental validity. It means that designers listen to users during the process of design provides valuable information to improve products. So, usability could be assessed through the use of diverse methods gathering subjective and objective information and results can show the usability levels.

Activity Theory

AT or Cultural Historical Activity Theory (CHAT) affirmed that "consciousness is formed through activity" (Wilson, 2008, p.120) and consciousness is expressed as "the product of an individual's interactions with people and artifacts or tools in the context of everyday practical activity" (Kaptelinin and Nardi, 2006). It is argued that consciousness or reality is subjective and is shaped by social and cultural understanding (Vygotsky, 1978). It means that individual understanding results of exchanges and experiences with tools and other individuals in the routine life in which they are engaged.

Additionally, AT suggests that consciousness and activity are united (Kaptelinin and Nardi, 2006) and activity is object-oriented but it cannot be detached from reality, which is objective in an expansive perception. Here, reality should be explained in terms of both natural science and cultural/social characteristics. It is because "the human subject is social in nature, shaped by culture, and influenced by language, acting with or through other people in organizations, groups, and communities" (Allen et al., 2011, p.780). At this point "the central concerns are the relationships between material action, mind, and society; the approach explores links between thought, behavior, individual actions and collective practices" (Blacker, 1993, p.867). So, practices are considered as arrays of human activities or routines patterns of such activities (Schatzki, 2002). Here, the unit of analysis is activity constituted by actions and operations, which are oriented towards the achievement of objects, goals, and conditions respectively (Leontév, 1978), as shown in Figure 2.

Nonetheless, Engeström (2001) suggests that activity should be enlarged to make possible its analysis considering different traditions or perspectives and individual and collective levels. Thus, these levels and perspectives originated different ways in how the activity should be scrutinized and three generations of AT have been proposed. Here, two of them are presented and used. The first generation refers the idea of mediation and is visualized schematically in a triangular model presenting the connection between stimulus and response. Figure 3 presents this model, which exhibits an activity culturally mediated and presented as "the triad of the subject, object, and mediating artifact" (Engeström, 2001, p.134). The mediating artifacts take the form of signs and language (abstract tools) (Vygotsky, 1978) and real and physical artifacts (material tools) (Leont'ev, 1978).

The second generation is built on the concept of "primeval collective hunt" (Leont'ev, 1981, cited in Engeström, 2001, p.134). Here, there is a "crucial difference between an individual action and a collective activity" (Engeström, 2001, p.134). So, the model was developed by Engeström (1987) and expressed that in a collective activity the subject(s) or individual (s) with the help of others (community) is motivated to act on objects facilitated by mediating artifacts (abstract and material tools). In these circumstances, that activity is controlled and shaped by cultural components, which include rules and norms (conventions) and division of labor (social division). Figure 4 shows this model that was named human activity system to differentiate it from the basic activity theory model, presented in Figure 3.

Additionally and in order to facilitate and guide the analysis of that activity system, five principles were enunciated and stated the present condition of AT (Engeström, 1987), as follows:

Figure 2. Activity, actions, and operations
Source: Adapted from Wilson, 2006.

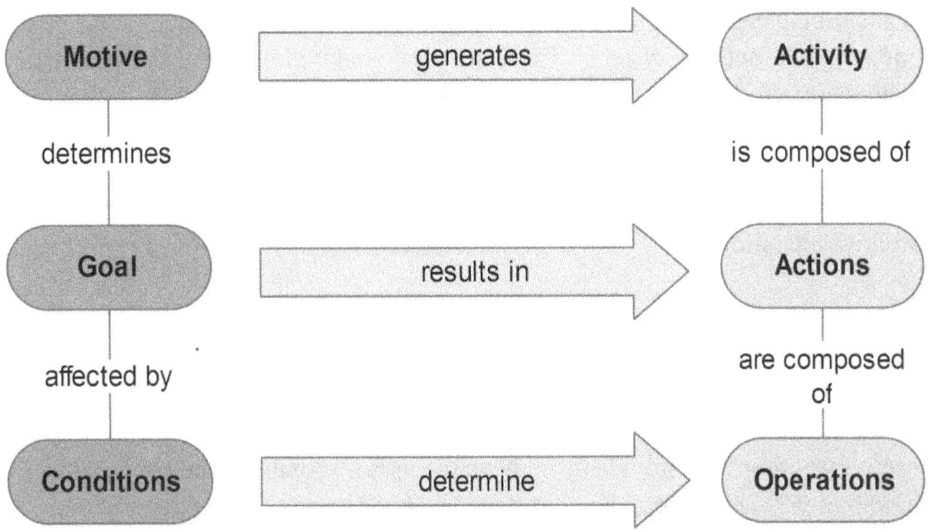

Figure 3. The basic activity theory
Source: Adapted from Vygotsky, 1978.

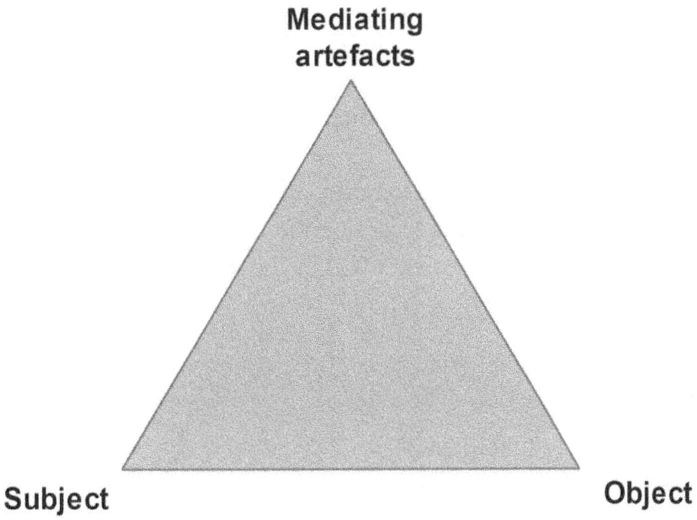

1. The first principle suggests that human activity system is collective, artifact-mediated and object-oriented. This provides the background to appreciate and understand, as a unit of analysis, the goal-directed actions, and operations at individual and collective levels.
2. The second principle refers to a multi-voicedness activity system. It denotes that there are culture and history in a community, which innovates to solve its problems through the negotiation between individuals in continuous interaction.

3. The third principle is historicity denoting an activity system that results of amassed transformations over time.
4. The fourth principle exhibits the tensions and contradictions as the principal source of change and development in the activity systems. These are generated within and/or between activity, systems provoking their transformation and advance, which are presented below.
5. The fifth principle exhibits the possibility of the activity system to expand their transformations. This suggests that the activity system is transformed consequently of change its primary object and motive. This also provides a broader perspective of possibilities in its trek and can be seen in its qualitative alterations.

Referring to the fourth principle, "contradictions are accumulating structural tensions within and between activity systems" (Engeström, 2001, p.137) that result in innovation. They can be seen as "the root of the causes of the problems" (Engeström, 2000, p.966) and are "recognized as frictions amongst the elements of the activity system" (Barab et al., 2002, p.80). Moreover, they can be found in diverse areas of the activity system and provide insights on the use of visuals within the organizations. In line, Engeström (1987, p. 87) shows that there are four levels of tensions and contradictions, presented in Figure 5. Primary inner contradictions (double nature) refer to those tensions within each constituent component of the central activity system. Secondary contradictions concern to those tensions between the constituents of the central activity system. Tertiary contradictions relate to those tensions between the object/motive of the dominant form of the central activity system and the object/motive of a culturally more advanced form of the central activity system. Quaternary contradictions apply to those tensions between the central activity system and its neighbor activity systems. Here, contradictions are seen as the source of transformation over time in the design of visuals in organizations. Initially, it suggests to focusing on the activity system perception in which visuals mediate that activity and are considered abstract tools in context.

In other words, contradictions could help to define attributes that should be included in the design of the visuals. For example, they could aid to define the elements to be incorporated in the forms of visuals referring to organizational requirements within the processes. It is considering the visuals as

Figure 4. The structure of a human activity system
Source: Adapted from Engeström, 1987.

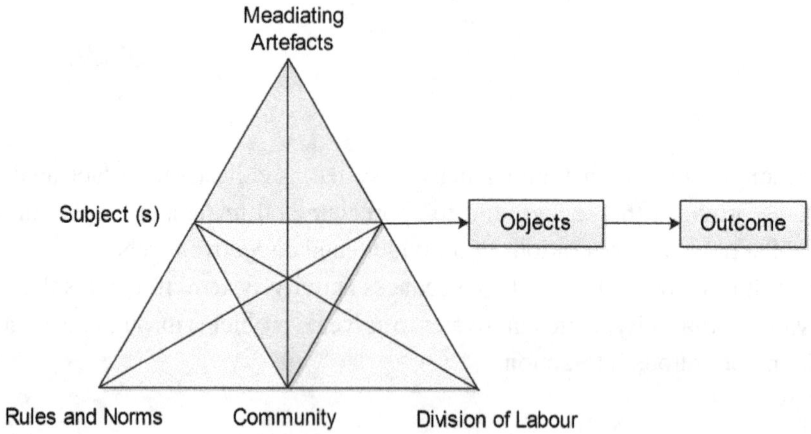

abstract tools incorporating suggestions in terms of security and safety and language aligned to the users or individuals who usually have diverse levels of culture. Additionally, tensions could assist on finding relationships between users, potential users and the achievement of objects or objectives with the visuals in the organizations. Moreover, contradictions could serve to put on perspectives the design of visuals, their use, and points of convergence. This means to combine attributes of use and methods to facilitate their design in order to minimize tensions in both use and design simultaneously. Furthermore, tensions could serve to know how the use of visuals at work supports the alignment of the activities described in Figure 1.

To put together and according to the theoretical departure, they put on evidence the lack of a methodology, which provides guidelines to design the used SWVs in manufacturing processes. It is considering that the principal objective of VF is collocated critical information in those areas where is essential and wanted. At the same level, diverse systems in VM have been developed in order to reinforce the role of individuals and their valuable contribution to the organizational values and understanding of the client needs through the use of visuals. So, the usability term is used to value the necessity to emphasize the importance of seeing the visuals as a product of a production system and their management and their procedures of design. In order to achieve usability, diverse methodologies have been adopted to evaluate the usability of products or during the design of products. This provides a gap in a guide in how the visuals should be designed for being used in the manufacturing processes. Here, usability should

Figure 5. Four levels of contradictions within the activity system
Source: Adapted from Engeström, 1987.

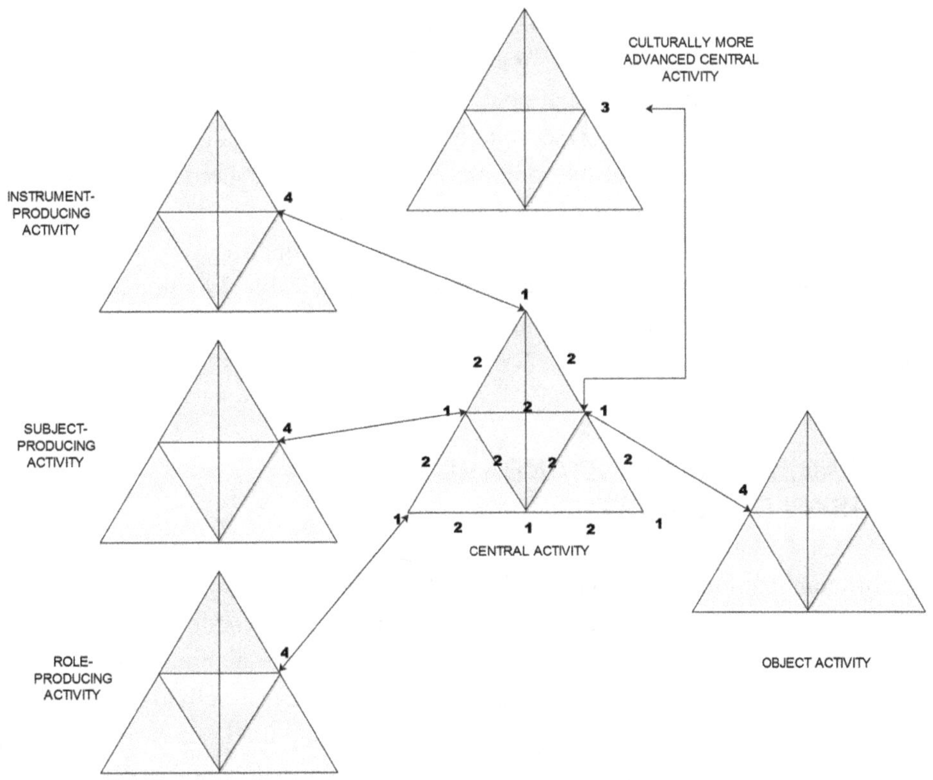

be reached at both product and methodological levels. Consequently, it is proposed AT as a conceptual framework that in past research served to achieve usability by providing advantages over other frames.

METHODOLOGY

Activity Theory is suitable as a framework and analytical tool for the research in the area of usability (Mwanza, 2002) and it "can be quite a powerful analytical tool and conceptual framework of inquiry" (Wilson, 2006, p.9). Particularly, it gives the chance to research the human activity in routine operation pondering the historical and cultural situation at individual and social levels (Wilson, 2008). This is because the "human mind emerges, exists, and can only be understood within the context of human interaction with the world, and this interaction, that is, activity, is socially and culturally determined" (Kaptelinin et al., 1999, p.28). The objective or object of the perception of visuals as an object-oriented activity helps to "understand not only what people are doing but also why they are doing it" (Kaptelinin, 2005, cited in Nardi, 2005, p.38). One form is discovering and framing the anatomy of actions and operations in the setting of study in succeeding and temporary moments of the activity in the study (Engeström, 2000). Similarly, it is important to understand the setting in which the activity is carried out. This permits to integrate as a whole each element of the activity system (Engeström, 2000) (see figure 3). Moreover, this integration permits to examine in detail the relationships and their mediation between the elements of the activity system permitting the microanalysis (Mishra et al., 2011). Additionally, historical and cultural characteristics are included into the integration facilitating their analysis as a whole.

However, it is important to consider and avoid confusions related to those activities object/oriented and those activities subject/oriented, the use of the terms object and goal and the language itself. Referring to the activities object/oriented and subject/oriented, the former is connected with the use of abstract and material tools achieving objects, and the latter, with social interactions between individuals interchanging information. Related to the terms object and goal, there are differences in terms of reaching activities and actions that are continuously transformed into one another (Bedny and Harris, 2005). Concerning to language, the development of a clear interpretation of each element should help on clarifying its use in context putting attention to the levels of the activity and the analysis. Here, a process could help to present in perspective the use of the visuals and the activity of perception.

In line, the methodology for creating methodologies is used. Table 3 exhibits 10 steps that should be done in order to develop a methodology, or in this particular case, it serves to develop the proposed checklist to evaluate the usability level of the SWVs used in manufacturing.

LEVEL OF USABILITY IN THE STANDARD WORK VISUALS (LUSWV CHECKLIST)

Following the structure presented in the last section and the Activity Theory, it is proposed the following checklist, which its principal objective is to evaluate the usability of the product and methodological levels of the standard work visuals employed in the manufacturing processes without forgetting the objectives of the visuals treated here. The flow diagrams present the sequence of activities in the processes; the procedures exhibit the phases needed for the activities and the one-point lessons showing critical operations that should be evaluated with rigid criteria. To do so, each step in the methodology defined in Table

Table 3. Methodology for creating methodologies

Step	Description
1. Define the direction	Specify the objectives for the process and identify who benefits from it.
2. Identify key issues	Identify key performance factors that affect the quality of the process
3. Put the process into context	Obtain a systematic overview to determine the scope, focus, and use of the process
4. Set criteria	Set criteria and determine outcomes that will be used to assess the quality of the process and its results
5. Inventory information and resources	Collect expertise in the use of the target process, including quality, quantity, timeliness, and the cost of relevant information
6. Logically order process	Organize the process into steps; include implicit feedback loops
7. Execute the methodology	Test the methodology, using it as a guide, not as a rule book
8. Assess each step	Collect data and measure performance in "real time" to improve future performance
9. Facilitate the process	Use facilitation, assessment, and management skills to help participants learn the process
10. Assess performance	Determine necessary changes in the methodology by analyzing the differences between the desired and actual outcomes

Source: Adapted from Smith and Apple, 2007.

3 is associated with each element of the activity system, presented in Figure 3, and the contradictions that could be presented in the relationships and mediations between same elements, as shown in Figure 4. Moreover, the checklist includes the psychological and socio-cultural adequacy. Particularly, the steps and their description are changed to adequate to the ergonomic perspective.

Consequently, Table 4 presents the proposed checklist to evaluate the level of usability in the standard work visuals used in manufacturing. It is suggested that subjective and objective information should be gathered to answer the checklist providing valuable information to the designer of visuals. So, diverse methods as interviews, questionnaires focus groups and observations should be employed to gather data. Once information is gathered, the following checklist should be used to evaluate the level of usability of the product and methodology. The information provided by the result should be used to improve the design of the visual and the method in how it should be designed. The answers to the statements offer areas of opportunity in the design and use of the visuals in context and in this way should be analyzed. In fact, it is considered that visuals could be adjusted consequently of the natural change of objects and motivations of the perception activity and the context in which activity is carried out. The principles of AT also exhibits various conditions of the system that have been included into this checklist.

CASES OF STUDY

Two cases of study are presented serving as examples in the design of standard work visuals. One is related to the medical industry and the other, the aerospace industry. It is important to remark that the analysis and results obtained from the visuals presented here are partial. It is because the participant organizations have decided to contribute with the research permitting the use of the visuals indicated in Appendices 1 and 2.

Table 4. The LUSWV checklist

Phase	Criteria on Visuals. Thick if Applicable
1. Guidance	___ Objective of visuals is obvious ___ Considerations of security and safety regulations are included ___ Scope of visuals is clear ___ The designer has experience conceiving visuals ___ Value of visual in the process is clarified
2. Key issues	___ The language is easy to read ___ Visuals are easy to follow ___ Visuals are easy to remember ___ Any user understands visual contents ___ Minimal training is required to master visuals ___ Minimal errors could be provoked ___ Figures and photographs are comprehensible ___ Colors are contrasting
3. Contextualization	___ Design of visuals is accorded to sector or industry ___ Language of visuals shows organizational culture ___ Design of visuals is particularized for the process or activity
4. Criteria	___ Visual is useful ___ Visual accomplishes its objective ___ Visual satisfies users ___ Designer understands her/his role ___ Designer is satisfied ___ Designer accomplishes organizational requirements ___ Visual clearly mediates operational work of individuals
5. Information and resources	___ Visuals exhibit experiences on their use ___ Diverse quality indexes are utilized to evaluate quality ___ The organization exhibits experience on designing visuals
6. Order and feedback	___ There is a logical order presenting the information ___ There are clear mediums to give feedback
7. Accomplishment	___ Formal tests of use are performed ___ Clear feedback is given to designer ___ User(s) is (are) empowered to give feedback ___ Feedback helps on improving visuals
8. Assessment	___ Quality of visuals is clearly stipulated and used ___ Improvements are included in visuals
9. Benefits	___ Visuals provide clear organizational benefits in their use ___ Visuals help on training workers ___ Visuals increase productivity ___ Visuals decrease errors ___ Visuals make clear organizational objectives ___ Visuals make clear expected outcomes
10. Comparison	___ Existence of a comparison between design and use of visuals ___ Existence of versions of visuals

Case 1: Medical Industry

Appendix 1 shows a flow diagram and it forms part of a medical process. For confidentiality reasons, the name of the company has been erased and the document is currently used in the industry. In order to show the use of the checklist, phase 1 is presented and Table 5 shows the conclusion of the analysis of the procedure.

Table 5. The LUSWV checklist for the flow diagram

Phase	Criteria on Visuals. Thick if Applicable
1. Guidance	___ Objective of visuals is obvious ___ Considerations of security and safety regulations are included _✓_ Scope of visuals is clear ___ The designer has experience conceiving visuals ___ Value of visual in the process is clarified

The result provides areas of opportunity in terms of usability. For instance, there are some areas of opportunity in the design of the visuals. The lack of the objective, the inclusion of regulations in the visuals and the absence of value suggest that designer should be trained on designing the visuals required and used in the manufacturing processes of the participant organization.

Case 2: Aerospace Industry

Appendix 2 shows an operating work instruction, or standard work visual named here, which contains both a procedure and a one-point lesson. The name of this visual is denominated Adjusting Pneumatic Presses and forms part of an aeronautical process. For confidentiality reasons, the organization denomination has been erased and the document is currently used in the industry. In order to show the use of the checklist, phase 1 is presented to both procedure and a one-point lesson in this industry. Table 6 exhibits the result of the analysis of the procedure, and the Table 7, the one-point lesson.

The result provides areas of opportunity in terms of usability. For instance, both visuals have the same areas of opportunity in phase 1. The lack of the objective and the inclusion of regulations in the visuals suggest that designer should be trained on designing the visuals required and used in the manufacturing processes of the participant organization.

Table 6. The LUSWV checklist for the procedure

Phase	Criteria on Visuals. Thick if Applicable
1. Guidance	___ Objective of visuals is obvious ___ Considerations of security and safety regulations are included _✓_ Scope of visuals is clear ___ The designer has experience conceiving visuals _✓_ Value of visual in the process is clarified

Table 7. The LUSWV checklist for the one-point lesson

Phase	Criteria on Visuals. Thick if Applicable
1. Guidance	___ Objective of visuals is obvious ___ Considerations of security and safety regulations are included _✓_ Scope of visuals is clear ___ The designer has experience conceiving visuals _✓_ Value of visual in the process is clarified

CONCLUSION

The LUSWV checklist is proposed in order to provide a guide for improving the design of the visuals. Their objectives are to put critical information available to individuals performing activities. To do so, three types of visuals are treated here and they form a crucial part of the LM through the concept Visual Factory. The flow diagrams, procedures, and one-point lessons served to put the basis to evaluate the usability levels of visuals used in two manufacturing organizations. One of them permitted to study a flow diagram, and the other, a procedure and a one-point lesson. The checklist permits to evaluate the usability at the product and methodological levels in a holistic way. This consists of 10 steps that similarly imply answering statements for accomplishing criteria related to the usability of both product and methodology. It is important to remark that expected outcome of its use is a guide in which designer should focus their efforts to improve visuals.

Their improvements are closely linked with the methodology and conceptual framework used to develop the mentioned checklist. For instance, the used methodology permits to make sure of what steps should be done in order to develop the proposed methodology. This similarly allows the improvements by developing specific procedures to achieve their objectives. In this particular case, the used methodology has allowed the inclusion of usability in the proposed methodology. On the other hand, the AT as the conceptual framework has been used in past research to define levels of usability in products. Indeed, the usability was decomposed under this frame into four chained activities: perception, involuntary and voluntary attention, discovering meaning and motor actions. This research has covered the elements of the perception activity guiding this study. Besides, these elements establish the core to develop criteria within each step that similarly has a clear objective to complete. Moreover, it has conferred the inclusion of the evaluation of the psychological and socio-cultural adequacy. However, only the step 1 was evaluated from the cases of study. Its outcome exhibits an area in which visuals should be improved. Particularly, it is suggested that designer should have knowledge of those objectives of visuals and regulations concerning to the processes that should be included.

Consequently, the use of the checklist has granted to uncover some recommendations and areas of future research. Referring to recommendations, the adoption of the checklist for evaluating visuals in diverse manufacturing sectors was mentioned. It could open the opportunity to boost them in context as a chained improvement. Another recommendation is that individuals are essential in those improvements because they use in context and could be capable to do it. For these reasons, they should be instructed using suitable mediums. In this line, the individuals should clearly see the strong relationships with the designers of visuals so that they can exchange information for their improvement and an organization change could happen. Another possible consequence is the decrease of the costs. Related to future studies, it is suggested to check the feasibility of the checklist in other manufacturing sectors revealing the possibility to pass its use to the service sector. This can stimulate the discovery of additional benefits by proving diverse configurations of the checklist in the form of customized models. These models should be developed using the criteria included in the checklist. Similarly, the models and the checklist can give the excuse to explore diverse perspectives and approaches. In conclusion, the checklist is the initial step in the development of one reliable tool that permits the evaluation of the usability of visuals used in the manufacturing processes. Similarly, the evaluation at the product and methodological levels serve to guide their improvement considering the user, the designer, and organizational requirements.

RECOMMENDATIONS

According to the results in the application of phase 1 of LUSWV checklist on three visuals utilized in manufacturing processes and personal experiences in the manufacturing sector, this checklist should be seen as an instrument that evaluates the product and methodological levels of usability in the visuals. Similarly, visuals should be identified as material and abstract tools utilized to mediate the perception activity included in the term usability. This activity is the initial activity from the chained activities corroborating usability of products. For these reasons, the checklist should support activities and tasks assisting the improvement of the visuals in context. It is recommendable not seeing it as a rule checklist. In addition, it is important to value each of the phases and criteria on it. For instance, this implies relevant aspects within the organizations regarded to operational issues, safety and safety regulations, user and organizational requirements, costs on gathering information for the improvement of visuals and psychological and cultural adequacy among other factors.

In this line, some additional recommendations are given in relation to the methodology of the LUSWV checklist, the strong relationship between the visuals and the users(s), the methods to gather data and the cost implied in the design of visuals. Initially, the methodology implies permanent feedback loops for evaluating usability and the perspectives of those connected with the design and use of visuals. It is seen through permanent adequacy of the visuals to the context, the listening to the user and the achievement of organizational objectives with their use (Mwanza, 2002). Consequently, the methodology could be considered as a subjective and objective element for checking the achievement of objectives in terms of grades of involvement in the Lean Manufacturing philosophy and Ergonomics.

Secondly, the connection between the user(s) and the visuals are detected in the elements of usability that should be included in the visuals in order to increase the quality of them. This supports the perspective on collocating the individuals as the most relevant element within the manufacturing processes by considering their viewpoints (Nielsen 1993) and their knowledge referring to the regulations applied to the production processes. Similarly, satisfaction in the use of visuals could be achieved, which consequently could influence in the effectiveness and efficiency of the user(s). Errors, accidents and quality problems could be decreased as a result of receiving feedback from individuals. So, this relationship should be exploited in order to increase the expected organizational outcomes and diverse methods could be used to gather that information.

Thirdly, diverse methods between the user(s), designer and organizations should be employed to give and receive feedback. These methods could definitely influence in the organizational culture and consequently, they permit and create environments in which information sharing is recognized as an asset and encouraged (Sonnenwald, 2006). Moreover, those methods could serve to receive suggestions from individuals, which could be included on visuals. For instance, the feedback or suggestions coming from individuals are appreciated and recognized in LM (Feld, 2000). This is also benefited with the use of diverse mediums for sharing information between individuals, which could decrease the organizational distance crossing organizational boundaries (Sonnenwald and Pierce, 2000). Thus, organizations should utilize diverse methods and mediums to exchange information between individuals so that visuals could be improved contributing with the production processes.

Finally, the designer and the organizations should recognize costs included into the design of visuals. They are seen in the diverse elements touched here. For instance, the use of the checklist in context is a cost that should be considered. It is because there are costs on gathering and analyzing information from the processes, answering the checklist and designing the visuals. However, these could be observed and analyzed as an investment, which should signify return from it (Brady, 2012). Since the costs of design and the improvement of visuals are seen as costs and investments at the same time, designers and organizations should confer a relevant importance within the manufacturing processes, appraise their economic value, and estimate their impact on subjective and objective issues.

FUTURE RESEARCH DIRECTIONS

Some directions in the use of the LUSWV checklist are enlisted in relation to the above exposed. Initially, it is important to check the feasibility of this checklist in other organizations. For these reasons, additional companies from the same sector can be invited to participate in the research so that this can be improved. It is expected that the entire checklist can be answered providing relevant information in the use of visuals. Later, it can be proved in other manufacturing sectors increasing its level of usability.

Next, the improvements could aid on putting on perspective its use and diverse scales could be developed to measure the levels of usability. In the same form, the checklist could serve as a framework to research the measurement of levels of usability in particular processes within or outside of the mentioned sectors. In addition, the checklist and their improvements could be translated to the service sector increasing its useful.

At the same time, the discovery of the benefits is a priority. It is important to uncover additional benefits from those mentioned in the current literature. This contributes with the generation of knowledge in the area of usability, Ergonomics, AT, LM, among other areas associated with this research. Particularly, those benefits connected with the user of the visuals and their designers.

Subsequently, it is expected to develop diverse models exhibiting multiple configurations of phases and criteria of the checklist in order to adequate the model to the context in which the research is done. This should mean the deeply knowledge of the significant variables within the checklist and models developed. Similarly, it also implies to move from qualitative to quantitative paradigms. The management of variables and their measurements should be consistent with the philosophy of LM, the continuous improvement and the development of tools for supporting manufacturing.

Finally, the probation of the useful of the checklist could open the opportunity to explore the mixture of frames passing from reductionist to systemic perspectives and from qualitative to quantitative approaches. For instance, the checklist initially includes the reductionist and systemic perspectives and afterward, the quantitative approach could be employed as mentioned previously. So, it is expected to explore other frames into the usability concept.

ACKNOWLEDGMENT

We would like to acknowledge to the Tecnológico Nacional de México, Instituto Tecnológico de Tijuana, Universidad Autónoma de Baja California, Universidad de Guadalajara and PRODEP for partially funding the study. We would also like to thank to the participant organizations for contributing with this study. The interpretations and views in this chapter, however, are solely those of the authors.

REFERENCES

Aceves-Gonzalez, C., Prado-León, L. R., & Avila-Chaurand, R. (2010). Factores ergonómicos en el diseño de indicadores visuales estáticos. *Revista de Diseño y Comunicación Visual, Septiembre* (6), 53-59.

Allen, D., Karanasios, S., & Slavova, M. (2011). Working with activity theory: Context, technology, and information behavior. *Journal of the Association for Information Science and Technology, 62*(4), 776–788. doi:10.1002/asi.21441

Ankrum, D. R. (1999). Visual ergonomics in the office — Guidelines. *Occupational Health & Safety (Waco, Tex.), 68*(7), 64–74. PMID:10081834

Anshel, J. R. (2006). Visual ergonomics in the workplace. *Professional Safety, 51*(8), 20.

Avila-Chaurand, R. (2013). *La usabilidad en el diseño de productos de consumo: un punto de vista desde la ergonomía.* Retrieved from https://es.scribd.com/document/172195315/LA-USABILIDAD-EN-EL-DISENO

Barab, S. A., Barnett, M., Yamagata-Lynch, L., Squire, K., & Keating, T. (2002). Using activity theory to understand the systemic tensions characterizing a technology-rich introductory astronomy course. *Mind, Culture, and Activity, 9*(2), 76–107. doi:10.1207/S15327884MCA0902_02

Bedny, G. Z., & Harris, S. R. (2005). The systemic-structural theory of activity: Applications to the study of human work. *Mind, Culture, and Activity, 12*(2), 128–147. doi:10.120715327884mca1202_4

Bevilacqua, M., Ciarapica, F. E., Mazzuto, G., & Paciarotti, C. (2013). Visual management implementation and evaluation through mental workload analysis. *IFAC Proceedings Volumes, 46*(7), 294-299.

Bilalis, N., Scroubelos, G., Antoniadis, A., Emiris, D., & Koulouriotis, D. (2002). Visual factory: Basic principles and the 'zoning' approach. *International Journal of Production Research, 40*(15), 3575–3588. doi:10.1080/00207540210140031

Blackler, F. (1993). Knowledge and the theory of organizations: Organizations as activity systems and the reframing of management. *Journal of Management Studies, 30*(6), 863–884. doi:10.1111/j.1467-6486.1993.tb00470.x

Bligård, L. O., & Osvalder, A. L. (2013). Enhanced cognitive walkthrough: Development of the cognitive walkthrough method to better predict, identify, and present usability problems. *Advances in Human-Computer Interaction, 2013*, 9. doi:10.1155/2013/931698

Brady. (2012). *Visual work handbook.* Brady Inc.

Dickson, E. W., Singh, S., Cheung, D. S., Wyatt, C. C., & Nugent, A. S. (2009). Application of lean manufacturing techniques in the emergency department. *The Journal of Emergency Medicine, 37*(2), 177–182. doi:10.1016/j.jemermed.2007.11.108 PMID:18722732

Durivage, M. A. (2016). *Practical Process Validation.* ASQ Quality Press.

Engeström, Y. (1987). *Learning by expanding: An activity-theoretical approach to developmental research.* Helsinki: Orienta-Konsuit.

Engeström, Y. (2000). Activity Theory as a framework for analyzing and redesign work. *Ergonomics, 43*(7), 960–974. doi:10.1080/001401300409143 PMID:10929830

Engeström, Y. (2001). Expansive learning at work: Toward an activity theoretical reconceptualization. *Journal of Education and Work, 14*(1), 133–156. doi:10.1080/13639080020028747

Feld, W. M. (2000). *Lean manufacturing: Tools, techniques, and how to use them*. CRC Press. doi:10.1201/9781420025538

García Criollo, R. (2000). *Estudio del trabajo: ingeniería de métodos*. Editorial McGraw-Hill.

Gómez-Bull, K. G., Hernández-Arellano, J. L., & Ibarra-Mejía, G. (2015). A proposed methodology for task analysis in ergonomic evaluations. *Procedia Manufacturing, 3*, 4756–4760. doi:10.1016/j.promfg.2015.07.573

Graif, M. (1991). *The visual factory*. Portland, OR: Productivity Press.

IEA. (2008). *EQUID Design process guidelines*. IEA Press.

Jiun, G. W., & Feng, C. J., Bin Kamaruddin, S., & Binti Muhammad, N. A. (2014). A visual board to facilitate production flow regulation at sequential processes. *International Journal of Industrial Engineering and Management, 5*(2), 85–94.

Jokela, T., Iivari, N., Matero, J., & Karukka, M. (2003). The standard of user-centered design and the standard definition of usability: analyzing ISO 13407 against ISO 9241-11. In *Proceedings of the Latin American conference on Human-computer interaction* (pp. 53-60). ACM. 10.1145/944519.944525

Kaptelinin, V. (2005). The object of activity: Making sense of the sense-maker. *Mind, Culture, and Activity, 12*(1), 4–18. doi:10.120715327884mca1201_2

Kaptelinin, V., & Nardi, B. A. (2006). *Acting with technology: Activity theory and interaction design*. MIT Press.

Kaptelinin, V., Nardi, B. A., & Macaulay, C. (1999). Methods & tools: The activity checklist: a tool for representing the "space" of context. *Interactions, 6*(4), 27-39.

Lehto, M. R., & Landry, S. J. (2012). *Introduction to human factors and ergonomics for engineers*. CRC Press.

Leont'ev, A. N. (1978). *Activity, consciousness, and personality*. Englewood Cliffs, NJ: Prentice Hall.

Lindlöf, L. (2014). *Visual management-on communication in product development organizations*. Chalmers University of Technology.

Lindlof, L., & Soderberg, B. (2011). Pros and cons of lean visual planning: experiences from four product development organisations. *International Journal of Technology Intelligence and Planning, 7*(3), 269-279.

Melton, T. (2005). The benefits of lean manufacturing: What lean thinking has to offer the process industries. *Chemical Engineering Research & Design, 83*(6), 662–673. doi:10.1205/cherd.04351

Mishra, J. L., Allen, D. K., & Pearman, A. D. (2011). Information sharing during multi-agency major incidents. *Proceedings of the Association for Information Science and Technology*, *48*(1), 1–10. doi:10.1002/meet.2011.14504801039

Morency, R. R., Rooney, E. F., & Foerster, D. R. (2003). A Methodology To Implement and Validate Ergonomic Improvements To Computer Workstations At LL Bean. *Advances In Industrial Ergonomics And Safety*, *V*, 319.

Mwanza, D. (2002). *Towards an activity-oriented design method for HCI research and practice* (Doctoral dissertation). Open University.

Nael, M. (2011). IEA EQUID Template for Cooperation between Product Designers and Ergonomists. *Human Factors and Ergonomics in Consumer Product Design: Methods and Techniques*, 261.

Niebel, B. W., Freivalds, A., & Osuna, M. A. G. (2004). *Métodos, estándares y diseño del trabajo*. Alfaomega.

Niebel, B. W. N., & Benjamin, W. (1996). *Ingeniería industrial: métodos, tiempos y movimientos*. Alfaomega.

Nielsen, J. (1993). *Usability engineering*. New York: AP Professional.

Parry, G. C., & Turner, C. E. (2006). Application of lean visual process management tools. *Production Planning and Control*, *17*(1), 77–86. doi:10.1080/09537280500414991

Prado-León, L. R., & Cárdenas-Vázquez, S. (2011). Lineamientos ergonómicos en el diseño de páginas Web. *Revista Ação Ergonômica, 2*(2).

Rossi, D., Bertoloni, E., Fenaroli, M., Marciano, F., & Alberti, M. (2013). A multi-criteria ergonomic and performance methodology for evaluating alternatives in "manuable" material handling. *International Journal of Industrial Ergonomics*, *43*(4), 314–327. doi:10.1016/j.ergon.2013.04.009

Schatzki, T. R. (2010). *Site of the social: A philosophical account of the constitution of social life and change*. Penn State University Press.

Smith, P., & Apple, D. K. (2007). *Methodology for creating methodologies*. Retrieved 03 17, 2017, from Faculty guidebook: http://www.pcrest3.com/fgb/efgb4/2/2_4_16.htm

Sonnenwald, D. H. (2006). Challenges in Sharing Information Effectively: Examples from Command and Control. *Information Research: an International Electronic Journal*, *11*(4), n4.

Sonnenwald, D. H., & Pierce, L. G. (2000). Information behavior in dynamic group work contexts: Interwoven situational awareness, dense social networks and contested collaboration in command and control. *Information Processing & Management*, *36*(3), 461–479. doi:10.1016/S0306-4573(99)00039-4

Steenkamp, L. P., Hagedorn-Hansen, D., & Oosthuizen, G. A. (2017). Visual Management System to Manage Manufacturing Resources. *Procedia Manufacturing*, *8*, 455–462. doi:10.1016/j.promfg.2017.02.058

Tjell, J., & Bosch-Sijtsema, P. M. (2015). Visual management in mid-sized construction design projects. *Procedia Economics and Finance*, *21*, 193–200. doi:10.1016/S2212-5671(15)00167-7

Vygotsky, L. S. (1978). *Mind in society: The development of higher mental process.* Cambridge, MA: Harvard University Press.

Weisner, K., & Deuse, J. (2014). Assessment methodology to design an ergonomic and sustainable order picking system using motion capturing systems. *Procedia CIRP*, *17*, 422–427. doi:10.1016/j.procir.2014.01.046

Wilson, T. D. (2008). Activity theory and information seeking. *Annual Review of Information Science & Technology*, *42*(1), 119–161. doi:10.1002/aris.2008.1440420111

Young, M. S., & Stanton, N. A. (2005). *Applying interviews to usability assessment. In The handbook of human factors and ergonomics methods* (pp. 29–1). CRC Press LLC.

KEY TERMS AND DEFINITIONS

Abstract and Material Tools: The mediators of an object-oriented activity, which takes the form of mental and physical artifacts.

Activity: A system of a human in which an individual works on an objective to reach a required outcome.

Activity Theory: The framework or meta-theory that helps to understand why and how individual(s) is (are) doing what in a cultural and historical context in which is (are) immersed.

Artifact: The mediators of an object-oriented activity taking the form of tools.

Lean Manufacturing: The holistic method which objective is to reduce the waste in a manufacturing system without ceding productivity.

Standard Work Visuals: The signs, labels, placards, displays, and other markings deployed in the physical workplaces exhibiting critical information so that individuals can use to carry out their activities.

Usability: The quality attribute in terms of ease to use and learnability of a human-made product and/or to the methods used to facilitate the use during the design process of a human-made product.

Visual Factory: The concept within the lean manufacturing that considers the collocation of visuals in the physical workplaces so that individuals can accede to relevant and critical information when it is solicited or required increasing safety and productivity.

APPENDIX 1: EXAMPLE OF THE FLOW DIAGRAM IN THE MEDICAL INDUSTRY

Figure 6. Flow diagram

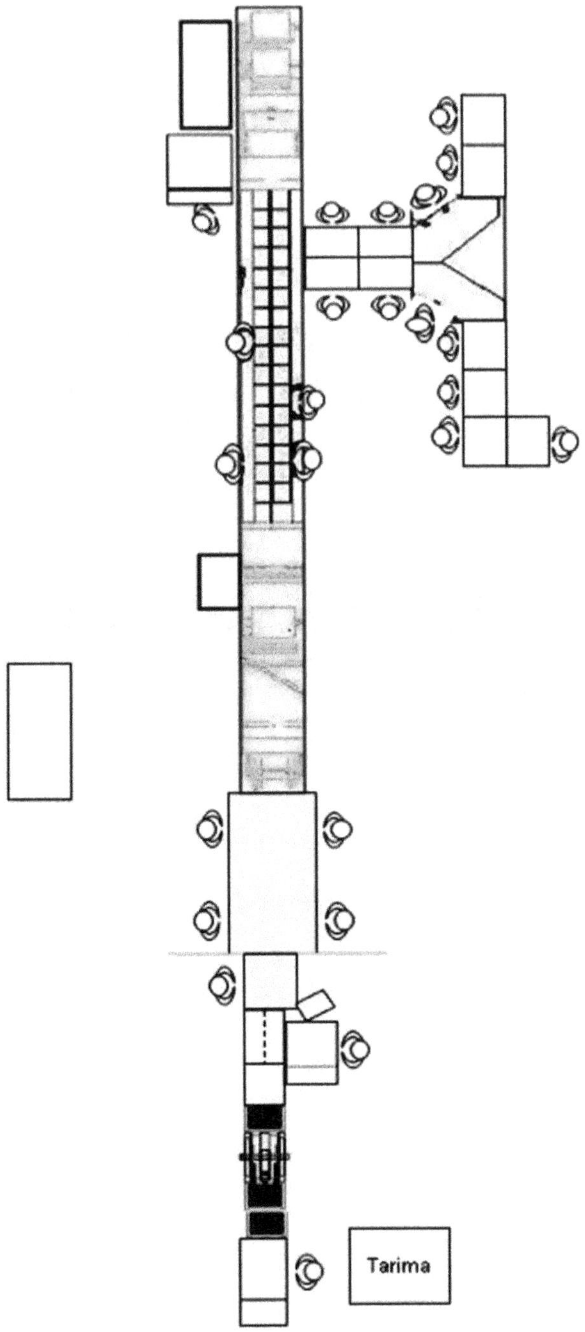

APPENDIX 2: EXAMPLE OF A PROCEDURE AND ONE-POINT LESSON IN AEROSPACE INDUSTRY

Figure 7. Procedure and one point-lesson

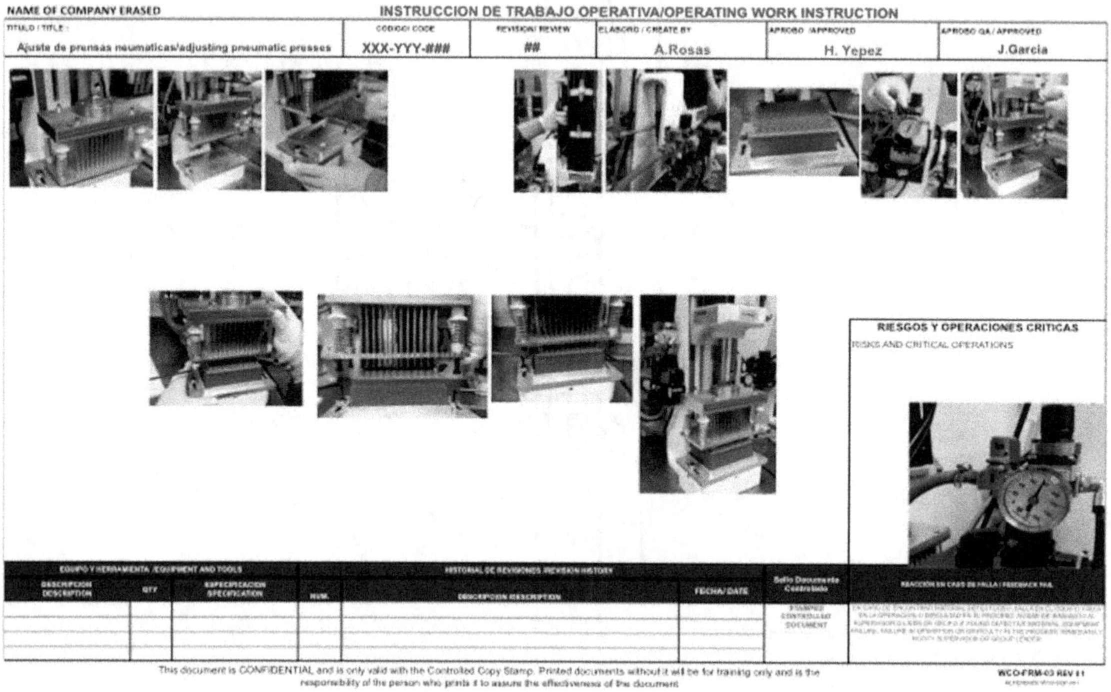

NAME OF COMPANY ERASED

INSTRUCCION DE TRABAJO OPERATIVA/OPERATING WORK INSTRUCTION

TITULO / TITLE:	CODIGO/ CODE	REVISION/ REVIEW	ELABORO / CREATE BY	APROBO / APPROVED	APROBO QA / APPROVED
Ajuste de prensas neumaticas/adjusting pneumatic presses	XXX-YYY-###	##	A.Rosas	H. Yepez	J.Garcia

RIESGOS Y OPERACIONES CRITICAS

RISKS AND CRITICAL OPERATIONS

EQUIPO Y HERRAMIENTA /EQUIPMENT AND TOOLS			HISTORIAL DE REVISIONES /REVISION HISTORY			Sello Documento Controlado	REACCION EN CASO DE FALLA / FEEDBACK FAIL
DESCRIPCION DESCRIPTION	QTY	ESPECIFICACION SPECIFICATION	NUM.	DESCRIPCION /DESCRIPTION	FECHA/ DATE		
						STAMPED CONTROLLED DOCUMENT	

This document is CONFIDENTIAL and is only valid with the Controlled Copy Stamp. Printed documents without it will be for training only and is the responsibility of the person who prints it to assure the effectiveness of the document

WCO-FRM-03 REV 11

Chapter 14
Usability Test and Cognitive Analyses During the Task of Using Wireless Earphones

Nora G. Bustamante
Autonomous University of Ciudad Juárez, Mexico

Aide Aracely Maldonado Macías
Autonomous University of Ciudad Juárez, Mexico

Adrian A. Durán
Autonomous University of Ciudad Juárez, Mexico

Juan Carlos Ortiz Nicolás
Autonomous University of Ciudad Juárez, Mexico

Andres R. Quiñones
University of Texas at El Paso, USA

ABSTRACT

The use of novel technologies can be complicated for some people; even products designed for everyday use can present problems. If a product is difficult to use, it causes frustration and prevents usage. The objectives of this study are to conduct a usability test, assess the mental workload, and identify potential human errors during the process of listening to music using wireless earphones from a mobile phone device to detect which operations were the hardest in order to suggest modifications to the design and to bring a better experience to the user. During the usability test some difficulties among users were detected and a hierarchical task analysis (HTA) was developed. NASA-TLX and SHERPA methods were applied. The use of these earphones resulted in an intermediate mental workload, and four types of human errors were identified. The methods used in this study helped to detect those difficult tasks and subtasks for the users.

DOI: 10.4018/978-1-5225-5234-5.ch014

Copyright © 2018, IGI Global. Copying or distributing in print or electronic forms without written permission of IGI Global is prohibited.

INTRODUCTION

Novel technologies introduce products that at first sight might seem appealing because people think that they will enhance the experience they were designed for. When an electronic product is purchased, the explicit properties that are stated in the package are what attracts the consumers' attention. Nevertheless, once the product is used at home or work, it can become a different and unpleasant situation. When these situations appear, issues such as usability become more important (Van Kuijk, 2008). Sometimes using a product can be more demanding than expected in the first place. A product that is difficult to use or takes too much time to figure out how to use it can lead to frustration. Evidence of this fact can be found in the work of Meuter, Ostrom, Bitner, & Roundtree, (2003), their studies have shown that computer anxiety is, in fact, a common occurrence. Moreover, one study found that 55% of Americans suffer from some degree of technophobia (Williams, 1994, cited by Meuter et al., 2003). Additionally, one-third of college students suffer from computer-related anxiety, yet technology anxiety can be extended in relation to technological tools in general (DeLoughry, 1993, cited by Meuter et al, 2003)).

Another issue equally important related with the design for the easiness of use of products is that initially, it must be stated what design is before an approach in error design is made. A purpose of design is to develop products that would respond to a person's needs in a satisfactory manner. The design is concerned with how things work, how they are controlled, and with the nature of the interaction between people and technology (D. A. Norman, 2003). On the latter basis, it can be understood that good designs will contribute to enhancing peoples' interaction with the products or systems; on the contrary, bad designs will affect peoples' lives and may cause frustration and errors; sometimes they can be subtle, but at other times, they can be fatal.

It is clear that the role of human error in the occurrence of accidents is obvious, and the consequences of not studying human errors as a risk factor can be catastrophic. Unfortunately, these errors are not always studied and classified to create safety parameters. All types of human errors, from the design phase to the poor interaction, result of using improper techniques, should be considered. The need for safety principles, and safe designs to prevent accidents and errors is more important due to the increasing development of modern technology (Ghasemi, Nasleseraji, Hoseinabadi, and Zare, 2013)

The solution to the challenges that new technologies represent is referred to as Human Centered Design. HCD is an approach that prioritizes human needs, limitations, capabilities, and behavior (D. Norman, 2013). User experience (UX) is not a property of the product, but the outcome of the human-product interaction, and therefore, it is dependent on the user (Desmet, 2007 in Ortíz Nicolas and Aurisicchio, 2011). The user brings to the interaction with the product a set of systems. These are studied to understand their effect on the assessment of UX. Accordingly, designs must satisfy them in the most complete manner, in order to become a successful design. Some scholars argue that good design starts with an understanding of Psychology and Technology, indicating what actions are possible, what is happening, and what is about to happen (D. Norman, 2013). In this process, task analysis carefully observes how the tasks are performed in order to design a product that would be usable. In the field of UX, it is also acknowledge the role of usability (Nicolas, Carlos, & Aurisicchio, 2011). An important tool to evaluate the usability of a product is usability test, which aims to improve the usability of the product that is being tested; another goal is to improve the process by which products are designed and developed, to avoid the recurrence of problems in similar products. These assessments should be done in the earlier design stages and through all the development process of a product (Dumas & Redish, 1999).

Additionally, cognitive methods can help designers detect opportunities where the product design can be improved by defining what tasks or subtasks may have mental workload and identifying the type of errors committed by users during this task. Analysts will be able to propose changes for better design of these tasks or products.

The general objective of this project is to conduct a usability test and cognitive analyses during the task of listening to music using wireless earphones. The cognitive analyses include develop a hierarchical task analysis to assess the mental workload and identify potential human errors during the process of listening to music using wireless earphones from their mobile phone device. Cognitive analyses will help recognize which tasks or subtasks may be difficult for users in order to suggest modifications to the design and to bring a better experience for them.

The use of these cognitive methods can help designers to determine what tasks or subtask represent problems to be performed. Correspondingly, detecting potential errors may suggest those changes that prevent the difficulties that users experienced, and reduce or eliminate frustration, to develop a more usable product. To a large degree, what makes a product usable is the absence of frustration when it is used, in order to be usable, it must be: useful, efficient, effective, satisfactory and accessible (Rubin & Chisnell, 2008).

LITERATURE REVIEW

The International Organization for Standardization (ISO) defines usability of a product as the extent to which specified users achieve specified goals with effectiveness, efficiency, and satisfaction in a specific context. The goal of a usability test is to uncover problems, not to demonstrate the existence of some specific phenomenon. Usability is 'the capability to be used by humans easily and effectively' (Shackel, 1981, 1991). Experience has shown that most major problems in usability tests can be uncovered with few participants (Dumas & Redish, 1999). The best results come from testing no more than five users; as with any human factors issue, however, there are exceptions, as stated by Nielsen, (2012): "For Quantitative Studies (aiming at statistics) tests, at least 20 users, for Card Sorting tests, at least 15 users per group, and for Eye Tracking tests, 39 users if you want reliable heat maps". Usability of a specific product can be evaluated in a specified context of use, so usability is a construct that is situated. It will vary with changing contexts and users (Van Kuijk, 2008).

Usability Testing is a research tool, with its roots in classical experimental methodology. The range of test one can conduct is considerable, from true classical experiments with large sample sizes and complex test designs to very informal qualitative studies with only a single participant. Each testing approach has different objectives, as well as different time and resources requirements (Rubin & Chisnell, 2008). Usability has five attributes: learnability, efficiency, memorability, errors and user satisfaction (Scholtz, 2004).

There are many methods for studying usability, the most basic and useful is user testing, which has three components (Nielsen, 2012):

- Choose representative users
- Ask the users to perform representative tasks with the design
- Observe what the users do, where they succeeded, and where they have difficulties with the user interface. Let the users do the talking

According to Rubin & Chisnell, (2008) an usable product or service must be:

- **Useful:** On how degree, a product or service, enables users to achieve their goals, and their willingness to use it again.
- **Efficiency:** How fast can the goal be accomplished accurately and completely, usually represented as a measure of time.
- **Learnability:** The training, time and effort required to learn to use the system.
- **Satisfaction:** Refers to the users' perceptions, feelings, and opinions of the product, usually assessed through written and oral questions.
- **Accessibility:** Defined as having access to the products needed to accomplish a goal, similar to usability, but it can be used when people with disabilities come into account. Considering accessibility for people with disabilities can clarify, and simplify, design for people who have temporary limitations.

More detailed usability issues provide specific design objectives:

- **Understandability:** Whether users understand what the system can do.
- **Operability or Supportiveness:** Supporting the users throughout the interaction and helping them to overcome problems that may occur.
- **Flexibility:** Enabling tasks to be carried out in different ways to suit different situations.
- **Attractiveness:** Encouraging user interest and motivating them to explore the system.

Basic elements of Usability Testing:

- Development of research questions or test objectives rather than hypotheses.
- Use of a representative sample of end users, which may or may not be randomly chosen.
- Representation of the actual work environment.
- Observation of end users who either use or review a representation of the product.
- Controlled and sometimes extensive interviewing and probing of the participants by the test moderator.
- Collection of quantitative and qualitative performance and preference measures.
- Recommendation of improvements to the design of the product.
- **Think Aloud:** Traditionally, thinking aloud studies are conducted with a psychologist or with user interface experts as experimenters who videotape the subjects and perform detailed protocol analysis. However, it is possible to run user test without sophisticated labs, simply by asking users to complete a test task and asking them to think out loud while they perform the tasks. Studies have shown that other evaluators with a minimum of training, will succeed in finding many usability problems (Nielsen, 1994).
- **Hierarchical Task Analysis:** HTA was initially developed in response to the need for greater understanding of cognitive tasks. In the 1960's, with a great level of automatization in the industry, the nature of workers' tasks was changing, the level of cognitive skills to perform an activity were increasing. In the original paper laying the approach for conducting HTA, Annett et al. 1971 make it clear that the methodology is based upon a theory of human performance (Nevil A Stanton, 2006). Contrary to traditional approaches, HTA had a different assessment approach than just

observing peoples' performance. Instead, it focused on representing system goals and plans. It is a very flexible method, which allows it to be used in many fields of human activities and systems. HTA analyzes not actions per se but goals and operations, the means to accomplish goals (Neville Anthony Stanton, Hedge, Brookhuis, Salas, & Hendrick, 2004). Charles Carver and Michael Scheier suggest that goals have three fundamental levels that control activities (D. Norman, 2013):

- Be goals, at the highest level, they determine how people act.
- Do goals, these goals determine plans and actions to be performed for an activity.
- Motor goals, specifying how the actions are performed.

A structured, objective approach to describing users' performance of tasks provides an understanding of tasks users need to accomplish a certain goal; these actions can be divided into multiple levels, or subtasks. In an HTA, first, the goal task must be identified. The next step is to describe in detail every step needed to accomplish the task, and the third step is to optimize these procedures. This will allow the task analyst to have different possible approaches to completing the tasks, which can help to optimize particular interactions (Hornsby, 2010). Most of the HF analysis methods require an HTA (Hilburn, in Harris, et al., 2005). These methods incorporate several cognitive elements such as monitoring, anticipating, predicting, and decision making (Alferez-Padron, Maldonado-Macías, García-Alcaraz, Avelar-Sosa, & Realyvasquez-Vargas, 2017). One of the more difficult aspects of HTA is to know when to stop the level of description, it is suggested to keep the number of immediate subtask to a small number (3-10). Annett et al. (1971), proposed that it is preferable to stop the analysis early than to continue it up to a point where it would not be useful (Nevil A Stanton, 2006).

Mental Workload (MWL) assessment is of crucial importance during the design and evaluation of complex systems. The increased role of technology and use of complex procedures has led to a greater level of demand (N. Stanton, Salmon, & Rafferty, 2013). MWL is defined as a multidimensional concept incorporating task and performance demands together with operator skills and attention. Assessments of MWL include measures of primary and secondary task performance, physiological measures, and subjective measures (N. Stanton et al., 2013). The mental workload of a task is the sum of mental demands placed upon the user, similar to physical workload or stress; if the effort is too great or cannot be sustained, the mental strain may result. According to ISO 10075, mental stress is defined as "any external influence impinging human beings and affecting them mentally" (Bridger, 2008)

Factors that affect mental workload are:

- **Time-Sharing:** Switching attention between two concurrent tasks.
- **Delays:** It can cause uncertainty in feedback.
- **Mental Models:** Can improve the operator's ability to understand current and predict future system behavior.
- **Absolute vs. Relative Judgment:** Absolute values can help to identify what actions can be done with the use of colors, green for a safe action, red danger or overload.
- **Memory Load:** If the information flow is too fast.
- **Recognition vs. Recall:** Presenting operators with a list of options is preferable than having to recall options from memory.
- **Controllability:** Ease of operation can reduce mental workload.
- **Error Tolerance:** Design systems that avoid illegal entries.

- **Environmental Design:** Optimize noise, lighting, and climate for the place where the activities are going to be performed.
- **Social Interaction:** Provide opportunities for communication and collaboration, and task sharing.
- **Time Pressure:** Workload can be reduced by relaxing the time standards. (ISO 10075, 1991).

National Aeronautics and Space Administration-Task Load Index (NASA-TLX) is a subjective workload assessment tool to rate the operator MWL (N. Stanton et al., 2013). It is a multi-dimensional scale designed to obtain workload estimates from one or more operators while performing a task or immediately afterward, it consists of six dimensions, which are: Mental Demand, Physical Demand, Temporal Demand, Performance Effort, and Frustration Level. Each dimension is rated on a scale, of 10 cm wide and divided into 20 intervals, with a reference of low and high on each side, where participants are asked to rate a task on this dimensions from a low to a high level. The ratings are converted to scores that vary from 0 to 100 (Bridger, 2008). The assumption is that some combinations of these dimensions are likely to represent "workload" (Hart, 2006). In the next step, each aspect of workload is compared to every other aspect of workload. A value of "1" is assigned to each aspect of workload every time it is rated as more important than the aspect to which it is being compared. On the third step, the ratings are multiplied by the weightings to give a set of adjusted workload rating. These are then summed and divided by 15 to give an overall TLX value for the task.

The years of research that proceeded subscale selection and the weighted averaging approach resulted in a tool that has proven to be reasonably easy to use and reliably sensitive to experimentally important manipulations over the past 20 years. Its use has spread far beyond its original application (aviation), focus (crew complement), and language (English)" (Hart, 2006).

The definitions of the scale items are as follows:

- **Mental Demand:** How much mental and perceptual activity was required (e.g., thinking, deciding, calculation, remembering, searching, etc.)?, Was the task easy or demanding, simple or complex, exacting or forgiving?
- **Physical Demand:** How much Physical activity was required (e.g., pushing, pulling, turning, controlling, activating, lifting, and carrying, etc.)? Was the task easy or demanding, slow or brisk, slack or strenuous, restful or laborious?
- **Temporal Demand:** How much time pressure did you feel due to the rate or pace at which the task or task elements occurred? Was the pace slow and leisurely or rapid and frantic?
- **Performance:** How successful do you think you were in accomplishing the goals of the task set? How satisfied were you with your performance in accomplishing these goals?
- **Effort:** How hard did you have to work (mentally and physically) to accomplish your level of performance?
- **Frustration Level:** How insecure, discouraged, irritated, stressed, and annoyed versus secure, gratified, content, relaxed, and complacent did you feel during the task?

Human Error is formally defined as all those occasions in which a planned sequence of mental or physical activities fails to achieve its intended outcome, and when these failures cannot be attributed to the intervention of some chance agency (Reason, 1990 in Stanton, 2013).

Psychologists have been investigating the origins and causes of human error since the dawn of the discipline. More recently, error research in the cognitive tradition has concentrated upon classifying errors with taxonomies and determining underlying psychological mechanisms have led to the classification of errors, driven errors, association activation errors, and loss of activation errors. (Neville A Stanton & Baber, 2002).

Human error occurs when human behavior and its influence on the system exceeds the limit. That is when wrong actions are performed (Tsang, 2006 in Alferez, 2017). HEI works on the premise that an understanding of an individual's work task and the characteristics of the technology being used allows indicating potential errors that may arise from the resulting interaction (Baber & Stanton, 1996). HEI techniques can be used during the design process to highlight potential design-induced error or to evaluate error potential in existing systems. The techniques usually describe potential errors, their consequences, recovery potential, probability and criticality, and offers associated design remedies or error reduction strategies (N. Stanton et al., 2013).

Norman first introduced a classification of errors in 1981 in slips and mistakes. Slips are defined as errors in which the intention is correct, but a failure occurring when carrying out the activities required. Mistakes arise from an incorrect intention, which leads to an incorrect action sequence, although this may be quite consistent with the wrong intention (Embrey, 2005).

- **Systematic Human Error Reduction and Prediction Approach (SHERPA):** The SHERPA method was created by Embrey in 1986, developed for use in the nuclear processing industry, and is probably the most used Human Error Identification method. Despite its origin in the nuclear processing industry, SHERPA taxonomy is generic and can be applied in any domain involving human activity such as, aviation, health care, and public technology; a series of validation studies have suggested that it is amongst the best and more reliable human error prediction tools available (N. Stanton et al., 2013) (Harris et al., 2005). The method consists of an error mode taxonomy linked to a behavioral taxonomy and is applied to an HTA of the task or scenario under analysis in order to predict potential human or design-induced error (N. Stanton et al., 2013).
- **SHERPA Behavior Taxonomy:** the assessment of the activity will be studied under the approach of what error can be done in each activity, the error can be classified under the external error mode taxonomy table in which actions are divided into four categories and an error code is given to each error mode, Figure 1. Errors are classified
- **Action (e.g. Pressing a Button or Pulling a Switch):** Errors in this category are classified as: operation too long/short; operation mistimed; operation in the wrong direction; operation too little/much misaligned; right direction on the wrong object; wrong operation on right object; operation omitted; operation incomplete; wrong operation on the wrong object.
- **Retrieval (e.g. Getting Information from a Screen or Manual):** Errors in this category are classified into: information not obtained; wrong information obtained; information retrieval incomplete.
- **Checking (e.g. Conducting a Procedural Check):** Errors in this category are classified into: check omitted; check incomplete; right check on the wrong object; wrong check on the right object; check mistimed; wrong check on the wrong object.
- **Selection (e.g. Choosing one Alternative over Another):** Errors in this category are classified into: selection omitted; a wrong selection made.

- **Information communication (e.g. Talking to Another Party) Errors in this Category are Classified Into:** information not communicated; wrong information communicated; information communication incomplete (Figure 1).

The advantages of SHERPA are that it provides a structured and clear approach to error prediction, giving a detailed analysis of potential errors. In comparison with other methods, studies have proved that is a reliable method, achieving the highest scores on evaluations for its performance. SHERPA's main disadvantages are that it is tedious and takes too long to complete it (Nevil A Stanton, 2006).

The technique works by indicating which error modes are credible for each bottom-level, based on the HTA, SHERPA involves eight steps:

Step 1: Conduct an HTA. The first step in a SHERPA analysis involves describing the task or scenario under investigation.

Step 2: Task Classification. Next, the analyst should take the first (or next) bottom-level task step in the HTA and classify it according to the SHERPA behavior taxonomy: action, retrieval, checking, selection and information communication.

Figure 1. SHERPA external error mode taxonomy

SHERPA external error mode taxonomy

Error code	Error modes based on behavioral taxonomy
A	ACTION ERRORS
A1	Operation is to long/short
A2	Operation is mistimed
A3	Operation in wrong direction
A4	Operation too little/much
A5	Misalign
A6	Right operation on wrong object
A7	Wrong operation on right object
A8	Operation omitted
A9	Operation incomplete
A10	Wrong operation on wrong object
C	CHECKING ERRORS
C1	Check omitted
C2	Check incomplete
C3	Right check on wrong object
C4	Wrong check on right object
C5	Check mistimed
C	Wrong check on wrong object
R	RETRIEVAL ERRORS
R1	Information not obtained
R2	Wrong information obtained
R3	Information retrieval incomplete
I	COMMUNICATION ERROR
I1	Information not communicated
I2	Wrong information communicated
I3	Information communication incomplete
S	SELECTION ERRORS
S1	Selection omitted
S2	Wrong selection made

D. Embrey

Step 3: Human Error Identification. The analyst uses the associated error mode taxonomy and domain expertise to determine any credible error modes for the task in question. For each credible error, the analyst should give a description of the form that the error would occur.

Step 4: Consequence Analysis. This involves determining and describing the consequences that the analyst thinks that can happen, associated with the errors identified in the previous step.

Step 5: Recovery Analysis. The analyst should determine the recovery potential of the identified error.

Step 6: Ordinal Probability Analysis. Once the consequence and recovery potential of error have been identified, the analyst should rate the probability of the error occurring, on a scale of low, medium, or high.

Step 7: Critical Analysis. Next, the analyst rates the criticality of the error in question on a scale of low, medium, or high.

Step 8: Remedy Analysis. The final stage in the process is to propose error reduction strategies.

Relationship Between Human Error and Design

Neville A Stanton & Baber, (2002) introduced the idea of predicting designer error by evaluating devices using human error reduction prediction approach (SHERPA). It is suggested that using HEI techniques as part of the design and evaluation process could help to make devices easier to use. In a study developed by Baber & Stanton, (1996), the use of human error identification (HEI) techniques was proposed as an alternative to observation studies for product evaluation. Many of the factors that affect mental workload are under the control of designers and ergonomists (Bridger, 2008).

Most industrial accidents are caused by human error: estimates range between 75 and 95 percent. The truth is that these accidents occur because of design problems when the percentage is so high, clearly other factors must be involved. While physical limitations are well understood by designers; mental limitations are greatly misunderstood (D. Norman, 2013).

In the airline industry, for example, human error is now the primary risk to flight safety. The high levels of automation offered a considerable advance in safety; however, new types of errors began to emerge on the flight decks. After the accidents of the Nagoya Airbus A 300-600, the Cali Boeing 757 accident, and the Strasbourg A320, in the early 90's, an exhaustive study of the pilot-aircraft interface on modern flight decks was commissioned by the Federal Aviation Administration (FAA). The report identified many major flight deck design deficiencies and shortcomings in the design process. Problems with flight deck interfaces were identified, confusing and unclear display symbology and nomenclature, among others. The report also criticized the flight deck design process, identifying a lack of human factors expertise on design teams, the application of little cognitive ergonomics and putting too much emphasis on physical ergonomics. Fifty-one specific recommendations came out of the report, including: 'The FAA should require the evaluation of flight deck designs for susceptibility to design-induced flight crew errors and the consequences of those errors as part of the type certification process' (Harris et al., 2005).

With the course of time and the advances of technology, new classifications and error types have emerged. Since the first studies in human error processes new technology devices are part of our daily lives, and sometimes, they represent a challenge in using them. Nowadays, even the use of a washer can be stressful for some people, as there are several functions included in a product that makes it difficult to use, and the intention to benefit from the product is not accomplished.

Some things that need to be considered when a designer starts a new project are: physical, cultural, semantic, and logical constraints: Physical constraints limit what can be done. They are more effective if they are easy to see and interpret. It prevents a wrong action from succeeding, only after it has been tried. An example that could be used, is that batteries have a very limited physical constraint, which makes it difficult to determine the proper orientation for the batteries. As a result, it would be better if designers developed a battery where orientation was irrelevant (D. Norman, 2013). Cultural constraints represent a set of actions permitted in a society, what can be accepted in one culture can be forbidden or frowned upon in another, there are general rules and information necessary for interpreting situations and for guiding behavior. Semantic constraints rely upon our knowledge of the situation and of the world. Like cultural constraints, they can change with time. Cultural and semantic constraints provide strong restrictions on what would make sense. New technologies can change the meanings of things. Logical constraints take the consistent relationship between the spatial or functional layout of components in the account to detect how they will affect or be affected by an action. Constraints are powerful clues, limiting the set of possible actions. Designers should take advantage of constraints and design products that would make error impossible to be committed instead of continuing with the *legacy problem* of using existing standards, even if they have usability, and continue doing things without taking the customers' needs into account (D. Norman, 2013).

Relevance of Human Factors in Product Development

Nowadays, professionals, such as industrial and software designers, are expected to have knowledge on ergonomic (human factors) matters and to put them in the center of the design process. A new culture has emerged as the HCD has become primordial in product development; this was very different back in the 1980's, where human factors were not particularly considered in companies making consumer products. The application of human factors has had a slow assimilation until now; before the total integration of human factors, in the early stages of product development, there was a time where the human factors specialist was asked to integrate some features in the product at a later product-creation process, resulting in superficial contributions. Now, companies see design as one of the few areas in which it is possible to have advantages over their competition. Good human factors are central in accomplishing excellence in design (Jordan, 2002).

The emotional design aims to evoke or prevent certain emotions, and it can be accomplished by an understanding of emotional processes. Considerable progress in the cognitive analysis of emotions has been accomplished in recent years, even though the emotional power of products has never been doubted, and designers have used their intuitions and artistic skills to exploit this appeal. A method was developed, based on the premises that facial expression and body language are universal, and by using animations and cartoon diagrams of emotional expressions (seven positive and seven negative emotions) (Desmet, 2002 in Norman, 2003). In addition, Desmet developed a methodology for aiding designers in reaching a desired product.

Usability has a new approach in design; now, customers expect products to be easy to use. Usability has evolved from being a satisfier to being a reason for dissatisfaction. People are no longer pleasantly surprised when a product is usable, but are unpleasantly surprised if it is difficult to use. Coincidently, the conception of human factors has changed to a problem-solving discipline, rather than a discipline that is positively increasing the market value of the products to which it contributes (Jordan, 2002).

The conception in which human factor should focus beyond usability in order to address the relationship between people and products or systems holistically to assess the pleasurability of products, challenges the profession with new approaches, perhaps the three major challenges are: understanding people holistically, linking product benefits to product properties, and developing methods and metrics for assessing product pleasurability. (Jordan, 2002).

METHODOLOGY

In order to achieve objectives of this study a usability Test, a Hierarchical Task Analysis, a NASA TLX Analysis, and a SHERPA Method, were conducted in order to determine the mental workload and to identify potential human errors

Steps for the Usability Test

Step 1: Literature and product review. The test administrators did a literature review to get familiar with the methods and analysis that were going to be used to run the evaluations, after the literature review, the earphones were analyzed to understand how they work.

Step 2: Creation of complementary surveys. Data recollection and evaluation forms were created for the three evaluations (usability, MWL, and EI), based on the target goal of using wireless earphones to listen to music from a mobile phone device.

Step 3: Pilot test. A pilot test was done to detect any possible mistakes. After corrections were made, another pilot test was conducted to be sure that the procedure would be correct.

Step 4: Contact the participants. A group of 10 participants, college students, ages 18-25, were asked to participate in the assessment based on their target user profile.

Step 5: Evaluation. Participants were evaluated in the classroom in sessions of 15 to 20 minutes. They were video recorded (with their prior consent).

Step 6: Data analysis. The surveys and videos were analyzed in order to obtain information for the methods.

Step 7: Design suggestions. Based on the results of the methods, changes on the design were proposed to improve the users' satisfaction.

Materials for the Usability Test

- Canon Power Shot camera, printed instructions, surveys, earphones, mobile device, and a camera to explain and practice the Think Aloud protocol.

Participants

- Test Moderator, Camera Operator, Participants (ten participants were involved in the test, college students, average age 18-25 years).

DATA COLLECTION

Procedure: Usability Test

First, the participants were instructed in the Think Aloud protocol by the test administrator. Then the test subjects were asked to perform the task, to listen to music using a pair of wireless earphones from a personal mobile device. Using the Think Aloud Protocol at all times; finally, a survey was conducted to evaluate the level of satisfaction of the usage of the earphones.

Step 1: Start to record the session.
Step 2: The test participants were thanked for their participation in the study.
Step 3: Participants were briefed on the overall procedure of the test and the Think Aloud Protocol.
Step 4: Participants were asked to practice the think-aloud protocol with a camera in order to get familiarized with the Think Aloud Protocol.
Step 5: Participants were asked to perform the activity being assessed, listening to music from a mobile phone device. The test administrator gave the instruction to perform each task to achieve the goal task.
Step 6: A survey was applied to every participant.
Step 7: Test administrator thanked the participant again for their participation.

Database Construction and Debugging

The videos were analyzed, in order to observe the time, it took the participants to perform each task, to establish the average time to rate the participants' performance Comments expressed on the Think Aloud protocols were taken into account to determine how they felt, and what they were thinking during the process. Surveys were used to determine the overall experience of using the earphones (general opinion, functionality, if they felt in control of the earphones all the time, what would they suggest improving them).

Hierarchical Task Analysis (HTA)

The seven-step procedural guideline (Neville Anthony Stanton et al., 2004) was used to do a HTA.

Step 1: Decide the purpose of the analysis. The purpose of the HTA was to determine if the participants had problems to achieve the goal to listen to music using wireless earphones from a mobile phone device and point where changes in the design can improve the users' satisfaction.
Step 2: Determine task goals and performance criteria. A list of the tasks and subtasks needed to be accomplished to achieve the goal activity was created.
Step 3: Identify sources of task information. Observation of the activity was the primary source of information, the users' manual of the earphones, as well as the usability test.
Step 4: Acquire data and draft decomposition diagram. Starting from the goal activity a series of tasks, subtasks and possible errors would be taken into account to develop the diagram Figure 2.
Step 5: Recheck the decomposition of the tasks. Take into account cognitive factors, such constraints, context, and other factors to assure the reliability of the diagram.

Step 6: Identify Significant Operations. Identify possible error related to input, action or lack of feedback, where impossible physical actions can be detected, the detection of these errors will help to take other actions.

Step 7: Generate hypothetical solutions to the performance problems identified in the analysis. Based on the previous analysis, design changes will be suggested to improve the usability of the wireless earphones.

NASA-TLX

The NASA-Task Load Index (NASA-TLX) preform was used and is shown in Figure 3 to assess the MWL experienced by participants in this study.

Step 1: The task under analysis to listen to music using wireless earphones was clearly identified.

Step 2: The HTA done previously, was used to identify the tasks that are necessary to perform the goal activity.

Step 3: The evaluation was performed with the group that participated in the usability test after a period of time.

Figure 2. HTA diagram

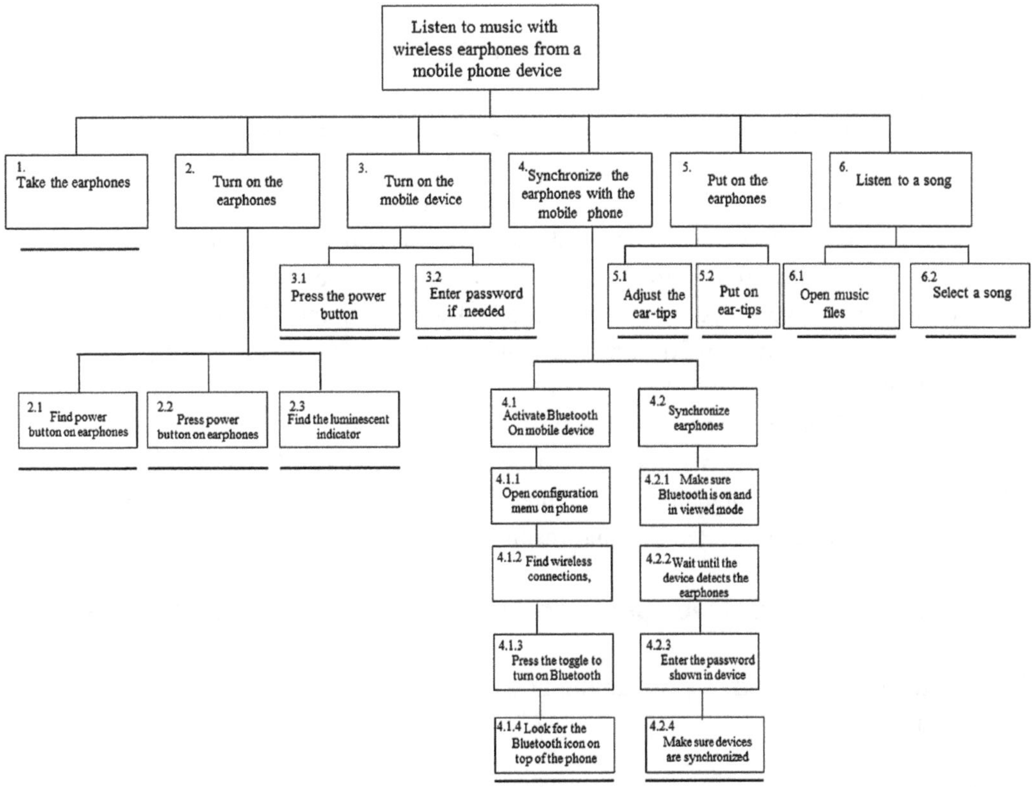

Figure 3. NASA-TLX rating scale

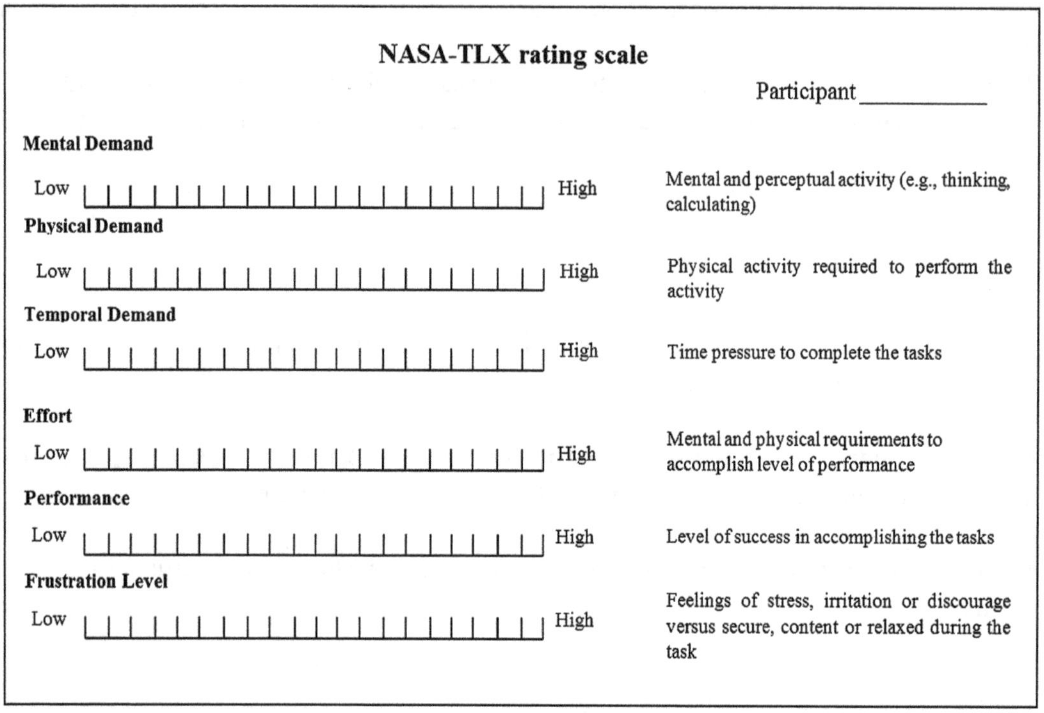

NASA-TLX rating scale

Participant __________

Mental Demand	Low – High	Mental and perceptual activity (e.g., thinking, calculating)
Physical Demand	Low – High	Physical activity required to perform the activity
Temporal Demand	Low – High	Time pressure to complete the tasks
Effort	Low – High	Mental and physical requirements to accomplish level of performance
Performance	Low – High	Level of success in accomplishing the tasks
Frustration Level	Low – High	Feelings of stress, irritation or discourage versus secure, content or relaxed during the task

Based on the paper-and pencil version of NASA-TLX

Step 4: The participants were briefed on the NASA-TLX technique, emphasizing the meaning of each dimension.

Step 5: In this step, participants should perform the task under analysis, but in this case, they had interacted with the product on the usability test previously.

Step 6: Weighting Procedure will begin with 15 pairwise comparisons of the six dimensions; to determine in which case they felt that a demand was stronger than the one being compared to, a value of one would be given and the sum of each dimension would be the weight.

Step 7: Rating Procedure. Participants will be presented with the interval scale for each of the dimensions, they will rate from 1 to 20 (low-high) and a value of five points will be given to each box (example temporal demand 12 times 5 = 60).

Step 8: TLX Score Calculation. The weight obtained on step six will be multiplied by the result in step 7 (in this case and an average of the ten participants was used) and then, the sum of those values will be divided by 15 to calculate the Weighted TLX it. To calculate the Average TLX the sum of the average values will be added and that result will be divided by six.

Systematic Human Error Reduction and Prediction Approach (SHERPA)

The SHERPA method was used to evaluate the possibilities of human error occurrence. This method involves eight steps:

Step 1: Conduct an HTA. On this first step, a description of every action needed to accomplish the task is described and divided into subtasks.

Step 2: Task Classification. Every step of the goal task is considered for error classification. In this study, four types of errors were detected: action, checking, retrieval, and selection.

Step 3: Human Error Identification. The next step is to describe the situation, and under which error code possible errors can be committed.

Step 4: Consequence Analysis. In this step, the consequences of the previous error outlined are described.

Step 5: Recovery Analysis. This step determines the recovery potential of the identified error.

Step 6: Ordinal Probability Analysis. Once the consequences and recovery potentials of the errors have been identified, a rating of the probability of the error occurring, in a scale of low, medium or high, will be logged.

Step 7: Critical Analysis. Next, the criticality of the error in question will be rated on a scale of low, medium, or high.

Step 8: Remedy Analysis. The final stage in the process, in which error reduction strategies are proposed.

RESULTS

Results of the Usability Test

Results on the usability test show that 100 percent of the participants had problems with turning on the earphones, as they did not see the luminal signifier when the earphones were turned on. Some participants did not synchronize the earphones with their mobile devices. Some participants did not see the power and volume controls. After the videos were analyzed, Table 1 shows that the results of the performance on the tasks evaluated were inadequate. Aside from the usability difficulties, some participants said that the sound quality was very good and that they liked the detachment of the earphones from the mobile device.

It is also important to understand that usability tests have limitations and that they do not guarantee success, or even prove that a product will be usable. Even the most rigorously conducted formal test cannot ensure with a 100 percent certainty that a product will be usable when released. The reason being, testing is an artificial situation, test results do not prove that a product will work because they depend upon the way in which the test was conducted, as well as the fact that participants are rarely full representatives of the target users. Because of this, testing is not always the best technique to use (there are other techniques designed to evaluate and improve products, in some cases, it is more effective to conduct an expert or heuristic evaluation of a product rather than test it). However, as part to an overall user-centered design approach, it is a very reliable tool to detect potential problems and how to resolve them (Rubin & Chisnell, 2008).

Results of Hierarchical Task Analysis

Results of the HTA are shown in Table 2. This analysis includes 6 tasks and 20 subtasks with their corresponding plan. Linear plans were used to determine the subtasks that needed to be performed to continue to achieve the goal task. The task which involved more actions was the one to synchronize the earphones to the mobile device via Bluetooth, with a total of 11 subtasks. Consequently, it was found to be the most time consuming and frustrating one.

Table 1. Usability measure indicators

Task	Turn on Earphones					Task	Synchronize Earphones With Device				
Participants	Time	0-9 sec.	10-13 sec.	14-17 sec.	18 + sec.	Participants	Time	15-36 sec.	37-58 sec.	59-80 sec.	81 + sec.
	Seconds	4 pts.	3 pts.	2 pts.	1 pts.		Seconds	4 pts.	3 pts.	2 pts.	1 pts.
1	23				x	1	36	x			
2	32				x	2	15	x			
3	74				x	3	234				x
4	15			x		4	84				x
5	14			x		5	122				x
6	53				x	6	171				x
7	9	x				7	86				x
8	23				x	8	174				x
9	15			x		9	96				x
10	6	x				10	145				x
		8	0	6	5			8	0	0	8
		Task #2 value			19			Task #4 value			16

40-31 points Optimum
30-21 points Adequate
20 - 11 points Inadequate
10 - 0 points Dreadful

Result of Mental Workload Analysis With NASA-TLX

As the results of the MWL analysis in Table 3 indicate, the information analyzed from the surveys concluded that a medium mental workload was experienced by the subjects and that the most demanding component was temporal demand, with the least demanding weighed component was the frustration experienced by the user. The Average TLX was 46.3, and the Weighed TLX 49.6.

The weighting procedure begins as a series of pairwise comparisons of the six defined dimensions, in order to determine which ones required of a higher mental demand. Once these dimensions are compared, an interval scale, ranging from 1-20, with 1 being the lowest, and 20 being the highest was presented to participants. These intervals represent five points each. The results of the pairwise comparisons obtained will then be multiplied by the points in the interval scale. The sum of all of these values will then be divided by 15, providing the Weighted TLX (748/15=49.9) In order to calculate the Average TLX, the total sum of the average values of the interval scale is calculated and then divide the result by 6 (278/6=46.3).

Results of HEI Analysis With SHERPA

As results of the HEI analysis, Table 4 shows 26 potential human errors that were identified in various tasks. This study showed that 50% of them were action errors (A), 26.92% were checking errors (C), 7.7% were retrieval errors (R), and 15.4% were selection errors (S). The task that had a higher probability of

Table 2. HTA

	Listening to Music From Your Mobile Device Using Wireless Earphones
1.-	Take the earphones.
2.-	Turn on the earphones.
2.1.-	Find the power button on the earphones.
2.2.-	Press the power button on the earphones.
2.3.-	Find the luminescent indicator.
3.-	Turn on the mobile device.
3.1.-	Press the power button on the mobile device.
3.2.-	If the mobile device is password locked, enter the password.
4.-	Syncchronize the earphones.
4.1.-	Activate Bluetooth.
4.1.1-	Open the configuration menu of the mobile device.
4.1.2.-	In “Wireless connections and Networks”, select “Bluetooth”.
4.1.3.-	Press the toggle to turn Bluetooth on.
4.1.4.-	Look for the Bluetooth icon on the top of your screen to confirm it is turned on.
4.2	Syncchronize the earphones.
4.2.1.-	Open the configuration menu of the mobile device.
4.2.2.-	In “Wireless connections and Networks”, select “Bluetooth”.
4.2.3.-	Make sure Bluetooth is turned on and that the device is configured as visible.
4.2.4.-	The device will search and display available devices on “Available Devices”.
4.2.5.-	Select the Bluetooth device you want to sync to your mobile device.
4.2.6.-	Enter the password that is shown in the mobile device screen.
4.2.7.-	If the synchronization is done successfully, the two devices will be connected.
5.-	Put on the earphones.
5.1.-	Adjust the ear-tips.
5.2.-	Put on the earphones.
6.-	Listen to a song.
6.1.-	Open the music/files application and search for a song.
6.2.-	Select the song

Table 3. NASA-TLX

NASA TASK LOAD INDEX									
	MD	**PhD**	**TD**	**Eff**	**Per**	**Fru**	**Weight**	**Average***	**R**
Mental Demand (MD)		1				1	2	50	100
Physical Demand (PhD)			1			1	2	24	48
Temporal Demand (TD)	1	1			1	1	4	60	240
Effort (Eff)	1	1				1	3	48	144
Performance (Per)	1	1		1		1	4	54	216
Frustration (Fru)							0	42	0
Total								278	748
Average TLX								46.3	
Weighted TLX									49.9

* Mean of ten participants

occurrence among the action errors was "keeping pressed the power button for less than 2 seconds". On the checking errors, there was a high probability of occurrence when subjects believed that the earphones had been turned on, as well of being unaware of the LED lights' power status.

About the criticality of an error occurrence, the one detected as a high criticality event was when mobile devices' password has being forgotten and it was classified as a Retrieval error; however, this event did not occur during the study. The rest of error modes presented low and medium levels of criticality.

Table 4. SHERPA method

Task Step	P	C	Error Mode	Description	Consequence	Recovery	Illustrative Remedies
1	L	L	A8	Earphones are not taken.	Earphones cannot be turned on.	Immediate	Indicate the test subject to take the earphones.
2	M	M	A8	Earphones are not turned on.	Earphones cannot be used while turned off.	Immediate	Turn on the earphones.
2	H	M	C1	Test subjects believe that earphones are already turned on.	They do not turn them on, and proceed with the experiment.	Immediate	Indicate that the earphones are not turned on to continue.
2.1	L	M	C2	Power button was not found.	Earphones cannot be turned on.	Immediate	Make the power button more visible.
2.2	H	M	A1	Earphones were pressed for less than 2 seconds.	Earphones do not turn on.	Immediate	Make the earphones more responsive to shorter presses.
2.3	H	M	C1	People were unaware of the LED's existence	They had no way to know if the earphones were turned on.	Immediate	Place the LED in a more visible place (such as, next to the power button).
3.1	L	M	C1	Power button is not found.	Mobile device cannot be turned on (if powered off).	Immediate	Include manuals that indicate where the power button is on the device.
3.1	L	M	A8	Power button is not pressed.	Mobile device cannot be turned on (if turned off).	Immediate	Lecture people about the importance of pressing the power button to use their mobile devices.
3.2	L	H	R1	Mobile device password is forgotten.	Mobile device cannot be accessed.	No recovery	Implement alternative methods to retrieve or disable mobile device passwords when they are forgotten.
4.1.1	L	L	C2	Mobile device configuration menu is not found	Bluetooth cannot be turned on, and earphones cannot be synced.	Immediate	Program the OS to make it so that the configuration menu is always visible on the home screen.
4.1.1	L	L	A8	Mobile device configuration is not selected.	Bluetooth cannot be turned on, and earphones cannot be synced.	Immediate	Inform your test subject of the importance of these following steps in order to connect the earphones.
4.1.1	L	L	S2	Different application is selected.	Unless the application is a bluetooth switch, bluetooth cannot be turned on, and earphones cannot be synced.	Immediate	Make the OS more accessible, by placing the name of the application under the application, and/or making the icons more representative of what the application is.
4.1.2	L	L	S2	Different option is selected.	Bluetooth cannot be turned on, and earphones cannot be synced.	Immediate	Space the options a little more, to make erroneous presses less likely.
4.1.3	M	M	R3	Test subject does not know how to turn on the "Bluetooth" option on their mobile device.	Bluetooth cannot be turned on, and earphones cannot be synced.	Immediate	Teach your test subject how to turn on the "Bluetooth".
4.1.4	L	L	C1	Test subject does not look for the indicator.	Test subject is not certain that the Bluetooth is turned on.	Immediate	Make the OS notify you that the Bluetooth is on, with a sound and/ or pop-up.

continued on following page

Table 4. Continued

Task Step	P	C	Error Mode	Description	Consequence	Recovery	Illustrative Remedies
4.2.1	L	L	C2	Mobile device configuration menu is not found	The earphones cannot be synced.	Immediate	Program the OS to make it so that the configuration menu is always visible on the home screen.
4.2.1	L	L	A8	Mobile device configuration is not selected.	The earphones cannot be synced.	Immediate	Inform your test subject of the importance of these following steps in order to sync the earphones.
4.2.1	L	L	S2	Different application is selected.	The earphones cannot be synced.	Immediate	Make the OS more accessible, by placing the name of the application under the icon.
4.2.2.	L	L	S2	Different option is selected.	The earphones cannot be synced.	Immediate	Space the options a little more, to make erroneous presses less likely.
4.2.3	L	L	A8	The mobile device is not visible.	The earphones might not be able to sync with the mobile device.	Immediate	Configure the mobile device as visible.
4.2.3	L	L	R1	The test subject does not know how to configure the mobile device to be visible.	The mobile device might not be visible, and the earphones might not be able to sync with the mobile device.	Immediate	Teach your test subject how to configure the mobile device as visible.
4.2.4	L	L	A1	The earphones turned off.	The earphones will not be able to sync with the mobile device.	Repeat steps 2.1 to 2.3, and try again.	Turn on the earphones.
4.2.5	L	L	A9	The earphones are not found by the mobile device.	The earphones will not be able to sync with the mobile device.	Immediate	Run the search again, and verify that the earphones are turned on.
4.2.6	L	L	A7	The password is incorrect.	The earphones will not be able to sync with the mobile device.	Go back to step 4.2.5	Type the password correctly the next time.
5.1	M	L	A5	Ear-tips are not placed correctly	The earphones may be uncomfortable, or might slip from the ear.	Immediate	Include instruction manual to indicate proper ear-tip placement procedure.
6.1	L	M	A9	Some people did not have any music on their mobile devices to select.	The earphones will not be able to play audio.	Immediate	Search for a song on the internet.
6.1	L	M	A9	Some people did not have an internet connection to look for a song.	The earphones will not be able to play audio.	Immediate	Connect the phone to the internet.
6.1	L	L	A6	Another application is selected.	The application chosen might not have any audio, or may not yield the desired results.	Immediate	Exit application, then select music.

P: Probability C: Criticality L: Low M: Medium H: High
Error Mode
A: Action C: Checking R: Retrieval I: Communication S: Selection

The use of these earphones resulted in a medium mental workload, and a diverse taxonomy of errors, mostly due to the fact that people had a hard time turning on the earphones, noticing they were on, and pairing them with the mobile device they were using. The methods used in this study helped in order to detect those difficult tasks and subtasks for the users. It is clear that changes on the design need to be done to improve the earphones.

DISCUSSION

The methods used in this study are well known for their efficiency and reliability, as they help to detect design problems by observing how people interact with them. But, as technology progresses, these methods should be updated, or adapted to perform certain evaluations, and have more accurate results. Studies show how a larger sample size yields better results than those obtained in studies with five participants (sets of 5 participants found 99% of the problems, other sets found 55%, with 10 participants, the lowest percentage of problems revealed was 80%, and with 20 participants to 95%) as Nielsen assures (Faulkner, 2003). The introduction of the concept of *new usability* positions it against existing approaches to usability. Emphasizes that the context of use is incredibly important to understand technology, there is a trend towards product integration and an increasing complexity of special domestic technology nowadays, (Thomas & Macredie, 2002). Studies based on the Think Aloud protocol, designed by Ericsson & Simon, highlight the need to include other factors to consider, due to the interactions with computers (Yang, 2003). It is important that the evaluations include all possible variants in order to achieve the best experience for the user of a product.

Common activities, such as watching a movie using a home entertainment device, can represent a cognitive challenge to complete a task efficiently, due to the high level of sophistication and complexity since the interfaces can be complicated. This will increase the mental workload and the possibilities of human error (Armenta- Hernández, Maldonado-Macías, García-Alcaraz, & Avelar-Sosa, 2016). It is therefore important to consider the impact of the context in usability in order to enhance UX, as previous research in this field has also suggested (Nicolas et al., 2011).

CONCLUSION

The goals of this study were accomplished in a favorable manner. From the results obtained by the methods, it can be concluded that the use of cognitive methods can help to identify and prevent design errors. Human error identification can complement traditional usability assessment. Correspondingly, detecting potential errors may suggest those changes that prevent the difficulties that users experienced, and reduce or eliminate frustration. The NASA-TLX method is probably the most commonly used subjective workload assessment technique, due to its simplicity in comparison to other methods, which is why it is used prominently in the actual working industry. Proper product design is a very important factor in the market, where even small differences in shape or design can result in entirely different products, functionally speaking. Because of this, manufacturers should aim to develop products that are, additionally of their intended purposes, easy to use. The challenge for the designer is to integrate different design techniques and physical and cognitive ergonomics to develop products that are usable and assure the users' full satisfaction.

DESIGN RECOMMENDATIONS

The recommended changes in this particular design of earphones device derived from this study are:

- The use of reflective controls.
- Reduce the time that the power button needs to be pressed to turn on the earphones.
- Changing the position of the LED light to a place where it is easier to see the status of the device, preferably near the controls of the earphones.
- Changing the position of the earphones' control panel to be further away from the earbud.

REFERENCES

Alferez-Padron, C., Maldonado-Macías, A. A., García-Alcaraz, J., Avelar-Sosa, L., & Realyvasquez-Vargas, A. (2017). Workload Assessment and Human Error Identification During the Task of Taking a Plain Abdominal Radiograph: A Case Study. In *International Conference on Applied Human Factors and Ergonomics* (pp. 108–119). Springer.

Armenta- Hernández. O. D., Maldonado-Macías, A. A., García-Alcaraz, J. L., & Avelar-Sosa, L. (2016). Analysis of mental workload and human error in the task of watching a movie using a home entertainment device. In Ergonomía Ocupacional. Investigaciones y soluciones (9th ed.; pp. 42–52). SEMAC.

Baber, C., & Stanton, N. A. (1996). Human error identification techniques applied to public technology: Predictions compared with observed use. *Applied Ergonomics*, *27*(2), 119–131. doi:10.1016/0003-6870(95)00067-4 PMID:15677051

Dumas, J. S., & Redish, J. (1999). *A practical guide to usability testing*. Intellect Books.

Embrey, D. (2005). Understanding human behaviour and error. *Human Reliability Associates*, *1*, 1–10.

Faulkner, L. (2003). Beyond the five-user assumption: Benefits of increased sample sizes in usability testing. *Behavior Research Methods*, *35*(3), 379–383. doi:10.3758/BF03195514 PMID:14587545

Ghasemi, M., Nasleseraji, J., Hoseinabadi, S., & Zare, M. (2013). Application of SHERPA to identify and prevent human errors in control units of petrochemical industry. *International Journal of Occupational Safety and Ergonomics*, *19*(2), 203–209. doi:10.1080/10803548.2013.11076979 PMID:23759191

Harris, D., Stanton, N. A., Marshall, A., Young, M. S., Demagalski, J., & Salmon, P. (2005). Using SHERPA to predict design-induced error on the flight deck. *Aerospace Science and Technology*, *9*(6), 525–532. doi:10.1016/j.ast.2005.04.002

Hart, S. G. (2006). NASA-task load index (NASA-TLX); 20 years later. *Proceedings of the Human Factors and Ergonomics Society Annual Meeting*, *50*(9), 904–908. doi:10.1177/154193120605000909

Hornsby, P. (2010). *Hierarchical task analysis*. Retrieved February 8, 2017, from https://www.uxmatters.com/mt/archives/2010/02/hierarchical-task-analysis.php

Jordan, P. W. (2002). *Designing pleasurable products: An introduction to the new human factors*. CRC Press.

Meuter, M. L., Ostrom, A. L., Bitner, M. J., & Roundtree, R. (2003). The influence of technology anxiety on consumer use and experiences with self-service technologies. *Journal of Business Research*, *56*(11), 899–906. doi:10.1016/S0148-2963(01)00276-4

Oritz Nicolas, J. C., & Aurisicchio, M. (2011). The scenario of user experience. DS 68-7: Proceedings of the 18th International Conference on Engineering Design (ICED 11), Impacting Society through Engineering Design, Vol. 7: Human Behaviour in Design, Lyngby/Copenhagen, Denmark, 15.-19.08. 2011.

Nielsen, J. (1994). Guerrilla HCI: Using discount usability engineering to penetrate the intimidation barrier. *Cost-Justifying Usability*, 245–272.

Nielsen, J. (2012). How many test users in a usability study. *Nielsen Norman Group, 4*(6).

Norman, D. (2013). *The design of everyday things: Revised and expanded edition.* Basic Books.

Norman, D. A. (2003). Designing Emotions Pieter Desmet. *The Design Journal*, *6*(2), 60–62. doi:10.2752/146069203789355444

Rubin, J., & Chisnell, D. (2008). *Handbook of usability testing: howto plan, design, and conduct effective tests.* John Wiley & Sons.

Scholtz, J. (2004). Usability evaluation. National Institute of Standards and Technology, 1.

Stanton, N., Salmon, P. M., & Rafferty, L. A. (2013). *Human factors methods: a practical guide for engineering and design.* Ashgate Publishing, Ltd. doi:10.1201/9781315587394

Stanton, N. A. (2006). Hierachical Task Analysis. In W. Karwowski (Ed.), International Encycopledia of Ergonomics and Human factors (pp. 3183–3190). Taylor & Francis.

Stanton, N. A., & Baber, C. (2002). Error by design: Methods for predicting device usability. *Design Studies*, *23*(4), 363–384. doi:10.1016/S0142-694X(01)00032-1

Stanton, N. A., Hedge, A., Brookhuis, K., Salas, E., & Hendrick, H. W. (2004). *Handbook of human factors and ergonomics methods.* CRC Press. doi:10.1201/9780203489925

Thomas, P., & Macredie, R. D. (2002). Introduction to the new usability. *ACM Transactions on Computer-Human Interaction*, *9*(2), 69–73. doi:10.1145/513665.513666

Van Kuijk, J. (2008) Make It Easy On Yourself, An 8-step guide to buying easy to use products. Dutch Ergonomics Magazine, volume 33, issue 1

Yang, S. C. (2003). Reconceptualizing think-aloud methodology: Refining the encoding and categorizing techniques via contextualized perspectives. *Computers in Human Behavior*, *19*(1), 95–115. doi:10.1016/S0747-5632(02)00011-0

KEY TERMS AND DEFINITIONS

Hierarchical Task Analysis: The classification of every action into tasks and subtasks needed to accomplish the goal task.

Human Error: Those occasions when something planned fails to reach the expected outcome.

Interaction: The relationship between the user and the product or system.

Mental Workload: The sum of mental demands experienced by a person at the same time.

NASA-TLX: A method used to assess the mental workload that a person is exposed to when they use a product or perform an activity in which the person will rate each dimension.

Performance: Level of success in executing or accomplishing a task.

SHERPA: A method used to detect a human error; it is linked to an error mode taxonomy of each and every action, as well as the consequences if the task is not achieved.

Think Aloud: A process where the person is asked to verbalize their thoughts, actions, and feelings that are experienced during the performance of the task.

Chapter 15
User Characteristics and Ergonomic Properties for Daily Objects Design

John A. Rey-Galindo
Universidad de Guadalajara, Mexico

Elvia Luz González-Muñoz
Universidad de Guadalajara, Mexico

Alicia Libertad Rizo-Corona
Universidad de Guadalajara, Mexico

ABSTRACT

An analysis of different objects of daily use was made, specifically four models of lemon squeezers, three models of can openers, and four models of clothespins. Such analysis is made focusing on the strength needed to complete the task and the dimensional relation between the hands and object grips that impact both in ease of use and user well-being itself, identifying the gap between young and older adults characteristics and capabilities and the features of the object. This comparison was supported by a dynamometric study to identify the maximum capabilities of strength application in two different grip forces very common in daily life. These two grip forces are the ground of the interaction with the analyzed products. In conjunction with the force, the registry made an anthropometric assessment of different anatomical points in the hands.

INTRODUCTION

Carry out most of the activities of daily living requires the help of products or objects. For instance, food preparation, hygiene or even displacement are activities in which a large number of objects are used. Objects that should act as facilitators and enable people to carry out their activities in an easy and effective manner. However, there is no much information on how the characteristics of the product respond to the needs of those who use them.

DOI: 10.4018/978-1-5225-5234-5.ch015

Copyright © 2018, IGI Global. Copying or distributing in print or electronic forms without written permission of IGI Global is prohibited.

Several analyses from the perspective of biomechanics and physiology have allowed recognizing that even though the products, in general, help people to carry out activities. There are some of them which could have a negative effect on the health of users due to sometimes products are not designed taking into consideration user's characteristics, as happens with some tools, which have been identified as risk factors for development of musculoskeletal disorders that affect workers (Daams, 1994; Resnick & Chaffin, 1995; Dong, et. al, 2007).

In the case of the use of daily products, the available information is limited. The repercussions of the characteristics of these objects on the users while interact with them are not clear and the situation becomes even more complex if it is recognized that many products that are offered in the market have similar functions. Just by taking a look at the shop windows, it is possible to find lots of products whose characteristics are different, but they are designed for the same activities. Often consumers do not know if its characteristics are adequate and what can be the repercussion of the use of these.

Another paramount point and which there is not much information is the case of the characteristics of users. It is not possible to know if the products are adequate, without having data of the capacities and limitations that the people for whom they are designed to have. This situation is not only problematic in relation with the evaluation of the products but also due to the lack of information. Thus, the decision-making of the designers is difficult and more likely to be inaccurate, causing the products to have problems of use and repercussions on user's well-being.

Ergonomics discipline allows the design to have a closer approach to users and confers to it useful tools to define the most important factors related to the characteristics of the people for whom the products are designed. In this way, designers can determine different aspects, with greater reliability so the requirements of use can be fulfilled. In the field of ergonomics has been developed the concept of ergonomic properties (Prado-León & Ávila-Chaurand, 2006), which consists of endowing the product with features that cover different users aspects. Thus, products are consistent with the requirements of use and could have less negative impact on user's health.

The ergonomic properties are six: 1) the anthropometric property, consists of adjusting the product dimensions to the measurements of the users, 2) the biomechanical property refers to harnessing and enhancing the strength that users have, through the form and mechanisms of products, 3) the anatomical property consists in adjusting the shape of the products to the anatomical area of users with which it will make contact. It is worth to highlight that not necessarily a well-designed anatomical object will copy the shape of the body, 4) the physiological property refers to the consideration of physiological functioning of the user body, considering that this is not affected by the use of the object by factors such as posture or contact with the body, 5) the psychological property, it is related to making the product comprehensible and clear in their its operation, taking into account the cognitive characteristics of the users and 6) the sensory adaptation, which refers to implement some aspects such as color and texture, to improve the use and understanding of the product.

In order to make effective the ergonomic properties of the products, it is necessary to identify the differences between users. This becomes even more complex when the characteristics of people who use the product are diverse, so it is essential identifying who will be the reference to design for. Thus, it is suggested working from the characteristics of those users whose capabilities are depleted or may require special conditions. It is expected that if the product is usable for them, people with a greater capacity will have not inconveniences when using it. In fact, older adults are becoming more relevant for designers, since the aging process reduces some capabilities, mainly of physical and sensorial nature.

This study focuses on the analysis of the capability of younger and older adults. This analysis is then related to the use of three products in three different models, which are used to carry out three activities of the daily life. To the study, can openers, clothespins, and lemon squeezers were chosen since they are widely used objects particularly for performing some instrumental activities of the daily life. Additionally, the study provides data on force application capabilities in two types of grips, as well as the anthropometry of hands of both groups of users. Therefore, the analysis of the products was focused on their anthropometric and biomechanical properties. An assessment of the dimensions and the amount of force required to carry out the task was made, complemented with a description of aspects such as their weight, material, and mechanisms.

It is expected that results of this study act as evidence of the need to analyze and redefine some characteristics of everyday products to better respond to the requirements of all users.

BACKGROUND

Mexico, as the rest of the world population, presents an outlook of constant aging. According to Manrique-Espinoza, et al (2013), 9% of the Mexican population in 2010, were older adults. Data indicate that annual growth rate in this sector, is 3,8% with a life expectancy of 76,4 years, although healthy life expectancy is 65.8 years. This shows a discouraging panorama considering that many of these older adults, due to their poor health condition, will present limitations in their daily life.

Manrique-Espinoza, et al., (2013) analyzed the functional state of 8874 older adults, dividing the sample into the urban and rural population in Mexico. The study was made by self-report employing the Katz scale of independence in activities of daily living and Lawton scale, to assess instrumental activities of daily living (see Table 1). As showed in the research 26,9% of the studied subjects had difficulties to perform the Basic Activities of Daily Living (BADL) and 24,6% presented difficulties in the Instrumental Activities of Daily Living (IADL).

Menendez, et al., (2015) found in a multicentric study about health, well-being and aging, that the proportion of people aged 60 years old and more who bespoke having difficulties to perform BADL were similar in all the studied cities, with a threshold limit value of 13,8% in Bridgetown and 23,7% in São Paulo, where the greatest proportion of surveyed people with difficulties to perform IADL was identified (40,3%). In Montevideo were found the lowest values of incapacity to perform BADL and IADL (16,8% and 17,0% respectively). Mexico City was included in the study and 19,4% of the participants were found to have difficulties with BADL and 28,6% with IADL. It is possible to identify that the number of people with difficulties to perform IADL is greater than those who presented difficulties with BADL, which could be explained by the greater motor abilities needed to perform these activities as well as the usage of diverse objects in their execution.

Among the Instrumental Activities of Daily Living, food preparation is of great importance due to its obvious relation to the basic need of nutrition. Several studies report that for older adults food preparation, as well as food shopping, is, in the same way, the expression of their independence and control (Andersson & Sidenvall, 2001; Sidenvall, Nydahl & Fjellström, 2001). Some people also report feeling to have a more normal and organized life when being able to prepare their own meals (Gustafsson, Andersson, Andersson, Fjellström & Sidenvall, 2003).

Table 1. Characterization elements of the functional state in the older adult

Activities	Manriquez-Espinoza, et al (2013)	Menéndez et al 2015
Basic Activities of Daily Living	Walk	Eat
	Take a shower	Going to the restroom
	Going to bed/Getting up	
	Get dressed	
Instrumental Activities of Daily Living	Meal preparation	Going out alone
	Food shopping	Getting into a vehicle
	Medication management	Housework
	Money management	Using the phone

Equally important, healthy eating is a determinant to Active Aging, and according to the policy framework on active aging by the World Health Organization (2002), comprised the ability to prepare meals. However, the changes in health conditions that come with age, as loss of vision and hearing, as well as osteoarthritis affecting hands, might limit elderly people's ability to prepare meals (WHO Ageing and Health report, 2015). These limitations may even lead to food insecurity at this stage of life (Lee & Frongilo, 2001).

During food preparation there is a wide range of obstacles to overcome by older people, such as schedule cooking time to fit with the time of the day when their health is at their best, maintaining positions for long periods, adjusting their cooking technique to fit their present capabilities and some more that are mediated by the kitchen's environment as well as the kitchenware used.

Many activities of food preparation depend on a great deal on how well does the tool or objects needed, such as food packaging (Hensler, Herren & Marks, 2015), fits user's capabilities and characteristics, facilitating the activity or being an obstacle to it. Some issues are at play as the shapes and angles of the grips on some objects, the materials that promote an easy grasp or the strength needed to complete a certain task. This accordingly to the functional characteristics of the hand as well as its changes in the different stages of life and the capability variations derived from this process.

The hand is composed of five different joint groups, distal radioulnar joint, which is the connection between the hand and forearm and allows supination and pronation movements. The wrist joint, that allows flexion, extension, ulnar deviation and radial deviation movements. The carpometacarpal joint, which presents a limited movement degree although, constitutes the support of the movement in the thumb distal zone, which is an essential element for pressure. The metacarpophalangeal joints, which connect the palm to the fingers and allow the extension, flexion, abduction, adduction and rotation movements of the fingers. Finally, interphalangeal joints that allow extension and flexion of finger phalanges (Ombregt, 2013).

Anatomic and Biomechanical analyses showed the great complexity around the functional aspects of the human hand (Schreuders, Brandsma, & Stam, 2014). Hand structure and the movements described are held by a group of extrinsic and intrinsic muscles. The first ones, which are the majority, are originated in the forearm, cross the wrist and insert in ligaments and bones in the hand. At the same time, intrinsic muscles are those which origin and insertion are in the hand itself, they enable adduction movement of the fingers and allow movements like power or spherical grip (Taylor & Schwarz, 1955).

Hand constitution is complemented by a complex system of nerve terminals, ligaments, joints, muscles and tendons (Ombregt, 2013; Reijnierse & Fernandes, 2013; Taylor & Schwarz, 1955), which are responsible for the wide range of movements which human beings are capable of, and for the force application, that enables the several grips employed in day to day activities.

Napier (1956) and later Landsmeer (1962) identified that grips could be understood as two basic forms of grasp; power grip and precision grip. The power grip is described as the flexion of fingers and palm over an object, making counter-pressure with the thumb in a plane similar to that of the palm (Napier, 1956). Within Cutkosky (1989) description, as a second form of power grip appears the so-called lateral pinch grip, in which the flexion of fingers and palm are present too, however, the pressure exerted by the thumb is not contrary to the palm, but in a lateral direction, based on the index finger. These two forms of grip are on which the present study is centered, as they are the core of the use of some everyday objects.

Several authors have assessed the force, according to the different types of grip (Delgado, 1990; Incel, Ceceli, Durukan, Erdem, & Yorgancioglu, 2002; Li & Yu, 2011; Mathiowetz, et. al., 1985; Parvatikar & Mukkannavar, 2009; Werle, et. al., 2009), equally important, it has also been recognized that these capacities vary over time which can affect the performance in some activities (Ranganathan, Siemionow, Sahgal, & Yue, 2001; Shiffman, 1992). This last issue takes on relevance since it proves the need to adjust the way of performing the activities and with them the characteristics of the objects to the needs of the elderly.

The aging of hands is due to a process of structural deterioration (bones, tendons, joints, muscles), combined with changes in neural control (Carmeli, Patish, & Coleman, 2003). A considerable decrease in strength has been recognized in the aging stage which appears to respond to a decrease in muscle mass probably due to loss of individual fibers volume, reduction in the total number of fibers, or a combination of both circumstances (Kirkendall & Garrett, 1998). In addition to muscle changes, the hands of older adults are much more likely to develop arthritis or osteoporosis, affecting their ability and movement (Carmeli, Patish, & Coleman, 2003).

In addition, aging is related to a decrease in the ability to maintain constant force application. Ranganathan, Siemionow, Sahgal, & Yue, (2001) mention that the high fluctuation in sustained strength in older adults may be due to the fact that their lower threshold of force is greater than that registered in young adults, therefore the discharge of decompression is greater. This can generate problems in the use of products that require sustained force application, making much more difficult to perform various activities.

MAIN FOCUS OF THE CHAPTER

Taking into account the functional characteristics of the hand, the grip shapes and the various changes that take place in the hand over time can allow designers to make more responsible decisions regarding the characteristics of the products proposed. Therefore, this study aimed to identify the agreement between the dynamometric capabilities and anthropometric characteristics of younger and older adults, in two types of grip, and the force application requirements in the use of three different products, present in instrumental activities of daily living.

METHOD

Participants

734 adult subjects of both sexes (422 older adults and 310 young adults) participated in the study. A not random sampling was made in 2 institutions, a university and a daycare center for older adults, with voluntary subjects. The proportion of subjects matches the population distribution.

Instruments

In order to measure the strength of palmar grip, the Baseline Hand Dynamometer 200 Lbs dynamometer was used; to measure the strength of pinch grip the Baseline Pinch Gauge 50 Lbs dynamometer was used. The evaluation of the resistance of the products was performed with the Force Gauge Model M5-200 Mark - 10 Series 5 dynamometer.

To measure the strength of palmar and pinch grip were used the Baseline Hand Dynamometer 200 Lbs and the Baseline Pinch Gauge 50 Lbs respectively. The evaluation of the resistance of the products was performed with the Force Gauge Model M5-200 Mark - 10 Series 5 dynamometer.

The anthropometers and calipers used in this research were constructed based on existing models such as the Martin anthropometer of Swiss origin and manufactured by GPM and the Harpender, with certain modifications that allow a greater ease of use. These instruments were tested and compared to the foreign models with an accuracy of 0.5 millimeters. A portable chair of adjustable height was used to measure the dimensions of the subjects in sitting position. The grip diameter was measured with a grip cone graduated in millimeters, specially designed for this population. To measure the feet, hands and head a small caliper (Glissier type) was used; body diameters and arm reaches were measured with the large and depth calipers. Product dimensions were assessed with a Vernier caliper WIN with a precision level of 0,05mm.

Procedure

To perform the dynamometry test, each of the participants was asked about any physical impediment to their participation; if responded negatively, they were asked to do some simple warm-up exercises and stretching of the hands during two minutes approximately. Then, were asked to adopt a posture of a relaxed arm, forearm flexed between 90 ° and 120 °, hand and wrist pronation without flexion or extension. This posture was used for lateral pinch grip and power grip. Subsequently, they were asked to apply the greatest force possible in power grip, using a Baseline hand dynamometer with the dominant hand. Once the maneuver was done, they were asked to go to the anthropometric measurement post. After a rest of approximately ten minutes the same procedure was done, this time with a lateral pinch grip, using the Baseline pinch gauge.

An anthropometric measurement of both hands was performed in each of the participants. They were asked to flex the forearm to about 90° while remaining with the arm relaxed and the hand in extended position, without a flexion or hyperextension of the fingers.

Pressure resistance of three different types of products used in daily activities was measured in the laboratory. The use of these products corresponds to the analyzed types of grips. Three models of can openers were evaluated in the opening of three different types of cans; as well as four types of clothespins and four types of lemon squeezers while squeezing a lemon. In the case of the can openers, eight measurements were made with each type of can. The number registered on each measurement was taken at the moment when the resistance of the can was defeated. For the clothespins, four evaluations for each were made, the data was registered at the point of greatest opening. As with the clothespins, four evaluations were made with each model of the lemon squeezers, data was obtained at the maximum closing point between grips while squeezing a lemon.

In addition to the resistance evaluation, the dimensions of grip areas of the products were taken. For can openers and lemon squeezers the dimensions of handle length, opening width of the grip at the point of force application and handle diameter at complete closure were analyzed. Then, a comparison with the anthropometric registry of the population was made. Finally, were described the materials, weight, and type of mechanisms in the objects.

Data Analysis

The information was gathered in the spreadsheet software Excel. The statistical analyses of data obtained from the exertion and anthropometric evaluations were analyzed with SPSS V17, as well as the analysis of the products

RESULTS

Participants Description

The sample consisted of 734 subjects, of both sexes, with ages ranging from 18 to 100 years. The highest percentage of subjects were females, as seen in Table 2.

Exertion Evaluation

Force exertion was compared between young and older adults in the two modalities; power grip and pinch grip, in order to find significant differences. Results showed that adults under 60 years old can apply greater force than those over 60 years of age (see Table 3).

The comparison of force application by gender showed that males have greater force capacity than females. These differences were statistically significant (see Table 4).

Simultaneously an anthropometric study of the participants was made, considering weight and height, as well as five anthropometric hand dimensions: hand length, palm length, total hand width, palm width and grip diameter.

Tables 5 and 6 present a comparison of dimensions by age group and by gender respectively. The Levene test indicates that equal variances are assumed in both groups.

The same dimensions were compared by gender. Significant differences between men and women in four dimensions were founded (See Table 6). Is particularly important to mention the difference in the dimensions of hand length and palm length.

Table 2. Demographic characteristics of the sample

Characteristic	Classification	Frequency	%
Gender	Female	471	64.2
	Male	263	35.8
Age	< 60	310	42.3
	≥ 60	422	57.7
Occupation	Student	178	24.3
	Administrative	119	16.2
	Operativo	35	4.8
	Retired	73	9.9
	Housekeeping	293	39.9
	Sales	34	4.6
	Agriculture workers	2	0.3
Obesity	Underweight	6	1.0
	Normal weight	164	27.8
	Overweight	242	41.0
	Obesity	170	28.8
	Morbid obesity	8	1.4

Table 3. Comparison of strength between young and older adult

Strength	Age	N	Mean	S.D.	Min.	Max.	Percentile			T-Test (p)
							5	50	95	
Power grip	< 60	176	38.92	14.38	14	76	21.7	34	60.00	10.477 (0.001)
	≥ 60	416	27.66	9.33	2.00	60.00	16.0	26.00	48.00	
Pinch grip	< 60	176	8.91	2.24	3	14	5.93	8.00	13.00	7.645 (0.001)
	≥ 60	418	7.33	2.33	2.50	18.00	4.00	7.00	11.50	

Table 4. Comparison of strength by gender (KgF)

Strength	Gender	N	Mean	S.D.	Min.	Máx.	Percentile			T-Test (p)
							5	50	95	
Power grip	Female	410	25.10	5.95	2.00	46.00	16.00	25.00	36.00	22.40 (0.001)
	Male	207	50.91	21.79	16.00	125.00	26.00	44.00	105.00	
Pinch grip	Female	410	6.80	1.723	2.50	12.00	4.00	7.00	10.00	19.03 (0.001)
	Male	184	10.01	2.25	3.00	18.00	6.12	10.00	13.00	

Table 5. Comparative anthropometric dimensions by age

Dimensions	Age	N	Mean	S.D.	Percentile			T-Test (p)
					5	50	95	
Weight (kg)	< 60	175	68.87	14.26	48.10	69.70	94.66	*
	≥ 60	415	68.47	13.42	49.70	67.00	90.84	
Stature	< 60	176	1638.38	88.16	1488.40	1633.50	1780.20	12.49 p<0.001
	≥ 60	418	1538.46	89.35	1417.95	1530.00	1708.05	
Hand length	< 60	176	175.28	11.34	157.00	175.00	194.15	5.43 P<0.001
	≥ 60	417	166.56	19.99	147.90	167.00	190.00	
Hand palm length	< 60	176	101.92	7.19	90.85	102.00	113.15	7.59 p<0.001
	≥ 60	417	94.88	11.38	82.00	95.00	109.00	
Hand breadth at thumb	< 60	176	93.61	10.20	81.55	93.00	105.15	*
	≥ 60	413	95.78	58.93	79.70	90.00	110.00	
Hand breadth at metacarpal	< 60	176	81.36	49.54	67.00	77.00	88.15	*
	≥ 60	413	81.44	59.45	68.70	75.00	88.00	
Grip diameter	< 60	71	40.34	3.58	33.00	40.00	46.40	8.19 p<0.001
	≥ 60	417	35.72	4.51	29.00	35.50	43.00	

*Not significance
Note: Weight is expressed in kg and dimensions in millimeters

Table 6. Comparative anthropometric dimensions by gender

Dimensions	Gender	N	Mean	S.D.	Percentile			T-Test (p)
					5	50	95	
Weight (kg)	Female	406	65.90	12.80	48.04	65.10	87.96	7.417 p<0.001
	Male	184	74.52	13.67	52.30	73.70	97.50	
Stature	Female	410	1525.19	76.21	1417.55	1521.00	1651.45	20.31 p<0.001
	Male	184	1663.60	78.14	1531.00	1669.00	1785.00	
Hand length	Female	409	164.22	17.18	147.00	166.00	182.00	10.66 p<0.001
	Male	184	180.09	15.83	161.00	182.00	195.75	
Hand palm length	Female	409	94.25	9.42	82.00	95.00	105.00	9.86 p<0.001
	Male	184	103.02	11.25	90.00	104.00	115.00	
Hand breadth at thumb	Female	406	93.78	46.13	79.00	90.00	110.00	*
	Male	183	98.12	56.72	81.20	96.00	106.80	
Hand breadth at metacarpal	Female	406	80.63	59.89	67.35	75.00	87.00	*
	Male	183	83.17	48.72	70.00	81.00	89.00	
Grip diameter	Female	359	35.42	4.42	29.00	35.00	42.00	8.19 p<0.001
	Male	129	39.11	4.32	32.00	39.00	47.00	

Note: Weight is expressed in kg and dimensions in millimeters

Characterization of the Objects

In the qualitative analysis of the objects, it was found that there are great similarities related to dimensional characteristics, however, in the case of the can openers, the point of force application is different by almost 10 cm between the can opener with the greatest grip width to the point of force application (B) and the can opener with the smallest grip width to the point of force application (C), which can impact on the force application itself. The weight was also very dissimilar, finding a difference of 92 grams between opener A and B. Although the twist and cut system is similar, the models A and C have a disc with a sharp disc that is in charge to realize the perforation and cutting, while model B has a blade for the same function (See Figure 1).

Likewise, with the clothespins, similar data was founded on weight, with a difference of only one gram between the heaviest models (B and D) and the lightest models (A and C). In relation to the material, models A, B and C were made of plastic and their mechanisms of metal. Model D is made of wood, with a metal mechanism. The pressure mechanism of the first three models (A, B and C) consists of a ring generating the resistance for closure; in model D, the mechanism is a metal spring. With respect to the measurements, model D is the longest and has the shortest distance between clamping zones, with a difference of 17.5 mm compared with model A (Figure 2).

Figure 1. Characterization of the three types of can openers: General description and mechanisms

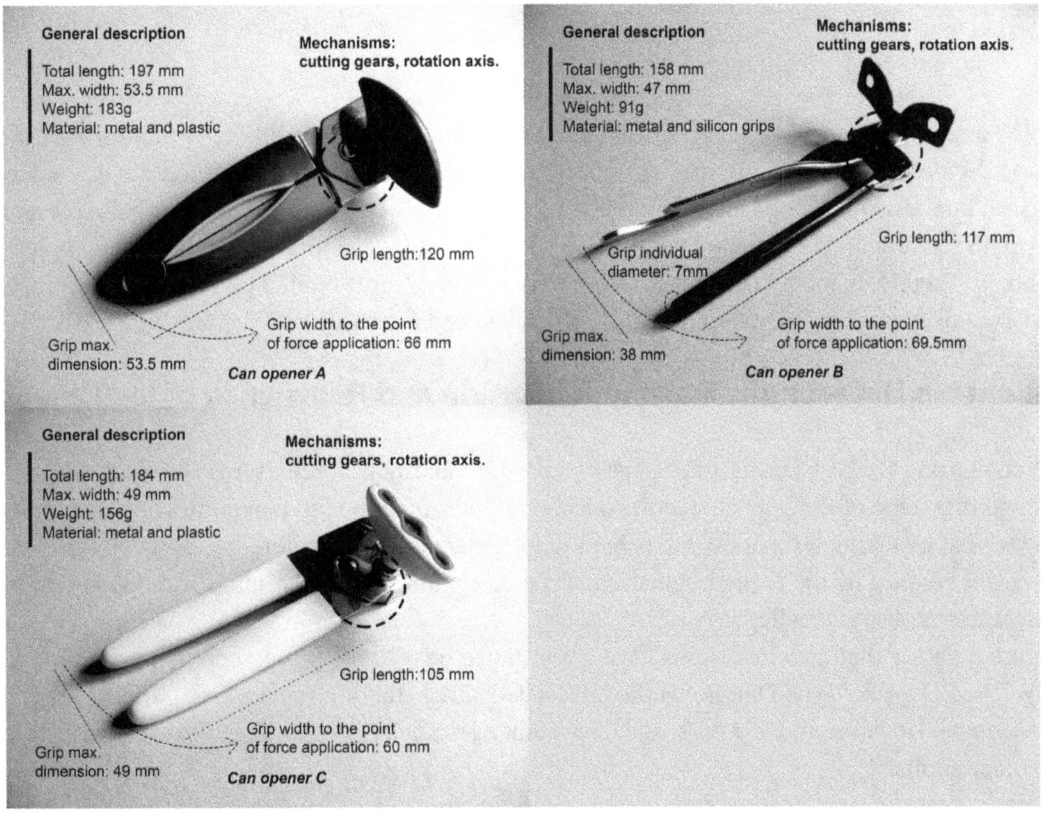

Figure 2. Characterization of the four types of clothespins: General description and mechanisms

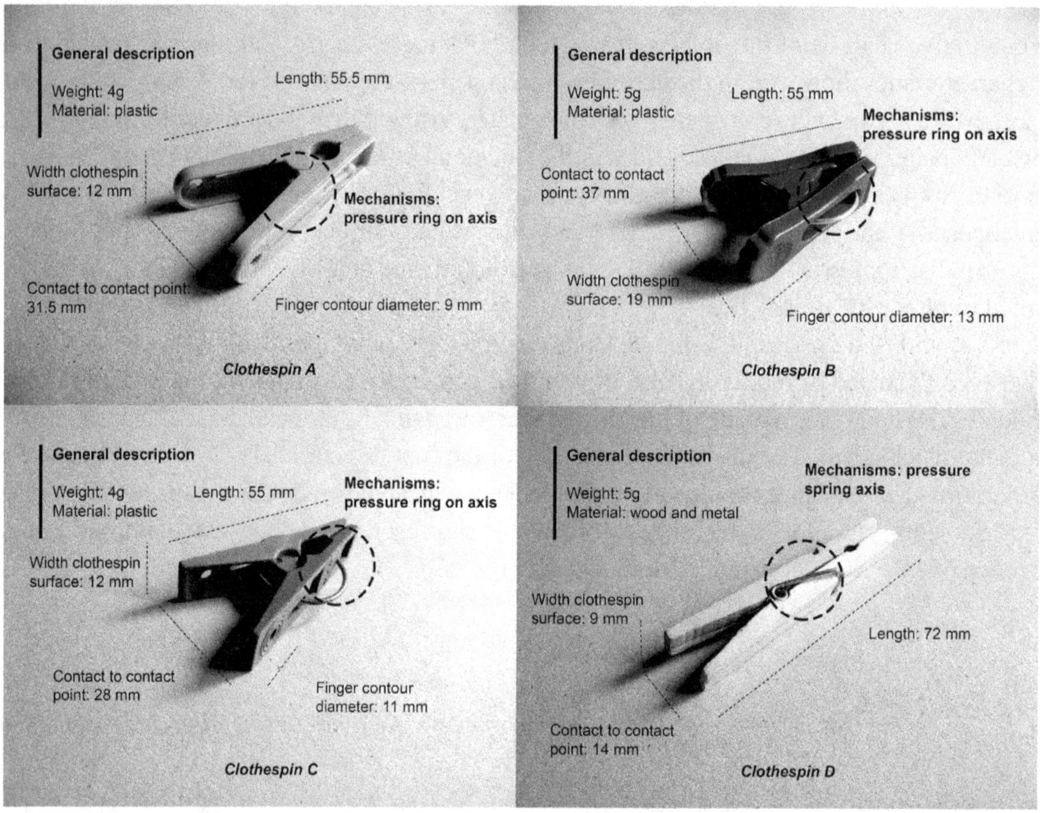

In the analysis of lemon squeezers, was found a difference of 108 g in the weight of model C and model D. Models A, B and C are made of plastic; model D is made of metal. Model A has an internal metal structure. The most noticeable difference between the models and perhaps the most relevant characteristic in relation to use, is the opening space between the two handles, when the lemon is placed, in the models C and D. In model C, the grip width to the point of force application is 152 mm, it represents a difference of 81 mm compared to model D which showed a measure of 71 mm (Figure 3).

Relationship Between the Maximum Exertion and Resistance of the Objects

When comparing the resistance of the objects with the maximum force exerted by both age groups, it was found that none of the openers require exertion higher than the 5th percentile, however in figure 4 is showed that by taking 50% of the 5th percentile of older adults as a reference, only the type B opener could easily be used by the entire population. Type A, requires a greater application of force, whereas type B, is the easiest to handle.

Figure 5 shows that the clothespins have a resistance near to the 5th percentile of the older adults, mainly C and D types. Type D presents the lowest resistance, therefore in terms of effort, would be the most adequate. However, models A, B, and C, present a resistance greater than 50% of the 5th percentile of the older adults.

Figure 3. Characterization of the four types of lemon squeezers: General description and mechanisms

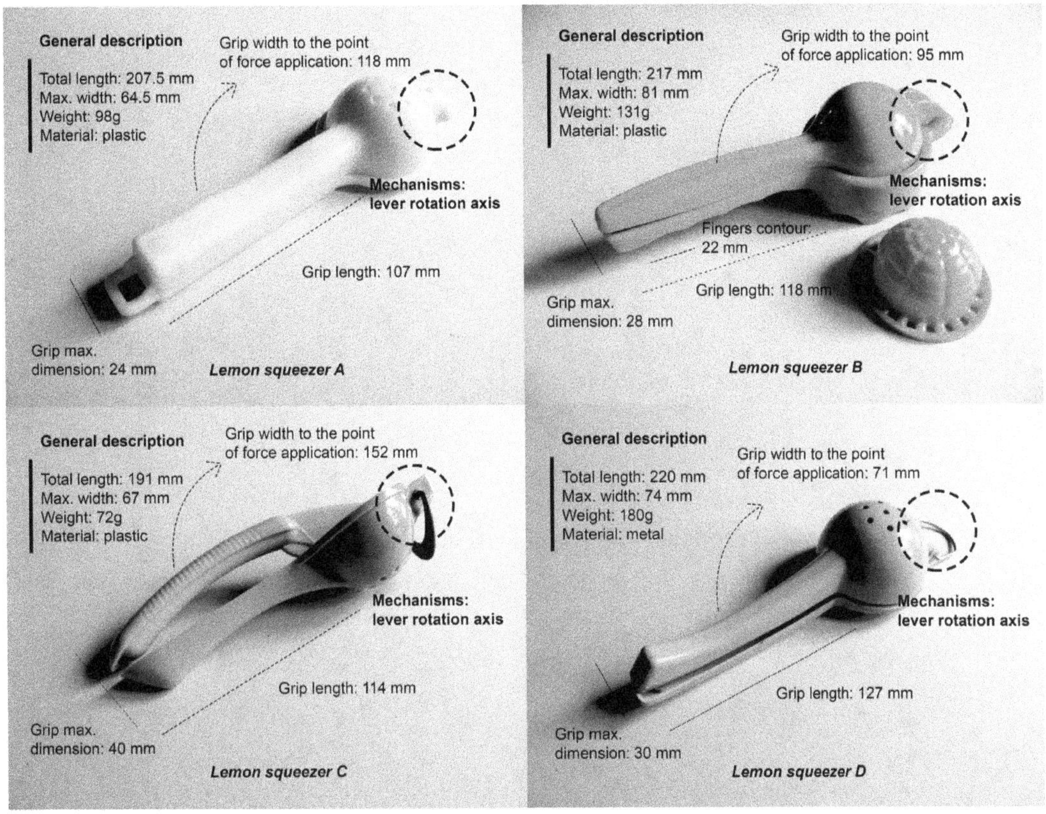

The analysis of the lemon squeezers showed that they do not require a higher effort than the 5th percentile of the older adults group, being Types B and C the most demanding models. Types A and D would be the most appropriate, since the resistance they present does not exceeds 50% of the effort of the 5th percentile for older adults.

Relation Between Hand Dimensions and the Use of Objects

The hand and palm length were compared to the size of the maximum opening space on the grip of the object at the point of force application while squeezing a lemon or opening a can, to verify if objects dimensions correspond to the anthropometric characteristics of the participants.

Regarding can openers, the point of force application varies from 60 to 69.5mm, which allows a correct holding because the dimension is smaller than the length of the hand of the 5th percentile.

At the same time, in the lemon squeezers, the relationship between the point of force application and the dimensions of the hand is different. Type A is the closest to the length of the palm (5th percentile = 82mm). The other three lemon squeezers have a grip width between 114 and 127mm while squeezing a lemon.

Figure 4. Comparison of the resistance of three types of can openers with young and older adults strength
Note: The reference is the strength that older adults can use, in the 50% of the 5th percentile.

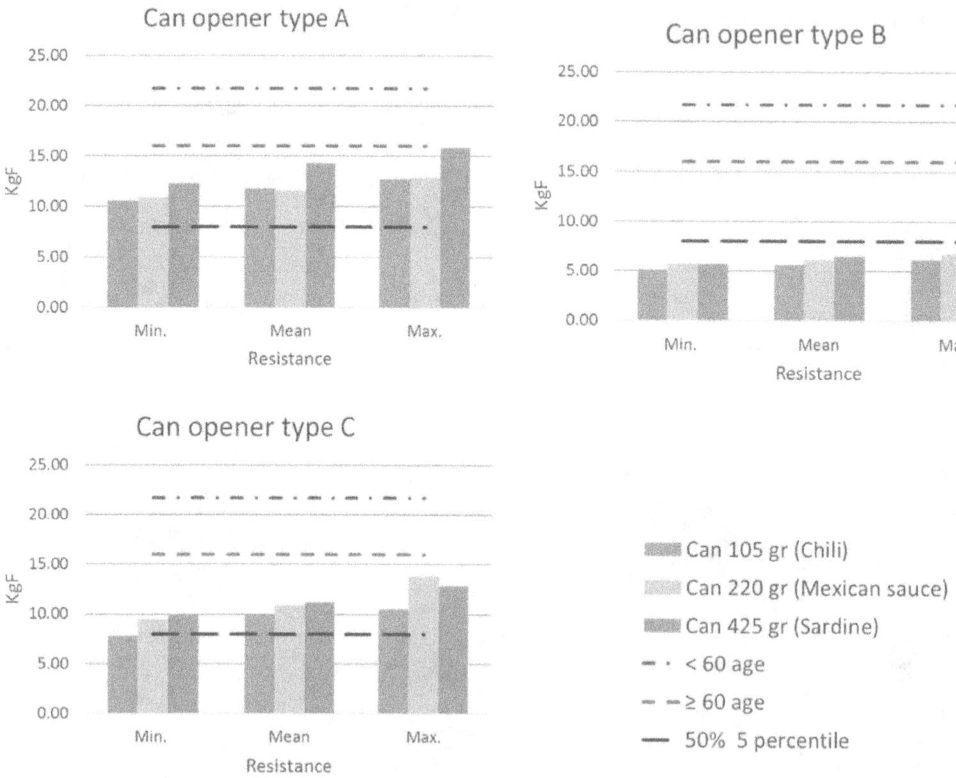

SOLUTIONS AND RECOMMENDATIONS

The results of the exertion tests showed that younger adults have a higher index of the two types of grip analyzed. The differences between younger and older adults were over 20% and 32% respectively, which agrees with that mentioned by Kallman, Plato, & Tobin, (1990); Stegink Jansen, et. Al, (2008). This difference could be explained by the loss of muscle mass within the aging process (Faulkner, Larkin, Claflin, & Brooks, 2007). Differences were also found between genders. It was found that women have a lower strength level than men, which is in line with previous studies (Stegink Jansen, et al., 2008). Therefore, according to this information, the best reference to design objects that require a force application, is the 5th percentile from older women population.

In regard to force application, several authors have discussed the importance of avoiding the use of maximum force, in order to take care of the functional hand structure. Lindstrom (cited by Freivalds, 2011) argues in particular, that for repetitive activities only between 33% and 50% of the force should be used. Carse, Thompson, & Stansfield (2011) analyzed what they called opening strategy and they found that older adults tend to use 40%, which can be a reference point to take design decisions. Furthermore, what was found in the present study shows that, in most cases, the products do not correspond to the characteristics and abilities identified in the participants.

Figure 5. Comparison of the resistance of four types of clothespins with young and older adults' strength
Note: The reference is the strength that older adults can use, in the 50% of 5th percentile.

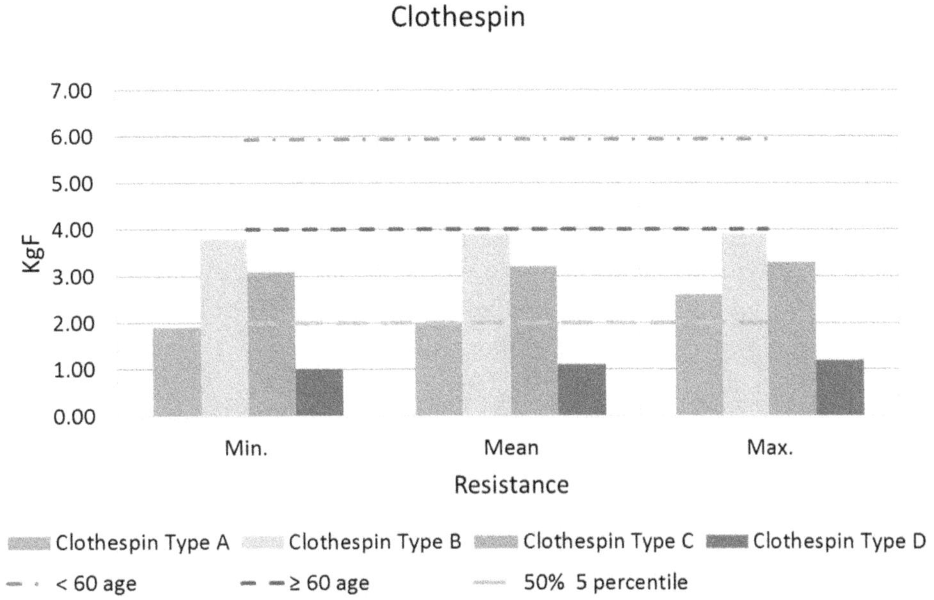

Figure 6. Comparison of the resistance of four types of lemon squeezers with young and older adults strength
Note: The reference is the strength that older adult can use in the 50% of 5th percentile.

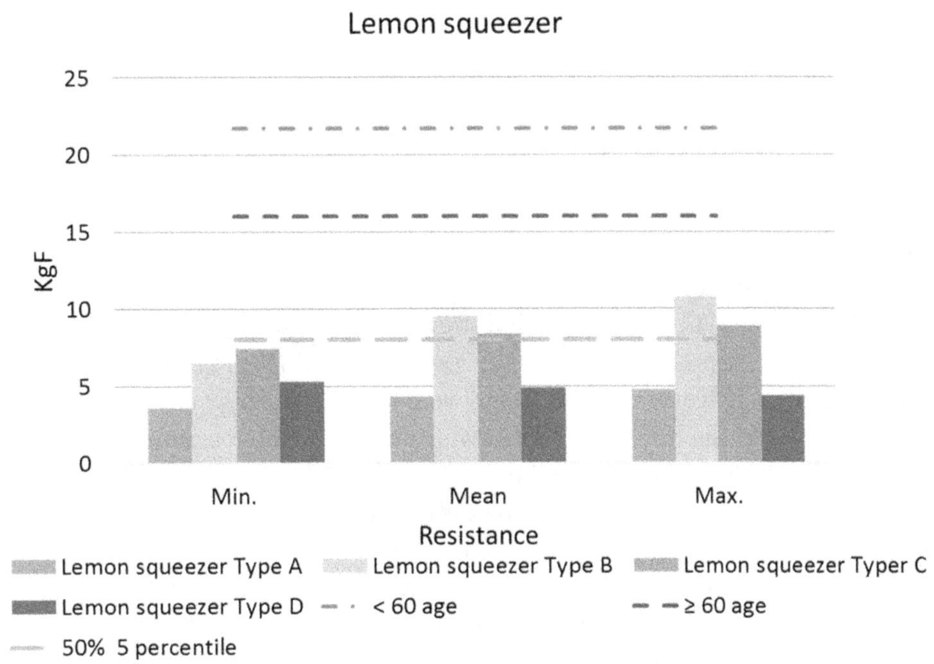

Figure 7. Comparison between hand dimensions and grip width to the point of force application while squeezing a lemon
Note: Dimensions are expressed in millimeters.

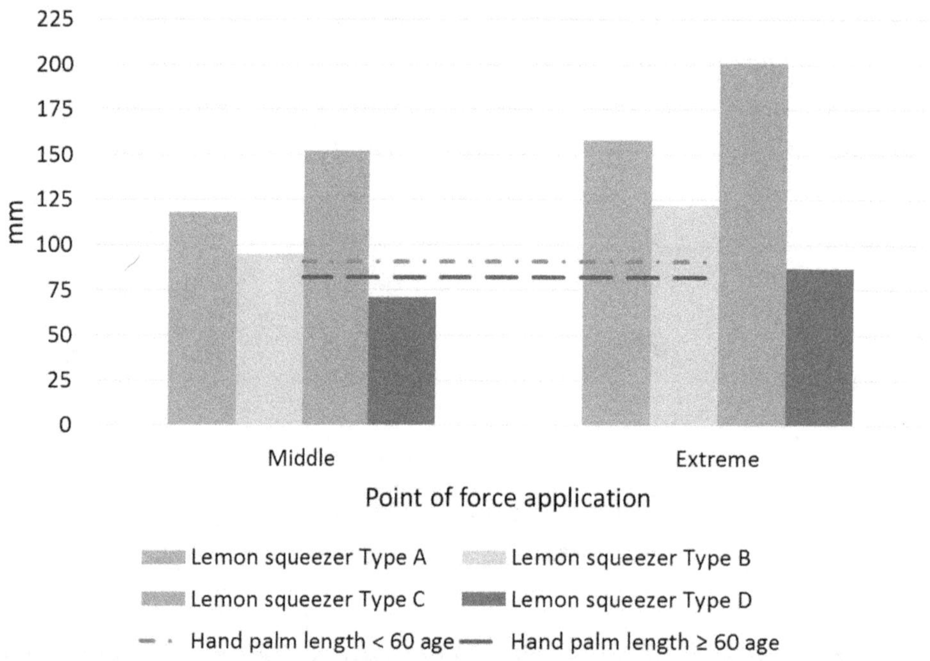

In the case of the can openers, only can opener B shows a level below 50% of the data obtained from the 5th percentile, the others required a higher level of force application. This result may be due to the type of mechanism, which is different from those in the other two can openers, rather than the differences in the levers formed by the length of the handles. However, this point requires further analysis.

In the case of the clothespins, the model which best fits to the 50% reference, is model D. The differences with the other three models may be due to several factors, such as the total length, which, in the case of model D is greater; however, it is possible that the main factor in the difference of the opening resistance is the mechanism, which is different to the ring that the other three models have.

The most effective lemon squeezers proved to be the models A and D. Although, they do not present similar dimensions, both have in common a metal structure, which seems to work much better in regard to force transmission, in comparison with the other two models which are made of plastic.

Significant differences in hand and palm length were found between younger and old adult populations, finding smaller dimensions in older adults, which has clear repercussions on the references that must be considered in the design of grips and other products of daily use (Holt, et al, 2013). Although this study does not have data that can specifically explain this difference, it is considered to be related to the secular trend (Pheasant, 2003) in the Mexican population.

The poor coincidence between the lemon squeezers measures, and the hand dimensions of the participants cannot only generate problems related to the ease of grasping the objects, but it impacts on the amount of force that can be developed (Oatis, 2009). Thus it is a relevant factor when defining how the shape and size of the grip should be. It is recognized that, in the case of lemon squeezers, dimensions prove to be the most difficult factor to use, even more than force application.

FUTURE RESEARCH DIRECTIONS

Having the reference on user strength capabilities and his anthropometric data is essential for making better design decisions. However, more detailed analysis of other aspects of the product is needed, such as the shape of the holding zones as well as the materials and textures in areas of contact. An example in relation to this is what happens with the analyzed can opener B when compared with the other models. It is the most effective in terms of strength application and adjustment to user hand dimensions, nevertheless, the poor contact area and the way of attachment can create problems of use.

Another needed research is to determine which of the aspects of the products have a greater impact on the force application, which will give to designers better tools to make decisions about the components and characteristics of the products.

Alternatively, a study related to user satisfaction could give light to what are the main issues to consider by the user while using them and make possible a future crossing of data with the information obtained in the present study.

Finally, it is important to develop a complete analysis of the products in the study in relation to the anatomical, physiological, psychological and sensorial adaptations (Prado-León & Ávila-Chaurand, 2006), in order to optimize them, and thus having a better understanding of various dimensions of the products.

CONCLUSION

The present study allowed to identify that most of the analyzed products, due to their characteristics, present problems of use in particular to the older adult population. In consequence, it is important to develop better mechanisms for analysis of the user needs to get accurate information about their abilities and capabilities to embed this information in the product conception process and then, to design more useful products.

It is essential that designers define the characteristics of the user taking in mind who could be the most vulnerable user while interacting with the product. The growing trend of the population of older adults and the reduction of their capacities makes them an increasingly important group considering it constitutes the main reference to define the characteristics of the objects to be designed.

Research in design requires a greater development to have more information when defining the characteristics of the products. In relation to this, Ergonomics has a vital importance in methodological terms and contributes to the design practice, pushing it to go further than merely working on the appearance of the product.

ACKNOWLEDGMENT

A special acknowledgment to Paulina Salas Gaxiola for her collaboration in the construction of tools used in the study.

REFERENCES

Cacha, C. (1999). *Ergonomics and Safety in Hand Tools Design*. Boca Raton, FL: CRC Press.

Carmeli, E., Patish, H., & Coleman, R. (2003). The aging hand. *The Journals of Gerontology. Series A, Biological Sciences and Medical Sciences*, *58*(2), 146–152. doi:10.1093/gerona/58.2.M146 PMID:12586852

Carse, B., Thompson, A., & Stansfield, B. (2011). A novel device for evaluating forces during the dynamic jar opening action—Do older and younger adults do things differently? *Medical Engineering & Physics*, *33*(4), 521–525. doi:10.1016/j.medengphy.2010.11.017 PMID:21212013

Cutkosky, M. (1989). On grasp choice, grasp models and the design of hands for manufacturing tasks. *IEEE Transactions on Robotics and Automation*, *5*(3), 269–279. doi:10.1109/70.34763

Delgado, S. (1990). Predictive values for grip and torque strengths using isometric strength testing unit (ISTU) hand attachments. *Journal of Disability*, *1*, 73–78.

Faulkner, J., Larkin, L., Claflin, D., & Brooks, S. (2007). Age-related changes in the structure and function of skeletal muscles. *Clinical and Experimental Pharmacology & Physiology*, *34*(11), 1091–1096. doi:10.1111/j.1440-1681.2007.04752.x PMID:17880359

Freivalds, A. (2011). *Biomechanics of the Upper Limbs*. Boca Raton, FL: CRC Press.

Gustafsson, K., Andersson, I., Andersson, J., Fjellström, C., & Sidenvall, B. (2003). Older women's perceptions of independence versus dependence in food-related work. *Public Health Nursing (Boston, Mass.)*, *20*(3), 237–247. doi:10.1046/j.0737-1209.2003.20311.x PMID:12716404

Hensler, S., Herren, D. B., & Marks, M. (2015). New technical design of food packaging makes the opening process easier for patients with hand disorders. *Applied Ergonomics*, *50*, 1–7. doi:10.1016/j.apergo.2015.02.002 PMID:25959312

Incel, N., Ceceli, E., Durukan, P., Erdem, R., & Yorgancioglu, R. (2002). Grip strength: Effect of hand dominance. *Singapore Medical Journal*, *43*, 234–237. PMID:12188074

Kallman, D., Plato, C., & Tobin, J. (1990). The Role of Muscle Loss in the Age-Related Decline of Grip Strength: Cross-sectional and Longitudinal Perspectives. *The Journals of Gerontology. Series A, Biological Sciences and Medical Sciences*, *45*, 82–88. PMID:2335723

Kirkendall, D., & Garrett, W. (1998). The effects of aging and training on skeletal muscle. *The American Journal of Sports Medicine*, *26*(4), 598–602. doi:10.1177/03635465980260042401 PMID:9689386

Landsmeer, J. (1962). Power grip and precision handling. *Annals of the Rheumatic Diseases*, *21*(2), 164–170. doi:10.1136/ard.21.2.164 PMID:14461996

Lee, J. S., & Frongilo, E. A. Jr. (2001). Factors Associated With Food Insecurity Among U.S. Elderly Persons: Importance of Functional Impairments. *The Journals of Gerontology. Series B, Psychological Sciences and Social Sciences*, *56B*(2), S94–S99. doi:10.1093/geronb/56.2.S94 PMID:11245369

Li, K. W., & Yu, R. (2011). Assessment of grip force and subjective hand force exertion under handedness and postural conditions. *Applied Ergonomics*, *42*(6), 929–933. doi:10.1016/j.apergo.2011.03.001 PMID:21439549

Manrique-Espinoza, B., Andersson, I., & Sidenvall, B. (2001). Case studies of food shopping, cooking and eating habits in older women with Parkinson's disease. *Journal of Advanced Nursing*, *35*(1), 69–78. doi:10.1046/j.1365-2648.2001.01823.x PMID:11442684

Manrique-Espinoza, B., Salinas-Rodríguez, A., Moreno-Tamayo, K. M., Acosta-Castillo, I., Sosa-Ortiz, A. L., Gutiérrez-Robledo, L. M., & Téllez-Rojo, M. M. (2013). Condiciones de salud y estado funcional de los adultos mayores en México. *Salud Pública de México*, *55*(Supl.2), S323–S331. doi:10.21149pm.v55s2.5131 PMID:24626711

Mathiowetz, V., Kashmann, N., Volland, G., Weber, K., Dowe, M., & Rogers, S. (1985). Grip and pinch strength: Normative data for adults. *Archives of Physical Medicine and Rehabilitation*, *66*, 69–74. PMID:3970660

Menéndez, J., Guevara, A., Arcia, N., León Díaz, E. M., Marín, C., & Alfonso, J. C. (2005). Enfermedades crónicas y limitación funcional en adultos mayores: Estudio comparativo en siete ciudades de América Latina y el Caribe. *Revista Panamericana de Salud Pública*, *17*(5-6), 353–361. doi:10.1590/S1020-49892005000500007 PMID:16053645

Mital, A., Kilbom, A., & Kumar, S. (2000). *Ergonomics Guidelines and Problem Solving*. Amsterdam: Elsevier.

Napier, J. (1956). The prehensile movements of the human hands. *The Journal of Bone and Joint Surgery*, *38*, 902–913. PMID:13376678

Oatis, C. (2009). *Kinesiology. The Mechanics & Pathomechanics of Human Movement*. Philadelphia: Lippincott Williams & Wilkins.

Ombregt, L. (2013). Applied anatomy of the wrist, thumb and hand. In L. Ombregt (Ed.), A System of Orthopaedic Medicine (pp. 102-111). Elsevier Health Sciences. doi:10.1016/B978-0-7020-3145-8.00072-7

Parvatikar, V., & Mukkannavar, P. (2009). Comparative study of grip strength in different positions of shoulder and elbow with wrist in neutral and extension positions. *Journal of Exercise Science and Physiotherapy*, *5*, 67–75.

Pheasant, S. (2003). *Bodyspace*. London: Taylor & Francis Ltd.

Prado-León, L., & Ávila-Chaurand, R. (2006). *Ergonomía y Diseño de Espacios Habitables*. Guadalajara: Universidad de Guadalajara.

Ranganathan, V., Siemionow, V., Sahgal, V., & Yue, G. (2001). Effects of aging and hand function. *Journal of the American Geriatrics Society*, *49*(11), 1478–1484. doi:10.1046/j.1532-5415.2001.4911240.x PMID:11890586

Reijnierse, M., & Fernandes, N. (2013). Hand and wrist. In I. Beggs (Ed.), *Musculoskeletal Ultrasound* (pp. 73–102). Lippincott Williams & Wilkins.

Schreuders, T., Brandsma, W., & Stam, H. (2014). Functional Anatomy and Biomechanics of the Hand. In M. Tuncay Duruöz (Ed.), Hand Function (pp. 3 – 21). London: Springer Science+Business Media. doi:10.1007/978-1-4614-9449-2_1

Shiffman, L. (1992). Effects of Aging on Adult Hand Function. *The American Journal of Occupational Therapy*, *46*(9), 785–792. doi:10.5014/ajot.46.9.785 PMID:1514564

Sidenvall, B., Nydahl, M., & Fjellström, C. (2001). Managing food shopping and cooking: The experiences of older Swedish women. *Ageing and Society*, *21*(02), 151–168. doi:10.1017/S0144686X01008121

Stegink Jansen, C., Niebuhr, B., Coussirat, D., Hawthorne, D., Moreno, L., & Phillip, M. (2008). Hand Force of Men and Women Over 65 Years of Age as Measured by Maximum Pinch and Grip Force. *Journal of Aging and Physical Activity*, *16*(1), 24–41. doi:10.1123/japa.16.1.24 PMID:18212392

Taylor, C., & Schwarz, R. (1955). The anatomy and mechanics of the human hand. *Artificial Limbs*, 22–35. PMID:13249858

Werle, S., Goldhahn, J., Drerup, S., Simmen, B., Sprott, H., & Herren, D. (2009). Age and Gender Specific Normative Data of Grip and Pinch Strenght in a Healthy Adult Swiss Population. *The Journal of Hand Surgery*, *34*(1), 76–84. doi:10.1177/1753193408096763 PMID:19129352

World Health Organization. (2002). *Active Ageing: A Policy Framework*. Madrid, Spain: WHO.

World Health Organization. (2015). *World report on ageing and health*. WHO.

KEY TERMS AND DEFINITIONS

Anthropometrics: The science that defines physical measures of a person size, form, and functional capacities. Applied to ergonomics, anthropometric measurements are used to study the interaction of people with tasks, tools, machines, vehicles, and other objects to determine the degree of adequacy.

Basic Activities of Daily Living: A group of activities related to self-care performed in the day to day that allows identifying the functional state of the subject by assessing them. Although there are many classifications, among them are personal hygiene, mobility, and eating.

Ergonomic Properties: Object characteristics in agree with user characteristics, capacities, and abilities. There are six recognized ergonomic properties: anatomic, anthropometric, biomechanical, physiological, psychological, and sensory properties.

Hand Strength: Used when referring to both grip force and pinch grip or to general strength in the hand.

Instrumental Activities of Daily Living: A group of activities performed in the day to day related to independence and control; they allow to identify the functional state of the subject by assessing them. There are several classifications but among them are using public transportation, shopping, preparing meals, and managing money.

Maximum Perceived Strength: Maximum strength possible to perform by the subject.

Older Adults: Adults over 60 years old.

Chapter 16
The Comprehension of Figurative Images of Food Items:
The Effect of Ergonomic Guidelines in Graphic Design

Lilia Roselia Prado-León
Universidad de Guadalajara, Mexico

Carlos Díaz de León Zuloaga
Universidad de Guadalajara, Mexico

Adrian Antonio Cisneros Hernández
Universidad de Guadalajara, Mexico

ABSTRACT

Graphic design is a discipline responsible for creating clear and concise visual messages that can enhance the understanding of information, easing the comprehension of the message when using different kind of images. In ergonomics, there have been studies about image comprehension, helping to set up and prove guidelines that enhance the efficacy and efficiency of symbols, icons, and pictograms. The goal of this research was to assess, within four stages, the comprehension of the figurative food illustrations and the effectiveness of ergonomic guidelines when used in the design process of these illustrations. The results showed that the familiarity and comprehension of some foods are limited due to the nutritional or cultural habits of the users. When comparing the graphic styles that use or lack of ergonomic guidelines, it is clear that the use of these guidelines helps the user to comprehend and identify better the graphical information presented.

DOI: 10.4018/978-1-5225-5234-5.ch016

Copyright © 2018, IGI Global. Copying or distributing in print or electronic forms without written permission of IGI Global is prohibited.

INTRODUCTION

Graphic design is a discipline which concerns itself with generating clear and concise visual messages. These messages are generated in conceptualized form, always taking the user into account, and always coming from a design perspective. While each message is mentally converted to an image which is true to life, some need to be transformed into abstractions, either for rapid understanding or to generate an image. For this reason, Abraham Moles describes the image "as a prop for visual communication that materializes a fragment of the visual space, not to mention the perceived universe" (Moles, 2007).

Likewise, Moles uses the term "iconicity" to describe a concept which expresses categories and levels of relationship, between an image and its corresponding real object. He postulates a decreasing scale of iconicity with which objects are classified according to their resemblance to reality, beginning at Level 12 and finally arriving at an incomprehensible abstraction at Level 0.

Another use of the process of graphic abstraction is the elimination of characteristic features of the objects to be portrayed. For this reason, Joan Costa mentions that the lower the level of iconicity, the higher the level of abstraction (Costa,1998). Therefore, at higher levels of iconicity, an image will conserve a greater proportion of its real-object properties. For example, at Level 5 on the iconicity scale (called "nonrealistic figurative image"), the object is still identifiable but its spatial relationships are altered. Such figurative images permit rapid identification of the objects portrayed; Hutchins, Hollan & Norman (1985) coincide with this viewpoint, affirming that the comprehension efficacy of an image can be attributed to a short articulation between the image and the real object, the shorter, the better.

Certainly, graphics do much more than simply give a good impression, they are an essential part of communication. As society becomes more visually oriented, graphics are becoming the principal channel of transmission of information to the end user (Robinson, 2009). Additionally, graphics can increase comprehension of textual information facilitating message processing and in some cases being more quickly perceived than isolated text (Rogers et al., 2000).

This study will focus on the design of recipes for dietary plans, in which figurative images provide a convenient alternative for representing the ingredients to be used. Thanks to their particular characteristics, these images make comprehension not only possible but also enjoyable, by providing a reader-friendly interface. In many cases, images work better than words to help the user identify ingredients and understand food preparation procedures (Prado, Diaz de Leon, Zambrano, Cisneros & Cárdenas, 2016).

BACKGROUND

Since decades ago, various studies have been made in the cognitive ergonomics field which helped to establish ergonomic principles or guidelines that allow for a better and faster comprehension of symbols and pictograms in various environments, especially in the case of pictograms which warn of danger (Cushman y Rosenberg, 1991; Hancock, Rogers, Schroeder y Fisk, 2004; Laughery, 2006; Jiamsanguanwong y Umemuro, 2014). Many of these ergonomic guidelines can be applied to other environments where the comprehension of an image's message is necessary in order to realize a particular action, as is the case in this study.

ISO norm 3864 (ISO, 2011) indicates that, in order for a graphic abstraction to be acceptable, it must be adequately understood by between 67% and 85% of all users (ANSI, 1987). Thus, Cognitive Ergonomics studies the understanding and retention of the meanings with diverse levels of graphic abstractions and with different objectives. For instance, it evaluates security pictograms which communicate dangers or risks efficaciously (Wogalter et al., 1997). It is worth to mention there are no universally effective graphic abstractions, however, Ergonomics must assess and produce guidelines in order to make them more comprehensible.

The use of figurative images generates a certain grade of inclusiveness at the moment in which a message is communicated, these images can serve to eliminate communication barriers, whether they be sociocultural barriers, language barriers, or barriers resulting from disparate educational levels (Wogalter y Silver, 1995), thereby allowing a wide range of users to comprehend complex information in a more efficient and effective manner.

The comprehension of abstractions has been evaluated using various methods, from multiple-choice answers to open-ended question exams (Wolf y Wogalter, 1998). The majority of the studies which have been undertaken to evaluate comprehension and recognition of graphic abstractions have focused on symbols or pictograms of danger or warnings (Mayer y Laux, 1989), leaving a wide-open field for the study of other types of graphic abstraction. Recently, signs and pictograms have been evaluated from a usability standpoint (Ng et al., 2011), which has proved effective for evaluating its comprehensibility.

For all of the above reasons, the focus of this study was to evaluate the comprehension of figurative images of food items, comparing the graphic styles which had been developed by Graphic Designers with a redesign based on the application of ergonomic guidelines, under the tutelage of experts in the field. The study was realized in four stages: (1) generation of the graphic styles to be used in the study, (2) assessment of comprehension for each of the four styles resulting from stage 1, (3) development of the redesign using ergonomic guidelines, and (4) comprehension assessment for the ergonomic redesign.

METHOD

Stage 1: Development of the Graphic Styles

Four Graphic Designers were given the task of creating nonrealistic images, based on level 5 of the iconicity scale, of the following cooking ingredients: corn, eggplant, milk, broccoli, beef filet, cooked beans, a bottle of cooking oil, orange, cheese, and radish, with the given task of applying a process of graphic abstraction. This process consists in simplifying the graphic representation; in so doing, it offers better expressiveness and clarity, in order to foster an adequate communication. Specific instructions were to arrive at a proposal which could unite all the necessary elements so that any user could recognize the illustration. Below, each of the styles of illustration that were developed is described, with examples.

Style 1: In a general sense, this style approximates reality, which is to say that it maintains a moderate level of abstraction with formal graphic features. Applying planar abstraction, the aforementioned elements are superimposed on each other, resulting in highlights, while the use of lighter and darker coloration similar to the objects' real colors generates a perception of depth, the cavalier oblique projection used for plane perspective generates the illusion of volume, and shadows are applied to locate the objects in space. Figure 1 is an example of this style.

Style 2: This style retains a rough proximity to the real objects, which is to say a moderate level of abstraction, but it exaggerates some graphic features, resulting in greater informality. Planar abstraction is implemented in most objects, where various planes superimpose themselves on each

Figure 1. Figurate images of style 1

other, resulting in the generation of highlights; grayscale colors were used between light and dark with the aim of adding depth, these colors are more brilliant than the corresponding real colors. Combined abstraction is implemented through the use of planes and contours. The cavalier oblique projection perspective is applied in the majority of the figures, but not in all. Most of the shadows do not interact with the planes corresponding to the various ingredients.

Style 3: This style can be observed in Figure 3. It generally maintains proximity to reality at a moderate level of abstraction, but exaggerates certain graphic features, allowing a sense of informality. Planar and combined abstraction are implemented, through the use of planes and contours in some of the examples. The superimposition of planes permits the generation of highlights and shadows, and

Figure 2. Figurate images of style 2

Figure 3. Figurate images of style 3

the colors aim to exemplify the true colors of the objects but using less saturation. The contour style is equivalent to the use of a fine paintbrush.

Style 4: In general, style 4 retains proximity to reality at a moderate level of abstraction, but exaggerates certain graphic features, allowing a sense of informality. Planar and combined abstraction are implemented, through the use of planes and contours in some of the examples. The superimposition of planes permits the generation of highlights and shadows, and the colors aim to exemplify the true colors of the objects but using less saturation. The contour style is geometric. The cavalier oblique projection perspective is applied in the majority of the figures, but not in all. Shadow is applied to only one ingredient, and color highlighting surrounds another ingredient.

Stage 2: Test of Graphic Style Comprehension

Participants

Undergraduate students working toward a Bachelor's Degree in Communications, specializing in Graphic Design, and those working toward a Bachelor's Degree in Architecture, 26 women and 26 men (n=52), ages 19-24. The distribution of participants was 13 for each graphic style.

Materials

Response form, which provides a record of participant responses in relation to each individual illustration:

- **Correct Response:** The participant responded with the correct name for the food item.

- **Incorrect Response:** The participant was unable to identify the item by name or claims to have forgotten it.

Four templates were created using a combination of the above styles.

Template A: (Fig.5) included the illustrations of the beans, corn, cheese, and radish, from Style 1, the bottle of oil and the grapefruit from Style 2, the eggplant and milk from Style 3, and the broccoli and beefsteak from Style 4.

Template B: (Fig. 6) took the illustrations of the oil and grapefruit from Style 1, the beans, corn, broccoli, and beefsteak from Style 2, the cheese and radishes from Style 3, and the eggplant and milk from Style 4.

Template C: (Fig. 7) took the illustrations of broccoli and beefsteak from Style 1, the eggplant and milk from Style 2, the beans, oil, grapefruit and corn from Style 3, and the cheese and radishes from Style 4.

Template D: (Fig. 8) used illustrations of eggplant and milk from Style 1, cheese and radishes from Style 2, broccoli and beefsteak from Style 3, and corn, beans, oil, and grapefruit from Style 4.

Procedure

Before beginning, an explanation was given to the participants regarding the content of the inquiry; they were told that their responses would be recorded in audio and they were given a consent form to sign. Once the forms were signed, demographic data was collected and recording was begun. A template with graphic illustrations of food items was shown to each participant, who was instructed to identify the illustrations as quickly as possible but without time limits, stating orally what each graphic represented, while the responses were recorded in audio, and the moderator noted correct responses and observations

Figure 4. Figurate images of style 4

Figure 5. Template A

Figure 6. Template B

on the response forms that had been provided. Each template was shown to thirteen different students, one time only.

Results

The database was formed using Microsoft Excel, version 13.13.3, and statistical analysis was accomplished using software SPSSS, version 22, both programs in Mac OS X.

Figure 7. Template C

Figure 8. Template D

Analysis of Results

Fisher's Exact Test was applied to compare the comprehension rates between the four styles and for each individual ingredient, independently. Only two of the illustrations presented a statistically significant difference between groups. The illustration of the bottle of oil in Style 3 had a greater frequency of recognition, in comparison to Style 1 (X^2 (1, n=26) = 7.721; p =0.015); in relation to the other two styles, no significant difference was shown. The illustration of cooked beans in Style 2 was more frequently recognized in comparison to Style 3; this difference was deemed significant between these two styles (X^2 (1, n=26) = 7.800; p = 0.015).

Since two of the styles (2 and 3) did present significant differences with respect to each other, a comparison to the total of the illustrations that had been recognized was made, according to style. Application of the chi-squared test revealed no significant differences in the frequency of recognition between styles (X^2 (3, n=520) = 2.889; $p = 0.409$). Table 1 illustrates recognition percentages by style; it can be observed that Style 2 results in a higher percentage of correct responses, with 73.8% comprehension, well ahead of Style 3 (67.7%).

Considering the aforementioned results, Style 2 was selected as the guiding graphic example to receive ergonomic treatment, due to the significant difference in recognition shown for the illustration of cooked beans, and to this style's percentage of general comprehension (74%), which comes the closest to the ANSI guidelines (85% recognition).

Stage 3: Ergonomic Redesign of the Style

In addition to obtaining a higher percentage of comprehension, Style 2 was able to encompass the most significant attributes of the ingredients, along with a certain formality in their visual representation, in order to be able to transmit an indicative message that can be clearly and precisely identified by the end users.

Even though it is recommended that in design methodology, ergonomic aspects be included right from the beginning of the process, and even though the intention of the graphic designers was to have their illustrations recognized by the vast majority of the people who saw them, a process of redesign of the style, based on ergonomic principles, was undertaken, the hypothesis being that this redesign would have a higher percentage of comprehension/recognition than the original design.

Based on the premise that, in the moment of conception of any graphic, the correct use of color, textures, and contrast must be taken into account, as well as the use of well-defined forms and contours so that the graphics are highly identifiable (Prado y Ávila, 2010), the following graphic recommendations were assigned in order to achieve illustrations which would be more recognizable and which, with their use of grayscale, might be able to conserve a greater quantity of the significant attributes of the ingredient.

The following is a list of the ergonomic guidelines that were applied in order to accomplish the task of ergonomic adaptation:

Guideline 1: Good distinction between figure and background; this was achieved by using closed figures whose limits were well-defined, either by a thick outline or by a figure which was completely filled in (see Figure 9)

Guideline 2: Apply perspective in order to generate volume; the recommendation was to generate all of the drawings in perspective, thereby imitating the visual image and the three-dimensionality of the object. The use of shadow, texture and the inclusion of details or highlights help to create this effect (see Figure 10).

Guideline 3: Horizontal alignment; the figurative images are aligned along a horizontal baseline in order to avoid the appearance of "floating" elements (see Figure 11).

Guideline 4: Graphic synthesis of commercial brands; those products which are better recognized under a particular brand name undergo a graphic synthesis which conserves a general appearance that would suggest that brand (see Figure 12).

Guideline 5: No more than four shades of gray; in order to distinguish the features of the ingredients, the use of a maximum of four shades of gray permits contrast between them (see Figure 13).

Table 1. Recognition percentages by style

Style		Comprehension	
		Yes	No
1	n	86	46
	%	65%	35%
2	n	96	34
	%	74%	26%
3	n	88	42
	%	68%	32%
4	n	92	38
	%	71%	29%

Figure 9. Good distinction between figure and background

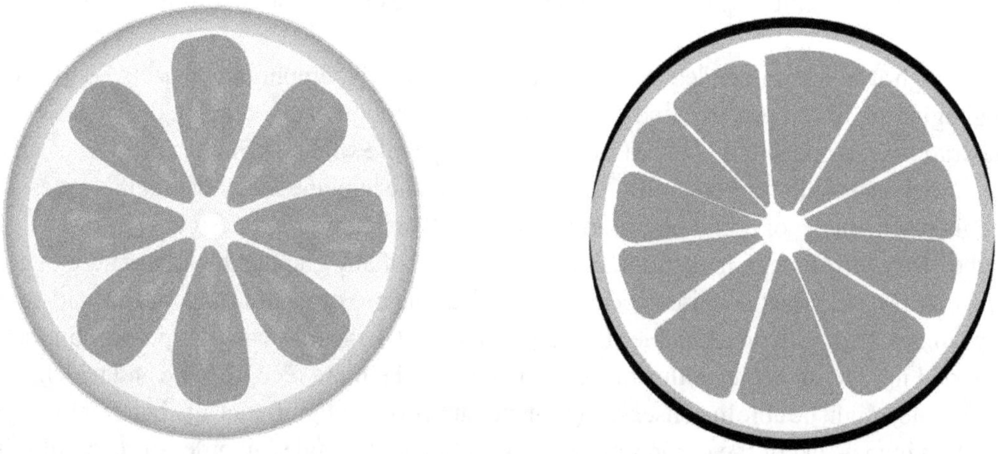

Figure 10. Apply perspective in order to generate volume

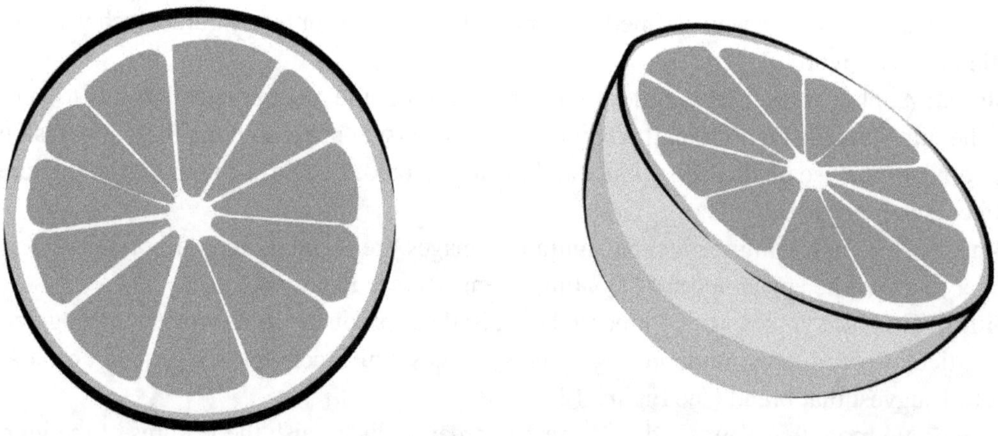

Figure 11. Use of horizontal alignment

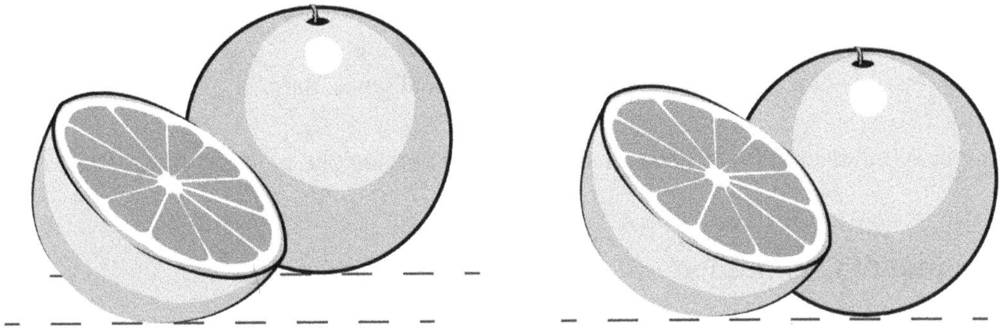

Figure 12. Graphic synthesis of commercial brands

Figure 13. No more than four shades of gray

Guideline 6: The use of special elements to reduce ambiguity; graphic syntheses of animals, for example, reinforce the mental picture of the type of ingredient; in this case, the graphic synthesis of a bull is superimposed behind the image of beefsteak (see Figure 14).

Guideline 7: Avoid color gradients, which can create visual noise, thereby hampering recognition (see Figure 15).

Guideline 8: Avoid the use of shadows that could cause problems in planar contrast (see Figure 16).

Stage 4: Test of Comprehension of the Ergonomic Design

To evaluate the redesign of Style 2 that was based on ergonomic criteria, the recognition test was reapplied, with the objective of comparing the illustrations of this style before (see Figure 17) and after (see Figure 18) ergonomic treatment.

Participants

Students working toward a Bachelor's Degree in design for the Department of Communications in Graphic Design at the University of Guadalajara, from first through seventh semesters; 28 were women and 24 were men (n=52), of ages ranging from 18 to 22 and a median age of 20 years old. The distribution: 26 participants for the "Style 2 by the Graphic Designer" (Style 2GD) group, and 26 for the "Style 2 with the application of ergonomic guidelines by experts" (Estilo 2WEL) group.

Materials

Response form, which provides a record of participant responses in relation to each individual illustration:

- **Correct Response:** The participant responded with the correct name for the food item.
- **Incorrect Response:** The participant was unable to identify the item by name, or claims to have forgotten it.

Figure 14. Use of elements to reduce ambiguity

Figure 15. Avoid color gradients

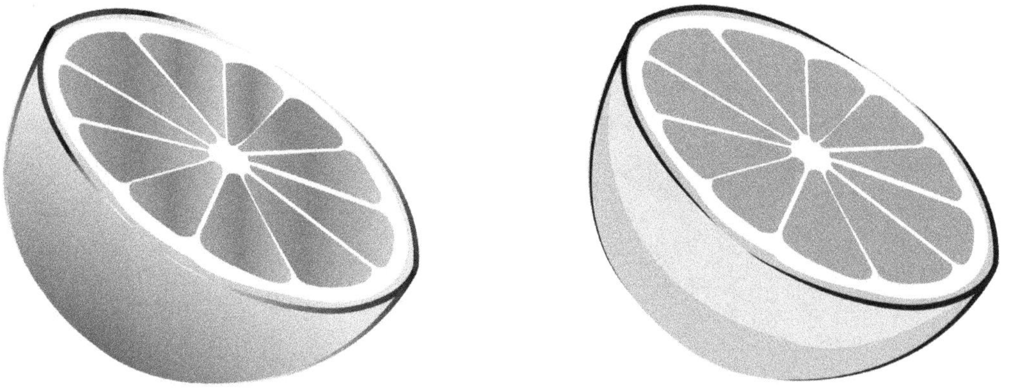

Figure 16. Avoid use of shadows

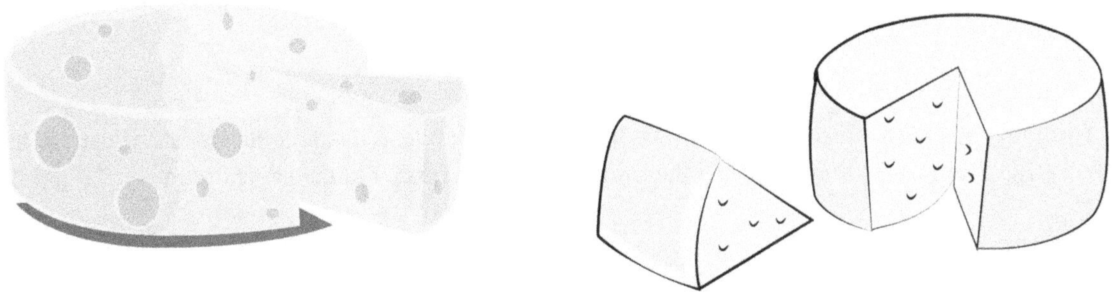

Figure 17. Style 2 without ergonomic guidelines

Figure 18. Style 2 with ergonomic guidelines

Two templates were created, one with the illustrations of style 2GD and one with the illustrations of style 2WEL. Both of the templates had the same ingredients placed in the same order.

Procedure

The procedure was exactly the same as Stage 2 with the only variation that each template was shown to 26 different students, one time only.

Results

The database was formed using Microsoft Excel, version 13.13.3, and statistical analysis was accomplished using software SPSSS, version 22, both programs in Operating System Mac OS X.

Analysis of Results

The chi-squared test was performed between the two styles, and for each of the ingredients, independently. Style 2WEL exhibited a significantly higher frequency of recognition than Style 2GD in the illustrations of the cooked beans (X^2 (1, n=52) = 9.665; p = 0.002) and the corn (X^2 (1, n=52) = 10.833; p = 0.001). Style 2GD had a higher frequency of recognition than Style 2WEL only for the illustration of Oil (X^2 (1, n=52) = 6.933; p = 0.019). In Table 2, it is possible to observe the frequencies of correct responses for the other illustrations.

As was the case in Stage 2, a comparison was made between the two styles, with regard to total number of recognitions. Upon application of the Chi Squared Test, significant differences in frequencies of recognition according to style were encountered (X^2 (1, n=520) = 5.735; p = 0.017), in which Style 2WEL (84%) far exceeded the percentage of recognitions of Style 2GD (75%), as can be observed in Table 3, which shows the respective recognition percentages.

Table 2. Frequencies of comprehension of each illustration, in Style 2 GD and Style 2WEL

Ingredient	Style	Comprehension				X^2	p
		Yes		No			
		n	%	n	%		
Beans	2-GD	10	38%	16	62%	9.665	.002
	2-WEL	21	81%	5	19%		
Eggplant	2-GD	21	81%	5	19%	0.433	.510
	2-WEL	19	73%	7	27%		
Milk	2-GD	25	96%	1	4%	1.020*	1.000
	2-WEL	26	100%	0	0%		
Broccoli	2-GD	25	96%	1	4%	0.000*	1.000
	2-WEL	25	96%	1	4%		
Meat	2-GD	24	92%	2	8%	2.080*	.490
	2-WEL	26	100%	0	0%		
Corn	2-GD	15	58%	11	42%	10.833	.001
	2-WEL	25	96%	1	4%		
Oil	2-GD	10	38%	16	62%	6.933	.008
	2-WEL	2	8%	24	92%		
Orange	2-GD	20	77%	6	23%	4.127*	.099
	2-WEL	25	96%	1	4%		
Cheese	2-GD	26	100%	26	100%	-	-
	2-WEL	0	0%	0	0%		
Radish	2-GD	20	77%	6	23%	1.209*	.465
	2-WEL	23	88%	3	12%		

Table 3. Recognition percentages, Styles 2GD and 2WEL

Style	Comprehension				X^2	p
	Yes		No			
	n	%	n	%		
2-GD	196	75%	64	25%	5.735	.017
2-WEL	218	84%	42	16%		

CONCLUSION

Even though ergonomics is integrated into the professional training of graphic designers, this study underlines the importance of multidisciplinary work, where the insertion of an ergonomics expert can make a significant difference in the comprehension of figurative images.

This study documents a real-time process of design methodology, in which the original proposals are evaluated and based on the results of the evaluation, a redesign is created and subsequently tested.

Stage 2, in which the graphic designers' styles were compared and contrasted in the evaluation process helped to select a baseline graphic style, achieved with the user testing. Consequently, that style was chosen for the subsequent ergonomic treatment, on the basis of its comprehension effectiveness. The determination of an illustrative style based on the participation of potential users permitted the establishment of general parameters for the graphic elements to be utilized, maintaining design coherence and fostering formal consistency between the various graphic elements of the different ingredients, as well as promoting faster recognition for the eye of the user.

However, a graphic configuration is not the only determining factor for its comprehension. The process of perception is intimately linked to the user's cognitive and sociocultural models. The cognitive theory explains how the exposure of an individual to different visual representations affects the manner in which that individual categorizes objects and images. Just so, human beings utilize mnemonic imprints to classify the objects or images that surround us. Such imprints, or prototypes, are visual representations which are stored in each person's memory and serve as representations of general classes or categories of objects. Categorization involves a process of comparison between new objects or images and the existing prototypes in the user's memory. These imprints or prototypes are the results of repeated exposure to objects, or images, over the course of time.

Accordingly, the comprehension tests demonstrated that in some cases, ingredient recognition can be impaired by the dietary habits of the participants. One concrete example was eggplant, a food item that was little comprehended in any of the styles, being so uncommon in the diet of the study participant population that the participants lacked the mnemonic imprint necessary for recognition of this ingredient.

Finally, the validation of the illustrative style, along with the recommendations that had been provided in the literature, permitted the development, under one unified guiding style, of a complete catalog of 169 illustrations of food items, 35 illustrations of containers, and 16 illustrations of actions. These illustrations will serve as the interface for each of the recipes included in the dietary plan, providing greater ease of learning and use, and improving procedural efficiency for the patient.

REFERENCES

ANSI 2535. (1987). *Criteria for safety symbols.* American National Standards Institute.

Costa, J. (1998). *La esquemática: visualizar la información.* Buenos Aires, Argentina: Paidós.

Cushman, W. H., & Rosenberg, D. J. (1991). Human factors in product design. *Advances in human factors/ergonomics, 14.*

Handcock, H. E., Rogers, W. A., Schroeder, D., & Fisk, A. D. (2004). Safety symbol comprehension: Effects of symbol type, familiarity, and age. Human Factors. *The Journal of the Human Factors and Ergonomics Society, 46*(2), 183–195. doi:10.1518/hfes.46.2.183.37344

Hutchins, E. L., Hollan, J. D., & Norman, D. A. (1985). Direct manipulation interfaces. *Human-Computer Interaction, 1*(4), 311–338. doi:10.120715327051hci0104_2

International Standards Organization (ISO). (1984). *International standard for safety colours and safety signs: ISO 3864.* Geneva, Switzerland: Author.

Jiamsanguanwong, A., & Umemuro, H. (2014). Influence of affective states on comprehension and hazard perception of warning pictorials. *Applied Ergonomics*, *45*(5), 1362–1366. doi:10.1016/j.apergo.2013.08.006 PMID:24045098

Laughery, K. R. (2006). Safety communications: Warnings. *Applied Ergonomics*, *37*(4), 467–478. doi:10.1016/j.apergo.2006.04.020 PMID:16759630

Liu, C. J., Kemper, S., & McDowd, J. (2009). The use of illustration to improve older adults' comprehension of health-related information: Is it helpful? *Patient Education and Counseling*, *76*(2), 283–288. doi:10.1016/j.pec.2009.01.013 PMID:19286343

Mayer, D. L., & Laux, L. F. (1989, October). Recognizability and effectiveness of warning symbols and pictorials. In *Proceedings of the Human Factors Society Annual Meeting* (Vol. 33, No. 15, pp. 984-988). Los Angeles, CA: SAGE Publications. 10.1177/154193128903301518

Moles, A. (2007). *La imagen comunicación funcional*. Trillas.

Ng, A. W., Lo, H. W., & Chan, A. H. (2011). Measuring the Usability of Safety Signs: A use of system usability scale (SUS). In *Proceedings of the International MultiConference of Engineers and Computer Scientists* (Vol. 2, pp. 1296-1301). Hong Kong: IAENG.

Prado, L. L. R., Diaz de Leon, Z. C., Zambrano, P. P.L., Cisneros, H. A. C., & Cardenas, V. S. (2016). *Evaluation of the usability of nutritional recipes: A comparison between a text-based interface and an interface based on ergonomic guidelines*. Paper presented at the The 6th International Conference for Universal Design, International Association for Universal Design, Japan.

Prado León, L. R., & Avila Chaurand, R. (2010). *Percepción Visual II. Aplicaciones Para el Diseño*. Editorial Universitaria.

Robinson, P. A. (2009). *Writing and Designing Manuals and Warnings 4e*. CRC Press. doi:10.1201/9781420069860

Rogers, W. A., Lamson, N., & Rousseau, G. K. (2000). Warning research: An integrative perspective. *Human Factors*, *42*(1), 102–139. doi:10.1518/001872000779656624 PMID:10917149

Wogalter, M. S. & Silver, N. C. (1995). Warning signal words: connoted strength and understandability by children, elders, and non-native English speakers. *Ergonomics, 38*, 2188-2206.

Wogalter, M. S., Sojourner, R. J., & Brelsford, J. W. (1997). Comprehension and retention of safety pictorials. *Ergonomics*, *40*(5), 531–542. doi:10.1080/001401397188017

Wolff, J. S., & Wogalter, M. S. (1998). Comprehension of pictorial symbols: Effects of context and test method. Human Factors. *The Journal of the Human Factors and Ergonomics Society*, *40*(2), 173–186. doi:10.1518/001872098779480433

KEY TERMS AND DEFINITIONS

Cognitive Ergonomics: A research field of ergonomics and human factors that its main focus is the mental processes of the user such as perception, memory, reasoning, and motor response. A field specialized in the study of the psychological interactions and cognitive process of the users with objects, information, or build environments.

Comprehension: Recognition of something through the expression of its meaning or name.

Dietary Plan: A set of scheduled meals in order to achieve different metabolic and health goals, such as weight loss, sugar control, muscle gain.

Ergonomic Guidelines: Different proved principles based on user characteristics and necessities that are used to apply ergonomics in the design process of the user-object-environment systems.

Figurative Images: An intermediate level of iconicity of an image (level 5 of 12) in were the object is still identifiable but some spatial relationships are altered; these images keep key characteristics of the real object. Figurative images can be classified as realistic and nonrealistic.

Food Items: Any edible element that can be used to prepare a meal or recipe.

Graphic Abstraction: The process of reducing visual elements, keeping the key features of the represented object in order to still be easily identified.

Graphic Design: A discipline which its goal is to create clear and concise visual messages.

Iconicity: The grade of abstraction of an image in relation with its realistic through its abstract representation.

Chapter 17
Understanding the Context of Design for Social Innovations:
A Methodological Case Study

Juan Carlos Ortiz Nicolás
Autonomous University of Ciudad Juárez, Mexico

Thomas Harrison
Imperial College London, UK

ABSTRACT

The chapter describes the applicability of designers to the space of finding solutions to social issues through the generation of social innovations. Based on a literature review, it was evident that design requires new methods to apply to the resolution of social issues, and as such, this chapter presents a new method that supports designers in the development of solutions within the design for social innovation (DfSI) framework. The proposed method consists of six stages: (1) identifying the challenge, (2) analyzing the system in which the challenge co-exists, (3) understanding the system using user-centered design techniques, (4) defining the design brief, (5) generating proposals, and (6) evaluating and concluding. The chapter also introduces a design case study that describes in detail the implementation of the design method. The results indicated that the method supports designers to structure their process when aiming to design for social innovation and in particular to define the potential solution, which is reflected in a design brief.

INTRODUCTION

Social innovation is significant to the addressing of local and global challenges that seek the common good and this may explain its current interest in academia and practice. Research has identified that there are differences between social and economic innovation, showing that social Innovations aims are often harder to measure and require thinking up-front to compare them. In the field of design, the influence of social innovation is acknowledged as Design for Social Innovation. At present, it is unclear how the differences between Social and Economic Innovation impact on the design process.

DOI: 10.4018/978-1-5225-5234-5.ch017

Copyright © 2018, IGI Global. Copying or distributing in print or electronic forms without written permission of IGI Global is prohibited.

This manuscript describes Social Innovation as the management of a social mess, a type of wicked problem associated with social mechanisms that are typically difficult to define and also difficult to resolve, however, the aim is particular, to create a common wealth and enhance society's capacity to act (Murray, Cualier-Grice & Mulgan, 2010). To address the latter challenges, up-front homework is undertaken in order to fully understand the social mess, including the contextual complexities that are discovered during the important Fuzzy Front End. Designing as a part of Economic Innovation, and designing as a part of Social Innovation are not identical, they require different methods to manage the different types of fuzziness and developing an open brief that cuts through the complexity in order to develop a clear context within which the designing can occur. The fact that each type of innovation requires particular approaches does not mean that there aren't overlaps exist in the undertake of specific design tasks, e.g. ideation, or that designers that focus on developing Economic Innovation cannot work on Social Innovation solutions and vice versa.

The aim of this chapter is to introduce a design method to support designers in the development of Social Innovation. The method focuses on supporting designers in defining the design brief when addressing a social challenge. Thus, the method focuses on the Fuzzy Front End of Innovation. The method is explained in detail and a case study is introduced to report its implementation.

To define the overall context of this research it is important to discuss basic concepts that help to frame this particular project. For instance, this manuscript addresses the following issues: wicked problems, the fuzzy front end of innovation, the development of an open brief, and theoretical issues in design methods.

BACKGROUND

This section demands conceptual and contextual clarity, especially in the definitions and characteristics of much researched Economic Innovation and its comparable peer, Social Innovation. The understanding of their differences will be achieved through a cross-comparison of the two concepts, starting with the more familiar Economic Innovation.

Differences Between Economic and Social Innovation

Generally, Innovation is seen as the successful exploitation of new ideas (Cox, 2005), and can also be defined as any departure from existing practice (Schumpeter, 1934). This is largely considered in terms of trends, such as consumer, organizational, or financial. This innovation process that takes a creative concept and makes it somehow concrete and pragmatic goes beyond simple creativity because innovation is relatively valueless if the idea remains in its abstract creative form. It is unusable, therefore has no applicable user.

The nature of the change or break in trend is typically what defines the type of the innovation, and these various types, include but are not exclusively Economic Innovation and Social Innovation (see Franz, Hochgerner, & Howaldt, 2012 for a thorough discussion). Economic Innovation seeks changes in consumer or organizational behavior within an economic system or network that may eventually lead to a financial reward. In comparison, Social Innovation seeks changes in Social Behaviour within a social system or network. This is introduced by the works of Schumpeter who suggests five methods by which innovation can occur with the outcomes being consumer market disruptions (Schumpeter, 1934); these are measured by financial changes experienced by the organization to characteristics such as profit and growth. Social Innovations in comparison generate, through trend-breaks, disruptions in social systems and ultimately lead to the delivery of social benefits. An example of such a Social Innovation would be the acceptance of household recycling during the 1990s where the benefits are often unquantifiable or immeasurable, and may be perceived to have no economic benefit or possibly seen to be uneconomic (c.f. landfill disposal of plastics).

The true definition of Social Innovation is perceived as being a fuzzy notion with other perceived differences between it and Economic Innovation as being directly linked to the involvement of stakeholders and differing activities (Djellal & Gallouj, 2012, p. 121). Djella goes on further to suggest that Social Innovation is perceived to be a wider topic than economic innovation is often seen as a lot closer to the users (Djellal & Gallouj, 2012). As such the usual go-no-go decision gateways as used in economic innovation would be difficult because there may be qualitative, rather than quantitative output characteristics (e.g. A business case may not have clear financial outcomes) (see Cooper & Kleinschmidt, 1986). On the other hand, in Economic Innovation the consideration, or assessment, aspect relies upon financial discussions and projections when considering the business case for continuing the development. These often do not exist in terms of Social Innovation, where the benefits are harder to quantify.

Wicked Problems, Messes, and Social Innovation

It is evident from the literature that different researchers and policy writers have looked into the problems associated with social issues and have generally come to a consensus that they can frequently be categorized as something called a wicked problem (Conklin, 2005; Rittel & Webber, 1973; see also Zwicky, 1969). These so called wicked problems or social messes came originally from studies into the complex area of social planning, another area of government policy (e.g. Environment, social care, homelessness, climate change, et cetera). In an example case of urban pollution, there are issues such as the frustrations aimed at the production of airborne car pollutants caused by a social trend in the use of petrol or diesel vehicles. In another example case of social care, there are issues about the intransigence in the resolution of urban homelessness within developed countries. Although not immediately obvious, many design problems are also seen to be characterized by wickedness (Lawson & Dorst, 2009, p.40) and resolution of these problems is a skill set already developed in design practitioners. It is, therefore, an important research area to look at how designers apply their skills to managing these social issues most effectively.

These so called wicked problems avoid delineation and understanding of causation; completely hiding their problematic nature (Rittel & Webber, 1973), and according to the philosopher Churchman are characterized as having thoroughly confusing outcomes. This is caused by unclear and complex information, large numbers of clients and decision makers with similarly large numbers of opinions, which are often contradictory (Churchman, 1967). The implication is that these problems are difficult to resolve. Social Messes, a type of wicked problem with a closer association with planning problems in social systems, are seen as being unclearly defined and ambiguous (Ackoff, 1981) and additionally are seen as being reactive to attempts to resolve them (Ritchey, 2011b). As such many social issues are also social messes. These are a form of wicked problem, which means that the relationship between determinants and the outcomes is clouded (Ritchey, 2011a). Ritchey more specifically states that they "defy efforts to delineate their boundaries and to identify their causes, and thus to expose their problematic nature" (Rittel & Webber, 1973; Conklin, 2005). This is typical because of emergence from an underlying highly complex or chaotic model, such as a social system.

At a holistic level, wicked problems are complex, ambiguous, and uncertain. Researchers have sought to categorize methods that manage the complex aspects of wicked problems (see Roberts, 2000). Roberts has suggested that solutions can be:

- Authoritative, where the complexity is reduced by limiting the number of stakeholders, thereby reducing the size of the system model;
- Competitive, where the ambiguity is reduced by allowing two or more teams to develop competing solutions based upon their own interpretations of the model;
- Collaborative, where the complexity is reduced through agreement upon the interpretations of the underlying system.

Although useful as categories, these methods are frequently used in parallel within the design process, for example where designers seek to simplify and prioritize the requirements (authoritative), retain parallel opportunities (competitive), and work with other groups to determine the best possible solution (collaborative).

Within design research, it is common to see all problems as wicked and 'indeterminate' due to the lack of specific focus of the subject of 'designing' (Buchanan, 1992); with design research, in general, seeking to improve the expertise of designers and effectiveness of methods associated with these wicked problems. Social messes and design problems require specific methods to discover understand and model the underlying behaviors of the users, and such a framework is User-centred Design. These methods and skills that fundamentally are in demand in design are those that allow them to manage the complexity, uncertainty, and ambiguity that describe wickedness.

Design, Innovation, and Wickedness

Designers are in their most basic form concept-generators (Yilmaz & Seifert, 2011) and are skilled in bringing together unstructured and mercurial environments in a meaningful way through the use of methods such as ideation. But the role of the designer is not clear in Social Innovation because the role of the designer is not clear in innovation generally. Several researchers have tried to address this including Manzini who seeks to understand the role of the designer within the social process (Manzini, 2014; Manzini & Coad, 2015). This obfuscates further the outcome of the economic innovation process that is typically focused upon making money. In comparison, the outcome of the design process could be far more varied, including the making of an elegant and timeless artefact. The role of the designer within innovation is important though, largely due to their non-creative skill set of being able to take the undefined and make it defined.

In real terms, Social Innovation, like Economic Innovation, can and does occur independently of design. There is no demand to have designers involved in the process; however, this is not to say that the skill-set of a designer is not important in the successful delivery of innovation, as they are seen to be solvers of problems. Complicating matters further is that Innovation and Design, although lazily used as similar terms, are very dissimilar activities that are often conjoined and aligned in their goals, but not always. Economic benefits of a nicely designed sculpture or a well designed park may add cultural value to a space within which people can enjoy say their lunch breaks, again are often hard to quantify.

Based on the work of Nigel Cross this manuscript argues that designers have skills to address wicked problems. Cross describes designerly thinking in terms of six key activities (Nigel Cross, 2006, p. 12).

These are: (1) dealing with uncertainty, (2) decision making on the basis of limited information (e.g. Decision making in uncertainty), (3) resolve ill-defined wicked problems by adopting solution-focusing strategies (e.g. Managing ambiguity and complexity), (4) employing creative/productive thinking, (5) using graphic or spatial modelling media (e.g. Structuring complexity). It is evident from the first three of these ways of thinking that they support working with wicked problems as designerly thinking involves working with uncertain problems that are complex and ambiguous (Lawson & Dorst, 2009, p. 33). As such designers are incredibly well placed for resolving them using methods and styles of thinking. Another method used by designers in navigating social messes is the framing, reframing, reformulation, and organization of ill-structured problems (Lawson & Dorst, 2009).

In addition to previously mentioned skills, designers are adept in building an understanding of social networks thereby partially unpacking the context of a social mess (Cross, 2006). This complicates the problem further with designers considering the subjective opinions and beliefs of each of these individuals (network-nodes). However, they still are capable of bringing together stakeholders during the early phases of the design activities into a coherent mass, which is most difficult during the Fuzzy Front End.

Design Briefing as Part of the Fuzzy Front End

The Fuzzy Front End is an important aspect of the innovation process as it embodies the initial undefined and ambiguous parts of the innovation process that organizations often fail to properly understand. This is in part because of the unstructured nature of these predevelopment activities (Cooper, 1994). The characteristics of the Front End make it dynamic, unstructured and with very little formalization (Murphy & Kumar, 1997). As such this requires a new skill set that is not commonly held within the risk-averse world of management where characteristics such as repeatability and dependability are rewarded. This leaves many economic and social organizations unprepared for the Fuzzy Front End and as such makes it more important as a research focus. The Fuzzy Front End is defined as being when an idea is initially considered as worthy of further consideration and ends when the business considers it suitable for development (Kim & Wilemon, 2002). Khurana and Rosenthal further describe this phase in a more holistic sense, as being an intersection between several key activities: the processing of complex data, managing conflicting organizational forces, coping with large amounts of uncertainty, and combining cross-functional desires and demands (Khurana & Rosenthal, 1998). These uncertainties, complexities, vagueness and incomplete information pose a threat to innovation models that currently exist within organizations (Cooper & Kleinschmidt, 1986).

In the design process of concept generation, there is a need to have a direction that evolves during the duration of the design process, and this is typically referred to as the brief. The briefing is important in Social Innovation because, like in Economic Innovation, the starting point of the endeavor could be very fluid and undefined. The briefing is a method used by the designer to structure and delineate the activities in opposition to the mercurial and complex nature of the environment (Lawson, 2005). The briefing process allows the gathering and structuring of customer needs and constraints, something that is seen as being a vital activity in under-determined design situations problems (Lawson & Dorst, 2009, p. 42). This structuring of inputs to the design process can originate from something so simple as a statement of intent or a mission statement that can be embryonic to the entire design process, in addition to being reinterpretable (Ambrose & Harris, 2009).

Some designers see the brief as simply the client's requirements (Ambrose & Harris, 2009). In many cases, designers do not deal with briefing development because the design task starts with the brief from the client (Crilly et al, 2009). Design problems, especially within the early phases of design and also the Fuzzy Front End, often cannot be clearly stated and so open briefs are vague, are not uncommon and vary in what they contain (Lawson, 2005). Lawson explains that one issue is that designers are adept at crafting the design problem and so prefer to be included from the very beginning of the project in order to develop a meaningful brief, from what could be as small as a statement of intent. The brief can often be only a few pages in length, however, the task of building the brief and adding the detail is what is of most value to the designer (Lawson, 1994). This process of determining what the actual problem is allows the designer to start from fuzziness and end with concreteness (Ambrose & Harris, 2009). In terms of categories, design briefs mainly come in two varieties: open briefs and closed briefs. Open briefing is the design method where the designer is able to have greater interpretation and designer assembles information from the design process such as requirements, constraints, and allows the requirements and desires to be absorbed in a more gradual rate following a slower rate of discovery (Franck & Sommaruga Howard, 2010). In contrast, closed briefing is where the outline, goals, and delimitations are clearly defined from the outset, and often both the user and designer are clear about what is to be delivered. Research related to briefing development has focused on Economic Innovation, thus research might ask, how effective is briefing in the context of Design for Social Innovation?

Briefing is one of the key aspects of designing and embodies the building up of knowledge about the situation. The designer can then look at this design situation in multiple ways and allows them to take 'stabs' at the design problem from many different perspectives (Lawson & Dorst, 2009). This, in turn, helps identify the design task that latter will be defined in a brief. Without a clear understanding of this design situation it reduces the effectiveness of the design process, and in the case of social problems, this may reduce the effectiveness of the overall design process. Understanding the situation or environment into which the design project has been placed is an important activity for the designer, if not because of the situation factors that originate from the environment directly (Kim & Wilemon, 2003). In response to these factors, the environmental discovery typically is described by periods of research. Examples of activities within this period include ethnographies, which aim to reduce to 'objective seperatedness' between the designer and the user (Guba & Lincoln, 1994; Christensen & Raynor, 2003). This provides the designer with a clear understanding of the day-to-day practices of the user and allows them to effectively frame the interactions with the future artefact.

The Fuzzy Front End in Social Innovation

This section combines two important concepts in Innovation that are characterised by uncertainty, complexity, and ambiguity, and these two concepts are the Fuzzy Front End and Social Innovation.

The concept of the Fuzzy Front End has been addressed to show its fundamental role in innovation in general and Social Innovation in particular. In the specific context of Design for Social Innovation, this manuscript has identified that there are many examples of design solutions and good practices to consider for social issues in designing (Papanek, 1985; Witheley, 1997; Manzini & Coad, 2015; Tromp and Hekkert, 2016; Ortíz Nicolás, 2016a). However, there are few studies that report a proper method to design them. The few examples that address this issue identify basic steps that should be considered; e.g. engagement, assessment, planning, implementation, evaluation, and termination (Margolin & Margolin, 2002). Other methods focus on similar areas such as behavioral change (Tromp, 2013), which is arguably similar. These methods can be improved if designers are supported in the Fuzzy Front End to identify relevant tasks within the context of addressing a social challenge, which is usually a wicked problem. This task later will be synthesized in a brief.

Theoretical Issues in Design Methods

To complete the literature review it was deemed important to have a brief report of relevant findings in the area of design methods.

Schön carried research into the field of design, and in particular focussed upon how people generally learn to design (Schön, 1987). He synthesized his findings in the concept of a 'reflective practitioner', which is a role embodying the importance of conscious reflection, especially within design practice. Schön's idea is in line with the synthesis that Daalhuizen (2014) has made in terms of the benefits that new method usage has in design practice:

- Extend the ability to design, e.g. based on methods that address particular issues, as in this case Design for Social Innovation.
- Enhance reflection on design, e.g. improving designer's expertise on the design activity
- Enhance learning to design, e.g. developing capabilities or awareness and strengthening of existing ones.

It is also relevant to mention that research into methods presents some novel challenges, for example, Dorst (2008) argues that design research aimed at directly improving the design process through methods often ignores the subtle differences between designers, differences between design contexts and the differences between the overall future design outcome. Improvements in methods focus upon generating efficiencies and increases in effectiveness for the inclusion of the people who implement them. Therefore, an area of opportunity to improve design methods is to simply pay more attention to the opinions of method users (Daalhuizen, 2014). A similar situation occurs in Social Innovation where it is important to increase the speed of Social Innovation (Mulgan, 2007), seemingly to reduce the risk of a failed delivery. Within Economic Innovation the speed of the innovation cycle is equally as important for the same reason (Kessler, & Chakrabarti, 1996).

Research Aim and Question

A common way to enhance design learning is based on improving process and methods (Schön, 1987; Blessing & Chakrabarti, 2009). In order to enhance design learning, these new methods have to be experienced and appreciated by designers (Schön, 1987). It is widely accepted that methods that stay purely in the theoretical domain are failures. This research, therefore, aims to develop a method that assists the designer in Design for Social Innovation and focuses on the early phases of design, and more specifically during the Fuzzy Front End of Social Innovation. The remainder of this article describes the development of a qualitative case study to show the effectiveness of such a method within a Social Innovation project.

MAIN CONTENT

This section describes the development of a design method that relies upon the ecology model during the early phases of the design project. This method was then adapted and used within a case study based in a municipality in Mexico City. The discoveries from the method are presented within this section and discussed in terms of the benefits to the design process.

Method: Developing a Method to Design for Social Innovation - DfIS

In general, designers have routinely shown an interest in the social impact that their work has on other people and society e.g. artefacts, architectural spaces, or services. This interest can be traced back to the Eighteenth Century when John Ruskin (Ruskin, 1849) published his seminal work The Seven Lights of Architecture, in which he proposed to start social change through Art and Architecture. Another designer that emphasized the relationship between social issues and design practice was William Morris, a central figure in the Arts and Crafts Movement. He stated that designers' main purpose was to save society, and the way he proposed to do it was through the creation of useful objects for individuals. Furthermore, he encouraged designers to take responsibility with the environment through the use of local materials and forgotten production techniques, that were seen not to pollute as much as the industrial revolution's great machines. This is something that is equally applicable in the modern era when considering the polluting effects of producing the rare earth metals that are required for modern computing devices. His techniques were generally created through formal patterns primarily inspired by nature in order to be more natural in their effects.

This research was initially missing a key aspect; a generalized method for Design for Social Innovation, which would be used as a basis for the on-going research. The literature review involved the inclusion of different types of literature sources as the focus of the study is largely based upon design practice and as such should not discount sources that are from design practice and not just academic design theory.

It was evident that Social Innovation literature does broadly acknowledge the positive role that design might have on the development of its outcomes (Murray, Cualier-Grice, & Mulgan, 2010; Cajaiba-Santana, 2014). However, it is rarely reported exactly how the design process exists within Social Innovation, in part because it is rarely discussed where design fits within the greater Innovation process. Additionally, innovation processes rarely refer to briefing, which is a cornerstone of the design process. With this in mind, four key methods were identified that are relevant to discussions on Design for Social Innovation (DfSI). Three of these aimed to consider the social impact that design solutions may have; the other being more generic but it has potential to contribute to fulfill the aims of Social Innovation. Table 1 introduces the two general steps of the identified methods.

For each of the methods in the Table 1 important phases were unpacked in each; these phases were the Preparation phase and the Design phase. The Preparation phase includes the activities that designers have to carry out in order to understand and define the broader design challenge. The Design phase includes standard activities that are undertaken to deliver a design solution. In this research, the focus is upon the first phase, before the typical design process when the brief has not been developed and agreed.

Each of the models evidently framed the innovation process in different ways, making them subtly dissimilar in the way they categorized the activities. Firstly, Margolin & Margolin (2002) propose six steps to social design, and these are: 1) Engagement, 2) Assessment, 3) Planning, 4) Implementation, 5) Evaluation and 6) Termination. There is little detail for each step, however, it can be inferred that the

Table 1. Models of social innovation

	Preparation	Design
Margolin & Margolin (2002)	• Engagement • Assessment	• Planning • Implementation • Evaluation • Termination
Tromp (2013)	• Phenomenon • Social Statement • Behaviours	• Strategy • Design Proposal
Standarization (2010)	Understanding Users	• Involve Users in the Design Process • Refine and Evaluate • Iterative Process • Includes UX • Multidisciplinary Approach
Murray et al. (2010)	Prompts, Inspirations, and Diagnoses	• Proposals and Ideas • Prototypes and Pilots • Sustainability • Scaling and Diffusion • Systemic Change

two first steps are oriented to engage with a particular social challenge, and assess the situation. The other four steps are standard activities of a design process. Secondly, Tromp (2013) proposes five design steps: 1) Phenomenon, 2) Social Statement, 3) Behaviours, 4) Strategy, and finally 5) Design Proposal. This method starts by defining the phenomenon of interest, and then designers can define what they want to contribute to society and establish a strategy that will deliver such a solution. The method has a direct focus on the design phase, which is of interest to this research and as with any method, it focuses on a particular issue, which in this case is behavior. Social Innovation indeed considers behavior but also other broader aspects, such as interactions among different groups, e.g. intra-, inter-, or extra- group (See Cajaiba-Santana, 2014). The latter is relevant because Cajaiba-Santana (2014) states that in Social Innovation it is fundamental to understand not only how people act but also how they give meaning to their actions. Thirdly, ISO 9241-210:2010(E) (Standardization, 2010) introduces principles of User Centred Design (UCD), which include: 1) The design is based upon an explicit understanding of users, tasks and environments, 2) Users are involved throughout design and development, 3) The design is driven and refined by user-centred evaluation, 4) The process is iterative, 5) The design addresses the whole user experience, and 6) The design team includes multidisciplinary skills and perspectives. UCD can support Social Innovation to develop an understanding of the reality of people involved in a social challenge. Furthermore, many of the principles suggested by UCD are relevant to improve a design solution, for instance, evaluation is becoming a standard activity in professional practice. Fourthly and finally, Murray et al. (2010) propose six steps to Social Innovation, and these are: 1) Prompts, Inspirations and Diagnoses, 2) Proposals and Ideas. 3) Prototypes and Pilots. 4) Sustainability 5) Scaling and Diffusion, and 6) Systemic Change. This method is aimed at Social Innovation and that as such it broadens its scope in comparison to other three methods that are suggested by designers. For instance, steps 4, 5 and 6 are aimed at inserting the solution and maintaining it, while the first three steps are linked to activities that designers also consider in the reviewed methods.

The literature review helped to identify the following issues. Firstly, there are many examples that show how design can contribute to Social Innovation. Secondly, there are methods that can stimulate

Design for Social Innovation (DfSI). Thirdly, many of the activities that are suggested on the four reviewed methods are aligned to activities that designers usually perform. Fourthly, the preparation phase is vaguer in comparison to the design phase (see Table 1), which may in part be because the activities within the early phases of design are not as obvious to predict, especially when defining open briefs. Therefore, designers need to further develop skills or have access to tools that help them carry out the preparation phase.

Introducing a Method to DfSI

The development of the two phases introduced in Figure 1 was informed by the methods reviewed in the previous section.

Figure 1 shows that the DfSI framework is described by two general phases, which are named Preparation and Design. In the first of the two phases, different activities are undertaken to understand the social challenge. This includes an approximation of the people that are affected by the selected social challenge. This information leads to a clearly defined design brief, which is traditionally the start of the design activity.

Identifying the Challenge

This first step focuses upon the selection of the phenomenon. In México and Latin America, there are many social challenges presenting such phenomena, it is, therefore, an issue to consider if the discipline of design can address them. For an example, some major social issues relating to health are difficult

Figure 1. DfSI: A method to design for social innovation

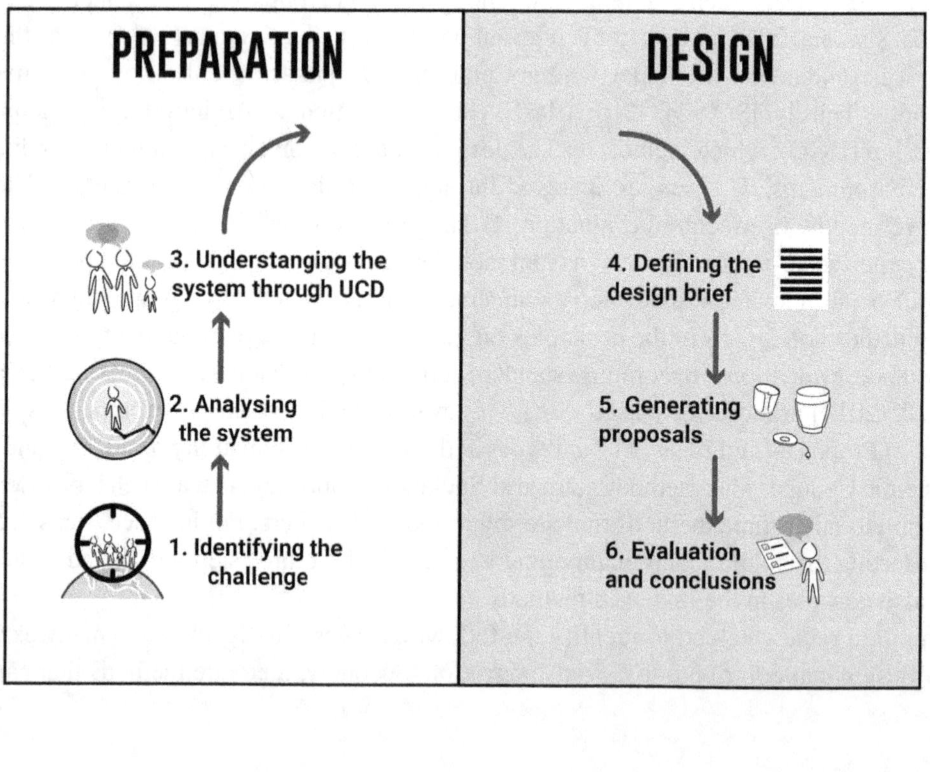

to be dealt with the design because of their size. As such two aspects that can help select a relevant phenomenon are: a group of people really demands a need and it is a phenomenon that is meaningful for the designer(s). Some strategies that can be used to select the challenge are documentaries and electronic media reviews. It is common that local media report current social challenges, which have the potential to become the phenomenon to work with. Another strategy is to select a community to which design specialists belong to, or have as one of their interests, in order to identify non-covered needs, for example, NGOs. Papanek (1985) has also suggested general subjects that have social relevance, among them are: design for people with disability, design for human life under marginal conditions or a popular suggestion, design for the third world. Furthermore, Whiteley (1997) mentions the Lucas Plan, which strongly emphasizes on social responsibility, an example of this: be accessible and useful to everyone in the community and not confined to fulfilling the needs of the selected few (Whiteley, 1997, p. 108).

Whiteley also argues that designers or the people affected by a social issue are the ones who should define their priorities to establish what is produced. This is instead of leaving that choice to the market's forces that are generally oriented to consumption. This supports the use of UCD techniques to define a relevant challenge to be addressed in a particular community. Another guide that could help to select a relevant social phenomenon is the National Economic Development Plan of each country as each country has particular and relevant issues to be addressed. It is important to keep in mind that Social Innovation does not intend to help people generally; it instead aims to generate solutions with the community to consider and include their reality.

Analysing the System

The objective of this step is to define where designers work. This is important because a potential solution will be inserted into a particular context that has specific dynamics. To map the context this research relied upon the ecology model suggested by Bronfenbrenner (1995). Based on the fact that ambiguity plays a role in the preparation phase there is reliance upon the ecology model to define the level of intervention at which designers can work (e.g. it is unclear how to structure it and previous research do not report tools that could help a designer to more fully define the design brief). The decision to use the ecology model is based around the issue that many social challenges are perceived to be wicked problems (Manzini & Coad, 2015; Ortíz Nicolás, 2016). Furthermore, Forlizzi (2007) suggests that environments can be described in terms of their relative scale and complexity; their inhabitants can be studied at various levels including individuals and small groups (micro level analyses), organizations or neighborhoods (meso level analyes), and regions or populations (macro level analyses). It is fundamental for DfSI to explore potential solutions that could be inserted at different levels. This implies that many potential solutions can be developed for a particular challenge (e.g. diabetes) and with Social Innovation being highly contextual the ecology model helps develop understanding about these contextual conditions. To structure those potential solutions, designers can rely upon the four systems of the ecology model (micro-, meso-, exo- and macro- systems). In other words, the ecology model can help designers decide at which level they will develop a solution.

The ecology model includes four interconnected levels: microsystem, mesosystem, exosystem, and macrosystem. Figure 2 shows how such a system analysis can be presented with the most immediate aspect of the system. The context presented is a summary of the larger part of the diagram due to its immediacy.

In order to more fully describe the system map as shown examples will be used and first an example for a microsystem would be a family unit. A mesosystem is related to the different interactions among

Figure 2. The ecology model

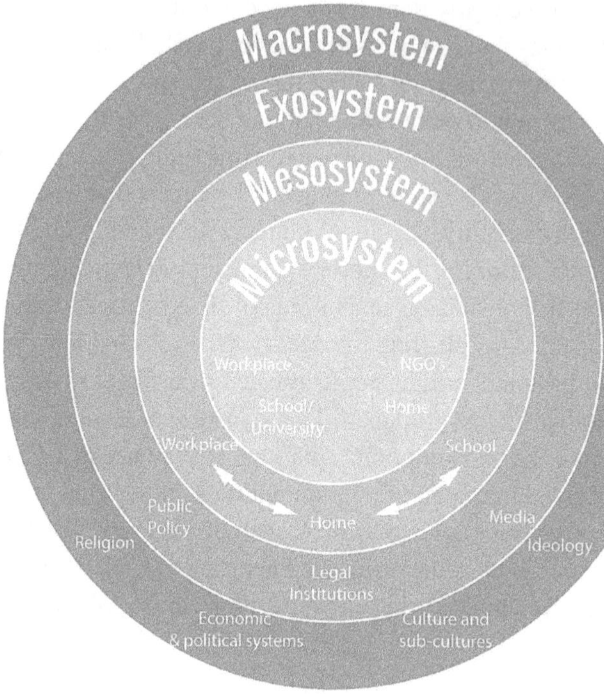

microsystems, e.g. between a family and a school. An exosystem would include upper social levels, for example, those in which a regular person does not usually access where public policies are developed and this influences a community. An example is a policy affecting public education, which affects both the micro- and mesosystems. Finally, the macrosystem connects and goes through all the sub-systems from low to high order, such as cultural values that affect the family, community, and public politics. For a detailed description of the ecology, model see Bronfenbrenner (1995). Mapping a social challenge based on the four systems of the ecology model helps visualize the different factors that influence it.

Furthermore, it supports designers to define the level at which they will work (e.g. micro-, meso-, exo- and macro-) and understand to some degree the dynamics of the context. It also helps designers understand that solutions work at different levels, for example, a solution aimed at the microsystem will not necessarily work at the exosystem.

Understanding the System (UCD Techniques)

Once the area of intervention has been defined (e.g. micro-, meso-, exo- or macro- systems), the next step is to use User-Centred Design (UCD) techniques to grasp people's reality, especially those who are directly affected by the problem. If the potential solution will be inserted in the microsystem is relevant to interview, observe, talk to the people who are directly involved in the social phenomenon. This helps comprehend the different influencers of a particular social challenge. It is important to point out that designers should first develop a strategy in order to use a UCD approach, which should at least consider the study's goal, data collection method, and data analysis techniques. UCD is also useful in questioning designers' beliefs or biases. They may believe that creating a particular solution, based only on their personal knowledge, will be good enough to address a particular social challenge. The dynamics of reality, however, make it really difficult to create plausible solutions without a deep understanding of context and people affected by a social challenge.

Defining the Design Brief

The fourth step of this method is to synthesise the acquired knowledge in a brief that describes the design task. The brief provides a guide to the design process. This step cannot be completed if the previous ones have not been concluded because they all inform the design brief. The general plan that is suggested is to establish the phenomena, then define the level of intervention (e.g. micro-, exo-, meso, or macro- systems) and finally understand the selected system based on UCD techniques. Based on the gained knowledge, designers can then more fully propose a particular design brief with a richer supporting argument. As Wendt (2015, p. 67) reports, framing problems and opportunities is a sense-making task. It involves finding meaning in the vast amount of observational data (e.g. a UCD technique). This research process generates a wealth of findings. Framing attempts to make sense of them and develop a coherent point of view for the designer to adopt.

Generating Proposals

Once the design brief has been generated, designers can use it to produce and explore alternative solutions. This describes aspects of a typical design process that can include sketching, mock-ups, and prototypes. The main issue in this aspect is to respect the spirit of the design brief. As in any design process, it is important to develop many solutions, this helps designers reflect on the pros and cons of each of them and select the one that fulfills the brief. At the same time it also helps designers to better understand the challenge that they are addressing (Wendt, 2015).

Evaluation and Conclusions

As in any iterative process, once a potential solution has been developed, it is expected to evaluate it with its potential users as a quality measure. This can deliver users' feedback and will help refine the solution before inserting it into the context of use. The evaluation process has to be prepared in advance to avoid pitfalls. Another important final step is to report the main lessons learned during the development of the project. This can help other designers or people interested in the development of similar projects.

Case Study: Social Innovation in Xochimilco

To offer more detail to the implementation of the method that was introduced in the previous section, a design project case will now be detailed. This project was the result of a university module delivered at the Design School of Industrial Design at the National University of Mexico (UNAM). The aim of the module was to implement Design for Social Innovation and was divided into 15 sessions of two hours per week. The first author assisted them with the implementation of the method using the structure shown in Figure 1. Three undergraduate students from the industrial design programme developed the design project. While the method initially seemed linear, when put into practice, it was very dynamic due to the project's complexity and the students' reflections on the method's implementation.

Identifying the Challenge

Within the challenge identification stage, the students expressed an interested in addressing one of the several issues that exist within their hometown community; the municipality of Xochimilco in Mexico City. In particular, the designers worked on the farming production in Xochimilco's Chinampas. A Chinampa is small floating land that farmers create over lakes and canals and at present, the Chinampas are used to cultivate flowers and vegetables. This farming technique has been used in Mexico City since pre-Hispanic ages for its effectiveness. However, its use has diminished over time.

Analysing the System

Once that the general phenomenon was defined the designers were introduced to the ecology model in order to assist in the analysis of the system. This helped them to develop a system map (see Figure 3), which offered a general overview of the identified relevant issues in the Municipality of Xochimilco. These included the number of inhabitants, the regions into which Xochimilco is divided, the products that are produced in the region and many of the cultural attractions and academic institutions. The map also includes information about particular farming activities, for instance, the students realized that Xochimilco is divided into three zones: the Mountainous zone, the Chinampera zone, and the Touristic zone. Farming was identified as a concern to all three zones and as such occurs was chosen to be the axis of the process.

By further utilizing the ecology model the students were able to decide to exclude the microsystem. Thus, they established to seek a social solution that could be embedded into the other interactions that affect farming in Xochimilco. Based on the gathered information from the different sources, e.g. news, records of Xochimilco's farming activities and informal observation, the students decided to work on the mesosystem. They identified an opportunity in the interactions that occur among canals, Chinampas, and farmers.

Understanding the System (UCD Techniques)

Having built and analyzed the system using the ecology model, which helped define the level of intervention at the mesosystem level the students prepared an interview to gain deep knowledge about farming in Xochimilco. Interviews used as a popular UCD technique to understand users. The gathered data allowed them to identify several challenges, for example, students identified that farmers needed a device to plough the Chinampas; tractors cannot be used because they are very heavy and can destroy the foundation structure. Figure 4 presents identified general problems and areas of opportunity, such as something that designers did not realized before of the interviews was the role that irregular settlings have in the pollution of water. How this is affecting the ecosystem of a landmark animal of Xochimilco, the Axolotl, was a key finding.

Designers also identified that the irrigation water was directly extracted from Xochimilco's canals. This water was polluted because the canals receive water from the sewages of irregular settlings that are located in Xochimilco, some of them in prohibited areas. At this point, designers acknowledged that canals' non-treated water was a central problem in the municipality of Xochimilco. The polluted water not only affected farming but also inhabitants' health. Based on designers' observations it was identified that locals used homemade filters to reduce water's pollution. Considering that a key issue

Figure 3. Xochimilco's systemic analysis (environmental awareness map)

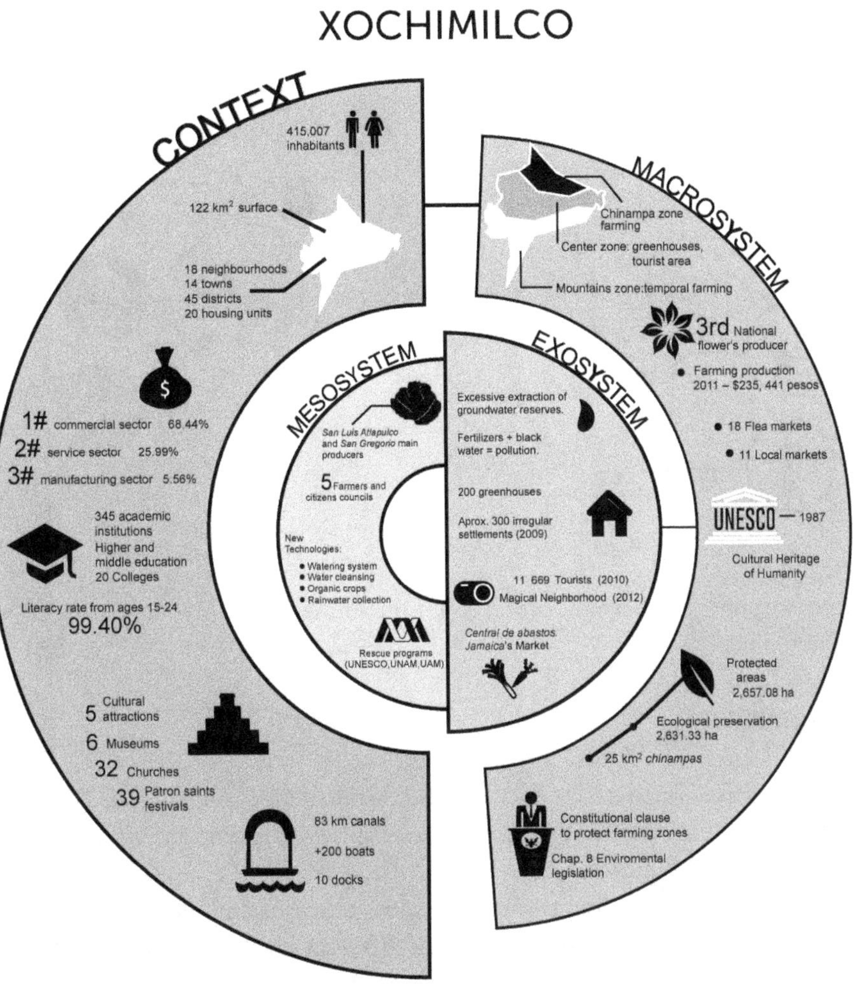

on the overall challenge was water pollution designers decided to further develop a water filter aim at irregular settlements

It was important to reflect that at this point there was a significant shift, which was strongly based on users' input. Students defined to focus on a particular phenomenon: farming in Xochimilco. However, based on the different activities that were undertaken they realized that there were other challenges that affected not only farming but also the ecosystems and its living flora and fauna. Considering that this was a key issue to be addressed, students decided to refocus from farming to cleaning polluted water.

Defining the Design Brief

Based on an extensive analysis of current problems in the municipality of Xochimilco, in particular in the Chinampas' zone the brief of the design project was defined as follows:

To design a filter that can be connected to sewers from irregular settlements that drain polluted water to Xochimilco's canals. The sewers are variable in diameter, which is a condition that the design would

Figure 4. Detected design challenges

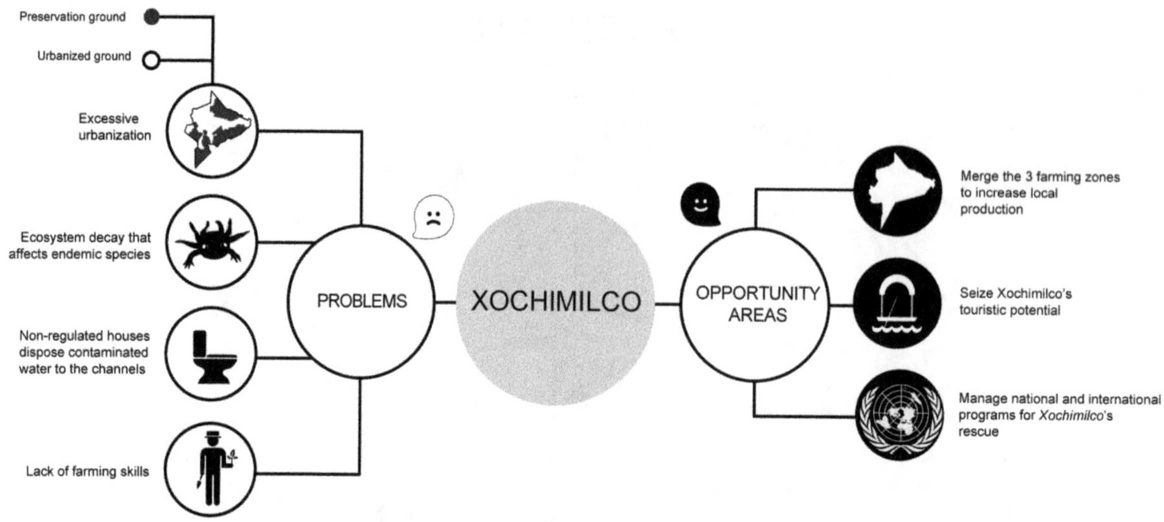

have to take into account. Furthermore, the filter's materials have to be easy to access, maintain and use, and cheap to replace.

Generating Proposals

Once the design brief was defined designers were basked into their comfort zone. They started to develop solutions that aligned to that brief. In this particular case, several solutions were developed and one of these options can be seen in Figure 5.

The filter, in Figure 4, included four trays with natural and artificial materials, for example, stones or plastic pellets. Each tray aimed to clean the water at different levels, from Residual to Biological. The Residual filter consisted of Tezontle, a porous and available volcanic rock that can filter large particles (e.g. litter, insects). The Biological level was based on materials that attract bacteria (in tray 3) and plants (tray 4). This idea also considered that the filter had to be cleaned and the trays could be easily removed once that the filter was located in place. Areas of improvement within this solution included the connector system; most sewage pipes are circular and consist of different diameters, and the solution has to adapt to this requirement. The design solution was further refined, however, it was decided to present this solution because as in any design process it is important to develop different ideas.

Evaluation and Conclusions

The filter that was developed as a solution to the selected challenge in Xochimilco was introduced to the community that was aimed to, people who lived in irregular settlements. The feedback was positive and their feedback was used to refine the solution. The solution, however, was not progressed beyond the design phase as it had fulfilled its educational purpose.

The case study aims to show how the method introduced in the previous section can be adopted by a group of designers when aiming to design for social innovation. Based on the analysis of existing meth-

Figure 5. One of the first design proposals

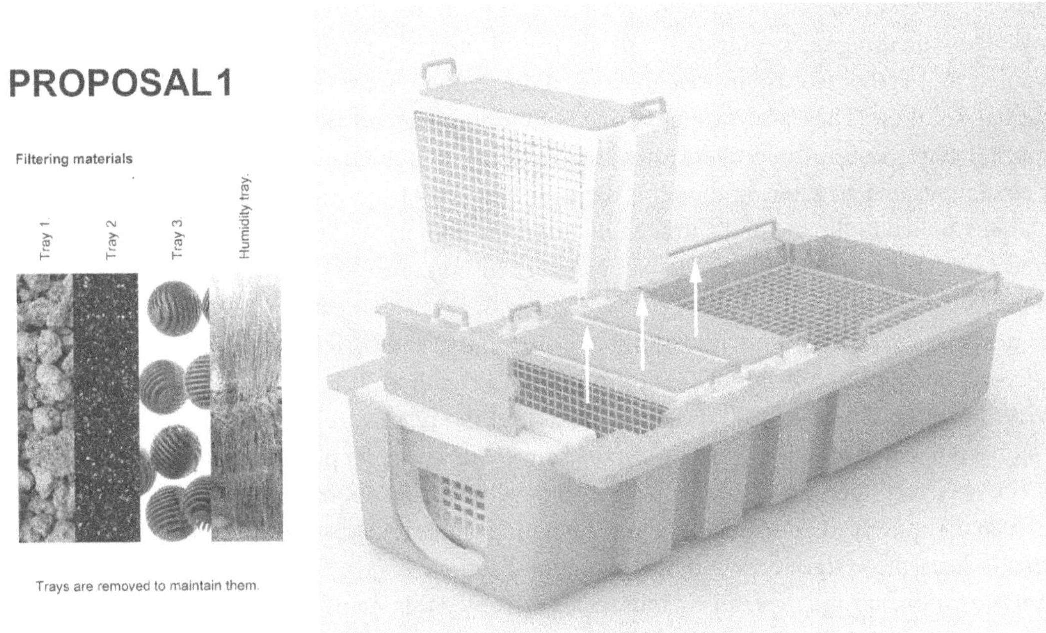

ods, it was identified that there is fuzziness in the definition of a design brief. It can be understood why a particular phenomenon has many potential solutions. To define the most relevant, within the terms of a design module, it relied upon the ecology model to define the level of intervention, in the reviewed cased it was the mesosystem. Then, it sought to grasp the complexity that affects the selected phenomenon through users' interviews. The gathered data redefined the focus of the project from farming to cleaning water. The particular case showed how designers identified different challenges and decided to focus on the one that they believed that was the most significant to the challenge at hand. It also showed the relevant role that designers can play in the implementation of design methods. In particular, the designer reported that to improve the method's implementation, it could be useful a set of examples of different social challenges and structured on the four systems of the ecology model.

DISCUSSION

The aim of this research was to introduce a method that assisted designers during Design for Social Innovation. The method focused on the early phases of design and more specifically during the Fuzzy Front End of Social Innovation. This aim was fulfilled based on the case study introduced. It details the different method's steps, and emphases on the preparation phase, which includes the Fuzzy Front End. It is acknowledged, however, that other means may be necessary to validate the introduced method. This is a first approach from industrial design to structure designers' thinking when developing social innovation projects.

It has been reported that methods consist of information – describing the structure of certain design phenomena – which needs to be taken in and be transformed by its user in order to be effective; a process that is called learning (Kolb, Boyatzis, Mainemelis, & others, 2001). The designers' opinion and experiences, however, are often ignored when design methods are evaluated (Dorst, 2008). This study acknowledges the designer's central role in the use of the DfSI method based on their use, guidance, and evaluation. In other words, the designers used the method, were guided when needed and delivered a particular solution. They also evaluated the solution and acknowledged the social influence that it had. Furthermore, areas of improvement were identified, e.g. collecting a set of examples of different social challenges and structuring them on the four systems of the ecology model. To further develop gain insights linked to the method's implementation the authors will structure this discussion based on the three aspects that impact on methods: designers (users), design task and the context in which the task is performed.

The users of the method were undergraduate students of industrial design and they could complete and deliver a solution as a result of the use of the DfSI method. It is acknowledged that the expertise of novice designers (Lawson, 2005) has an impact on the results; for instance, they are starting their professional education and have very little experience within design practice. Other designers, e.g. professionals, may rely upon particular strategies to develop a concept once that it has been defined (See Cross, 2004). Similarly, their experience may help them analyze data more effectively, develop creative solutions or keep the systemic view in the potential solution.

In relation to the design task, this manuscript has argued that many social challenges align well with wicked problems (Farrell & Hooker, 2013), for instance, the novel designers who participated in the module identified different potential solutions to the particular challenges they were dealing with. Thus, it may be relevant to consider in more detail how to address wicked problems and the Front End of Innovation (Koen et al., 2001; Harrison & Aurisicchio, 2011). It is important to report that the introduced method may not be useful when closed problems are involved because the first steps aim to identify the potential challenge to be addressed based on an open design brief. In this line, the method has potential to address wicked problems. For instance, in the DfSI method, the exploration of different design opportunities was based first on the ecology model, considering that a particular solution may be relevant only for one of its four systems. A second step was to study the system at which the potential solution was going to be inserted through UCD techniques. This helped understood the complexity of the system (Hassan, 2014). A strong difference of the DfSI method in comparison to others is that in many cases the design process starts with the design brief. In the method used in this study, novel designers defined it based on a particular challenge, prior to analyzing and selecting a particular level of the context in order to understand it fully. This implies that they developed skills to evaluate and select the relevant phenomena and improved their skills in reflection and analysis.

The design method helps students define the project and develop a solution to a particular social challenge. The first author supported novice designers in the understanding of the ecology model (see Step 2 in Figure 1) and in the use of the User Centred Design techniques (see Step 3 in Figure 1). These are fundamental tasks that required particular skills. As such, a condition of the use of the DfSI method is to first hold a basic knowledge of the ecology model and UCD techniques. It is relevant to mention that the use of UCD brings knowledge of what is to be changed (Hassan, 2014; Wendt, 2015). As the context becomes more complex in understanding, UCD techniques can help identify what is required within that particular messy context. This, of course, demands developing design skills to explore, observe and research. Additionally, all the projects were carried out by groups of two to three students. This helped undertake all the activities that involved the DfSI method and supported the reflective practice (Schön, 1987) that is relevant in this type of projects.

Design methods are not strict instructions or recipes that a person can automatically follow to create and deliver the best solution for the task at hand (Daalhuizen, 2014). Nevertheless, they are useful tools that designers rely upon to learn design. Thus, design methods can be considered tools that are embodied by designers to help them become professional designers. The latter is of great interest to academia, considering that every person can engage in designerly behaviors (Cross, 1993). Furthermore, involving designers in the evaluation of a particular method has shown positive results (Ortíz Nicolás, Aurisicchio, & Desmet, 2013).

A relevant aspect that is frequently ignored in Design for Social Innovation is the measurability of the delivered solution. However, the methods suggested within such frameworks identify the need to identify measurable social benefits in order to evaluate the success of programmes. Non-measurable social benefits are problematic because they are difficult to pass through go-no-go decision gateways as used in the Fuzzy Front End (Cooper & Kleinschmidt, 1986). As such non-measurable Social Innovations are difficult to defend due to the lack of clarity over success measurements. The measurability of such social benefits is important to design because it justifies the quality of the design process, as laid out in the brief. These delivered social benefits are non-financial, often qualitative, and sometimes difficult to quantify. In the example of the Child Happiness Index (see UNICEF Innocenti Research Centre, 2007), it would be easy to provide arguments on what components define child happiness, and ultimately what the eventual measurement would mean in terms of social realities. It would also be in terms of how it relates to other more definable measures of unhappiness (e.g. recorded numbers of self-harm). Another key aspect of measuring Social Innovation is the individual and social attitudes towards the benefits themselves. Social behavior has always been difficult to predict largely for the issue that people have beliefs and opinions towards outcomes and the utility of these outcomes, even when they are not financial.

It is obvious that financial benefits are easy to measure because they are quantifiable and as such objectively agreeable by the majority of people within the system. On the other hand, social benefits are often qualitative, often making the interpretation of their benefits difficult to agree objectively and by a majority. When they are quantitative, these measures are also difficult to agree in terms of how beneficial they are to the society. As an example, many people have no interest in the concept of saving the environment because their subjectively more trustworthy set of opinions and beliefs may suggest that the environment does not need saving. As such, the measurement of the benefits may be more complicated than the delivery of the benefits. And those social issues that do exist within society may be obfuscated largely due to conflicts of opinions such as different interest groups having different beliefs and priorities. In addition, social problems may often have unquantifiable economic costs making them difficult to prioritize within a political and economic context. This manuscript has pointed out the relevance to measure the impact of a Social Innovation. It is therefore important that design methods oriented to enhance Social Innovation consider how to measure their successes.

FUTURE RESEARCH DIRECTIONS

The steps of the DfSI method have been introduced in some detail allowing other designers to readily adopt and use it. It is understood, however, that the given descriptions may not answer all the designers' questions that they may have when implementing it. The authors are, thus, developing supporting tools to implement the method and in particular are focusing upon the ecology model as a generative idea tool. Designers can then consider the different levels of the ecology model within a design challenge.

This will allow them to propose ideas that can become design opportunities. The tool aims to introduce ideas for each level based upon a particular design challenge, and in this way further studies will aim to facilitate the implementation of the method.

Thus, further research can also define the skills or benefits that designers will gain when addressing Social Innovation. For instance, it has been identified that designers' skills can be enhanced through the development of DfSI projects having some typical issues, e.g. considering users as active agents, considering the impact of the context, managing the ambiguity and uncertainty within the context, and stimulating research for design (Ortíz Nicolás, 2016).

Another relevant future discussion concerns the identified commonalities between the Fuzzy Front End for Social and Economic Innovation in a greater context. These include information gathering, the problem solver is similar in both cases as a financial market full of consumers is essentially a different type of social system, albeit one with mainly economic innovations. In social systems, the choice of methods is more likely to be more specific to social issues, for example delivering the common good.

Social innovation has acknowledged the role that design may have on its development in at least two areas: creativity and the development of solutions, e.g. an artefact, service (Grimm, Fox, Baines, & Albertson, 2013; Van der Have & Rubalcaba, 2016). It is up to designers to insert their discipline in a broader context, i.e. Social Innovation. This insertion may be a perfect environment to stimulate inter- and multidisciplinary work, a long desired dream to design (see Papanek, 1985).

CONCLUSION

The DfSI method aims to enhance the capacity of both designers and users to act to current social challenges. The results of this research indicate that the DfSI method shows potential to frame social challenges and deliver solutions that aim to tackle those challenges. In a similar line, the ecology model is a useful tool that supports design decisions. In Social Innovation, the context is incredibly important and any tool that helps the designer, or indeed innovator, to understand the context and situation will be improving the efficiency of the process.

Many Social Innovation challenges are wicked problems, thus existing that deal with this type of problems in Economic Innovation may be relevant to consider in Social Innovation.

Social Innovation can benefit from Industrial Design and vice versa; designers can develop particular skills when undertaking this type of projects, e.g. Trans-disciplinary work. It may also be a clear action to create solutions that surpass the oriented market ones.

REFERENCES

Ackoff, R. L. (1981). *Creating the corporate future: Plan or be planned for*. University of Texas Press.

Ambrose, G., & Harris, P. (2009). *Design thinking*. London, UK: AVA Publishing.

Blessing, L., & Chakrabarti, A. (2009). *DRM, a design research methodology*. London, UK: Springer. doi:10.1007/978-1-84882-587-1

Bronfenbrenner, U. (1995). The bioecological model from a life course perspective: Reflections of a participant observer. In P. Moen & G. H. Elder Jr., (Eds.), *Examining Lives in Context: Perspectives on the Ecology of Human Development* (pp. 519–618). Washington, DC: American Psychological Association. doi:10.1037/10176-017

Buchanan, R. (1992). Wicked problems in design thinking. *Design Issues*, *8*(2), 5–21. doi:10.2307/1511637

Cajaiba-Santana, G. (2014). Social innovation: Moving the field forward. a conceptual framework. *Technological Forecasting and Social Change*, *82*, 42–51. doi:10.1016/j.techfore.2013.05.008

Christensen, C. M., & Raynor, M. E. (2003). *The innovator's solution: Creating and sustaining successful growth*. Boston, MA: Harvard Business Press.

Churchman, C. W. (1967). Guest editorial: Wicked problems. *Management Science*, *14*(4), B141–B142.

Conklin, J. (2005). Wicked problems and social complexity. In *Dialogue mapping: Building shared understanding of wicked problems* (pp. 1–25). Chichester, UK: John Wiley & Sons.

Cooper, R. G. (1994). New products: The factors that drive success. *International Marketing Review*, *11*(1), 60–76. doi:10.1108/02651339410057527

Cooper, R. G., & Kleinschmidt, E. J. (1986). An investigation into the new product process: Steps, deficiencies, and impact. *Journal of Product Innovation Management*, *3*(2), 71–85. doi:10.1016/0737-6782(86)90030-5

Cox, G. (2005, May). *Cox review of creativity in business: Building on the UK's strengths*. London, UK: The Stationary Office.

Crilly, N., Moultrie, J., & Clarkson, P. J. (2009). Shaping things: Intended consumer response and the other determinants of product form. *Design Studies*, *30*(3), 224–254. doi:10.1016/j.destud.2008.08.001

Cross, N. (1993). Science and design methodology: A review. *Research in Engineering Design*, *5*(2), 63–69. doi:10.1007/BF02032575

Cross, N. (2004). Expertise in design: An overview. *Design Studies*, *25*(5), 427–441. doi:10.1016/j.destud.2004.06.002

Cross, N. (2006). *Designerly ways of knowing*. Basel, Switzerland: Birkhäuser.

Daalhuizen, J. J. (2014). *Method usage in design: How methods function as mental tools for designers* (Unpublished PhD dissertation). Delft University of Technology.

Djellal, F., & Gallouj, F. (2012). Social innovation and service innovation. In H.-W. Franz, J. Hochgerner, & J. Howaldt (Eds.), *Challenge social innovation: Potentials for business, social entrepreneurship, welfare and civil society* (pp. 119–137). Berlin, Germany: Springer-Verlag. doi:10.1007/978-3-642-32879-4_8

Dorst, K. (2008). Design research: A revolution-waiting-to-happen. *Design Studies*, *29*(1), 4–11. doi:10.1016/j.destud.2007.12.001

Farrell, R., & Hooker, C. (2013). Design, science and wicked problems. *Design Studies*, *34*(6), 681–705. doi:10.1016/j.destud.2013.05.001

Forlizzi, J. (2007). The product ecology: Understanding social product use and supporting design culture. *International Journal of Design*, *2*(1), 11–20.

Franck, K., & von Sommaruga Howard, T. (2010). *Design through dialogue: A guide for architects and clients*. Chichester, UK: John Wiley & Sons.

Franz, H.-W., Hochgerner, J., & Howaldt, J. (Eds.). (2012). *Challenge social innovation: Potentials for business, social entrepreneurship, welfare and civil society*. Berlin, Germany: Springer-Verlag. doi:10.1007/978-3-642-32879-4

Grimm, R., Fox, C., Baines, S., & Albertson, K. (2013). Social innovation, an answer to contemporary societal challenges? Locating the concept in theory and practice. *Innovation (Abingdon)*, *26*(4), 436–455. doi:10.1080/13511610.2013.848163

Guba, E. G., & Lincoln, Y. S. (1994). Competing paradigms in qualitative research. In N. K. Denzin & Y. S. Lincoln (Eds.), *The Handbook of Qualitative Research* (pp. 105–117). Thousand Oaks, CA: Sage.

Harrison, T., & Aurisicchio, M. (2011). Understanding the front end of design. In *Proceedings of the International Conference on Engineering Design, ICED 2011*. Technical University of Denmark.

Hassan, Z. (2014). *The social labs revolution: A new approach to solving our most complex challenges*. Oakland, CA: Berrett-Koehler Publishers. doi:10.9774/GLEAF.9781783531455_5

International Organisation for Standardization. (2010). *Ergonomics of human-system interaction: Part 210: Human-centred design for interactive systems*. Geneva, Switzerland: ISO.

Kessler, E. H., & Chakrabarti, A. K. (1996). Innovation speed: A conceptual model of context, antecedents, and outcomes. *Academy of Management Review*, *21*(4), 1143–1191.

Khurana, A., & Rosenthal, S. R. (1998). Towards holistic "Front ends" in new product development. *Journal of Product Innovation Management*, *15*(1), 57–74. doi:10.1016/S0737-6782(97)00066-0

Kim, J., & Wilemon, D. (2002). Focusing the fuzzy front-end in new product development. *R & D Management*, *22*(4), 269–279. doi:10.1111/1467-9310.00259

Kim, J., & Wilemon, D. (2003). Sources and assessment of complexity in npd projects. *R & D Management*, *33*(1), 15–30. doi:10.1111/1467-9310.00278

Koen, P. A., Ajamian, G., Burkart, R., Clamen, A., Davidson, J., D'Amore, R., & Wagner, K. (2001). Providing clarity and a common language to the "Fuzzy front end". *Research Technology Management*, *44*(2), 46–55. doi:10.1080/08956308.2001.11671418

Kolb, D. A., Boyatzis, R. E., Mainemelis, C., & ... (2001). Experiential learning theory: Previous research and new directions. *Perspectives on Thinking, Learning, and Cognitive Styles*, *1*(8), 227–247.

Lawson, B. (1994). *Design in mind*. Amsterdam: Architectural Press.

Lawson, B. (2005). *How designers think: The design process demystified*. Amsterdam: Architectural Press.

Lawson, B., & Dorst, K. (2009). *Design expertise*. Amsterdam: Architectural Press.

Manzini, E. (2014). Making things happen: Social innovation and design. *Design Issues*, *30*(1), 57–66. doi:10.1162/DESI_a_00248

Manzini, E., & Coad, R. (2015). *Design, when everybody designs: An introduction to design for social innovation.* Cambridge, MA: MIT Press.

Margolin, V., & Margolin, S. (2002). A "social model" of design: Issues of practice and research. *Design Issues*, *18*(4), 24–30. doi:10.1162/074793602320827406

Mulgan, G. (2007). *Social innovation: What it is, why it matters and how it can be accelerated.* London, UK: The Young Foundation.

Murphy, S. A., & Kumar, V. (1997). The front end of new product development: A canadian survey. *R & D Management*, *27*(1), 5–15. doi:10.1111/1467-9310.00038

Murray, R., Cualier-Grice, J., & Mulgan, G. (2010). *The open book of social innovation.* London: National Endowment for Science, Technology and the Arts.

Ortíz Nicolás, J. C. (2016). Diseñando el cambio. La innovación social y sus retos. *Economía Creativa*, *06*, 9–34.

Ortíz Nicolás, J.C. (2016a). Innovación social y diseño, una propuesta metodológica. *2° Coloquio Internacional, Las facetas de la evaluación.* Ciudad de México: Universidad Autónoma Metropolitana.

Ortíz Nicolás, J. C., Aurisicchio, M., & Desmet, P. (2013). Designing for anticipation, confidence, and inspiration. In *Proceedings of the 6th international conference on designing pleasurable products and interfaces* (pp. 31–40). ACM. 10.1145/2513506.2513510

Papanek, V. J. (1985). *Design for the real world: Human ecology and social change.* New York: Academy Chicago Publishing.

Ritchey, T. (2011a). About Fritz Zwicky. In *Wicked problems–Social messes.* Berlin, Germany: Springer-Verlag. doi:10.1007/978-3-642-19653-9_9

Ritchey, T. (2011b). Wicked problems and genuine uncertainty. In *Wicked problems–Social messes* (pp. 19–29). Berlin, Germany: Springer-Verlag. doi:10.1007/978-3-642-19653-9_3

Rittel, H. W. J., & Webber, M. M. (1973). Dilemmas in a general theory of planning. *Policy Sciences*, *4*(2), 155–169. doi:10.1007/BF01405730

Roberts, N. (2000). Wicked problems and network approaches to resolution. *International Public Management Review*, *1*(1), 1–19.

Ruskin, J. (1849). *The seven lamps of architecture.* Chichester, UK: John Wiley & Sons.

Schön, D. A. (1987). *Educating the reflective practitioner: Toward a new design for teaching and learning in the professions.* Chichester, UK: Jossey-Bass.

Schumpeter, J. A. (1934). *The theory of economic development.* Cambridge, MA: Harvard University Press.

Tromp, N. (2013). *Social design* (Unpublished PhD dissertation). Delft University of Technology.

Tromp, N., & Hekkert, P. (2016). Assessing methods for effect-driven design: Evaluation of a social design method. *Design Studies, 43*, 24–47. doi:10.1016/j.destud.2015.12.002

UNICEF Innocenti Research Centre (2007). *UNICEF: Child poverty in perspective: An overview of child well-being in rich countries*. UNICEF Innocenti Report Card 7.

Van der Have, R. P., & Rubalcaba, L. (2016). Social innovation research: An emerging area of innovation studies? *Research Policy*, *45*(9), 1923–1935. doi:10.1016/j.respol.2016.06.010

Wendt, T. (2015). *Design for Dasein: Understanding the Design of Experiences*. Thomas Wendt.

Whiteley, N. (1997). *Design for society*. London: Reaktion Books.

Yilmaz, S., & Seifert, C. M. (2011). Creativity through design heuristics: A case study of expert product design. *Design Studies*, *32*(4), 384–415. doi:10.1016/j.destud.2011.01.003

Zwicky, F. (1969). *Discovery, Invention, Research - Through the Morphological Approach*. Toronto: The Macmillan Company.

KEY TERMS AND DEFINITIONS

Design Brief: A document that provides a direction and scope to the task being completed; ranging from "open and ambiguous" to "closed and specific".

Design for Social Innovation: Design approach that aims to enhance the common good.

Design Method: A particular approach to address a design task. In a way it guides designers to generate solutions.

Ecology Model: Scalable theory of the context that deconstruct it on four systems: micro-, exo-, meso-, and macrosystem.

Fuzzy Front End (FFE) of Innovation: Work that is performed before the organization has begun to take the idea seriously; this is both in terms of resources and management attention.

Social Innovation: Type of innovation that aims to enhance the common good.

Wicked Problem: A multifactorial challenge where it is difficult to define its cause, and commonly experienced as design problems.

Chapter 18
A Creative Teaching Strategy to Generate Concept Designs and Their Possible Application in Tourism

Juan Manuel Madrid Solórzano
Autonomous University of Ciudad Juárez, Mexico

Ercilia Loera Anchondo
Autonomous University of Ciudad Juárez, Mexico

Porfirio Peinado Coronado
Autonomous University of Ciudad Juárez, Mexico

Ludovico Soto Nogueira
Autonomous University of Ciudad Juárez, Mexico

ABSTRACT

The aim of this chapter is to present the activities and results from a didactic strategy developed to generate concept designs that can be perceived as possible objects to be manufactured. This strategy is directed towards university students on an intermediate level with low performance in the industrial design major. Similarly, this chapter analyzes the importance of creating concept designs for promoting touristic destinations through a logo. The methodology for the present study is of qualitative nature and employs action research. Some methods stemming from a focus on user-centered design and creativity procedures were utilized. The collage technique and a semi-structured interview were the tools utilized on focus groups. It was concluded that the pedagogical proposal contains activities that assist the student in generating ideas in a systemized way. These actions can be applied with different purposes such as creating a design with the aim of promoting a product and a touristic destination.

DOI: 10.4018/978-1-5225-5234-5.ch018

Copyright © 2018, IGI Global. Copying or distributing in print or electronic forms without written permission of IGI Global is prohibited.

INTRODUCTION

Students with low academic performance in the Industrial Design (ID) major at Autonomous University of Ciudad Juárez, Universidad Autónoma de Ciudad Juárez (UACJ), show deficiencies towards generating concept designs that reflect aspects regarding function, usability, aesthetics, and manufacturing at an abstract level. These minimum requirements are essential to conceive the possibility and viability of a design proposal. The students' inability to create concept designs is present due to three reasons: first, students show deficiencies in their competence to associate and produce ideas according to collected data about the consumer; thus, they cannot structure coherent ideas that can be depicted visually. Second, there is a lack of training in research tools that are frequently used to obtain an understanding of consumers´ necessities; these instruments include surveys and interviews, which require the use of statistics for their design and validation. Students need to be knowledgeable in the methodological aspect in order to avoid mistakes due to intuition in the data analysis. The third reason, is that students do not have guidelines for the selection of ideal manufacturing materials and processes to demonstrate the viability of a concept design.

The didactic strategy in the present study has the intention of responding to the problem faced by an artisan association in the community of Creel, a town located in the highlands of the state of Chihuahua in Mexico. The purpose is to improve the design of products currently offered by artisans in order to break into a segment of the market or town's population which presently shop in supermarket chains, such as Wal-Mart or Soriana.

In view of the foregoing, the design of a pedagogical proposal for the conceptualization of industrial products should fulfill four qualities: (a) explore the expectation of consumers through the use of qualitative research tools, (b) offer a procedure for the analysis and interpretation of collected data in a population under study, (c) orient towards the proposition of a considerable number of sketches, and (d) provide students strategies to elaborate guidelines for the development of design proposals which express relevant and coherent features of the manufacturing process. Consequently, the objectives of the present study are the following:

1. To propose a creative didactic strategy for students to generate design proposals that, in general, fulfill functional, aesthetic, structural, and manufacturing requirements, through a pilot test with learners in fourth and fifth semesters.
2. To identify the advantages and disadvantages of the didactic strategy that the students carried out.
3. To analyze the importance of concept design in the creation of a logo as part of destination branding in order to promote a touristic destination.

This didactic proposal must not be considered as a design method; it is a reference to be used by teachers who encounter similar problems with students who struggle to create sketches. It is a procedure that intends to provide a general vision regarding concept designs that are viable for the solution of a problem that befalls a social group. Therefore, this strategy does not include activities planned to satisfy requirements for commercialization, transportation, sustainability, legal affairs, and regulated issues.

BACKGROUND

In view of the previously stated, it becomes clear that concept design is a key step in the manufacturing process of a final product. This fact leads to address concepts and theories that will provide the theoretical basis for the subject under study.

Concept Design

The term of concept design can be conceived from different perspectives. For example, for some authors, it is an idea that suggests its utility to solve a problem. It also must show manufacturing, functional, and aesthetic requirements for its physical creation (Rodgers & Milton, 2011; Keinonen & Takala, 2010). Moreover, most professors at UACJ have defined "concept design" as an idea, expressed in a sketch, which represents a notion of how to satisfy a necessity, and with the possibility of being manufactured (Lom, Macías, Madrid, Martínez, & Soto, 2009). The previous definition arises because of their working experience in the local twin-plant industry, in which a design idea should meet two features: the capacity of satisfying a necessity and the viability of being manufactured, leaving under-valued aesthetic requirements of a product. Therefore, for the present study and from the perspective of the industrial design field, a concept design is defined as an idea directed towards solving a problem in relation to a service or an artifact. Similarly, it should show an approximation of the aesthetic, functional, structural, and manufacturing requirements, as well as those of usability, to such an extent that it can be considered as an idea to be perfected with the application of a design methodology (didactic strategy) that will be discussed below.

Didactic Strategy

Nowadays, the learning process of university students is different to that of years ago. It finds itself influenced by scenarios presented by the so-called "information society" (Rodríguez & Pérez, 2015). For example, there are different means of electronic communication to satisfy the necessity of knowledge; however, the student does not experience the application of the acquired knowledge within a context of fluctuating situations such as those requiring decision making, which improves their personal and intellectual growth. One of the procedures to encourage that the student becomes a builder of their own knowledge in changing environments is through the means of didactic strategies; besides being beneficial to the learner, these become an important support for teachers when they wish to develop a well substantiated, dynamic, and constructive class (Criollo, Romero, & Ruiz, 2017). This type of teaching leads to a harmonious learning, as well as to a disposition and concentration that facilitate the assimilation of the topics; therefore, the student will be active at all moments of the academic activities carried out inside and outside of the classroom. This will provide foundations for knowledge built from not only the instructor's explanation, but also from the student's attitude (Rivera, Cuartas, & Renteria, 2011).

The use of didactic strategies in different disciplinary areas is widespread. For example, they have been put into practice when developing critical thinking abilities in midwife students and when connecting theory with practice (Rankin & Brown, 2016). They also have been employed in accounting and finance students to improve the self-directed learning ability with the use of Communications and Information Technologies (Mulet, Domínguez, & Flores, 2015). Didactic strategies have been implemented to raise the grade point average in high school students in the chemistry subject (Ramos et al., 2016). In the design discipline, a didactic strategy was applied for graphic design students to comprehend the theory and application of semiotics based on the formulation of a real design problem; the steps consisted of reading theory of semiotics, collecting data from a population under study, data analysis and generation of design proposals. The results show that most of the students expressed that the activity enabled them to gain an understanding of semiotics and that they had better communication with the teacher (De la Cruz & Mejía, 2017). Hence, a didactic strategy allows the teacher to decode complex topics for students, to relate the concepts with real life applications, and to improve student/teacher communication.

Collage and Focus Group

One of the central aspects to consider in the didactic proposal is to explore the people`s expectations regarding a product. In order to carry out this task, the methodology proposed by Costa, Schoolmeester, Dekker, and Jongen (2003) will be used. The authors combine two qualitative methods: the focus group and the collage technique which are referenced frequently in the literature as appropriate to obtain information about the consumers´ necessities (Maguire & Bevan, 2002); another advantage is its ease of use (Designkit, n.d.). This strategy has also been employed in graphic design to create promotional images for touristic places based on identification of tourists´ perceptions (Prebense, 2007). Likewise, these two tools are considered suitable for stimulating memories or thoughts through the interaction with other people and the use of images, situation that does not take place when answering a survey (Bruseberg & McDonagh-Philp, 2002).

Grounded Theory

A crucial activity within the didactic proposal is the development of relevant meanings drawn from collected data. Grounded Theory will be used for this purpose, According to Cuñat (2007), Grounded Theory is a qualitative research method which, through induction, produces an explanatory theory of a determined phenomenon. The author points out that one of the steps consists of the classification of the collected data in categories; each category is configured with words, phrases or images that are similar, different or show a degree of consistency; next, when comparing the categories, an area of interest can be identified to explore. Creswell (1998) adds the theory that results from the interrelation of categories can take the form of hypotheses, propositions, or a visual picture.

Forced Relationships and Combination Techniques

Galindo (2009) proposes a didactic intervention to orient students towards generating sketches through the use of Forced Relationships and Combination techniques; these are implemented on students without experience in creativity methods. The Forced Relationships technique relies on creating sketches based on observing objects which are shown randomly; the results presented by the author describe that the generation of ideas is more fluid and with less tension from the students. The Combination technique consists in manipulating the concept designs originated through Forced Relationships, with the purpose of unfolding other proposals. The goal is to obtain a high quantity of ideas from which the most viable are selected, even though some proposals might not be able to be manufactured. Immediately after the last step, the teacher and the student select the sketches together; in this way, they discard those that do not fulfill the functional requirements of the designed object, as well as those that do not satisfy the environmental conditions in which the object will be used, and those that are less likely to be assembled. Afterwards, there is another selection filter where a scale model of at least three chosen proposals is made; the least likely to fulfill the functional requirement are eliminated.

Logo and Branding

The main issue to consider in this study as an essential part of a product to manufacture is the one related to its image. The creation of a logo and a brand is not only intended to distinguish the product from others, but also to promote the place where it is manufactured, in this case the community of Creel, Chihuahua. For this reason, the following concepts are presented below:

Logo

According to Becer (2002), logos "are defined as symbols that carry a brand or emblem characteristic, which serve to promote a product, company, or service" (as cited in Denecli, 2015, p. 321). However, to Martins (2000), the logo, besides being a symbol, can be a trademark or a visual element; the author identifies it as a component of the brand that helps to recognize it (as cited in Efer, 2017).

In addition to the definitions that have emerged, the diversity of logos that have been created has resulted in a more detailed classification of the concept. For example, Rivers (2008) identified "...five basic types of logo: text only, image only, text and image, abbreviation only and abbreviation and image" (as cited in Tero, 2012, p. 65). Nonetheless, these logos must not be perceived as a type of design that will work by itself to be present in the consumers. In this sense, for Kelly (2017), the logo is a crucial element when it comes to creating branding; the author highlights the importance of not treating the logo as an isolated component, but to integrate it as part of "visual structures used to create meaning".

Branding

Branding is a concept that has evolved throughout history based on a variety of changes. Silva (2011) in her review of *Qué es el branding?* (What is Branding?) by Healey, affirms that the brand implies a "promise of satisfaction" and has gone through transformations due to the social, cultural, political, and economic changes that have had effects on the commercial culture or the way to do commerce. Considering the previous, companies have evolved and have focused on being innovative, as well as reaching excellence and quality. Now, it can be observed that branding is applied to "corporations, destinations, and public figures" Silva (2011).

Efer (2017) highlights the importance of branding by associating it in an inherent way with the aspects of merchandise like "value, service and function" which are aimed towards having a role in the life of the consumers including their habits, attitudes, moral standards, manners, etc.

This view reveals the influence of branding on the consumer; therefore, its design has required the incorporation of elements as the ones presented by company brands. According to Tero, the specific components of a brand of this kind are the following: an image as a component for the most efficient and the fastest manner to recognize the brand; this image replaces the brand most of the times. The other component is the name of the brand, and lastly, the slogan. The author also affirms that acknowledging these three elements, it is how most logos correlate to brands.

In relation to a touristic destination, the brand also plays an important role in its promotion; in this case the concept which has been coined is "destination brand". Ritchie & Ritchie (1998), (as cited in Blain, Levy, and Brent, 2005, p. 329) define it in the following way:

A name, symbol, logo, word mark or other graphic that both identifies and differentiates the destination; furthermore, it conveys the promise of a memorable travel experience that is uniquely associated with the destination; it also serves to consolidate and reinforce and recollection of pleasurable memories of the destination experience. (p. 103: bold added)

As can be seen, the key verbs in the definition are identify and differentiate, which intend to maintain a solid association of the brand with the place in the visitor´s mind. Similarly, experience is also an essential term for the authors, due to the fact of how our connection with a touristic destination involves all our senses; consequently, these good or bad experiences we have will be associated with any vacation spot that relies on a brand. They eventually will end up consolidating our image regarding that place.

METHODOLOGY

The implementation of the proposed didactic strategy included the participation of nine male and nine female students. They were enrolled in fourth and fifth semesters in the ID major, showing low performance in classes.

In the first part of the five-step process, this group of students was instructed in a classroom for a period of two hours. In this class, they learned how to apply the collage method to subsequently implement it to the participants outside of college grounds. In this session, the students simulated the population under study: people who buy handcrafts. By the end of the learning session, the students decided that the characteristics of the sample under study were the following: a) subjects were friends or family, for accessibility reasons b) participants who frequently shopped in supermarkets and c) they had acquired a handcraft at some point. Likewise, it was agreed that three to six people would be invited to form one focus group. Taking into consideration that there were eighteen students, eighteen groups were formed. The learners concluded that the maximum number of participants would be six, which facilitated control over the group. The testers also decided that step 1 would not last longer than two hours, as to accommodate the schedules of the participants in the focus group. It should be noted that it was possible for one focus group to form two teams; this led to the potential creation of two collages per group. The teams were principally made up by family members who displayed the previously mentioned characteristics; furthermore, the participants were mostly females with various academic degrees.

Step 1: Data Gathering

The first step of the didactic proposal was the application of the collage method in a focus group in the interest to explore the expectations of a population under study –in this case, people who buy handcrafts–. This method aimed to register beliefs related to any condition that resulted in frustrations, desires, unease, shortcomings, etcetera. It is important to mention that the students applied the method with their participants during the weekend and off university grounds; the average time for the creation of the collage was an hour. The four activities present in step 1 were as follows.

Activity 1

The student explained the objective of the study and the activities to be performed to the participant sample. He/she also requested authorization to record the collage presentations in the second-to-last activity.

Activity 2

Each student formed teams so that all focus groups made at least one collage. Students provided magazine cutouts that participants would then glue on a sheet of a flip chart. The directions for the construction of the collage were the following: (a) each team member was required to paste images that represented positive experiences on the right side of the sheet and negative experiences on the left side; (b) the group, as one, picked images that represented both positive and negative meanings and then pasted them in the center.

Activity 3

Each team gave a presentation explaining the reasons and criteria for the elaboration of the collage. In this phase, the student recorded the participants' explanations and annotated comments on them. It is worth noting that Costa et al. (2003) propose making a transcription of the recording. This task was not performed due to the magnitude of work that would result in writing each team member's dialogue verbatim; it would have taken a considerably large amount of time.

Activity 4

The student concluded the activities with its focus group by asking open questions to every participant with pursuance to clarify any doubts respecting what was mentioned during the presentation; the student also took notes about the relevant points in the testimonials. Finally, the student closed the session with a word of thanks.

In Figure 1, an example of a collage developed by four participants (three females and a male) is displayed; this one shows a larger number of images on the right side, which symbolizes a positive attitude towards handcrafted objects, referring to them as unique and elegant; however, on the left side they display the fragile nature of these.

Step 2: Categorization

In the following step, the student conducted the analysis and interpretation of the gathered data in the collage method. The Grounded Theory method was used for the analysis of the recordings and notes taken on the first impressions of the focus groups. The student carried out two, two-hour long sessions within a week, in which four activities were performed. Each of the activities is described below:

Activity 1

The gathered information was organized into categories with the purpose of finding recurring patterns or similarities. Each category was assigned a label, or title, for easy recognition; for example, "decorating

Figure 1. Collage made by a focus group

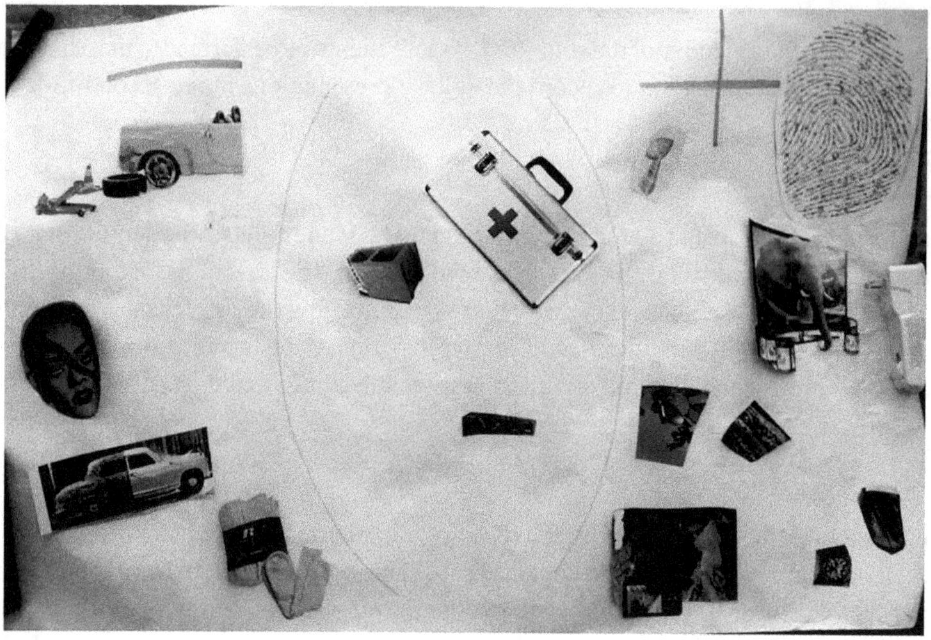

home." This category was created due to the prominent presence of phrases in the data that referenced the decorative utility of handcraft in a home.

Activity 2

The student linked elements in one category with elements from others with a view to create a comprehensive compilation of short phrases that depicted ideas, meanings, needs, wants or frustrations from the participants. For example, some people said, "A handcraft is made with local materials," others said, "Handcrafts are easily broken," "A haversack is hand sewn."

Activity 3

The phrases from the students were grouped once again into categories; subsequently after linking each of the newly created sets of expressions, students created a description which sought a finding or a problem that is solvable through the industrial design discipline. Another set of information that was derived from the formation of categories relates to the location or objects in the surroundings that affect that situation, also referred to as the 'context'. Lastly, it was necessary to explain the implication, or the proposal to the finding or the problem. The following is an example of a phrase that contains a finding, a context, and an implication: "An adult does not buy a handcraft if it is not useful in a home space, whether it be in the bedroom or the study room." For the previous example, the finding is "An adult does not buy a handcraft if it is not useful in a home space," the context is, "In the bedroom or the study room," and the implication is: "Thus, it must lead to the proposal of a handcraft that has a decorative and an additional usage value."

Activity 4

In the next step, the student related and compared the previously created sentences in order to specify the challenge to be designed or solved; in other words, define the design problem. Bearing in mind the previous example, the design problem was the following: "The handcraft must contain two values of usage: one as a home decoration and the other as an artifact that can be used (finding); but it should also display its own handcraft characteristics for the study room or the bedroom (context)." It should also be taken into consideration that since it would be manufactured with local materials, handcraft should show shapes or motifs that identify the region or community where it was crafted. By meeting the characteristics that are mentioned in the findings and context parts, it can be concluded that a handcraft that possesses a decorative utility at home must be designed (implication). This is one of the various features that students came up with. After specifying the design problem, the student then proceeded to propose a concept design that could give an approximate solution to the given problem; for example: "Designing a desk lamp with Tarahumara identity." Afterwards, the student created a hierarchized list of criteria in the interest of evaluating the sketches selected to be designed in step 3, which will be used to select the most feasible to develop. The listing included short sentences that reflected the functional characteristics as well as aesthetic requirements of the concept design. These criteria were formulated from data gathered in the focus groups: How does the product work? How and where will it be used? What shapes and forms must the product have to be identified as Tarahumara handcrafting? and What materials must be used? From these questions, some characteristics of the possible object to be created arose: 1) usage of green, brown and blue colors that represent nature, 2) placement of an opening in the packaging that allows the consumer to feel the texture of the handcraft, 3) usage of natural local materials for its manufacturing, 4) dimensions appropriate for a desk and for lighting small spaces, 5) usage of appropriate shape and form that identifies it as Tarahumara handcrafting, among other characteristics. For hierarchizing criteria, the students set in first place that which is considered the most viable for providing a solution to the given design problem.

It is important to clarify that, as mentioned before, in Grounded Theory, the result of the categorization process is the generation of a theory, or hypothesis. In the case of the industrial design field, the outcome relates to the phrases that describe a problem, a context that describes the elements affecting that problem, and the implication, which is the proposal to the problem. Therefore, based on the information that was generated, the most relevant hypotheses that concern marketing in different objects are presented.

1. For the creation of packaging that can protect, transport, safeguard, highlight and improve the visual aspect of a Tarahumara handcraft, it is considered the market target, which is international and domestic tourists of female and male genders within the age range of twenty-five to fifty-two years old.
2. The use of a higher variety of color in handcrafts will have a greater acceptance from a higher number of users; by consequence, the variety of colors that identify the craftsmen's town will result in a higher probability of product placement in supermarket chains.
3. Consumers ranging from eighteen to thirty years old wish to buy products used frequently in their day-to-day life with decorations made by craftsmen. Thus, in order to fulfill the usage and decorative values, the craftsmen must incorporate materials, colors, and symbols that identify their culture in the crafted objects.

4. The young consumer, which ranges from twenty-one to twenty-seven years old, identifies that a product with a brand or logo represents reliability of purchase, that is, they perceive that the product is of good quality and is protected by warranty.

Step 3: Creation of Sketches

As it was mentioned before, Galindo (2009) proposes the techniques of forced relationships and combination for the creation of sketches. The objective is that the student does not create them without a concrete direction. In this step, only the Forced Relationships technique was used. The Combination technique was eliminated since, as the author affirms, it is complex: it has a great number of repetitive alternatives generated, and it increases the time spent in selecting viable sketches. It is important to mention that in the intention of saving time and effort in the present study, the Forced relationships technique underwent three modifications. The first modification concerned removing the step in which developing functional models are formed, as to minimize students' expenses; the second dealt with the items displayed to students: physical objects were substituted by digital images due to the accessible nature of these. The images had to allude to different concepts among themselves (insects, flowers, tools, etc.) to avoid projecting similar alternatives; the last one, the process of producing combinations was thought as a complementary possibility to improve a final proposal, by fusing it with any of the physical attributes of the eliminated ones.

In this case, six digital images were displayed randomly with the help of a projector. Each student created five sketches for each of the six pictures shown, which resulted in thirty illustrations per student. The time allotted to present an image and create the five drawings was twenty-five minutes. Collectively, one hundred and fifty minutes, equating to two hours and a half, were dedicated for the creation of sketches. Two class sessions of two hours each were used to carry out the task. In the first session four images were shown: two of them referred to insects, and the other two to furniture. In a following session, fifty minutes were dedicated to present the rest of the images, one alluding to a gem and the other one to a flower. The rest of the time was allocated to elaborate on sketches that students considered appropriate.

Step 4: Selection of Sketches

Once each student completed the required number of sketches, they selected the one they considered most suitable to solve the problem. This step consisted of four exercises:

1. Each student placed the sketches over their table for easy visualization.
2. Next, each learner grabbed the listing of hierarchized criteria that was originated in the last activity of step 2, with the intent to organize the sketches in the following way: according to the student's own judgment, the one considered as most adequate to satisfy the primary criteria was ranked in first place, and the least feasible was ranked in last place. The time assigned for this exercise was two hours.
3. After organizing the sketches, the first three were selected; then, students had to infer whether it would be possible to combine components from the selected proposals for the configuration of a better idea; otherwise, one of the three was selected.

4. Finally, it was required that they elaborated, in a speedy manner, on the final sketch by providing color, shade, a brief description of the object, and any other attribute that highlighted the physical product's appearance.

Step 5: Preliminary Evaluation

This stage involves performing a preceding assessment of the manufacturing process that restates necessary changes that must be conducted on the chosen sketch and, thus, evaluating it as a potentially manufactured object. An industrial engineer with experience in industry assessed the students in their decisions. The allocated time for this step was four hours that were divided into two, two-hour long sessions in one week. The process undertaken consisted of the following actions:

1. **Exploded-View Drawing of the Conceived Object:** Students rapidly drew each component of the object.
2. **Hierarchizing the Components:** A classification of pieces was made based on the following question: Which components would be taken into consideration first to begin manufacturing the concept? Afterwards, those that began with the manufacturing of the artifact were placed as first on a list, leaving the parts that finished the assembling in last place. Then, the students were asked to imagine being in a workshop where they crafted their proposal from the beginning of the process to the end, which included tools, equipment and machinery to be used. The students verbalized materials and methods for the construction of the product to clarify thoughts and reaffirm taken decisions. The purpose of these activities was to propose the changes made to the pieces by the advisor and the student. The changes were mainly related to the geometry of the components, to assure that it would consume the least amount of material, workmanship, equipment, and machinery for manufacturing. Therefore, the student made a new sketch with the suggested changes.
3. **Standardization of Components:** The student standardized forms and dimensions of components to increase probability of low-cost production.
4. **Material Exploration:** Through this action, students performed visits to regional businesses to identify the most commonly used materials for the manufacturing of products whose designs were similar to the schemed proposal. Through this process, a benchmark was set to acknowledge the most accessible and affordable materials in the local market. The previous was done by observing photographs taken of products that were identified in handcrafts shops and supermarkets; subsequently, the students, aided by the engineer, analyzed the gathered information to select the manufacturing materials. With the results from actions 2, 3 and 4, the student made the pertinent modifications to the sketch created in action 2. This last activity was undertaken at a separate time from the two assigned sessions for step 5 during the same week.

Upon completion of the five steps, students were asked to elaborate on their final sketch on paper by hand in order to be evaluated by specialists in the manufacturing area; however, students proposed moving their sketch to a digital format, arguing that this is the way they presented final projects in other subjects of their major. A concept design for a lamp is shown in Figure 2 shows the sketch of the lamp and Figure 3 shows the same object after completing the didactic strategy.

Figure 2. Example of a concept design proposal
Created by Manuel Heriberto Ramírez López.

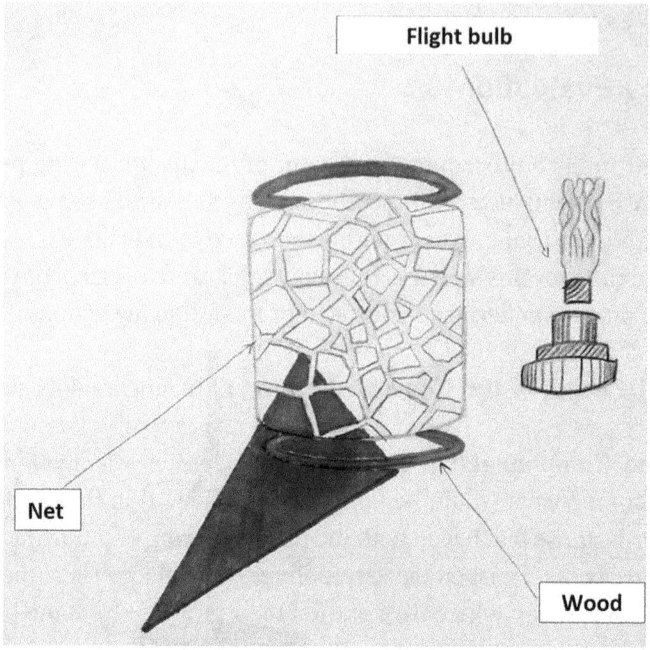

Finally, with the purpose of assessing whether the didactic strategy had fulfilled its objective, teachers from the ID major were invited. One industrial engineer and two industrial designers, with work experience at twin plants, were in charge of evaluating the thirty concept designs created by students. The activity lasted a session of two hours. At the beginning of the session, the professor explained to the evaluators the aim of the task and the problem regarding the artisan community in Creel. Afterwards, the instructor asked the specialists to observe carefully each of the sketches. Then, based on their own parameters and in a consensual way, they selected the sketches that met the characteristics that a competent concept design must possess. Lastly, a resolution was reached and comments regarding the taken decision were given.

Estimated Costs

During the execution of the pilot test in step 5, one of the students proposed developing an activity to generate an estimated cost of production of the object with the possibility to be manufactured. Even though this point was not contemplated in the didactic strategy by the professor, the suggestion was accepted. In this way, the student would identify whether his concept design would be suitable to manufacture the number of objects requested by the market segment. In addition, this task would help him to determine whether the price was adequate to compete with existing similar products in the region. The object shown in Figure 2 (b) was used to calculate the estimated cost.

In this case, the student was asked to repeat the activity in step 5, which consisted in imagining how the complete object from his concept design would be assembled, but in a more thorough way in which the manufacturing time for each task was also taken into account. Once the previous exercise was com-

Figure 3. Sketch of the lamp after completing the didactic strategy
Created by Manuel Heriberto Ramírez López.

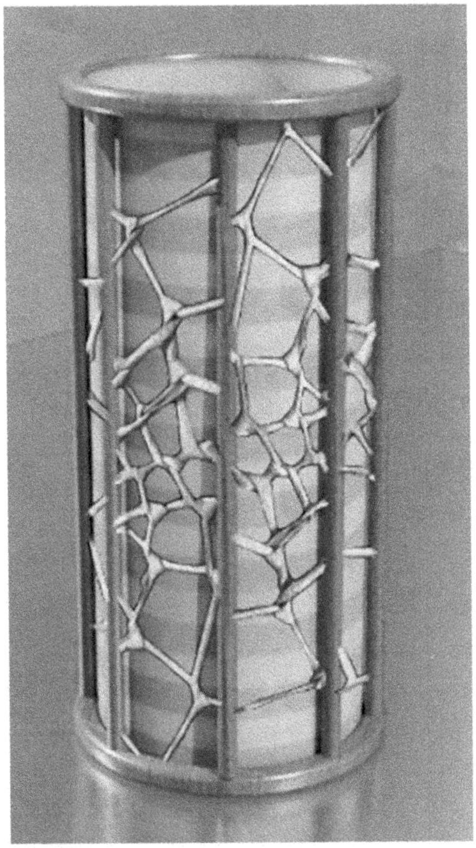

pleted, the student concluded that a person could produce forty units in a period of four weeks. Then, in a list, he elaborated on the production consumables or materials needed to manufacture the forty pieces, as shown in Table 1. As can be seen, two main columns were created: Fixed Cost and Variable Cost. On the top left square of the same table (Fixed Direct Cost), the price of consumables that do not change in a monthly basis and that are directly related to the assembly of the concept design were noted down. For the purpose of this exercise, it is important to notice that the costs were calculated and shown in Mexican currency.

Next, in the lower left square (Indirect) those that were not directly related to assembly, such as pencils, advertising, among others, were placed. As can be seen, the Variable Cost column is also classified as direct and indirect; the supplies and utilities of the manufacturing process whose price could change at an established time were set in this category. It is important to mention that the concepts of Variable Cost and Fixed Cost which were used by the student are not necessarily linked to the definitions that could be found in the literature regarding project costs. He gained an understanding of both concepts from the professor's explanation while carrying out the activity. However, it was considered as appropriate to show the result of the idea for didactic purposes.

Table 1. Estimated costs to produce the object in Figure 2 (b)

Fixed Cost				Variable Cost	
(Direct)				(Direct)	
Jigsaw	1,995	Screws	2,160	Pine plywood	3,300
6 manual machine presses	754	Nails	3,750	Varnish	7,225
Sandpaper	750	Tape	198	Cow leather	1,650
Work benches	4,300	Lightbulbs	30,000	Vegetal paper	7,500
Scissors	275	Sockets	3,000		
Miter saw	4,845	Paint gun	1,615		
Sanders	1,330	Circular saw	1,150		
Nail gun	645	Cutters	36		
Soldering iron	160	Drill	1,035		
Glue	1,310	Welding wire	275		
Total			$71,584	Total	$19,675
(Indirect)				(Indirect)	
Advertising	1,000	Invoices	300	Water	250
Water glasses	100	Work coats	75	Electricity	400
Screwdriver	24	Glasses	60	Gas	50
Heavy-duty waistbands	380	Pencils	87		
Extension cord	275	Compressor	3,950		
Cellphone	7,000	Computer	10,000		
Rent	2,000	Salaries	4,000		
Air ducts	3,000				
Total			$31,251	Total	$700

Source: Created by authors.

RESULTS

In relation to the students' perceptions about the didactic proposal, they evaluated the steps of the strategy in a positive way, recognizing it was useful for their academic preparation. However, over half of them thought it was necessary to reduce the time of the didactic strategy. Although some techniques could be useful, they mentioned they would not employ this method in their professional practice because of the time factor.

When analyzing the results particularly, some students commented that step 1 was very dynamic, to such a degree that it motivated the studied population to explain their necessities and to propose solutions in an open way, contrasting other methods used previously.

Concerning step 2, students judged that the activities of creating categories and phrases used up too much time; thus, they suggested creating similar actions that replaced them and were developed simultaneously with step 1. In contrast, the creation of criteria lists to evaluate proposals was well perceived. Students defined it as a visual guide as a reminder of what they must focus on when making required changes to the sketch; in fact, five students designed their own format to express phrases. Figure 4 and

Figure 5 show two formats for the elaboration of a box design to store handcrafts. Figure 4 shows a triangular box and Figure 5 shows a rectangular box. In both parts, two columns can be seen: the right column shows the most important phrases of the hierarchized list that must be fulfilled by a sketch. It can be observed how students added drawings as a manner to link the verbal expressions to physical attributes. In the left column, they named the category to which the phrases and images belong.

Relating to step 3, students considered that the allotted time to sketch was insufficient. In the same way, they indicated that it was not necessary to create as many sketches when visualizing images. They suggested that it would be better to stop the creation process when similar drafts appeared and create new combinations based on the existing sketches. Once this was done, the combinations could be once again hierarchized with the criteria guide derived in activity 4, from step 2. On the other hand, opinions regarding the technique of presenting images as a guide to stimulate sketching were positive. In this sense, students commented on how this activity enabled them to create sketches in a concrete direction; however, they questioned the selection of presented images: What would have happened if the images contained different geometric forms to the ones presented? Students concluded that the results would

Figure 4. The first criteria list based on the students' proposals

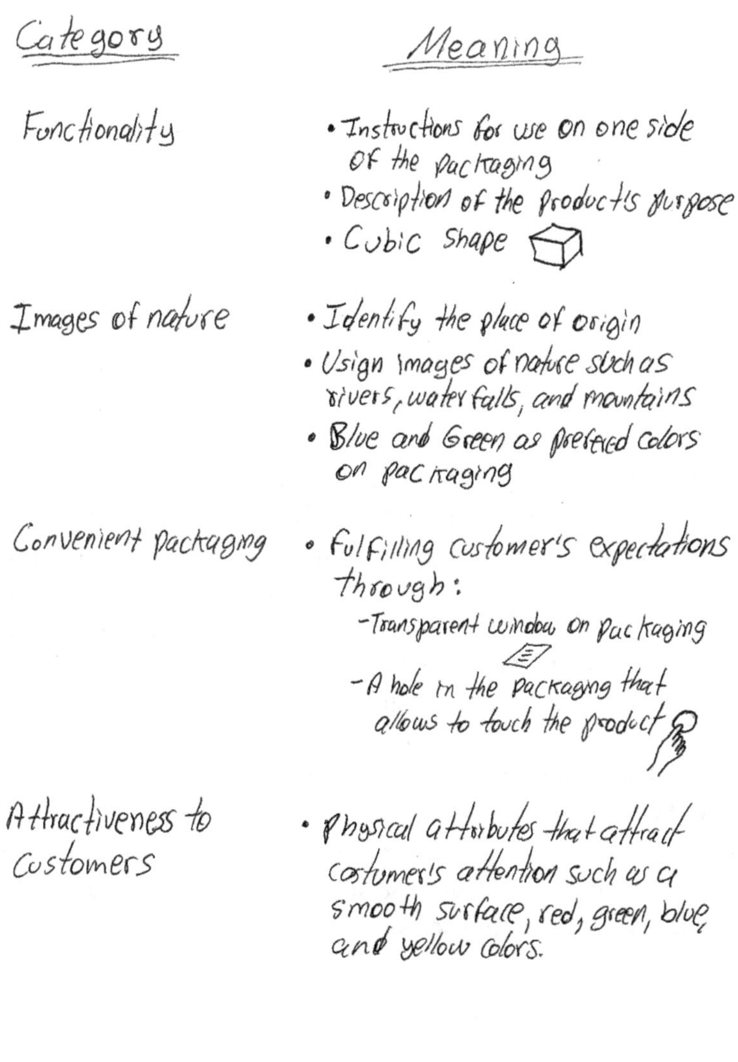

Figure 5. The second criteria lists based on the students' proposals

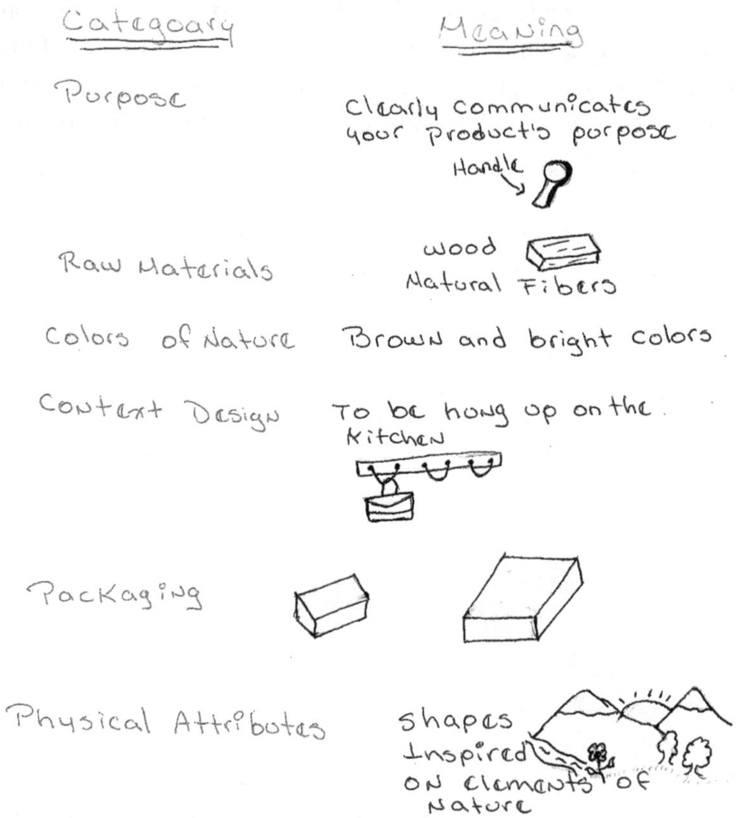

have probably been different. On average, eleven different alternatives emerged from each student in the pilot test.

Regarding step 5, in the beginning it was thought that tasks at this point would be some of the most complicated, due to the supposition that students were not acquainted with multiple manufacturing processes; however, the previous knowledge acquired in wood and metal workshops facilitated the job. Most of the participants in the pilot test commented that this step was perceived as feedback; it allowed them to notice the flaws in the design proposal. It was also mentioned that the last action, which is the material exploration, was an exercise that they would implement in school projects.

Concerning the evaluation conducted by professors in the ID major, experts coincided that five proposals out of eighteen were perceived as feasible to manufacture and had high probabilities of satisfying the necessities of the user; therefore, they concluded that they were trustworthy to use, practical, had a design that generated interest, and looked resistant. Other six sketches were described as possible to manufacture, resistant, safe to use, practical, and convenient, but did not show an attractive style that reflected the designer's intervention. Regarding the seven remaining proposals, evaluators explained that flaws were observed in safety of use, convenience, resistance, and the geometry of one of the components which would complicate the manufacturing process; two of them were graded with a style that kindled interest. Thus, it was considered that these seven designed images would not be able to satisfy the costumer expectations. Taking this into account, it was concluded that the objective of the didactic strategy was not completely fulfilled because most of the proposals failed to generally satisfy one or more of the conditions related to the user requirements, as well as functional, structural, and manufacturing considerations.

Regarding the economic aspect for the manufacturing of the object in Figure 2 (b), the following results are obtained. The student concluded that the estimated costs were high to place the product in supermarket chains. This activity made the student reflect and think about modifying the appearance of his design in such a way that the product was competitive in the market. For this reason, it is suggested that this exercise is conceived as a reference point to demonstrate the viability of a concept design.

Finally, when the participants presented the collage that they were assigned, they mentioned that they would like to see a brand, as this is linked to the quality of a product. At that moment, the following question arose: Who would the consumer contact in the case of a defective product?

Moreover, as with the brand, students expressed that it was necessary to include a logo and place it in the product with the intent to generate an identity with a physical location. Although students did not specify the type of logo, as referred to by Rivers (2008), they pointed out the importance of one: A handcraft must give an idea of where it comes from, tell a story of the place it originates from, who makes it, and conveys a specific environment. In this sense, Norman (2004) mentions how objects must tell a story to be well perceived, and this is sought through a logo or a brand.

Bearing in mind that a logo or brand would make handcrafts more attractive, it is reasonable to believe that more people would buy them; therefore, the community of Creel, which is the context in this case, would be benefited. The favorable conditions in economic terms would not only have an impact on its artisans, but also on the region through the creation of a logo and a brand that promote it. These visual structures are essential components in what is known as destination branding.

CONCLUSION

According to the results, any type of didactic strategy similar to the one presented in this work must be taken into account with the following criteria: 1) simplicity in the processing of information, and 2) generation of criteria by the teacher to allow the student to create a high number of non-repetitive sketches. In conclusion, even though the didactic strategy did not accomplish the expected results, this provided redeemable activities that can be used in any part of the development of an industrial product.

The collage technique in a focus group facilitates the interaction and empathy between the researcher and the population under study; as a result, a barrier-free dialog is established to communicate wishes, frustrations, and possible solutions to the studied phenomenon in a fluid, live way. The technique of Forced Relationships that provides a strategy to stimulate sketching, temporarily drives students away from stereotypes and avoids the loss of patience in the first attempts to transfer ideas into images or vice versa. The implementation of the didactic strategy makes students reflect about the way in which they learn the best, in this case having to do with metacognitive processes which will be useful in their learning process.

In a similar way that the didactic strategy can be applied to the creation of a sketch for a product, its implementation can be directed towards different fields such as tourism. The design of images as part of conceptualizing logos and brands to promote a touristic destination is an essential element for attracting visitors. Logos have a relevant role when it comes to choosing a destination; they "effectively stimulate awareness and communicate desired attributes" (van Riel and van den Ban 2001, as cited in Blain, et al., 2005, p. 332).

Hence, this research indicates the importance of having the proper didactic strategies to be implemented in the classroom to obtain key concepts, ideas and meanings that will be used in the creation of designs and later in the elaboration of a logo or brand. These images will transmit the meaningful aspects of a place by consolidating its characteristics in customers and tourists. In the present study, the elaboration of collages in focus groups, the Grounded Theory method, and Forced relationships technique were used to elicit designs to make a product more attractive and functional.

Lastly, the effort of designing this pedagogical tool originated to aid the percentage of students that show deficiencies in their ability to associate and produce ideas based on gathered information. Taking notice of this, it is important to point out that it is the teacher's duty to provide the conditions to foster learning and provide strategies and techniques oriented towards a student-centered teaching.

RECOMMENDATIONS

As the development of a didactic strategy is the main concern in this study, the suggestions in this section are related to the steps that comprise it. In order to satisfy the first step of the pedagogical proposal that deals with the application of the collage method in a focus group, it is recommended to adhere to Costa et al.'s (2003) procedure. It is necessary to include two people performing two crucial tasks in the focus group: one person in charge of moderating the discussion and another keeping track of recording and note taking.

For step 2, it is necessary to investigate the pertinence of using techniques to organize and process knowledge in a graphical manner, rather than using the Grounded Theory method. For example, other alternatives would be the use of concept maps, spider graphs, and cause-effect diagrams. In the same way, as it was previously mentioned, in this phase the time spent to accomplish step 2 is a concern. According to learners, the activities carried out at this point take too much time. Since students live in a fast-paced society, finding a quick solution to a problem has become a habit, and it creates a similar expectation in their academic lives; in this case, it does so in the terrain of creativity and in the information process. Learners are discouraged when they notice that it is a relatively a long and slow process to analyze, interpret information, and generate sketches.

Another work to develop regarding the improvement of this didactic proposal is that of developing an activity in which consumers evaluate the visual appearance of the concept design. The objective is to determine whether the physical attributes of the handcraft are attractive for the market segment. This type of information is useful, as it provides a view about how the designed object is perceived compared to other products (McDonagh, Bruseberg, & Haslam, 2002). Based on this, students would make the pertinent changes in their proposals to increase their probability of being more competitive in the market.

ACKNOWLEDGMENT

Special thanks to all eighteen students who participated in the implementation of the didactic strategy and for the disposition they displayed to share every result for each step of the process. We are equally grateful for the cooperation of the professors who participated in the evaluation of the students' work.

REFERENCES

Blain, C., Levy, S. E., & Brent, J. R. (2005). Destination Branding: Insights and Practices from Destination Management Organizations. *Journal of Travel Research, 43*(4), 328–338. doi:10.1177/0047287505274646

Bruseberg, A., & McDonagh-Philp, D. (2002). Focus groups to support the industrial/product designer: A review based on current literature and designers' feedback. *Applied Ergonomics, 32*(1), 27–38. doi:10.1016/S0003-6870(01)00053-9 PMID:11827136

Costa, A., Schoolmeester, D., Dekker, M., & Jongen, W. (2003). Exploring the use of consumer collages in product design. *Trends in Food Science & Technology, 14*(1-2), 17–31. doi:10.1016/S0924-2244(02)00242-X

Creswell, J. W. (1998). *Qualitative Inquiry and Research Design: Choosing among Five Traditions.* Thousand Oaks, CA: SAGE Publications.

Criollo, M., Romero, M., & Ruiz, T. F. (2017). University students' self-efficacy for learning how to conduct research. *Psicología Educativa, 23*, 63-72. Retrieved 04 04, 2017, from http://web.a.ebscohost.com

Cuñat, R. J. (2007). Aplicación de la teoría fundamentada (Grounded Theory) al estudio del proceso de creación de empresa. *Decisiones basadas en el conocimiento y en el papel social de la empresa: XX Congreso anual de AEDEM.* Palma de Mallorca: Asociación Española de Dirección y Economía de la Empresa (AEDEM). Retrieved 03 17, 2015, from Dialnet: https://dialnet.unirioja.es/servlet/articulo?codigo=2499458

De la Cruz, L., & Mejía, G. M. (2017). Reflective didactic strategy to integrate semiotic theory and creative practice in graphic design education. Art. *Design & Communication in Higher Education, 16*(1), 83–97. doi:10.1386/adch.16.1.83_1

Deneçli, S. (2015). The Effect of Consumers' Attitudes towards Logos or Emblems on Brand Preference. *Journal of Marmara University Social Sciences Institute/Öneri, 11*(44), 319-335. doi:.10.14783/od.v11i44.5000080017

Designkit. (n.d.). *Methods.* Retrieved 05 24, 2016, from http://www.designkit.org/methods/25

Efer, O. O. (2017). 8. Industrial Design: The Roles and Factors of Aesthetics, Modeling, Styling, Product Brand and Branding in Design/Design Education. *Review of Artistic Education, 14*(1), 186–199. doi:10.1515/rae-2017-0024

Galindo, R. V. (2009). Creatividad en el diseño industrial. Experiencias con dos técnicas creativas. *Revista Digital Universitaria, 10*(12), 1-14. Retrieved 04 20, 2016, from http://www.revista.unam.mx/vol.10/num12/art83/int83.htm

Keinonen, T. K., & Takala, R. (2010). *Product Concept Design: A Review of the Conceptual Design of Products in Industry.* Springer Science & Business Media. Retrieved 04 10, 2017, from https://books.google.com.mx

Kelly, M. (2017). Analysing the Complex Relationship between Logo and Brand. *Place Branding and Public Diplomacy, 13*(1), 18–33. doi:10.1057/pb.2016.3

Lom, H., Macías, M., Madrid, S., Martínez, T., & Soto, N. (2009). *Áreas integrales de diseño industrial: Percepción, metodologías, manufactura y ergonomía*. Ciudad Juárez, Chihuahua, México: Universidad Autónoma de Ciudad Juárez.

Maguire, M., & Bevan, N. (2002). User Requirements Analysis: A Review of Supporting Methods. In *17th World Computer Congress* (pp. 25-30). Montreal, Canada: Kluwer Academic Publisher. Retrieved 05 24, 2016, from http://www.usabilitynet.org/papers/WCC_UserRequirements.pdf

McDonagh, D., Bruseberg, A., & Haslam, C. (2002). Visual Product Evaluation: Exploring Users´ Emotional Relationships with Products. *Applied Ergonomics*, *33*(3), 231–240. doi:10.1016/S0003-6870(02)00008-X PMID:12164507

Mulet, M. T., Domínguez, M. M., & Flores, E. F. (2015). Didactic Strategy with the Support of the ICT for the development of professional Skills in the Licentiate in Accounting and Finance in the Model Semiprescencial. *Pedagogía Universitaria, 20*(1), 71-81. Retrieved 06 04, 2017, from http://web.a.ebscohost.com

Norman, D. (2004). *Emotional Design: Why we love (or hate) Everyday things*. New York, NY: Basic Books.

Prebense, N. K. (2007). Explorign Tourists' Images of a Distant destination. *Tourism Management*, *28*(3), 747–756. doi:10.1016/j.tourman.2006.05.005

Ramos, E. Z., Torres, L. O., Guzmán, J. M., Salazar, M. N., González, R. G., Espinosa, D. H., . . . Aviña, K. P. (2016). Didactic strategies in teaching-learning: In respect to the study of nomenclatureof organic chemistry in students of the Atotonilco Regional High School. *Educación Química, 27*, 43-51. Retrieved 06 05, 2017, from http://www.sciencedirect.com

Rankin, J., & Brown, V. (2016). Creative teaching method as a learning strategy for student midwives: A qualitative study. *Nurse Education Today, 38*, 93-100. Retrieved 06 04, 2017, from www.elsevier.com/nedt

Rivera, P. B., Cuartas, M. C., & Renteria, Y. P. (2011). *Facultad de Educación y Humanidades: La importancia del uso de estrategias de aprendizaje*. Retrieved 06 04, 2017, from Universidad Católica Luis Amigó: http://www.funlam.edu.co/uploads/facultadeducacion/53_LA_IMPORTANCIA_DEL__USO_ESTRATEGIAS_DE_APRENDIZAJE.pdf

Rodgers, P., & Milton, A. (2011). *Product Design*. London: Laurence King Publishing. Retrieved 04 10, 2017, from https://www.ebscohost.com

Rodríguez, A. V., & Pérez, E. V. (2015). Didactic strategy to stimulate the personality education. *Educación Médica Superior, 29*(2), 340-350. Retrieved 06 04, 2017, from https://www.ebscohost.com

Silva, B. Y. (2011). Reseña de "¿Qué es el branding?" [Review of the book ¿Qué es el Branding?]. *Revista del Centro de Investigación. Universidad La Salle, 9*(35) 43-44. Retrieved 05 08, 2017 from http://www.redalyc.org/articulo.oa?id=34218346006

Tero, M. (2012). The Logotype, Fundamental Unit for a Brand's Visual Identity. *Scientific Bulletin of The Petru Maior University of Targu Mures*, *9*(1), 64–68.

KEY TERMS AND DEFINITIONS

Collage Technique: A procedure consisting of using a range of different images from magazines or other printed sources to visually express feelings or experiences to evaluate or describe how a product is perceived.

Creativity Technique: A procedure that generates innovative ideas that result in design proposals to solve a problem.

Didactic Strategy: An action plan that comprises methods, techniques, and organization of instructions to achieve an intentional, reflexive, conscious, and self-directed learning in any given teaching-learning environment.

Focus Group: A qualitative research technique in which a small group of people discuss a specific topic under the guidance of a moderator who asks predetermined questions. The researcher gathers information about the participants´ attitudes, experiences, and beliefs to be analyzed and interpreted later.

Functional Requirements: The minimum desirable qualifications a concept design should fulfill to ensure the proper operation of a product. These include mechanics, reliability, versatility, endurance, and finish.

Manufacturing Considerations: The minimum desirable qualifications a concept design should satisfy to ensure the means and methods for its manufacturing.

Pilot Test: The implementation of an initiative or activity for the first time with the purpose of learning about its viability and ponder the possibility of a subsequent improvement.

Structural Requirements: The minimum desirable qualifications a concept design should meet to ensure protection of the product´s parts and mechanisms as well as to provide functional stability to its structure.

Twin Plants: Subsidiary companies of foreign firms generally located in the United States. The sister company in Mexico receives materials or components to be assembled. Once the product is finished, it is sent back or exported to another country.

User Requirements: The minimum desirable qualifications a concept design should satisfy to ensure that a consumer will interact with the future product in a correct manner. These requisites include practicality, safety, repair, handling, anthropometry, ergonomics, and perception.

Section 5
Additive Manufacturing and Technology

Chapter 19
Methodology to Apply Design for Remanufacturing in Product Development

Karina Cecilia Arredondo-Soto
Universidad Autónoma de Baja California, Mexico

Rosa María Reyes-Martínez
Instituto Tecnológico de Ciudad Juárez, Mexico

Jaime Sánchez-Leal
Instituto Tecnológico de Ciudad Juárez, Mexico

Jorge De la Riva Rodríguez
Intstituto Tecnológico de Ciudad Juárez, México

ABSTRACT

This chapter discusses the definitions of environmental design, design for remanufacturing (DfR), as well as a case study to demonstrate the convergent point between these topics. Many products that are currently remanufactured were not designed with this objective, generating a complicated process that requires that the manufacturing engineers develop in a corrective way, modifications in the original design of the products related to its components and the process. The case study analyzes a product that was not originally designed as remanufacturable. The decision was made to develop a reconstruction process that fulfilled the characteristics of remanufacturing. Finally, in applying the DfR, it was possible to expedite the remanufacturing so that material planners do not fear to run out of good parts and have to order the purchase of new product, decrease the use of assembly details that are purchased at a high price, and thus save on the cost of remanufacturing. Analyzing this case and applying DfR implied a savings of 37% compared to the initial process that did not apply this tool.

DOI: 10.4018/978-1-5225-5234-5.ch019

Copyright © 2018, IGI Global. Copying or distributing in print or electronic forms without written permission of IGI Global is prohibited.

INTRODUCTION

The design of a product directly influences the way a value chain will be managed; the product innovation starts with the design phase and requires the development of generations with greater advantages: it is necessary to modify subsequent generations in order to facilitate the assembly processes, meet functional requirements, reduce costs and offer a more competitive product due to the high quality involved. Until a few years ago, when designing a product, logical thinking was to favor a design that would facilitate the manufacturing process, thus creating the DMA (Design for Manufacturing and Assembly). However, the inclusion of new industrial standards such as ISO 14000 and OSHAS 18000 added to the interest of companies to incorporate a sustainable vision has led to the association of innovative concepts to the DMA, emphasizing Green Design or Ecodesign, as well as others approaching to Design for EoL (End-of-Life), strategies among which the Design for Remanufacturing stands out.

Once a product is designed it goes through the life cycle stages as shown in figure 1. In the last phase the product final disposal is made, for this purpose, there are different options such as those included in figure 2. These options recover part of the value of the product previously considered as waste. The remanufacturing is evaluated as the most complete option under the concept of saving everything that can be recovered, by preserving the original geometry of the components, assuring the customer the resulting product meets the design specifications and therefore with the functional requirements, in addition, it offers a guarantee of quality of a new product to a lower price. And is that few concepts integrate so many desirable and attractive topics in one such as this, for example new business opportunities, high return on investment, job creation, technology development, green industry, low prices and sustainable manufacturing.

The remanufacturing process includes initial inspection, disassembly of components, cleaning of components, assembly (this phase includes machining operations, reconstruction of parts and change of critical components by new ones) and final test (Guide, 2000; Hatcher, Ijomah, & Windmill, 2014). Therefore, it is not only important the easy to assemble the product, but also facilitate the process of disassembly. It is necessary to reduce the time of disassembly and generate the least possible damage to the components, which implies a more efficient process. At this regard, Design for Remanufacturing has been addressed from the following approaches:

- Design for Disassembly (DfD).
- End-of-Life Decision Making in product design (EoLDM).
- Environmental Design (Design for Environment), which has been treated indistinctly to Eco-design or Green Design (GD) in the context of remanufacturing.

Figure 1. Life cycle stages

Figure 2. End of life strategies

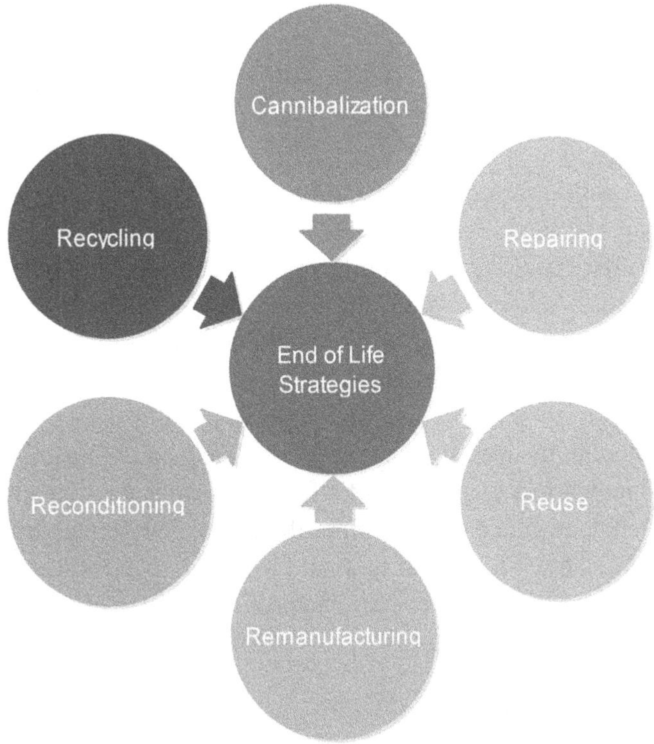

This paper discusses the definitions of Design for Remanufacturing and its associated strategies as DfD, EoLDM and GD; differences and similarities are considered because they are frequently confused with each other. Brewer & Arnette (2017) identify a lack of empirical testing for many keys Design for X (DfX) techniques, therefore the importance of analyzing the practices of the companies that apply these concepts. DfR activities are exemplified in the case study by describing the methodology to apply the design for remanufacturing in the development of products for an Original Equipment Manufacturer (OEM). Due to the novelty of the subject, many products that are currently remanufactured were not designed with this objective, generating a complicated process that requires that the manufacturing engineers develop in a corrective way, modifications in the original design of the products related to its components and the process, all this work promotes new generations of products (Matsumoto et al., 2016).

BACKGROUND

Design for Sustainability (DfS) promotes product innovation, in this regard, Ceschin & Gaziulusoy (2016) propose four levels: (a) product innovation level, (b) product-service system innovation level, (c) spatial-social innovation level and (d) socio-technical system innovation level. The product innovation level, is the design approach focused on improving existing or developing completely new products, the Green Design, and the Ecodesign belong to this innovation level. The DfS also incorporate economic, environmental and social aspects; Table 1 includes some of the main design techniques classified from the sustainability approach. This section addresses design activities considering the ecological approach, describing and highlighting their differences. Table 1, shows the taxonomy of design from sustainability approach.

Design for Environment (DfE)

It can be considered that within the ecological approach, DfE is the concept that implies more design activities; it implies that new products are designed to efficiently use resources at each stage of the product lifecycle; the primary goal is sustainability, defined as responsibility toward future generations (Rybaczewska-Blazejowska, Masternak-Janus, and Gierulski, 2016). The United States created the Environmental Design Program through the Environmental Protection Agency (EPA) in 1992 to (a) identify safer chemical products to use them as an alternative and (b) define and use best practices in production processes (cleaner, more efficient practices, and technologies). In this regard, the Life Cycle Assessment (LCA) is the most used tool in the measurement of the environmental impact.

Schöggl, Hofer, & Baumgartner (2017) identified nine key categories for sustainable product development: resource efficiency, resource consumption, the use of low-impact materials, optimization of the EOL phase, health and safety aspects, transport and logistics, social and ethical aspects, decrease of environmental pollution, and economic efficiency and profitability; the concept of DfE is also identified as Green Design.

Ecodesign or Green Design (GD)

The ISO standard 14062 on integrating environmental aspects into product design and development, defines eco-design as "… the integration of environmental aspects into product design and development (ISO, 2002). The objective is to increase the environmental performance, to achieve it the following model is proposed in ISO standard 14062: (1) Planning; (2) Conceptual design; (3) Detailed design; (4) Testing/prototype; (5) Production/launch on the market; (6) Product review (Goepp et al., 2014). In this regard, the EU is working to meet the energy efficiency targets scheduled for 2020. Due to the demands of the world market for more efficient products from an environmental perspective, the EU through its eco-design legislation seeks to eliminate less efficient products from the market.

Table 1. Taxonomy of design from sustainability approach

Economic Approach	Ecological Approach	Social Approach
Design for Procurement	Design for Environment (Green Design and Ecodesign)	Design for Ergonomics
Design for Manufacturing and Assembly	End-of-Life Decision Making in Product Design	Design for Social Responsibility
Design for Logistics	Design for Remanufacturing	
Design for Supportability	Design for Disassembly	

Source: (De los Rios & Charnley, 2016; Arnette, Brewer, & Choal, 2014; Brewer & Arnette, 2017).

Design for Remanufacturing

The design of the original product can determine the level of utilization that can be achieved by applying remanufacturing operations. DfR refers to the different design strategies applied to the product in order to incorporate features to facilitate the remanufacturing process (Matsumoto et al., 2016). DfR requires minimize damage to parts to be reused and isolate expected damage to removable and replaceable parts Shu & Flowers, 1999). According to Ijomah et al., (2007) DfR is a design activity focused on managing the factors involved in remanufacturing related to product design, its objective is to overcome the technical barriers to apply the remanufacturing process.

DfR involves changes that can focus on the product or the process. It is desirable the changes be made in the product design phase using less quantity and variety of material. However, many of the applied changes are related to the process; operational practices are improved through maintenance functions, efficient management, stream segregation, better material handling, inventory control, and training. Another strategy is the technological change: layouts changes, increased autonomy, improved equipment and new technology (Goepp et al., 2014; Hatcher, Ijomah, & Windmill, 2014).

Design for Disassembly (DfD)

OEMs and Independent Remanufacturers (IR) are involved in activities to repair and remanufacture products in the circular economy and to recover valuable materials. For this purpose, the products will have to be designed from the origin for the disassembly. The design process should consider how components will be separated, how the product will be updated for future applications (if it is desirable and possible), and what the component parts could become below (Chiodo & Ijomah, 2014). The key to successful DfD implementation lies in maintaining flexibility, separation of components should be easy as well as access to them (Soh, Ong, & Nee, 2016).

The disassembly process is the first operation performed on the products are remanufactured only after the inspection process where it is determined whether the product should be disassembled or not (Rybaczewska-Blazejowska, Masternak-Janus, & Gierulski, 2016). According to Chiodo & Ijomah (2014) with the aim of facilitating DfD it is necessary to consider: reduce the variety in quantity and type of components, standardize the use of fasteners and simplify the design to make components easily separable.

METHODOLOGY

The case study company is a remanufacturer of turbine components; it is divided into working cells and it is determined in a general way the common processes necessary for the reconstruction of the different components. The measurement of the performance of the remanufacturing process within the company is called Material Metric Discarded (MMD) Vs. Materials Induced (MI) to the remanufacturing process, and it takes the amount (in monetary terms) of the material discarded during the remanufacturing process and compares it as a percentage of the amount of material induced to the remanufacturing process.

The improvement project validation manual (internal document of the company) states the financial benefit of reducing the amount of material discarded is given by its variable cost, which is 84% of the fixed cost of the discarded material. It is also established that it must be verified the reduction in the material discarded will have a directly proportional benefit (in terms of costs) in another process of the company. In summary, the company states for every dollar saved in discarded material costs, the benefit of having saved that dollar should also be reflected in another area of the business. To cite examples: the increase in sales, or, the reduction in costs of raw material. In order to reduce costs for discarded material and for the purchase of raw material to replace this material, it is necessary to apply a methodology allows defining the problem, quantifying it and proposing solutions contributing to modify the original design of the product to facilitate remanufacturing, this through the recovery operations for products development. The methodology applied to design for remanufacturing through the development of operation of the product consists of six steps which are explained below.

Step 1: Defining the Problem

According to the internal procedures of the company, the quality module of its Enterprise Resource Planning System is used to declare and process all discarded material, regardless of origin or reason. This is the first problem for this research work since this process does not systematically distinguish the cause by which the material is being discarded. It may be due to some inherent condition of the part, which is perfectly within the scope of this research project; as well as maybe some defect in the reconstruction process, which is totally out of reach as it is a quality control aspect.

Step 2: Quantifying the Problem

In order to determine if the data that could be extracted from the Enterprise Resource Planning System in a systematic way would serve for this research work, a randomly selected sample of 288 transactions was taken and the following information was registered: serial number, code of cause, inspector's employee number, rejection text, process operation (from which it is being rejected), and work cell.

Step 3: Technical Personnel Survey

A survey of seventeen questions about the material disposal process was also generated, focusing mainly on the following aspects of the process:

- Acceptance criterion available (published).
- Acceptance criterion acquired by experience (not published).

Step 4: Repeatability and Reproducibility Study

A Study of Repeatability and Reproducibility by Attributes is carried out, in order to know if there is a considerable difference in the acceptance criterion between different workgroups. In order to carry out the study, a set of fourteen identical pieces was chosen, which were then evaluated by a group of people who together determined the acceptable and not acceptable pieces. Subsequently, the set was evaluated between three different employees to know the percentage of error in their own evaluation, as well as between them. The simple format for repeatability and reproducibility studies by attributes published by the Automotive Industry Action Group (AIAG) was used to record the results and to quantify error percentages.

Step 5: Reject Sampling Study

The company is divided into five work cells and a total of eight manufacturing engineers, each one with a specific number of parts exclusively in their charge. Therefore, it was decided to maximize the allocated resources and to study the rejections of discarded material based on the parts that most contributed in terms of costs to the measurement of material rejected by manufacturing engineer. To collect the data, it is necessary specifying the sample size, what will be measured and where the data will be taken.

Step 6: Results Analysis

The analysis of the results is carried out considering: the analysis of the material review process in the company, the business needs and the analysis of evaluation criteria and methods.

CASE STUDY

Step 1: Defining the Problem

The company requires reducing the amount of discarded material during the remanufacturing process while reducing the amount of components purchased as raw material. These components serve as substitutes for the discarded; therefore, the company needs to carry out a project to define the aspects intervening for the acceptance or rejection criterion of components. It is also important to determine the congruence of these criteria, as well as to define and stratify the causes of rejection. Finally, different options must be analyzed to modify the original design and to facilitate the remanufacturing process.

Step 2: Quantifying the Problem

Considering the total population of transactions used in a 12 month period to discard material, the 288 transactions used in the sampling yielded the following data: 274 of them were initiated when the piece was in the process of its disposal inspection, which represents 95% of the sample. It is worth noting the analysis and interpretation of this sampling made it possible to use of the entire population of transactions of discarded material for the other planned studies without the need to segregate it from a material that has been discarded for quality defects.

Step 3: Technical Personnel Survey Results

The survey of the technical staff was answered by 10 inspectors and nine engineers. The survey questions were intended to find out what staff was based on when applying their evaluation criteria; they affirmed the Acceptance Criteria Handbook and the own accumulated experience of each employee are the two most frequently used knowledge sources during the evaluation of the product. In second place was the level of training (documented) that has the inspector and the participation of the manufacturing engineer (during the disposal process). And lastly, the least used resource is to follow some review process of the material to be discarded.

Acceptance Criteria Handbook. Although the importance of the Acceptance Criteria Handbook has been set out as one of the two most frequently used criteria sources, it should be highlight in several of the responses it was concluded that the perception is that it is not entirely up to date and that the process for doing so is slow, almost impossible to have it updated. With the exception of one person interviewed, all others state that the Acceptance Criteria Handbook is not the only technical reference document used, but that they are a good reference to start from the experience acquired by the staff conducting the inspection. These other references used are:

- Other technical documents (drawings, reports, specifications).
- The accumulated experience of each employee.
- The capacity of the repair process.
- History of external rejections, by the end customer.

Own acceptance criteria. One problem found following the survey is the documentation of the criterion of self-acceptance; one has been created based on experience, the consultation of third parties, plans, reports of failure, specifications; if you get to document most of the time will be in the instructions of process and without endorsement of Design Engineering, who has maximum responsibility for the product. The technical staff accepted they are relatively consistent in documenting their acceptance criteria in the documents they control, in a less frequently way they also communicate their criteria verbally by means of electronic mail.

Development of new repairs. One-third of the manufacturing engineers surveyed agreed that they never spent their time developing new repairs that could extend the scope of repairs and thus increase the number of rebuilt parts.

Step 4: Repeatability and Reproducibility Study

Evaluation of the acceptance criteria. The results of the repeatability and reproducibility study were that the inspector one accepted three pieces that had to be rejected and had no differences between his evaluations. Inspector two rejected an acceptable piece, accepted a piece he should have rejected, and in three trials had differences in his evaluations. And finally, inspector three rejected three acceptable pieces, accepted a piece that he should have rejected and had no differences between his evaluations. The tabulation of the results of the repeatability and reproducibility study by attributes is available in table 2.

Step 5: Reject Sampling Study

Reject sampling study results. From the sample taken described in Table 3, it was possible to stratify the information in an appropriate way and focus on the areas of greatest opportunity, the time for the data collection was 12 months and the manufacturing engineering is in charge.

Stratification by cause codes. Based on the financial impact, 84% of the discarded material is typified under the following cause codes during the disposal inspection:

Table 2. Repeatability and reproducibility study results

Population		Operator 1		Operator 2		Operator 3		Y/N	Y/N
Sample	**Attribute**	**Try 1**	**Try 2**	**Try 1**	**Try 2**	**Try 1**	**Try 2**	**Agree**	**Agree**
1	Pass	Pass	Pass	Pass	Pass	Fail	Fail	N	N
2	Pass	Pass	Pass	Pass	Pass	Fail	Fail	N	N
3	Fail	Fail	Fail	Fail	Pass	Fail	Fail	N	N
4	Fail	Fail	Fail	Fail	Fail	Fail	Fail	Y	Y
5	Fail	Fail	Fail	Pass	Fail	Fail	Fail	N	N
6	Pass	Pass	Pass	Pass	Pass	Pass	Pass	Y	Y
7	Pass	Fail	Pass	Fail	Pass	Fail	Fail	Y	N
8	Pass	Pass	Pass	Pass	Pass	Pass	Pass	Y	Y
9	Fail	Pass	Fail	Pass	Fail	Pass	Pass	Y	N
10	Fail	Pass	Fail	Fail	Fail	Fail	Fail	N	N
11	Pass	Pass	Pass	Pass	Pass	Pass	Pass	Y	Y
12	Pass	Pass	Pass	Pass	Pass	Pass	Pass	Y	Y
13	Fail	Fail	Fail	Fail	Fail	Fail	Fail	Y	Y
14	Fail	Fail	Fail	Pass	Fail	Fail	Fail	N	N
Percentage appraiser score			100%		78.57%		100%		
Percentage vs. attribute			78.57%		64.29%		71.43%		
Screen percentage effective score Screen percentage effective score vs. attribute			57.14% 42.86%						

Table 3. Measurement plan

Factor	Metric	Source of Data	Sample Size	Additional Data
Component condition	Defect code Rejection Text Rejected part Amount rejected Operation rejected	Rejection Module	10 random rejects of the 5 most impact part numbers chosen by and manufacturing engineer.	Additional information that may exist in the traceability system for serialized parts.
Part costing	Cost Cost vs. cause code Value of the material Code cost	Cost-of-order report of discarded material	25% of the rejections analyzed by the manufacturing engineer.	
Effectiveness of the Acceptance Criteria Handbook	Handbook vs. condition vs. cause code Disposal process	Rejection module vs. Acceptance Criteria Handbook	10 random rejects of the 5 most impact part numbers chosen by and manufacturing engineer.	
Technical staff experience	Employee Manufacturing Engineer Other staff	Rejection module	10 random rejects of the 5 most impact part numbers chosen by and manufacturing engineer.	Name of any other employee who participated in the evaluation of the part.

A: Fractured or broken.
B: Erosion, excess of.
C: Functional test, does not comply.
D: Critical dimension, out of tolerance.

Stratification of material discarded by code of cause A. In terms of money the families of products most discarded by fractures and/or rupture are: combustion liners (30.5%), turbine diffusers (14.4%), turbine blades (11.3%), injectors (9.7%), cones compressor rotor (8.4%) and multiple fuel (5.9%). The study of this group of information also showed that the Acceptance Criteria Handbook is the most used technical reference, with 38%.

Stratification of material discarded by a code of cause B. The product families most discarded by erosion are combustion liners (28%), valves (21.3%), combustion housing (13.6%), stators (12.5%) and injectors (11.4%). The results of the study also showed that the Criterion Acceptance Handbook was used a third of these times, but also widely used other technical references as well as experience.

Stratification of material discarded by code of cause C. Material discarded for not functioning properly is mostly made up of fuel injectors (60.7%) and this is because its functional test is the flow test according to the planes of engineering of the respective models.

Stratification of material discarded by code of cause D. The families of material most discarded for being out of dimensional tolerance are number 2 bearing housings (43.7%), compressor housings (16.5%), combustion liners (16%) and rotor cones compressor (15.5%).

Stratification by working cell. The three cells that contribute 95.9% of the amount of material discarded are static (69.7%), rotating (16.8%) and housing and brackets (9.3%).

Continued with the stratification, now by cell of individual way has that in the static work cell,the families of product that most impact the metric of discarded material are: segmented nozzles (19%), diaphragms (14%), turbine booms (12%), compressor booms (11%), injectors (6%), combustion liners (6%) and nozzle shells (6%). For the rotating work cell, the most discarded product families are variables, compressor disks, compressor cones, central screws, turbine diffuser, and power turbine shaft compressor.

Finally, the stratification of the working cell and housings shows the product families that are most discarded and these are: diffuser (21%), compressor housings (20%), combustion housings (13%), support No.2 (13%), turbine collector (13%) and nozzle housing (9%).

Step 6: Results Analysis

Analysis of the Material Review Process in the Company

Following the first sampling, where it was validated 95% of the transactions made to discard material were made during the layout inspection, it is valid to declare only 5% of the discarded material is identified as such after the piece has been accepted for the reconstruction process and for the business it is no longer valid to think the material is discarded because of its unacceptable risk condition, but because of a deficiency in the inspection of disposition, when not being able to evaluate properly the piece, or, for a defect of process originated independent to the characteristics of the piece.

Like any production process, the accepted material after the layout inspection to the reconstruction process, in the event of a defect being detected, is rejected by an inspector and later by a Material Review Board (MRB) process, it is reviewed and depending on the type of defect it is discarded, repaired or accepted. There is no material review process for discarded material during layout inspection. Then, if the Discounted Material Metrics vs. Induced Material to the reconstruction process of being able to reduce would represent a great opportunity for the company, in terms of money the amount could be significant.

Business Needs

The multiple answers obtained from the survey being interpreted as business needs can be summarized in the following six points:

- Certification process of all technical personnel who makes the material disposal.
- Have only one source of acceptance criteria.
- Create a simplified feedback process to keep the Acceptance Criteria Handbook updated.
- Development of new repairs to key strategies.
- Do not discard material for cosmetic reasons only.
- Create a material review process for the material disposal process induced in the rebuilding process.

Analysis of the Evaluation Criteria and Methods

It is appropriate to recognize respect to the perception about the disposition process in in the company, it can be said that the quality of the inspection of disposal by company's technical staff is considered as good. Following the recommendations published by the A.I.A.G. for the interpretation of the results obtained when using its simple format for studies of repeatability and reproducibility by attributes, the necessity to train or re-train the technical personnel was exposed for the following reasons:

- It is unacceptable inspector 2 has not been able to repeat its own evaluations, regardless of whether or not the evaluation was correct.
- There were differences between the three inspectors who participated in the Study of Reproducibility and Reproducibility by attributes as a whole and this since the percentage of error as a whole was 57.14% having as a recommendation the A.I.A.G. That is 100%.

From the rejection sampling, a data was revealed, 32% of the total amount of material discarded was based entirely on the Acceptance Criteria Handbook. If it starts from the premise that this handbook is the only official document published by the Design Engineering department that should be used to accept or reject material, the answers obtained from the survey are reinforced where the problems related to the evaluation criterion can be summarized in the following points:

- Lack of unification of criteria.
- Use of informal means to communicate acceptance criteria.
- Documentation of new criteria under a single process.
- Failure to share responsibility for the disposal process.

Specifically, together with the points listed and especially the last point is evident the lack of processes:

- Review of material.
- Development of repairs.
- Documentation of acceptance criteria.

Rejection Sampling: The Static Work Cell

The different stratifications made with the information extracted from the sampling of rejections leads to an easy conclusion, the static work cell represents in itself 70% of the monetary amount of material discarded. The concise 70% data is contundent, in the different analyzes presented it is not surprising that the families of products that make up this work cell appear and reappear in all the different stratifications that were presented.

SOLUTIONS AND RECOMMENDATIONS

In order to the execute the improvement implementation the Material Review Board was created beside the Material Review Process (MRP), to take the first step in establishing the processes that are needed in the reconstruction process, this MRP was implemented as described in Figure 3. This process has as main objective to reduce the amount of discarded material, and secondly to serve as a receiver mechanism for those parts that require some new repair (development) and to give feedback of the acceptance criteria and, if necessary, provide the appropriate follow-up to obtain its publication in the Acceptance Criteria Handbook (See Figure 3).

Measurement and Validation of the Improvement. Respect to the Discarded Material Metric vs. Induced Material to the reconstructed process, there is a change between before and after the implementation of the new material review process. The result obtained implies the average of the material discarded as a percentage of the material induced to the reconstruction process was halved, from 13% to 6%

To discard this reduction in the Metric of Material Dismissed vs. Material Induced to the Reconstruction Process was due to a reduction in the volumes of production, figure 14 likewise represents a before and after the amount of material induced to the process of reconstruction on its own; as Figure 4 indicates, there was no significant change in production volumes. Analyzing the amounts of material discarded, in figure 15 it can be observed after the months of implementation of the new material review process the monthly amounts of discarded material begin to decline, explaining the significant reduction of the Metric Material Discarded vs. Material Induced to the Reconstruction Process (See Figure 4).

Looking at Figure 5 shows the amounts of material discarded corresponding to the month 6 is similar to the obtained after the 13th month. Due to the above, a variance analysis was performed on the set of amounts of material discarded before and after the implementation of the material review process. Analysis of variance shows that both populations are statistically significant using a 90% confidence interval, described visually by the mustache chart in Figure 6.

The company value initiatives or projects related to the reduction of discarded material and the requirement to demonstrate the financial benefits attributable to them. Figure 7 represents the average cost of the 25 part numbers that historically make up 50% of the amount of material discarded. It can be seen after 7th month a reduction in its average cost was seen as a percentage again, thus confirming that the results of the implementation of the material review process are being reflected in the cost of material of the company in the last point of the material chain.

Figure 3. New process for material disposal

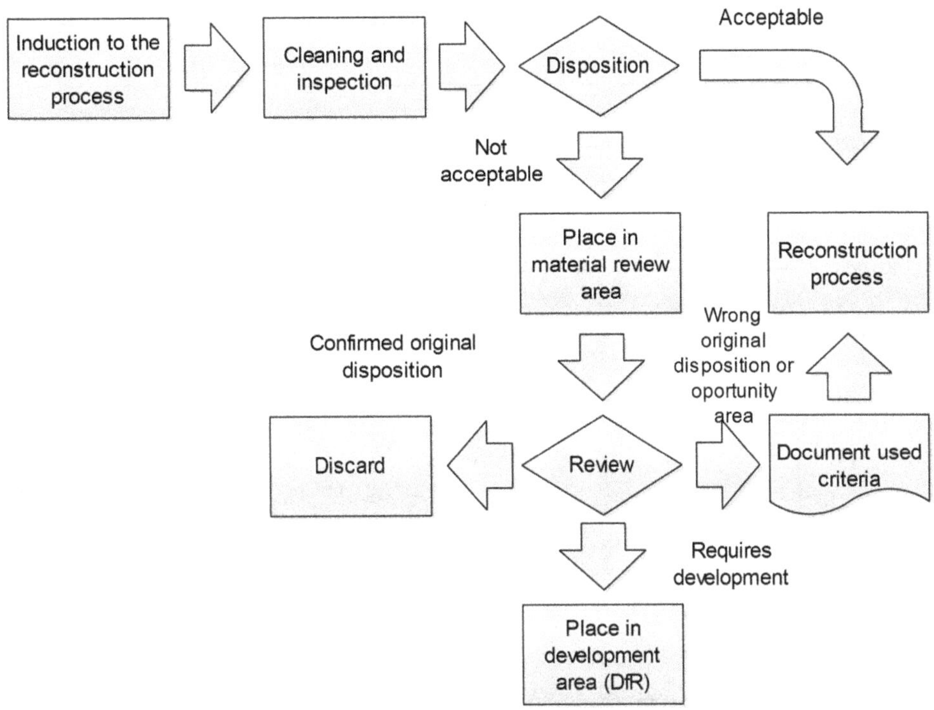

Figure 4. Induced material to the reconstruction process

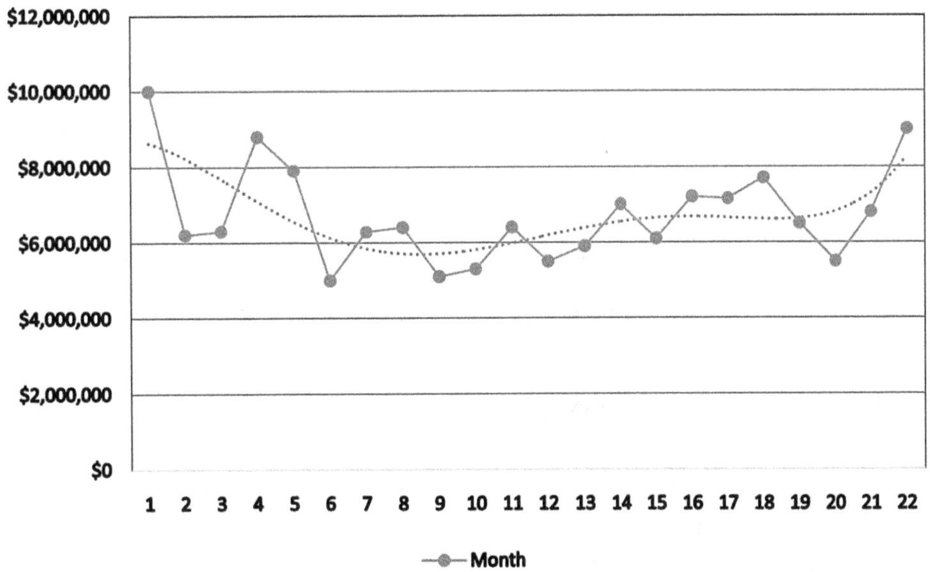

Figure 5. Total of discarded material

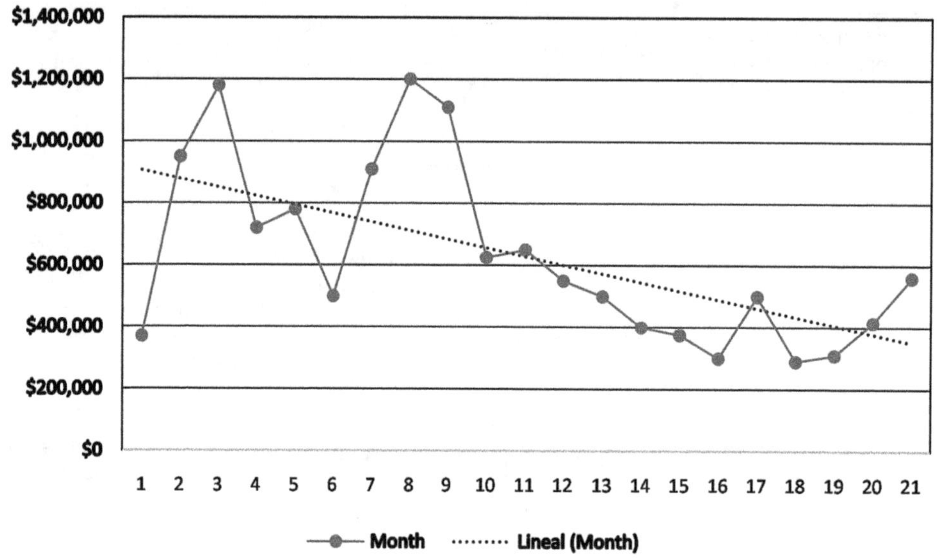

Figure 6. Analysis of variance

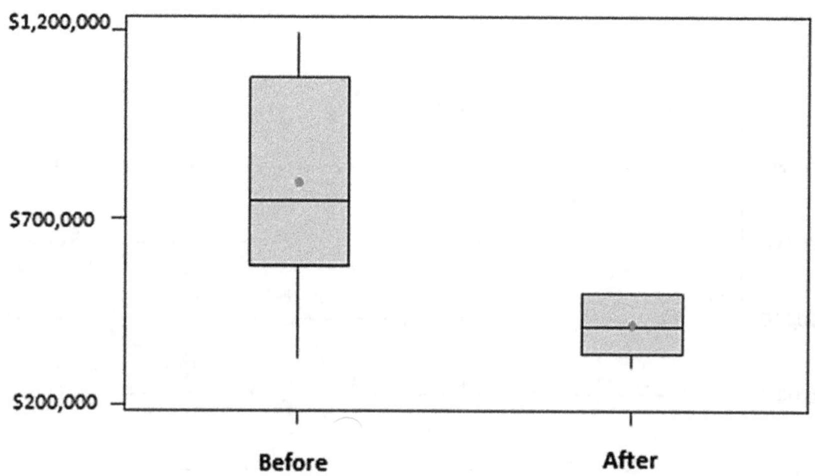

FUTURE RESEARCH DIRECTIONS

Currently, the existing theory in Product Design is abundant from the functional approach of the product. Regarding the ergonomic perspective, an interest has arisen in determining the relevant aspects to consider. On the other hand, the environmental aspects have taken center stage, which has resulted in the increase in the number of research projects related to Ecodesign. However, the theoretical aspects of design for remanufacturing are still scarce; the methods in this field are based on reducing the technical

Figure 7. Mean cost of the 50% of discarded material

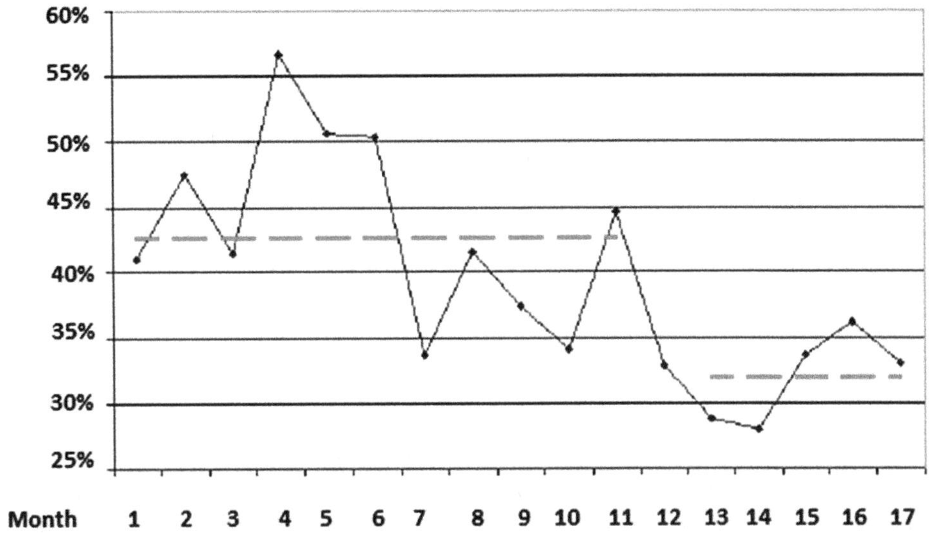

barriers applying the remanufacturing process. Remanufacturing considers the human factor fundamental to the success of the process, due to the slowness in the learning curve of practitioners, the future trend is linked to increasing the number of studies that include applications on the practice of design for different activities; With the aim of determining what is being done, what is possible to improve in this and evaluating the results in the implementation of these proposals.

CONCLUSION

Applying the design for remanufacturing in products that were not designed to be remanufactured implies an additional effort to recover the greater amount of components. If the product is considered not acceptable to be remanufactured due to its characteristics, through the review process it is possible to identify three options:

- The layout was wrong and the product can be remanufactured.
- The original arrangement is correct and it is not possible to remanufacture.
- Finally, a third option is if remanufacturing is not possible in its current state, product development activities are carried out, which implies new manufacturing operations (cleaning, heat treatments, machining, among others). These activities are proposed and managed by remanufacturing development engineers who design processes to make remanufacturing possible. For this it is necessary to evaluate from the economic perspective the feasibility of carrying out these additional activities.

As a result of this research, it was possible to identify a simple solution to reduce the amount of material that would not have a revision process to be discarded, replaced with new material and the profit margin of the company would be affected. It also identifies business needs to be implemented in a medium-term and in a formal way as the ongoing process of development of repairs and feedback and improvement of evaluation criteria. A recommendation that can be made after the analysis is that the static work cell could benefit if the areas of injectors and combustion liners were detached from it, and then, based on data, focus financial and human resources to solve the problems of greater opportunity.

This research received no specific grant from any funding agency in the public, commercial, or not-for-profit sectors.

REFERENCES

Arnette, A. N., Brewer, B. L., & Choal, T. (2014). Design for sustainability (DFS): The intersection of supply chain and environment. *Journal of Cleaner Production*, *83*, 374–390. doi:10.1016/j.jclepro.2014.07.021

Brewer, B., & Arnette, A. N. (2017). Design for procurement: What procurement driven design initiatives result in environmental and economic performance improvement? *Journal of Purchasing and Supply Management*, *23*(1), 28–39. doi:10.1016/j.pursup.2016.06.003

Ceschin, F., & Gaziulusoy, I. (2016). Evolution of design for sustainability: From product design to design for system innovations and transitions. *Design Studies*, *47*, 118–163. doi:10.1016/j.destud.2016.09.002

Chiodo, J. D., & Ijomah, W. L. (2014). Use of active disassembly technology to improve remanufacturing productivity: Automotive application. *International Journal of Computer Integrated Manufacturing*, *27*(4), 361–371. doi:10.1080/0951192X.2012.667151

De los Rios, I. C., & Charnley, F. J. (2016). Skills and capabilities for a sustainable and circular economy: The changing role of design. *Journal of Cleaner Production*.

Goepp, V., Zwolinski, P., & Caillaud, E. (2014). Design process and data models to support the design of sustainable remanufactured products. *Computers in Industry*, *65*(3), 480–490. doi:10.1016/j.compind.2014.02.002

Guide, V. D. R. Jr. (2000). Production planning and control for remanufacturing: Industry practice and research needs. *Journal of Operations Management*, *18*(4), 467–483. doi:10.1016/S0272-6963(00)00034-6

Hatcher, G. D., Ijomah, W. L., & Windmill, J. F. C. (2014). A network model to assist 'design for remanufacture' integration into the design process. *Journal of Cleaner Production*, *64*, 244–253. doi:10.1016/j.jclepro.2013.09.021

Ijomah, W. L., McMahon, C. A., Hammond, G. P., & Newman, S. T. (2007). Development of robust design-for-remanufacturing guidelines to further the aims of sustainable development. *International Journal of Production Research*, *45*(18-19), 4513–4536. doi:10.1080/00207540701450138

ISO. (2002). *ISO/TR 14062: Environmental Management - Integrating Environmental Aspects into Product Design and Development*. ISO.

Matsumoto, M., Yang, S., Martinsen, K., & Kainuma, Y. (2016). Trends and research challenges in remanufacturing. *International Journal of Precision Engineering and Manufacturing-Green Technology, 3*(1), 129-142.

Rybaczewska-Błażejowska, M., Masternak-Janus, A., & Gierulski, W. (2016). Inverse problem of life cycle assessment (LCA): Its application in designing for environment (DfE). *Management, 20*(2), 224–241. doi:10.1515/manment-2015-0062

Schöggl, J. P., Baumgartner, R. J., & Hofer, D. (2017). Improving sustainability performance in early phases of product design: A checklist for sustainable product development tested in the automotive industry. *Journal of Cleaner Production, 140*, 1602–1617. doi:10.1016/j.jclepro.2016.09.195

Shu, L. H., & Flowers, W. C. (1999). Application of a design-for-remanufacture framework to the selection of product life-cycle fastening and joining methods. *Robotics and Computer-integrated Manufacturing, 15*(3), 179–190. doi:10.1016/S0736-5845(98)00032-5

Soh, S. L., Ong, S. K., & Nee, A. Y. C. (2016). Design for assembly and disassembly for remanufacturing. *Assembly Automation, 36*(1), 12–24. doi:10.1108/AA-05-2015-040

U.S. EPA Design for Enviroment from United States Environmental Protection Agency. (2017). *Reports and Publications of the Design for the Environment and Safer Choice Program.* Retrieved June 27, 2017 from http://www.epa.gov/saferchoice/design-enviroment-programs-initiatives-and-projects

KEY TERMS AND DEFINITIONS

Automotive Industry Action Group (AIAG): AIAG is a not-for-profit association where auto industry members collaborate to develop common global standards for quality.

Design for Manufacturing and Assembly (DMA): It is a set of tools to ease the assembly process. Some considerations include ensuring minimizing the number of jigs and fixtures, reduction in the number and variety of cutting tools, minimization of tolerance stack up, and assembly mismatch.

Design for Remanufacturing (DfR): DfR is a design activity focused on managing the factors involved in remanufacturing related to product design. Its objective is to overcome the technical barriers to apply the remanufacturing process.

Material Metric Discarded (MMD): The amount (in monetary terms) of the material discarded during the remanufacturing process.

Materials Induced (MI): The amount of material induced to the remanufacturing process.

Original Equipment Manufacturer (OEM): An OEM is the original manufacturer of products who designed and proposed the specifications, subsequently participates in reverse logistics activities to obtain end-of-life products and remanufacture them.

Product Development: In the context of an OEM remanufacturing its own products, it is a process of generating recovery operations on products were not originally designed to be remanufactured, resulting in new product generations; it is used synonymously with DfR.

Chapter 20
Selection of Prototyping Process and Part Orientation for Virtually Manufactured Gears

Divya Zindani
NIT Silchar, India

Kaushik Kumar
Birla Institute of Technology, India

ABSTRACT

This chapter explores transformation of a virtually produced gear to a physical product using rapid prototyping technique. The same can be segregated into two parts. The first being selection of the best possible rapid prototyping technique using dynamic mechanical analysis and the next was optimization of the process parameters of the selected rapid prototyping technique considering tensile strength as the criteria. The results indicated that fused deposition modeling process provides the best solution amongst the techniques studied. Moreover, as orientation of the product in rapid prototyping techniques plays an important role in providing the strength and also accuracy of the product profile, from results it was also ascertained that strength and profile accuracy of the product was optimized with horizontal position.

INTRODUCTION

The most prominent challenge faced by product designers is to be able to present a tangible product. Despite advancements in 3D CAD technologies, such situations usually end with the product designer being asked: "Can you prove it?" Additive manufacturing or in the context of this work, rapid prototyping, has come as a shot in the arm. With the ability to present a tangible representation of the concept in the form of a physical object within short product lifecycles and even shorter development times, this concept is gaining importance and popularity with designers and engineers.

DOI: 10.4018/978-1-5225-5234-5.ch020

Copyright © 2018, IGI Global. Copying or distributing in print or electronic forms without written permission of IGI Global is prohibited.

In the present work, an arrangement is required to be made to transform a virtual object into reality using Rapid Prototype (RP) technique in order to produce a virtually manufactured gear. The database of the output product i.e. the virtual gear serves as the input files for RP machine and plastic gears are manufactured.

RAPID PROTOTYPING TECHNIQUES

Rapid prototyping or 'Layered manufacturing technology' describes a wide range of techniques wherein 3-dimensional objects are constructed in laminated form. Though diverse, each of these shares the same set of fundamental parameters.

First, the object to be fabricated must be described in terms of an accurate 3-dimensional design representation. This must then be reformatted to describe the object in terms of a number of slices with finite thickness. This "slice information" is then used to fabricate the appropriate number of slices from the desired material. These slices are assembled to form the solid object. In practice, a number of the current technologies combine the slice fabrication and assembly processes by using a previously deposited slice as a template for the deposition of subsequent slices. Hence Rapid prototyping (RP) or layered manufacturing technology got its existence to fill the need in the manufacturing industry for development of representative or functional prototypes of objects normally manufactured in large quantities by tooled processes (Kai and Fai 1997). Models and prototypes manufactured with RP technology not only for visualization purposes but also to build functional parts. It is an important technology as it has potential to reduce the manufacturing lead time of the product up to 30–50% even when the relative part complexity is very high (Massod et al. 2001).

However, many RP techniques are available and the output of these processes also differs. Therefore, it is essential to select the appropriate RP technique for the generation of the gears. The quality of rapid prototyped parts mainly depends on careful selection of process variables. Hence, identifying the RP process parameters that affect the quality of processed parts is of utmost importance. In recent years, eminent researchers such as Masood, S.H. and his team (Massod et al. 2000, Massod et al. 2003, Omar et al. 2015 and Omar et al. 2016) have explored a number of ways to improve the mechanical properties and part quality using various experimental design techniques and concepts.

DIFFERENT RP PROCESSES

Various RP processes have been developed and also being developed in order to improve the realization of a virtual product to the extent possible. In this case as the product to be produced is of a non-metallic material hence out of all the RP techniques available only Stereolithography (SL), Selective Laser Sintering (SLS), Fused Deposition Modeling (FDM) and Laminated Object Manufacturing (LOM) have been studied.

STEREOLITHOGRAPHY

Stereolithography (SL) is a popular RP process in which intricate parts of a plastic monomer are directly built by photo polymerization process with the model constructed using a computer-aided design (CAD) package (Kai and Fai 1997). The process of SL involves; modeling of part with a CAD package to generate 3D solid model; conversion of 3D solid model into standardized triangular language (STL) file format to create volumetric mesh and creation of support structure; slicing of STL format of 3D solid model to provide a series of cross-sectional layers; exporting the sliced model to the computer of stereolithography apparatus (SLA); building the support structure and the part layer by layer (from bottom to top) over a vat of specially designed liquid resin with a helium–cadmium or argon laser, which traces the outline of the planar sections and solidify the resin in SLA; removal of support structure to get the green part; post-curing of green part to undergo final polymerization in a post-curing apparatus (PCA), which is either a controlled furnace or an ultraviolet oven. The dies made through SL process are subjected to high tension due to high injection pressure. Strength is crucial in case of rapid tooling since the parts have to withstand pressures during the test of fitment and when they are used as a die for injection molding.

SELECTIVE LASER SINTERING

Selective laser sintering (SLS) is an additive manufacturing technique that uses a high power laser (for example, a carbon dioxide laser) to fuse small particles of plastic, metal (Direct Metal Laser Sintering), ceramic, or glass powders into a mass representing a desired 3-dimensional object. The laser selectively fuses powdered material by scanning cross-sections generated from a 3-D digital description of the part (for example from a CAD file or scan data) on the surface of a powder bed. After each cross-section is scanned, the powder bed is lowered by one layer thickness, a new layer of material is applied on top, and the process is repeated until the part is completed. Compared to other methods of additive manufacturing, SLS can produce parts from a relatively wide range of commercially available powder materials. These include polymers such as nylon, (neat, glass-filled or with other fillers) or polystyrene, metals including steel, titanium, alloy mixtures, and composites and green sand. The physical process can be full melting, partial melting, or liquid-phase sintering. In many cases large numbers of parts can be packed within the powder bed, allowing very high productivity. Additionally, the selective laser sintering (SLS) process makes parts that have greater stability than SL and selective laser sintering (SLS) parts generally do not lose their shape or post-cure over time.

FUSED DEPOSITION MODELING

Fused deposition modeling (FDM) is an additive manufacturing technology commonly used for modeling, prototyping, and production applications. The technology was developed by S. Scott Crump in the late 1980s and was commercialized in 1990. FDM works on an "additive" principle by laying down material in layers. A plastic filament or metal wire is unwound from a coil and supplies material to an extrusion nozzle which can turn on and off the flow. The nozzle is heated to melt the material and can be moved in both horizontal and vertical directions by a numerically controlled mechanism, directly controlled by a computer-aided manufacturing (CAM) software package. The model or part is produced by extruding small beads of thermoplastic material to form layers as the material hardens immediately after extrusion from the nozzle. The FDM process constructs three-dimensional objects directly from CAD data. A temperature-controlled head extrudes thermoplastic material layer by layer. The FDM process starts with importing an STL file of a model into the preprocessing software. This model is oriented and mathematically sliced into horizontal layers varying from +/- 0.127 to 0.254 mm thickness. A support structure is created where needed, based on the part's position and geometry. After reviewing the path data and generating the tool paths, the data is downloaded to the FDM machine. Several materials are available, with different trade-offs between strength and temperature properties like Acrylonitrile Butadiene Styrene (ABS) polymer, Polycarbonates, Polycaprolactone, Polyphenylsulfones and Waxes. A "water-soluble" material can be used for making temporary supports while manufacturing is in progress, this soluble support material is quickly dissolved with specialized mechanical agitation equipment utilizing a precisely heated sodium hydroxide solution.

LAMINATED OBJECT MANUFACTURING

The first commercial Laminated Object Manufacturing (LOM) system was shipped in 1991. LOM was developed by Helisys of Torrance, CA. The main components of the system are a feed mechanism that advances a sheet over a build platform, a heated roller to apply pressure to bond the sheet to the layer below, and a laser to cut the outline of the part in each sheet layer. Parts are produced by stacking, bonding, and cutting layers of adhesive-coated sheet material on top of the previous one. A laser cuts the outline of the part into each layer. After each cut is completed, the platform lowers by a depth equal to the sheet thickness (typically 0.002 to 0.020 in), and another sheet is advanced on top of the previously deposited layers. The platform then rises slightly and the heated roller applies pressure to bond the new layer. The laser cuts the outline and the process is repeated until the part is completed. After a layer is cut, the extra material remains in place to support the part during the build. The excess material supports overhangs and other weak areas of the part during fabrication. The cross-hatching facilitates removal of the excess material. Once completed, the part has a wood-like texture composed of the material layers. Moisture can be absorbed by the paper, which tends to expand and compromise the dimensional stability. Therefore, most models are sealed with paint or lacquer to block moisture increase.

SELECTION OF APPROPRIATE TECHNIQUE FOR THE PRESENT STUDY

The ability to evaluate and determine the best part building orientation for different rapid prototyping (RP) processes is important for building a satisfactory part/prototype. It is also an essential step towards the identification of the most suitable RP process for a given RP application.

In the present study, primary criteria for the selection of the RP technique is based on the fact that built plastic gears will have the optimum combination of strength, dimensional accuracy and surface finish. It is quite easy to ascertain the accuracy of the RP product on last two aspects, namely, dimensional accuracy and surface finish. Achievable accuracy, surface finish, manufacturing time and building cost are significantly affected by the differences in layering methods of various RP techniques. However, strength analysis is not that straightforward especially when the product is a complicated object like gears. As a result, considerations for the selection of suitable orientation that relates to above-mentioned criteria vary with different RP processes.

Therefore, before proceeding further with the manufacturing of gears by any one of the RP techniques, Dynamic Mechanical Analysis was performed on the test specimens made from three most popular and most widely used processes utilizing non-metallic materials, namely, Stereolithography, Laminated Object Manufacturing, and Fused Deposition Modeling. The output results of the analysis are compared and the best option is selected. Once the RP technique is selected, evaluation of best part building orientation is considered for that particular RP method.

EXPERIMENT SCHEME FOR DYNAMIC MECHANICAL ANALYSIS

The objective of the work was to find out the best possible RP process for the fabrication of virtually manufactured gears. The test specimens are made as shown in Figure 1. Dynamic Mechanical Analysis is performed on the test specimens (dimension 12.5 mm X 6 mm X 3 mm) made from each of the processes.

The samples of stereolithography (SL) are prepared on SLA -250 / 50 machine while the samples of laminated object manufacturing (LOM) are produced by LOM 1015Plus machine. Dimension SST 768 machine is employed to manufacture the samples of fused deposition modeling (FDM). In the dynamic mechanical analysis, the strength of the material produced by RP processes is determined under varying temperatures at which the product may operate during its uses. Thus the sample is subjected to a sinusoidal stress (σ) of varying frequencies. The temperature is varied during the test. The resulting strain (ε) is also sinusoidal with respect to the frequency (Figure 2) and in case of elastic materials, it occurs exactly in phase with the stress. However, when internal molecular motion occurs at the same frequency as the applied stress, the material exhibits viscoelastic characteristics and the strain response lags slightly behind the stress. The phase angle is expressed as an angle δ. The modulus of a viscoelastic material (E*) is a complex quantity,

$$E^{*} = E' + E'' \tag{1}$$

Figure 1. Dynamic Mechanical Analysis test specimens

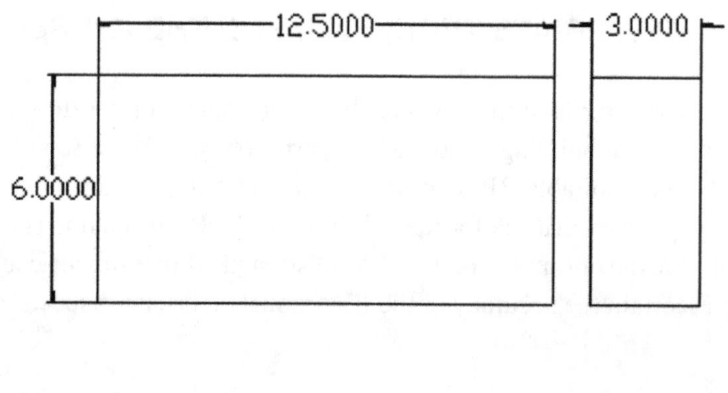

where E' is the storage modulus or in-phase component, and E'' is the loss modulus or out of phase component.

Tangent of the phase angle (δ) can be given by

$$\tan\delta = E'' / E' \tag{2}$$

Since the material used for the present study exhibits viscoelastic behavior to some extent, DMA is carried out with varying frequency range.

EXPERIMENTATION

Dynamic mechanical analysis is performed in 3 point bending mode, the specimens were clamped between strain gauges and subjected to small sinusoidal strain (static strain of 0.2% and dynamic strain of 0.1%) at the frequency of 1 Hz. The measurements were carried out at 20–1200C temperature range with a rate of heating 5^0C/min and the oscillation amplitude is set at 15 μmm. The machine used in this measurement was DMA Q800, TA Instruments, USA. The data was collected and was subsequently analyzed with the help of proprietary software DMA V4.2.

RESULTS AND DISCUSSION

The dynamic mechanical properties of products are significantly dependent upon the type of materials and processing techniques. Polymeric materials exhibit more than one relaxation region or so-called transition over a wide range of temperature during dynamic thermo-mechanical analysis. The variations of dynamic mechanical properties (storage modulus, loss modulus, and tan δ with temperature for different samples prepared by various prototyping processes are graphically presented in (Figures 3, 4 and 5). In Figure 3, a variation of the storage modulus made by SL, LOM, and FDM is measured as a function of temperature. It is observed from the graph that the decrease in storage modulus with temperature occurred in all three cases. However, maximum storage modulus was observed in case of the samples prepared by FDM and the minimum by those prepared by LOM process.

Figure 2. Sketch for storage modulus and loss modulus

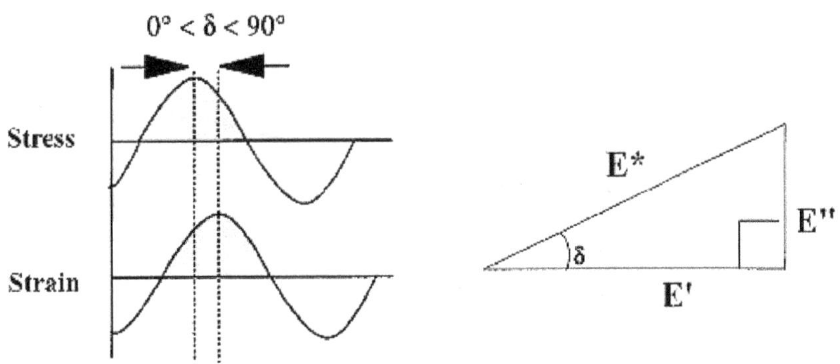

Figure 3. Variations of storage modulus with temperature for different samples prepared by FDM, SLA, and LOM

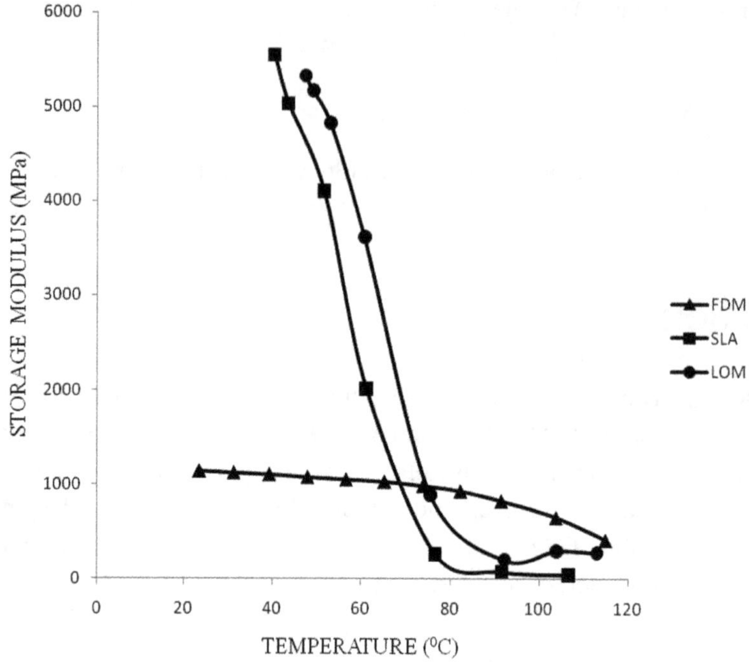

Figure 4. Variations of loss modulus with temperature for different samples prepared by FDM, SLA, and LOM

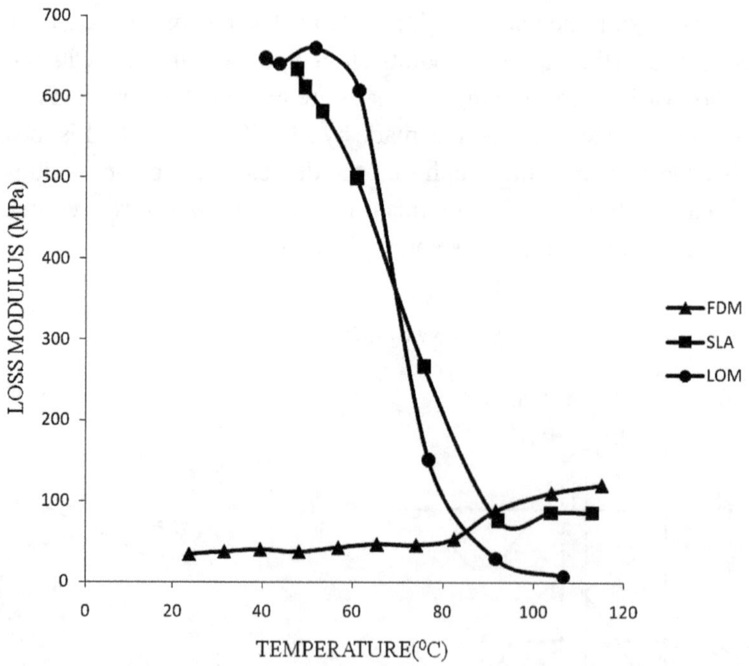

Figure 5. Variations of tan δ with temperature for different samples prepared by FDM, SLA, and LOM

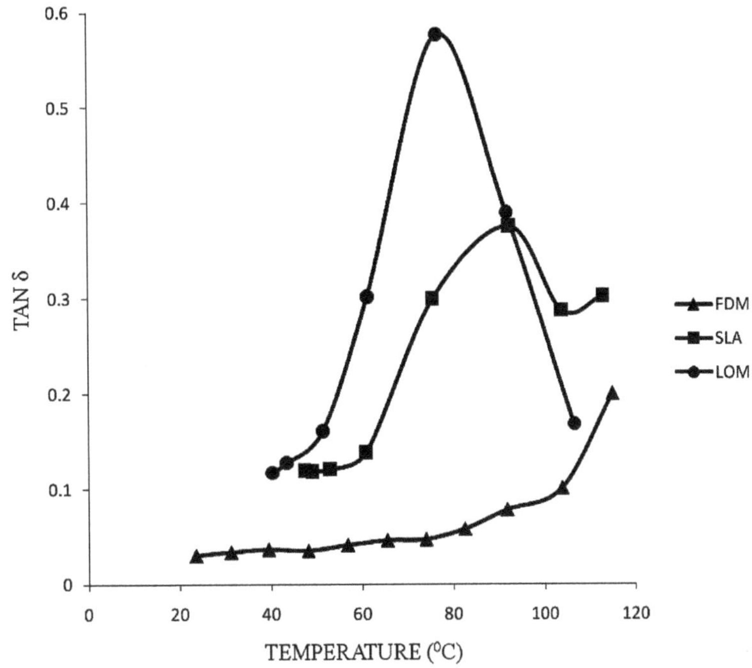

As the basic material used in case of LOM is paper, which has comparatively lower strength, the sample shows minimum storage modulus. Whereas the sample prepared by FDM is thermoplastic in nature and has very high interlayer integrity when it is processed layer by layer. But in the sample prepared by SL, although thermosetting in nature, the interlayer adhesion is poor due to incomplete curing of the interior portion of the sample and results in lower storage modulus compared to material prepared by FDM. It can be inferred that the FDM sample gives the maximum storage modulus up to 800C which can be assumed to be the maximum working temperature for gears made from plastic-type materials. The same procedure is followed for the loss modulus of the samples made by three processes as well. If the mechanical energy is not stored elastically, it may be lost and converted to heat through molecular friction (that is viscous dissipation) within the material. The extent to which this happens is reflected by the magnitude of loss modulus, higher the value of loss modulus, higher is the loss of mechanical energy. Figure 4 exhibits the variation of the Loss modulus for SL, LOM, and FDM samples, respectively as a function of temperature. It is revealed that the loss modulus is the minimum for the FDM sample over the range of temperature. So it can be inferred that the FDM sample exhibits more dynamic strength as it has the minimum loss modulus amongst the three processes considered. In fact, its loss modulus may even reach as low as zero around 800C indicating almost elastic behavior at the maximum working temperature of a plastic type of gears. The ratio of loss modulus to storage modulus (E"/ E') is measured as the mechanical loss factor or tan δ. The variation of tan δ as a function of temperature for different formulated samples is shown in Figure 5. As shown in Figure 5, tan δ, which corresponds to the damping properties of the material, is found to be maximum in case of the sample prepared by FDM and minimum in case of the sample generated by SL. As the material used in FDM is in nature of thermoplastic, having a more flexible polymeric backbone, it allows more movement of macromolecule chain with an increase in temperature. Whereas the material used in SL is thermosetting in nature and the rigid cross-linked backbone does not allow larger long chain movement with an increase in temperature. Hence, Fused Deposition Modeling is selected as the process for fabrication of plastic type gears for its better damping character as well as higher modulus even at a higher working temperature (80^{0}C).

PART DEPOSITION ORIENTATION FOR FDM

So FDM has been identified as the most suitable RP technique for the present study hence now evaluation of best part building orientation is required to be estimated. It is a very important factor as it affects strength, build time, support structure, dimensional accuracy, surface finish and cost of the prototype. A number of process specific parameters and constraints are required to be considered while making this decision. Determining the optimal part deposition orientation is a difficult and time-consuming task as one has to trade-off among various contradicting objectives (Pham and Demov 2001) like part strength, surface finish and build time. There have been several attempts (Alexander et al. 1998, Frank and Fadel 1994, Chang et al. 1995, Lan et al. 1997, Xu et al. 1997, McClurkin and Rosen 1998, Pham et al. 1999, West et al. 2001, Goyal et al. 2000, Mishra and Thirumavalavan 2014, Phatak and Pande 2012, Paul and Anand 2011 etc.) to determine suitable part deposition orientation for different objectives like accuracy, build time, support structure etc. The build time of a prototype can be reduced by depositing it with thicker layers; however, this may lead to the rough surface. Moreover generation of curved surfaces, if not oriented properly, gets deformed due to phenomena are known as Staircase Effect. Staircase Effect is so called because of the step formed between two adjacent layers when the layers are built. Staircase Effect in a build will result not only in very rough surface finish but also inaccurate surface profile. However, deposition using thin layers increases built time drastically. In the present work, average part surface roughness and the build time of the part are considered as two objectives and are minimized by minimizing their weighted sum.

SELECTION OF FDM PROCESS PARAMETERS TO MAXIMIZE STRENGTH OF THE PRODUCT

The FDM machine used for the present experiment is SST 768 with installed proprietary software Catalyst EX. The strength of FDM processed component primarily depends on input process parameters. They are defined as follows:

1. **Orientation:** Part build orientation or orientation refers to the inclination of a part in a build platform with respect to X, Y and Z axes. X and Y-axes are considered parallel to build platform and Z-axis is along the direction of the part build.
2. **Layer Thickness:** It is the thickness of the layer deposited by the nozzle and depends upon the type of nozzle used i.e. the height of each layer of material extruded to produce a part. In the present machine, it may be either 0.010 or .013 inch (0.2540 or 0.3302 mm). The resolution will affect build time and surface finish i.e. 0.010 will be a smoother finish but will take a longer time to build, the default value being 0.2540 mm.

3. **Model Interior:** It establishes the type of fill to be used for the part. It refers to the way in which the part is filled. It controls the density of the part. There are two types of build styles in the machine used.
    a. "Solid" fills the interior part completely. It will build a part with a stronger interior.
    b. "Sparse" minimizes the material volume and build time by leaving gaps. It utilizes a unidirectional raster.

Support style refers to the portions of a model that extends outward to prevent the part from collapsing during the building process. It is of four types.

1. "Basic" is the standard raster tool path support structure. It supports all part features with small support raster curves.
2. "Sparse" reduces the amount of support material volume. It uses less amount of material than basic support.
3. "Surround" fill the surround of small features or small parts. It is used for all parts which are small in nature and requires more support due to the geometry.
4. "Break-away" is similar to "Sparse", but it contains discrete boxes and it is easier to remove than other three types of support structures.

The objective of the study is to determine a combination that may lead to a product with maximum strength with dimensional accuracy maintained within a proper range. Accordingly, a set of experiments is carried out using tensile test specimens produced by FDM having varying parameter setting as mentioned above.

EXPERIMENTATION

In order to make tensile test specimen, a 3D model of the same is developed using CREO software (Figure 6) and is saved as 'stl' file. The dimensions of the test specimen are chosen according to ASTM D 638 (Figure 7).

Figure 6. 3D model of tensile test specimen drawn using CREO software

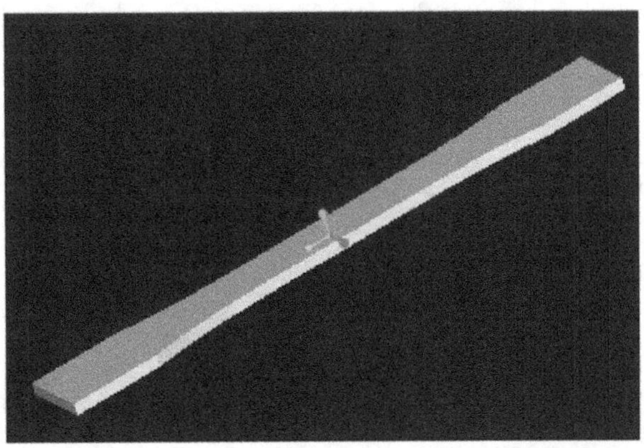

Figure 7. The Tensile model with dimensions (As per ASTM D 638)

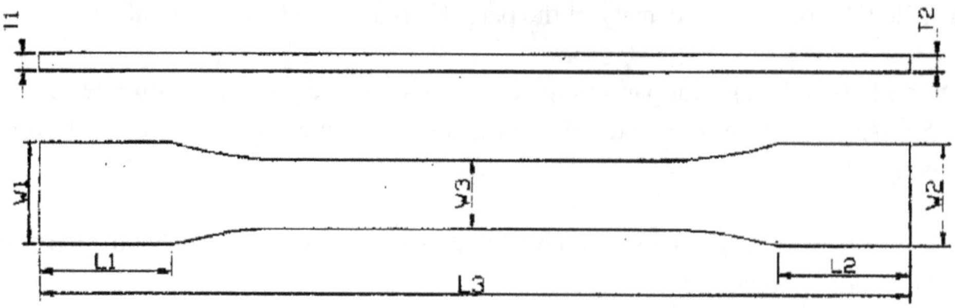

The 'stl' versions of the model are loaded on the FDM machine and an attempt is made to create eight different RP models on the FDM machine with three different parameters such as orientation, layer thickness, and model interior. Table 1 shows the variation in parameters along with build time, build volume and support volume required for these eight samples. However, these values are given by the software even before the samples are actually made by FDM.

Three different orientations are the obvious choice i.e. horizontal, vertical and standing, however, for standing orientation, due to heavy vibration of the machine and very small base, the samples fell before completion of the product. This implies that this condition should be avoided. So two different orientations were considered by the software (Catalyst EX) and are shown in Figure 8. For each of the horizontal and vertical orientations, four samples were created by FDM.

Next, Instron universal-testing machine-Model 3360 was used to find the tensile strength of all these samples at 1.0 mm/min crosshead velocity.

Table 1. The various parameters and their variations used for experimentation

Code No.	Orientation	Layer Thickness	Model Interior	Build Duration (Hrs.)	Model Material Required (cc)	Support Material Required (cc)
H1	Horizontal	0.2540	Solid	2.133333	36.13	12.31
H2	Horizontal	0.2540	Sparse	1.7	26.03	12.31
H3	Horizontal	0.3302	Solid	1.416667	36.07	15.4
H4	Horizontal	0.3302	Sparse	1.4	26.27	15.4
V1	Vertical	0.2540	Solid	3.45	34.55	8.59
V2	Vertical	0.2540	Sparse	2.683333	29.29	8.59
V3	Vertical	0.3302	Solid	2.166667	34.26	10.27
V4	Vertical	0.3302	Sparse	2.616667	34.29	10.27

Figure 8. The various orientations (Horizontal and Vertical) as seen by the Catalyst EX software

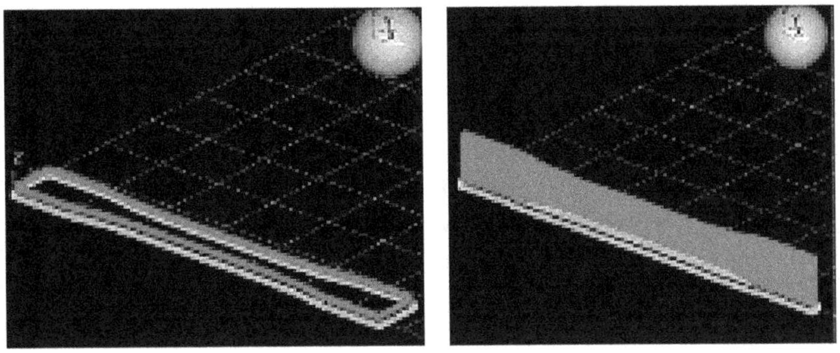

RESULT AND DISCUSSION

Tensile Test Results

The output results of the tensile test are presented in Table 2. The table also includes the dimensional variation of different test specimens.

Figure 9 shows the ultimate tensile stress of sample prepared in both the direction one in horizontal another in a vertical position. Figure 9 also indicates that the in both the cases sample H1 and H3 gives maximum ultimate tensile stress compare to sample H2 and H4 as the model interior is sparse which increases the interlayer gaps and hence the stress decreases. Moreover, it was also observed that the between sample H1 whose layer thickness is 0.2540 and sample H3 having layer thickness 0.3302, H1 shows more stress compare to H3. These results can be attributed to the fact that as layer thickness decreases the interlayer void space also decreases results in more dense structure compare to a material having more layer thickness. Similar kind of results also observed in the sample prepared in the vertical direction. The solid model gives more stress compared to the sparse structure. But the variation in stress between samples with various layer thicknesses is not very much as the contact area between layers is less in the sample prepared in a vertical position.

Table 2. The results of the tensile testing along with variation in dimensions of all samples

Sl. No.	Sample Code	Ultimate Stress (MPa)	Ultimate Strain (%)	Modulus (MPa)	Variation in Dimension (%)	
					Max.	Min.
1.	H1	20.32	4.75	635.40	5.63	0.11
2.	H2	13.76	4.25	439.30	5.00	0.11
3.	H3	18.59	5.33	559.80	8.43	0.00
4.	H4	12.76	4.08	418.45	7.81	0.63
5.	V1	17.78	7.17	665.10	7.52	1.69
6.	V2	8.55	40.42	572.80	7.12	1.25
7.	V3	18.94	13.42	665.45	8.80	0.42
8.	V4	15.85	10.42	651.00	6.48	0.95

Figure 9. Ultimate tensile stress of sample prepared in horizontal and vertical position

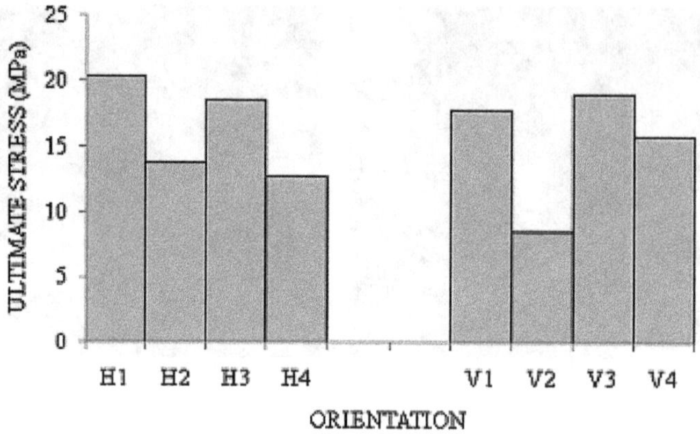

Figure 10 shows the variation of ultimate tensile strain for samples prepared in a different orientation. It was observed from data that the vertically oriented sample shows much higher strain compare to sample prepared in a horizontal orientation. As in the tensile strength measurements the test specimen drawn in the direction which is parallel to the layer orientation in case of sample horizontal oriented and perpendicular to the layer orientation in case of sample vertically oriented. Now when the sample is drawn in tensile strength machine the layer which is in machine direction restrict more and do not allow stress transfer rapidly whereas in case of vertical sample since the layer-oriented opposite direction, during tensile drawing the interlayer may slip on over other and stress may transfer to different interface and allows more extension which is shown in graphs.

Figure 11 depicts the tensile modulus of all samples. It is obvious that the solid sample H1 and H3 show higher modulus compare to the sparse sample H2 and H4. But the modulus of H3 sample is found to be lesser than the sample prepared in H1 condition as the layer thickness is higher which increases the interlayer gaps and hence the strength decreases. Similar kinds of results were observed in modulus in case of the sample prepared in a vertical condition. As the area of the layer is less and number of the interface of the layer is high the change in modulus of the sample prepared in vertical condition with

Figure 10. Ultimate tensile strain of sample prepared in horizontal and vertical position

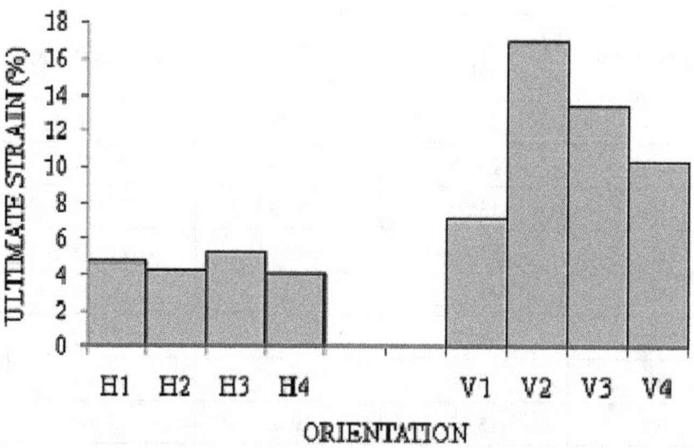

Figure 11. Modulus of sample prepared in horizontal and vertical position

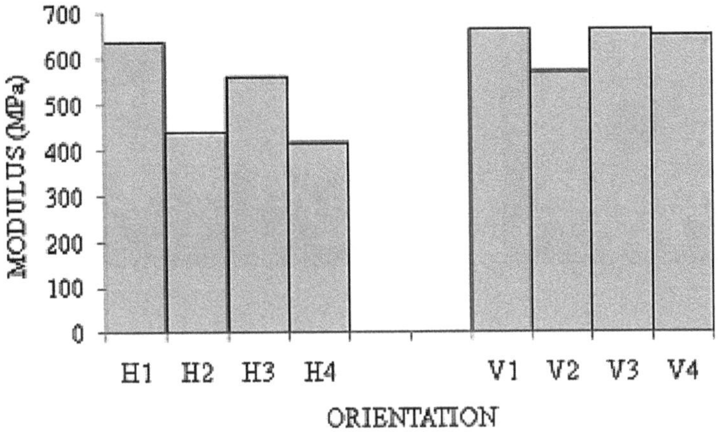

different parameter within the small elongation is not a considerable amount. Though solid sample in vertical orientation shows higher modulus compare to sparse sample but differences in value are not considerable.

CONCLUSION

There are two parts in this work. The first being selection of the best possible rapid prototyping technique using dynamic mechanical analysis and the next was optimization of the process parameters of the selected RP technique considering tensile strength as the criteria.

The samples prepared using SL, FDM and LOM, subjected to dynamic mechanical analysis indicated that the storage modulus for liquid-based processes (SL) is less due to the incomplete curing of the resin. It occurs as the resin remains in contact with the laser for a shorter time and even with post-processing also the curing is never 100% complete. This results in lower storage modulus. In LOM paper are glued together using an adhesive which also shows lower strength. Hence in such situation, FDM, wherein an extruded wire creates the model, is strongest of all these within the specified temperature range.

The orientation of the product in RP techniques plays an important role in providing the strength and also the accuracy of the product profile. Form the results two orientations were found to be suitable H1 and V4. V4 orientation results in highest yield Stress, ultimate strain etc. but the curved profile of tooth of the gear would be improper due to 'staircase effect'. So H1 parameters, although offers lesser strength, is more proper considering the product profile. Therefore H1 condition was chosen for gear generation by FDM and depicted in Figure 12.

Figure 12. Spur gear created using FDM

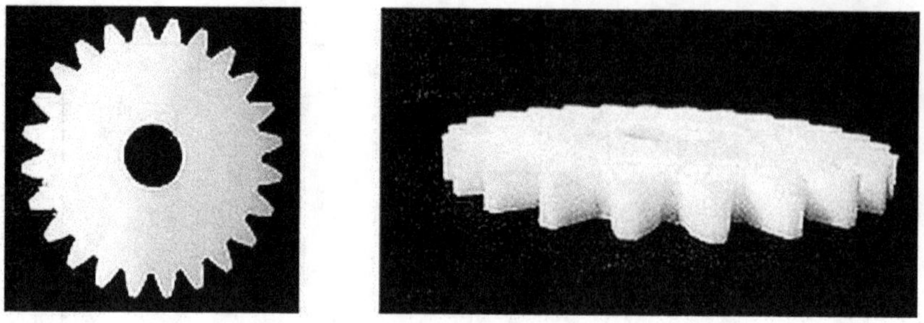

FUTURE WORK

The work can be extended for generation of other gears and also evaluation of metal RP for generating metallic products.

ACKNOWLEDGMENT

The authors sincerely acknowledge the comments and suggestions of the editors and reviewers that have been instrumental for improving and upgrading the paper in its final form.

REFERENCES

Alexander, P., Allen, S., & Dutta, D. (1998). Part orientation and build cost determination in layered manufacturing. *Computer Aided Design, 30*(5), 343–356. doi:10.1016/S0010-4485(97)00083-3

Cheng, W., Fuh, J. Y. H., Nee, A. Y. C., Wong, Y. S., Loh, H. T., & Miyajawa, T. (1995). Multi-objective optimization of part-building orientation in stereolithography. *Rapid Prototyping Journal, 1*(4), 12–23. doi:10.1108/13552549510104429

Frank, D., & Fadel, G. (1994). Expert system based selection of the preferred direction of build for rapid prototyping. *Journal of Intelligent Manufacturing, 6*(5), 334–339.

Goyal, A., Sudha, A., Kumar, A., & Rao, P. V. M. (2000). Optimum part orientation for rapid prototyping based on estimated number of adaptive slices. *Proceedings of International Conference on E-manufacturing,* 247–253.

Kai, C. C., & Fai, L. K. (1997). *Rapid Prototyping, Principles and Applications in Manufacturing. John Wiley & Sons .*

Lan, P.-T., Chou, S.-Y., Chen, L.-L., & Gemmill, D. (1997). Determining fabrication orientation for rapid prototyping with stereolithography apparatus. *Computer Aided Design*, *29*(1), 53–62. doi:10.1016/S0010-4485(96)00049-8

Masood, S. H., Rattanawong, W., & Iovenitti, P. (2000). Part build orientations based on volumetric error in fused deposition modeling. *International Journal of Advanced Manufacturing Technology*, *16*(3), 162–168. doi:10.1007001700050022

Masood, S. H., Rattanawong, W., & Iovenitti, P. (2001). A volumetric approach to part-built orientations in rapid prototyping. *Journal of Materials Processing Technology*, *119*(1-3), 348–353. doi:10.1016/S0924-0136(01)00924-4

Masood, S. H., Rattanawong, W., & Iovenitti, P. (2003). A generic algorithm for part orientation system for complex parts in rapid prototyping. *Journal of Materials Processing Technology*, *139*(1–3), 110–116. doi:10.1016/S0924-0136(03)00190-0

McClurkin, J. E., & Rosen, D. W. (1998). Computer-aided build style decision support for stereolithography. *Rapid Prototyping Journal*, *4*(1), 4–13. doi:10.1108/13552549810197505

Mishra, A. K., & Thirumavalavan, S. (2014). A Study of Part Orientation in Rapid Prototyping. *Middle East Journal of Scientific Research*, *20*(9), 1197–1201.

Mohamed, O. A., Masood, S. H., & Bhowmik, J. L. (2015). Optimization of fused deposition modeling process parameters: A review of current research and future prospects. *Advances in Manufacturing*, *3*(1), 42–53. doi:10.100740436-014-0097-7

Mohamed, O. A., Masood, S. H., & Bhowmik, J. L. (2016). Optimization of fused deposition modeling process parameters for dimensional accuracy using I-optimality criterion. *Measurements*, *81*, 174–196.

Paul, R., & Anand, S. (2011). Optimal Part Orientation in Rapid Manufacturing Process for Achieving Geometric Tolerances. *Journal of Manufacturing Systems*, *30*(4), 214–222. doi:10.1016/j.jmsy.2011.07.010

Pham, D. T., & Demov, S. S. (2001). *Rapid Manufacturing: The Technologies and Applications of Rapid Prototyping and Rapid Tooling*. London: Springer-Verlag London Limited. doi:10.1007/978-1-4471-0703-3

Pham, D. T., Demov, S. S., & Gault, R. S. (1999). Part orientation in stereolithography. *International Journal of Advanced Manufacturing Technology*, *15*(9), 674–682. doi:10.1007001700050118

Phatak, A. M., & Pande, S. S. (2012). Optimum Part Orientation in Rapid Prototyping Using Genetic Algorithm. *Journal of Manufacturing Systems*, *31*(4), 395–402. doi:10.1016/j.jmsy.2012.07.001

West, A. P., Sambu, S. P., & Rosen, D. W. (2001). A process planning method to improve build performance in stereolithography. *Computer Aided Design*, *33*(1), 65–79. doi:10.1016/S0010-4485(00)00064-6

Xu, F., Wong, Y. S., Loh, H. T., Fuh, J. Y. H., & Miyazawa, T. (1997). Optimal orientation with variable slicing in stereolithography. *Rapid Prototyping Journal*, *3*(3), 76–88. doi:10.1108/13552549710185644

KEY TERMS AND DEFINITIONS

Dynamic Mechanical Analysis: A technique to study the viscoelastic nature of materials especially polymers. The applied stress is sinusoidal in nature and is usually accompanied with varying temperature and frequency.

Fused Deposition Modeling: An additive manufacturing process that builds a part from thermoplastic filament by extruding and heating the filament.

Laminated Object Manufacturing: A rapid prototyping technique where a part is built from layers of papers.

Layer Thickness: The thickness of the layer deposited by the nozzle for creating the solid.

Model Interior: The type of fill to be used for the part (i.e., sparse or solid).

Part Orientation: The inclination of the being built with respect to the X, Y, and Z axes of a RP machine.

Rapid Prototyping: The process of making physical prototypes for scale down physical model obtained from CAD data.

Selective Laser Sintering: An additive manufacturing technique in which high-power laser source is used to fuse the powdered material to obtain three dimensional physical prototypes.

Stereolithography: One of the 3D printing techniques of producing prototypes by converting liquid plastics to solid plastics.

Tensile Modulus: The ability of material to withstand change in length when under tension.

Ultimate Tensile Strain: The strain at the point of failure.

Ultimate Tensile Stress: The maximum stress a material can sustain while being stretched before breaking.

Chapter 21
Thermal Power Sector Sustainability:
A Framework for Sustainable Supply Chain Management

Suchismita Satapathy
KIIT University, India

Jitendra Narayan Biswal
KIIT University, India

ABSTRACT

Sustainable supply chain management (SSCM) practices in thermal power plants is dependent on mostly three pillars: social factor, economic factor, and environmental factor. So, in this chapter, sustainable supply chain management of Indian thermal power sector is evaluated. Artificial neural network (ANN) method is implemented to measure whether the benefits of sustainable supply chain management are achieved after practices of sustainable supply chain management in Indian thermal power sector. This chapter also designs a framework by QFD (quality function deployment) method to find solution for some unsatisfactory measures (inputs in sustainable factors) that are not achieved against outputs. As sustainable supply chain management practices in thermal power plants are influenced by a significant number of interrelated enablers and barriers, the drivers or enablers of SSCM are taken as the design requirement to improve SSCM in thermal power industries, and the most important driver is prioritized against the unsatisfied measurands in thermal power sector.

INTRODUCTION

Energy or power is one of the most rudiments of modern society. A life without energy is a life without accessing the opportunities by the people provided by the modern world. In a modern society indispensable social services are heavily reliant on a reliable source of energy supply, known as electrical energy (Electricity). Without electricity, it is not possible to function effectively the state of health and education

DOI: 10.4018/978-1-5225-5234-5.ch021

Copyright © 2018, IGI Global. Copying or distributing in print or electronic forms without written permission of IGI Global is prohibited.

of a developing country. Lacking in access to electricity entails growing in problems of health hazards, diseases and improper education. Power is the major contributor and vital source for economic development of any developing country across the globe. This is why accessing energy for all in a modern developing and developed countries is the basic need and prime importance in the global scenario. However, for safe and reliable generation of power through using fossil fuels like coal, is the only way to provide electricity at affordable cost compared to other forms of energy. Being coal as cheapest, it will continue for next 2-3 decades as backbone of generating electricity in global context. With regards to this, Indian economy depends on its power sufficiency making a suitable leading position in world economy. India mostly depends on thermal power plants using coal as fuel, the only fossil fuel easily available in large quantity in the country, to maintain its energy requirement. It has been seen that, energy generated by the coal based thermal power plants is about 50% of the total electrical energy produced in the country (Rao and Kumar, 2014). Growing use of coal in thermal power plants confronts increasing in environmental issues. Thermal power plants using coal as fuel are treated as one of the major sources of environmental pollution across the globe. It causes degradation of land use, air pollution, water contamination, health hazards, greenhouse gas emission, waste generation, degradation of aquatic life, scarce of natural resources etc. (Adham et al., 2015). Increasing serious negative impacts on environmental as well as economic and societal issues by thermal power plants have been faced major challenges since one or two decades (Govindan et al., 2013). Increasing in consumption of natural resources in unsustainable manner, by the thermal power plants results consensus regarding awareness for environmental issues due to arising of incessant pressure from government, society, customers and stake-holders needs to be focused (Tyagi et al., 2015). Owing to this it needs to incorporate and implement sustainable supply chain strategy to deal with various problems and issues associated with supply chain to reduce environmental and societal hazards (Jayaram and Avittathur, 2015; Mathiyazhagan et al., 2014;

Diabat et al., 2014). UN Brundtland Commission defined sustainability is "development that meets the needs of the present without compromising the ability of future generations to meet their own needs" (World Commission on Environment and Development, 1987). According to the Environmental Protection Agency (USA), sustainability is a principle in which "everything that we need for our survival and well-being depends, either directly or indirectly, on our natural environment. To pursue sustainability is to create and maintain the conditions under which humans and nature can exist in productive harmony to support present and future generations". However, in order to implement sustainable supply chain management (SSCM) practices in thermal power plants seems like a difficult task owing to its dependence on a number of various interrelated factors. These factors which can be categorized mainly into two groups, likely enablers and barriers. Enabler is one that encourages SSCM adoption and the barrier is one that offers hindrance to SSCM.

Interdependence and knowledge regarding of these influential factors will enable the decision makers of the organization to consider the best possible suitable strategy to reduce the negative impacts of barriers and to implement the benefits of the enablers. So, managerial implications of SSCM practices in thermal power plants and understanding their mutual interdependence is important to face the challenges, barriers and the global competition to achieve sustainable success.

LITERATURE REVIEW

The power sector is the biggest customer of coal, it depends on coal for power generation. As coal is the maximum polluting material, so its transportation and storing is an important point. All over the world pollution is a headache and enviornmental condition is the focused issue for discussion. So a lot of efforts are taken by government and research units. Media always debates and asks for suggestions and new developments that can resolve this problem. So, researches are undergoing on this to find a better way to use coal without creating environmental pollution. The current researchers have realized that, environmental and societal issues must be considered such that the greatest revolutions in human thought, uniting the entire world in a fight against the emissions which are produced during industrial activities(Dubey et al., 2015). Particularly the last decade has seen an increased pressure to broaden the accountability of the industries beyond economic performance, for shareholders to sustainability performance, for all stake holders (Labuschagne et al., 2005). Consequently, an increased interest was exhibited by organizations in addressing sustainability in their supply chains, which has been described as Sustainable Supply Chains Management (SSCM) that incorporates the triple bottom line of sustainability (Walker and Jones, 2012). SSCM is the management of raw materials and services from suppliers to manufacturers/service providers to customers and back with the improvement of the social and environmental impacts explicitly considered (Grzybowska, 2012).

Green or environmental aspect of SSCM focuses on minimization of the adverse environmental consequences of various activities of supply chains where as the second one, social aspect of SSCM ensures ethical as well as decent working conditions of various stakeholders including the suppliers. The third component of the SSCM, the economic aspects ensures local economic generation through purchasing from local suppliers (Walker and Jones, 2012).

SSCM implementation in industries is influenced by a number of factors and there is no exception for the thermal power industries. These influential factors broadly can be divided into two categories; enablers the factors that encourage SSCM adoption and barriers that hinder SSCM adoption.

Enabler as defined in layman's term is "an entity that makes it possible or easy". Therefore, enablers for sustainable supply chains are process that can drive a supply chain to being sustainable (Hussain, 2011).

Scanning of contemporary literature reveals that there exists a number of studies that focused exclusively either on the enablers (Grzybowska, 2012; Faisal, 2010; Walker and Jones, 2012; Muduli and Barve, 2013; Diabat et al., 2014) or barriers (Bhattacharyya, 2010; Walker and Jones, 2012; Ageron et al., 2010). However, this study provides a framework to study both the categories factors simultaneously, so that it will be easy to identify the relative importance of an enabler with respect to a barrier. This in turn will enable the organizations to identify their degree of strength in handling a particular barrier.

SSCM is at its nascent stage in India; moreover, its adoption amongst Indian thermal power sector is limited to only few big companies. Hence, an analytical approach such as AHP has been proposed in this to study to analyze various influential factors of SSCM instead of statistical approach that depends upon bigger sample sizes. AHP developed by Saaty (1970), is a simple mathematical method based on elementary operations with matrices and relies on decomposition of the problem into objective, criteria, sub-criteria and alternatives, pair-wise comparison of elements of each level with respect to its immediate upper level factor using a nine-point scale and generation of priority vector (Muduli and Barve, 2015). It has the ability to accommodate qualitative attributes in an organized manner and can be used to structure a system and its environment into mutually interacting elements and then to synthesize them by measuring and ranking the impact of these elements on the entire system (Sambasivan and Fei, 2008) with the objective or goal (SSCM implementation) occupying position at the top level of the hierarchy, various criteria and sub-criteria occupy positions in the subsequent level of utility.

RESEARCH METHODOLOGY

Thermal industries are in a urge to get profit, fullfill demand and show maximum benefit which they deserve to achieve by implementing SSCM in their regular practice. So to find the measuring criteria which are not helping in achieveing the goal of SSCM implementation in thermal power industries, a standard questionnaire is designed that consists of 34 questions for economic performance, 37 social questions and 22 questions for Environmental factors.Then 400 questionaire is send to diferent Indian Thermal industries by post, mail and personal contact and the respondents are advised to respond each item of the Questionnaire in a five -point Likert-type scale (1 = totally disagree, 2 = partially disagree, 3 = no opinion, 4 = partially agree, 5 = totaly agree).The details of items in the questionnaire are given in Appendix.1. Among them 272 responses received and the response rate is more than 65%. Then, the datas are feed for statistical analysis like Principal component analysis to find the items or questionnairre falling under three dimensions or pillers of SSCM like Social, Environmental and Economic factors. Then further ANN method is used to measure the successive implementation of SSCM under the three pillers of SSCM (i.e. Social, Environmental and Economic factors) against benefits achieved due to implementation of SSCM. After getting those measurands that are not help full in achieving goal of SSCM, a frame work is designed and the most important design requirements are suggested, which can improve the SSCM practice in thermal power sector.

RESULT AND DISCUSSION

The collected data against three pillers of SSCM in Indian thermal industries are subjected to various statistical analyses such as factor analysis and Kaiser–Meyer–Olkin (KMO) test. Factor analysis on 272 useful responses has been conducted using principal component method followed by varimax rotation via SPSS 17.0. After analysis it is found that only 11 items are coming under 3 dimensions of Social factor, 13 items under 6 dimensions of environmental factors and 10 items coming under four dimensions of economic factor of SSCM. Percentage of total variance explained was found to be 71%, 73.3% snd 70.9% respectively which is an acceptable value for the principal component varimax rotated factor loading procedure (Johnson and Wichern, 2002). The internal consistency of the actual survey data was tested by computing the Chronbach's alpha (α). The value of alpha for each dimension is shown in Tables 1 and Table 4 and Table 7. The value of KMO, which is a measure of sampling adequacy, was found to be 0.55 and 0.57 and 0.61 indicating that the factor analysis test has proceeded correctly, and the sample used is adequate as the minimum acceptable value of KMO is 0.5 (Othman and Owen, 2001). Therefore, it can be concluded that the matrix did not suffer from multicollinearity or singularity. The result of Bartlett test of sphericity shows that it is highly significant (significance = 0.000) which indicates that the factor analysis processes is correct and suitable for testing multidimensionality (Othman and Owen, 2001). The factors coming after factor analysis under Social dimension are Workplace/Internal, Institutions/Systems Community/External, the factors coming under environmental factors are Air, Water, Land, Materials, Mineral and energy resources, Institutions/systems and the factors coming under economic factor are Economic performance, Financial health, Market and structure and Institutions/Systems .

The social dimension of sustainability relates to the human capital of the supply chain. Improving sustainability with respect to the social dimension involves developing and maintaining business practices that are fair and favorable to the labor, communities, and regions touched by the supply chain. Social performance indicators are grouped into three categories. (1) Workplace: Refers to the internal human resources, i.e., those who work within the supply chain. (2) Community: Refers to all people outside of the supply chain, including those who are directly and indirectly affected by the chain's performance. (3) Institutions/Systems: Refers to the internal and external systems, procedures, and values that relate to the social dimension.

Table 1 shows the constructs Workplace, community and Institutionsor systems are coming under Social factors of SSCM. Table 2 mentions the items coming under social factors.

The term environment characteristically refers to the natural environment. Improving environmental sustainability means reducing the ecological footprint of the supply chain of the three dimensions, the environmental aspect of supply chain management has been studied the most. Handfield et al. (2002) identify more than 50 environmental performance indicators by which suppliers can be evaluated; Labuschagne et al. (2005a) also identify many factors related to environmental sustainability. The environmental factors divided into six categories. (1) Air: Refers to the local impacts, such as carbon monoxide emissions, as well as global impacts, such as ozone depletion. (2) Water: Refers to both quality and quantity impacts, e.g., toxic discharges as well as total usage. (3) Land: Refers to how much land is used, how it is used, and the impacts of the use, such as soil pollution. (4) Materials: Refers to quantity of material used, the type of material used, and the potential effects of that material. (5) Minerals and energy: Refers to the use of non-renewable mineral and energy resources. (6) Institutions/Systems: Refers to the values, procedures, and systems both internal and external that relate to the environment.

Table 1. Social factors

Group	Item	Factor 1	Factor 2	Factor 3	Chronbach's Alpha
Workplace/Internal	15	0.788			0.635
	17	0.630			0.575
	23	0.724			0.524
	36	0.551			0.505
Community/External	1		0.536		0.506
	24		0.588		0.565
	32		0.652		0.506
Institutions/Systems	22			0.534	0.514
	26			0.566	0.549
	27			0.708	0.588
	29			0.601	0.590

Table 2. Naming of constructs

Item	Name
15	External Social Criteria
17	Health
23	Regulatory Public Service
36	Stake holder Empowerment
1	Initial Social Criteria
24	Sensory stimuli
32	Supporting Community Projects
22	Supporting Educational
26	Cultural properties
27	Economic Welfare Growth
29	Contractual Stake holders

Table 3 shows the dimensions like Air, Water, Land, Material, Minerals and energy and Institutions coming under the umbrella of Environmental factor. Table 4 shows the items of Environmental dimensions.

Table 4 explains the name of items that are coming under Environmental Factors.

The economic dimension of the supply chain refers to the profits earned by the members of the chain as well as the economic benefits realized by the host nations, regions, and communities of those members. Economic factors are sorted into four categories. (1) Economic performance: Refers to the ability of the firm to carry out its operations as well as the market value of the firm. (2) Financial health: Refers to well-being and longterm viability of the firm with respect to financial resources. (3) Market and struc-

Table 3. Environmental factors

Group	Item	Factorl	Factor2	Factor3	Factor4	Factor 5	Factor6	Chrobanch's Alpha
Air	9	0.650						0.558
	10	0.625						0.530
	13	0.658						0.625
Water	14		0.635					0.581
	15		0.648					0.526
Land	8			.761				0.617
	21			0.604				0.557
Materials	5				0.588			0.523
	16				0.656			0.559
Mineral and energy resources	12					0.557		0.624
	19					0.646		0.545
Iinstitutions/Systems	3						0.689	0.508
	17						0.630	0.537

Table 4. Naming constructs

Item	Name
9	Environmental Planning
10	Environmental Assignment
13	Resource Consumption
14	Consumption raw material.
15	Consumption water
8	Environmental Identification.
21	Recyclability
5	Process Adoption.
16	Pollution Produced
12	Environmental Performance
19	Production waste
3	End Pipe Controls
17	Production polluting agents

ture: Refers to health of the market and the configuration of the supply chain. (4) Institutions / Systems: Refers to the internal and external systems, procedures, and values that relate to the economic dimension.

Table 5 and Table 6 refers to the dimensions of Economic factors and naming of the constructs.

Table 6 shows the name of constructs that are coming under economic factor of SSCM.

The main benefits (output)of Sustainable supply chain management in thermal power sector are found by expert opinion are shown in Table 7.

These Benefits are considered as Outputs and all the items coming under three pillers of SSCM (i.e. Social, Environmental and Economic factors) all together are considered as inputs of SCM. As it is predicted that how much the power sector industries are benefited by practicing SSCM . ANN method

Table 5. Economic factors

Group	Item	Factor 1	Factor 2	Factor 3	Factor 4	Chrobanch's Alpha
Economic performance	25	-0.752				0.747
	31	0.677				0.718
	34	0.762				0.783
Financial health	2		0.592			0.708
	18		0.591			0.780
Market and structure	11			0.548		0.796
	20			0.738		0.789
Institutions/Systems	6				0.575	0.744
	30				0.665	0.753
	33				0.565	0.538

Table 6. Naming constructs of economic factor

Item	Name
25	High operating cost
31	Cost investment for waste treatment
34	Engage in environmentally friendly disposal
2	Improvement in Cost reduction of output
18	Depth of supplier pool
11	Less Transportation cost per unit
20	Increase in Market share
6	New use of technologies
30	Reduce energy usage during production
33	Cost investment for disposal of hazardous wastes

Table 7. Output achieved on successive implementation of SSCM in thermal power sector

N°	Name
1.	Customer satisfaction
2.	Supplier's innovation capacity
3.	Quality trust
4.	Upstream Supply chain risk management
5.	Optimal inventory
6.	Flexibility
7.	Reduced Supply chain cost
8.	Trust in suppliers
9.	Waste management

is implemented to find that whether the benefits of Sustainable supply chain managenent are achieved by the measured indicators and what are the items need improvement.

Artificial Neural Network (ANN) is a computational model based on the structure and functions of biological neural networks. Mittal et al. (1998) discussed about neural network and explained that neural network can effectively exploit and represent the non-linear relationship between the consumer satisfaction and their perception of the service; it can be used for modeling of a customer's decision making. For effective evaluation and measurement of quality, neural networks based on back propagation algorithm are widely used to model qualitative and intangible aspects of different service sectors. Nordmann and Luxhoj (2000) have applied for forecasting of service problems in aircraft structural component grouping. Satapathy (2014) has applied ANN to test the service satisfaction in electricity service. Satapathy et.al.(2012) have developed a systematic assessment of the sustainability of water services provided to the consumers in rural, urban and municipality area in India by neural network method.

NETWORK PARAMETERS

The back propagation module of a neural network package Nanuet Pro version 2.3 is used for the training and testing of survey data. The network parameters are taken as follows: Input layer with 34 nodes (all items of social, environmental and economic factors taken after factor analysis), one hidden layer with ten nodes and an output layer with a single node. A single question regarding overall customer satisfaction is considered as the output. Similarly for 9 (benefits) outputs of SSCM the similiar process is followed .As per the software recommendations, the number of nodes in the hidden layer (H) is decided by the relation as: $H = 2\surd\,(I + 1)$ (1)

Normalization of raw data was carried to obtain values between 0 and 1 for expressing all data in a common scale. Learning rate and momentum parameter are set at 25% and 20% respectively during the training phase. The numbers of correct outputs were noted till the root mean square error (RMSE) is minimized to a reasonable value.

SENSITIVITY ANALYSIS

Sensitivity analysis is used to study the impact of changes in service performance along the various items (inputs) of SSCM measuring practices against Output (i.e. benefits of SSCM practice). The model was run by varying the learning parameter, momentum parameter and number of cycles till RMSE is minimized. Both the learning parameter and momentum parameter are set at 0.25 and 0.20 to obtain the best results. The model is run for different number of cycles for diferent outputs till RMSE of 0.01is achieved. Training of the network is stopped at this point. In order to find the robustness of the proposed model, sensitivity analysis was carried out. Sensitivity analysis is used to find the impact of SSCM practices measures on nine outputs or benefits achieved after implementations.

The inputs in the test samples are varied one at a time systematically, up and down $\pm 10\%$ from its base value holding other items at their original values. The scaled change in output is calculated with the current input increased by 10% and the current input decreased by 10%. The scaled change in output is given by: (Scaled output for 10% increase in output - Scaled output for 10% decrease in output) / 2.

Thus, the results obtained are the scaled output change per 10% change in input. The calculation is repeated for every input and every fact and then average across all the facts, yielding a single mean scaled change in output for each input criterion. Table 8 shows the ANN prediction results of all measurand inputs (i.e factors of SSCM) with outputs (benefits of SSCM). The result of Table.8 shows that items like 12,1,5,2,12 and 31 are common items of three factors Social, Environmental and Economic factors as all the ouputs are shoing negative scores. Table.8 shows the items ranked as per the maximum negative score. After these common items some items like 19, 23, 26, 27, 3 etc.are also showing negative scores. But the common negative items (i.e, Environmental Performance, Initial Social Criteria, Recyclability, Process Adoption, Improvement in cost reduction output, Consumption of water and Cost investment for waste treatment) must be taken care for overall improvement. Table 8 shows the measured indicators of SSCM.

Common Items selected from the nine outputs of Sustainable Supply chain management are 12, 1, 21, 5, 2, 15 and 31 with negative scores. So, these are the items need to focus for improvement.

Table 8. Measured indicators of SSCM with respect to outputs (Benefits of SSCM)

Dimension	Item No.	Customer Satisfaction	Supplier Innovation Criteria	Quality Trust	Optimal Inventory	Trust of Suppliers	Flexibility	Reduced Cost	Upstream SCM	Waste Management
Social Factors	15	-0.0275	-0.03	-0.2765	-0.01	-0.08	-0.25	-0.075	-0. 65	-0.08
	17	0.0102	-0.08	0.395	0.15	-0.005	-0.03	0.105	0.0395	-0.006
	23	0.61	-2.283	-0.34	-0.045	1.77	0.35	0.01	0.31	1.237
	36	1.7	-0.0725	0.2825	-0.13	0.31	-0.05	-0.19	0.0825	-0.21
Workplace/ Internal	1	-3.08	-0.76	-0.51	-0.31	-0.276	-0.23	-0.29	-0.43	-0.31
	24	-0.065	-0.26	-0.0475	0.01	0.435	0.25	0.055	-0.157	0.3735
	32	0.275	-0.125	0.275	0.135	0.045	-0.1	0.04	0.11	0.0045
Institutions/ Systems	22	-0.495	0.02	0.2	0.06	0.455	-0.05	0.055	0.22	0.155
	26	0.07	-4.5	0.205	-0.03	-0.22	-0.05	0.06	0.1205	-0.022
	27	-0.05	0.125	-0.69	-0.075	-0.01	-0.3	0.02	-0.479	-0.31
	29	10.86	0.245	-0.086	0.185	0.08	1.05	0.015	-0.025	0.08
Air	9	-0.137	0.0825	-0.2	0.03	0.225	0.2	-0.065	-0.08	0.125
	10	0.2975	12.61	0.14	-0.110	0.165	0.3	-0.04	0.234	0.265
	13	0.2875	0.14	0.143	0.075	0.005	0.2	0.015	0.043	0.103
Water	14	0.255	0.2325	0.1	0.01	0.075	-0.05	-0.025	0.21	0.175
	15	-2.37	-0.29	0.105	0.06	0.155	0.1	0.03	0.103	0.155
Land	8	-0.27	-0.3775	0.095	-0.105	0.134	-0.2	-0.07	0.025	0.134
	21	-0.51	-0.3775	-0.1	-0.49	-0.41	-0.35	-0.22	-0.31	-0.41
Materials	5	-0.775	-0.215	-0.525	-0.57	-0.719	-0.35	-1.05	-1.037	-0.87
	16	0.015	-0.082	0.01	-0.05	0.05	0.3	0.035	0.009	0.05
Mineral and energy resources	12	-14.29	-0.3275	-0.535	-0.62	-0.43	-0.29	-0.86	-0.335	-0.13
	19	0.256	1.106	-0.019	0.48	1.23	-0.35	-0.06	-0.0039	1.23
Institutions/ System s	3	0.115	-0.95	-0.7	0.015	0.012	-0.1	0.15	-0.017	0.012
	17	-0.315	10.34	-0.13	0.205	-0.032	-0.05	0.095	-0.027	-0.132
Economic performance	25	0.77	-0.17	-0.08	0.01	0.05	-0.05	0.02	0.18	0.0075
	31	-0.5	-0.09	-0.07	-0.02	-0.06	-0.2	-0.01	-0.27	-0.26
	34	-0.127	0.2	1.1	0.02	0.11	-0.05	-0.04	2.105	0.19
Financial health	2	-0.9	-0.28	-0.34	-0.185	-0.234	-0.25	-1.025	-0.3544	-0.5734
	18	-0.037	-0.11	-0.13	0.001	0.09	0.115	-0.085	-0.113	0.29
Market and structure	11	0.256	0.324	-0.24	-0.05	-0.07	0.3	0.075	-.024	-0.37
	20	0.028	0.022	0.213	-0.01	-0.021	0.125	-0.012	0.2313	-0.02
Institutions/ Systems	6	-0.031	-0.21	0.012	0.03	-0.4	-0.11	-0.10	0.112	-0.14
	30	-0.046	-0.13	0.008	-0.305	-0.062	-0.24	-0.07	-0.03	0.262
	33	-0.12	0.16	0.12	-0.072	-0.035	0.123	-0.03	0.011	-0.135

(The negative score for average scaled change in output scores per 10% variation in inputs is the norm).

QUALITY FUNCTION DEPLOYMENT

Quality Function Deployment (QFD) is a structured approach to defining customer needs or requirements and translating them into specific plans to produce products to meet those needs. Anwar et al. (2010) have focused on the application of QFD on a utility service company to improve its quality of service through designing the house of quality matrix. Delgado et al. (2007) have proposed customer satisfaction index scores measured through application of QFD. Satapathy (2013) have demonstrated what a utility customer might reasonably expect in his electric service to identify and rank the system design requirements to improve services in Utility sector. Common Items selected from the nine outputs of Sustainable Supply chain management are 12,1,21,5,2,15 and 31 with negative scores are considered as Customer requirement/customer needs shown in Table.9. Enablers or drivers for SSCM in thermal power sector are the design requirements of QFD. So, a questionnaire is designed for drivers of SSCM and factor analysis is done to find important design requirements.The collected data for barriers and drivers of SSCM in Indian thermal industries are subjected to various statistical analyses such as factor analysis and Kaiser–Meyer–Olkin (KMO) test. Factor analysis on 146 useful responses has been conducted using principal component method followed by varimax rotation via SPSS17.0. After analysis it is found that only 21 items are coming under 16 items under 9 enablers. Percentage of total variance explained was found to be 74.3% which is an acceptable value for the principal component varimax rotated factor loading procedure (Johnson and Wichern, 2002). The internal consistency of the actual survey data was tested by computing the Chronbach's alpha (α). The value of alpha for each dimension is shown in Tables 10. The value of KMO, which is a measure of sampling adequacy, was found to be 0.65 and 0.57 indicating that the factor analysis test has proceeded correctly, and the sample used is adequate as the minimum acceptable value of KMO is 0.5 (Othman and Owen, 2001). Therefore, it can be concluded that the matrix did not suffer from multicollinearity or singularity. The result of Bartlett test of sphericity shows that it is highly significant (significance = 0.000) which indicates that the factor analysis processes is correct and suitable for testing multidimensionality (Othman and Owen, 2001). The factors comingafter factor analysis SSCM drivers are Top Management Vision & Initiative, Employee Pressure, Expected Economic Benefits, Expected Risks of Non-Adoption, Government Policies, Academic Involvement, Competitors Action, Customer's expectation.

Drivers

1. Top Management Vision and Initiative
2. Employee Pressure
3. Expected Economic Benefits
4. Expected Risks of Non-Adoption
5. Government Policies
6. Academic Involvement
7. Competitors Action
8. Customer's Expectation
9. Stakeholder's Pressure

Table 9 shows that after factor analysis the following 16 items are coming under nine dimensions (drivers).

Table 10 shows the design requirement to satisfy customer needs.

Table 11 shows the Correlated matrix between customer ratings and design requirement.

Figure 1 shows HOQ for SSCM in thermal power sector of India and ranks most important design requirement.

The HOQ (matrix) is the mainly acknowledged form of QFD. HOQ is constructing from these major components as explain below.

- **Customer Needs (What's):** A structured list of requirements derived from expert's feedback
- **Design Requirements (How's)**: A structured set of relevant and measurable characteristics/ services which are required for fulfilling what's.
- **Planning Matrix (Left Matrix):** Gives customer/expert perceptions observed in surveys. It includes the relative importance of requirements. Interrelationship matrix (centre matrix): Gives the expert's perceptions of interrelationships between design requirements and customer needs. An appropriate scale is applied and illustrated using symbols or figures. Filling this portion of the matrix involves discussions.

Table 9. Driver factor analysis

Dimension	Item	Factor1	Factor2	Factor3	Factor4	Factor5	Factor6	Factor7	Factor8	Factor9	Chronbach's Alpha
Top management vision and initiative	12	0.939									0.838
	15	0.623									0.804
Employee pressure	8		0.829								0.94
	9		0.825								0.861
	17		0.556								0.805
Expected economic benefit	3			0.871							0.883
	1			0.701							0.809
Expected risk of non adoption	5				0.912						0.835 0.843
	6				0.576						0.893
Government policies	17					0.521					0.856
	2					0.901					0.804
Academic involvement	11						0.536				0.946
	20						0.887				0.878
Competitor's action	7							0.797			0.944
Customer expectation	13								0.876		0.777
Stakeholder's pressure	10									0.933	0.881

Table 10. Design requirement

N°	Item	Name
1	12	Employee pressure
2	15	Voice of customers
3	8	Investors pressure
4	9	Media pressure
5	17	Government rewards and incentives
6	3	Competitor's actions
7	1	Top management vision and relation
8	5	Government regulatory requirements
9	6	Supplier's green initiatives
10	17	Government rewards and incentives
11	2	Customer's expectations
12	11	Pressure from labour union
13	20	Compliance of emission, ash utilization, and regulatory norms
14	7	Other stakeholders (such as NGOs)
15	13	Corporate Social Responsibility
16	10	Pressure from local community

Table 11. Correlated matrix between SSCM measurands and driver's of SSCM

	CR	1	2	3	4	5	6	7	8	9	10	11	12	13	14	15	16	Customer Rating
1	5	●	O	O	O	O	O	Δ	Δ	O	Θ	Θ	Θ	Θ	Θ	Δ	Δ	7.1
2	2	●	●	●	Δ	●	●	●	Δ	Θ	Δ	●	●	Δ	●	O	O	3.6
3	5	Δ	Δ	O	Θ	●	●	●	Δ	O	●	●	O	Θ	Δ	O	O	6.43
4	3	O	Δ	O	Θ	O	O	●	Δ	●	●	●	Δ	Δ	Θ	Θ	Θ	4.2
5	2	●	O	O	Δ	Θ	●	O	●	Δ	●	●	O	Θ	●	Θ	Θ	4.33
6	4	●	●	O	●	●	●	●	●	O	●	●	O	Θ	O	O	O	6.43
7	4	●	●	O	Θ	●	●	●	●	O	●	●	Δ	●	Θ	O	Δ	5.4
		11.3	12.2	27.5	18.5	16.01	21.8	11.5	19.5	25.02	15.2	18.1	22.57	18.7	19.2	25.4	23.2	
Revised rating		15.1	15.3	29.9	23.6	20.8	26.8	14.5	21.9	28.3	18.1	20.05	24.7	20.5	20.7	27.6	24.6	
Normalised		0.505	0.51	1	0.78	0.69	0.89	0.48	0.73	0.94	0.605	0.67	0.826	0.67	0.69	0.92	0.82	
Ranking		14	15	1	7	10	4	16	8	2	13	12	6	11	9	3	5	

- **Design Correlation (Top) Matrix:** Used to identify where design requirements support or impede each other in the system or product design Customer Rating= Zi +1/n-1($\sum B_{ij}$ Zj), where

Bij is ith initial customer rating and Zj denote the relationship between customer need and customer rating respectively n number of customer needs Bij [B1ij, B2ij].

Figure 1. HOQ for SSCM in thermal power sector of India
Symbols for QFD: 8 Strong O, 6 Moderate Θ, 4 Week Δ, 2 Very week •Then HOQ is designed in Table 13.

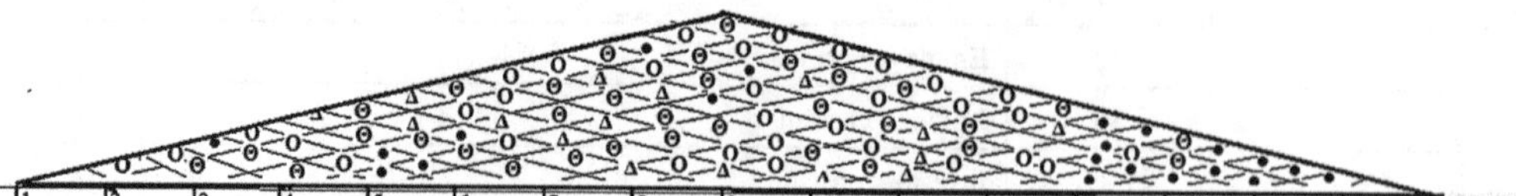

	CR	1	2	3	4	5	6	7	8	9	10	11	12	13	14	15	16	Customer's rating
1	5	•	O	O	O	O	O	Δ	Δ	O	Θ	Θ	Θ	Θ	Θ	Δ	Δ	7.1
2	2	•	•	•	Δ	•	•	•	Δ	Θ	Δ	•	•	Δ	•	O	O	3.6
3	5	Δ	Δ	O	Θ	•	•	•	Δ	O	•	•	O	Θ	Δ	O	O	6.43
4	3	O	Δ	O	Θ	O	O	•	Δ	•	•	•	Δ	Δ	Θ	Θ	Θ	4.2
5	2	•	O	O	Δ	Θ	•	O	•	Δ	•	•	O	Θ	•	Θ	Θ	4.33
6	4	•	•	O	•	•	•	•	•	O	•	•	O	Θ	O	O	O	6.43
7	4	•	•	O	Θ	•	•	•	•	O	•	•	Δ	•	Θ	O	Δ	5.4
		11.3	12.2	27.5	18.5	16.01	21.8	11.5	19.5	25.02	15.2	18.1	22.57	18.7	19.2	25.4	23.2	
Revised rating		15.1	15.3	29.9	23.6	20.8	26.8	14.5	21.9	28.3	18.1	20.05	24.7	20.5	20.7	27.6	24.6	
Normalised		0.505	0.51	1	0.78	0.69	0.89	0.48	0.73	0.94	0.605	0.67	0.826	0.67	0.69	0.92	0.82	
Ranking		14	15	1	7	10	4	16	8	2	13	12	6	11	9	3	5	

The individual rating for each design requirement is obtained from the centre matrix using the following relation:

Design requirement = 1/n $\sum$ Aij Xj

where Aij and Xj denote the relative importance of the ith characteristics with respect to the jth customer need in the relationship matrix and the importance of jth customer needs (customer ratings) and n is the number of customer needs. Figure 1 shows the HOQ designed for improving the SSCM practice by suggested design requirement. It is found that the most important design requirement is 3 (Investors pressure), 9 (Media pressure) and 15 (Corporate Social Responsibility).

CONCLUSION

The implementation of SSCM is a comprehensive and difficult phenomenon, due to its dependency on a number of factors. Those power sector industries implemented SSCM in their practice, they are concerned to see how much they are benefited after this good practice. Such that small power industries also learn lesson from their practice. The output factors or benefits of SSCM are interconnected and their strong relation need to be understood for effective SSCM execution. Hence, an attempt has been made in this paper to suggest about few items that are not able to satisfy SSCM implementation criterion (goal/objective) and showing negative score. These items are coming under SSCM inputs or performance measuring criterion of thermal power plants operating in India. The suggested items are Environmental, Performance, Initial Social Criteria, Recyclability, Process Adoption, Improvement in cost reduction output, Consumption of water and Cost investment for waste treatment. These items needs improvement. the HOQ made by QFD show that the thermal power industries must consider the design requirement like System for waste water and energy recycling ranked first, Waste disposal of water for generating electricity by biomass ranked 2nd and new power plants shall also promote adoption of clean coal and clean power generation technologies ranked 3rd for successfull practice. As these items will be taken care the SSCM implementation or practice will be successfull and the power sector industries will earn maximum benefits from them. Hence the policies must be framed by Government in such a manner that these seven points must be taken care.

FUTURE WORK

As SSCM is a very important concept for all types of Industries.S o in future more studies can be done on different industries. Waste management and barriers must be taken care for providing maximum benefit to Indian Thermal power sectors. More private and public Thermal power industries must be included for more clarity in SSCM concept. Other optimization techniques like DEA, SEM must be added to find more results.

REFERENCES

Adham, K. N., Siwar, C., Bhuiyan, M., & Hossain, A. (2015). *An overview of Malaysian government initiatives on sustainable consumption and production practices*. Academic Press.

Ageron, B., Gunasekaran, A., & Spalanzani, A. (2012). Sustainable supply management: An empirical study. *International Journal of Production Economics*, *140*(1), 168–182. doi:10.1016/j.ijpe.2011.04.007

Anwar, R.M., Masud, A.K.M., Abedin, F., & Hossain, M.E. (2010). QFD for utility services: A case study of electricity distribution company DESCO. *Int. J. Quality and Innovation*, *1*(2), 184–195.

Bhattacharyya, S. C. (2010). Shaping a sustainable energy future for India: Management challenges. *Energy Policy*, *38*(8), 4173–4185. doi:10.1016/j.enpol.2010.03.045

Delgado, J., & Saraiva, P. A., & Almeida, A.T.de. (2007). Electrical power delivery improvement in Portugal through quality function deployment. *Ninth International Conference, Electrical* 10.1109/EPQU.2007.4424223

Diabat, A., Kannan, D., & Mathiyazhagan, K. (2014). Analysis of enablers for implementation of sustainable supply chain management–A textile case. *Journal of Cleaner Production*, *83*, 391–403. doi:10.1016/j.jclepro.2014.06.081

Diabat, A., Kannan, D., & Mathiyazhagan, K. (2014). Analysis of Enablers for implementation of sustainable supply chain management – A textile case. *Journal of Cleaner Production*, *83*, 391–403. doi:10.1016/j.jclepro.2014.06.081

doi: 10.1016/j.jclepro.2014.06.081

Dubey, R., Gunasekaran, A., Wamba, S. F., & Bag, S. (2015). Building Theory of Green Supply Chain Management using Total Interpretive Structural Modeling. *IFAC-Papers Online*, *48*(3), 1688–1694. doi:10.1016/j.ifacol.2015.06.329

Faisal, M. N. (2010). Sustainable supply chains: A study of interaction among the enablers. *Business Process Management Journal*, *16*(3), 508–529. doi:10.1108/14637151011049476

Govindan, K., Khodaverdi, R., & Jafarian, A. (2013). A fuzzy multi criteria approach for measuring sustainability performance of a supplier based on triple bottom line approach. *Journal of Cleaner Production, 47*, 345–354. doi:10.1016/j.jclepro.2012.04.014

Grzybowska, K. (2012). Sustainability in the supply chain: Analyzing the enablers. In Environmental Issues in Supply Chain Management (pp. 25-40). Springer-Verlag.

Handfield, R., Walton, S. V., Sroufe, R., & Melnyk, S. A. (2002). Applying environmental criteria to supplier assessment: A study in the application of the Analytical Hierarchy Process. *European Journal of Operational Research, 141*(1), 70–87. doi:10.1016/S0377-2217(01)00261-2

Jayaram, J., & Avittathur, B. (2015). Green supply chains: A perspective from an emerging economy. *International Journal of Production Economics, 164*, 234–244. doi:10.1016/j.ijpe.2014.12.003

Johnson, R. A., & Wichern, D. W. (2002). *Applied Multivariate Statistical Analysis*. Upper Saddle River, NJ: Prentice-Hall.

Labuschagne, C., Brent, A. C., & Claasen, S. J. (2005a). Environmental and social impact considerations for sustainable project life cycle management in the process industry. *Corporate Social Responsibility and Environmental Management, 12*(1), 38–54. doi:10.1002/csr.76

Labuschagne, C., Brent, A. C., & Van Erck, R. P. G. (2005). Assessing the sustainability performances of industries. *Journal of Cleaner Production, 13*(4), 373–385. doi:10.1016/j.jclepro.2003.10.007

Mathiyazhagan, K., Govindan, K., & Noorul Haq, A. (2014). Pressure analysis for green supply chain management implementation in Indian industries using analytic hierarchy process. *International Journal of Production Research, 52*(1), 188–202. doi:10.1080/00207543.2013.831190

Mittal, V., Ross, W. T., & Baldasare, P. M. (1998). The asymmetric impact of negative and positive attribute-level on overall satisfaction and repurchase intentions. *Journal of Marketing, 62*(3), 33–48. doi:10.2307/1251801

Mohammed, H. (2011). *Modelling the enablers and alternatives for sustainable supply chain management* (Ph.D. Thesis). Concordia Institute for Information Systems Engineering, Canada.

Muduli, K., & Barve, A. (2013). Developing a Framework for Study of GSCM Criteria in Indian Mining Industries. *APCBEE Procedia, 5*, 22–26. doi:10.1016/j.apcbee.2013.05.005

Nordmann, L. H., & Luxhoj, J. T. (2000). Neural network forecasting of service problems for aircraft structural component grouping. *Journal of Aircraft, 37*(2), 332–338. doi:10.2514/2.2599

Othman, A., & Owen, L. (2001). Adopting and measuring customer service quality (SQ) in Islamic banks: a case study in Kuwait finance house. *International Journal of Islamic Financial Services, 3*(1), 1-26.

Rao, A. B., & Kumar, P. (2014). Cost implications of carbon capture and storage for the coal power plants in India. *Energy Procedia, 54*, 431–438. doi:10.1016/j.egypro.2014.07.285

Saaty, T. L. (1970). *Optimization in integers and related external problems*. McGraw-Hill.

Sambasivan, M., & Fei, N. Y. (2008). Evaluation of critical success factors of implementation of ISO 14001 using analytic hierarchy process (AHP): A case study from Malaysia. *Journal of Cleaner Production, 16*(13), 1424–1433. doi:10.1016/j.jclepro.2007.08.003

Satapathy, S. (2014). ANN, QFD and ISM approach for framing electricity utility service in India for consumer satisfaction. *International Journal of Services and Operations Management, 18*(4), 404–428. doi:10.1504/IJSOM.2014.063243

Satapathy, S., Patel, S. K., Mahapatra, S. S., & Biswas, A. (2013). A framework for electricity utility service by QFD technique. *International Journal of Services Sciences, 5*(1), 58–73. doi:10.1504/IJSSCI.2013.057659

Satapathy, S., Patel, S. K., & Mishra, P. D. (2012). Discriminate analysis and neural network approach in water utility service. *International Journal of Services and Operations Management, 12*(4), 468–489. doi:10.1504/IJSOM.2012.047954

Shen, L., Muduli, K., & Barve, A. (2015). Developing a sustainable development framework in the context of mining industries: AHP approach. *Resources Policy, 46*, 15–26. doi:10.1016/j.resourpol.2013.10.006

Tyagi, M., Kumar, P., & Kumar, D. (2015). Analysis of interactions among the drivers of green supply chain management. *International Journal of Business Performance and Supply Chain Modelling, 7*(1), 92–108. doi:10.1504/IJBPSCM.2015.068137

Walker, H., & Jones, N. (2012). Sustainable supply chain management across the UK private sector. *Supply Chain Management, 17*(1), 15–28. doi:10.1108/13598541211212177

KEY TERMS AND DEFINITIONS

ANN: Artificial neural network is a computational model based on the structure and functions of biological neural networks. Information that flows through the network affects the structure of the ANN because a neural network changes—or learns, in a sense—based on that input and output.

HOQ: House of quality is a voice of customer analysis tool and a key component of the quality functional deployment technique.

QFD: Quality function development is a structured approach to defining customer needs or requirements and translating them into specific plans to produce products to meet those needs. The "voice of the customer" is the term to describe these stated and unstated customer needs or requirements.

SSCM: Sustainable supply chain management is a business issue affecting an organization's supply chain or logistics network in terms of environmental, risk, and waste costs. There is a growing need for integrating environmentally sound choices into supply-chain management.

APPENDIX 1

Table 12 shows the appendix 1 information.

Table 12. Appendix 1

Item	Scale				
Question (Social Criterion)	**1**	**2**	**3**	**4**	**5**
Internal Social Criteria					
Employment practices					
Disciplinary and security					
Employee contracts					
Equity labor sources					
Diversity					
Discrimination					
Flexible working arrangements					
Job opportunities					
Employment compensation					
Research and Development					
Career development					
Health and Safety incidents					
Health and Safety practices					
External social criteria					
Local communities influence					
Health					
Education					
Housing					
Service infrastructure					
Mobility infrastructure					
Supporting educational					
Regulatory and public services					
Sensory stimuli					
Security					
Cultural properties					
Economic welfare and growth					
Grants and donations					
Contractual stakeholders					
Procurement standard					
Partnership screens and standards					
Supporting community projects					

continued on following page

Table 12. Continued

Item	Scale				
Question (Social Criterion)	**1**	**2**	**3**	**4**	**5**
Consumers education					
Other stakeholders influence					
Decision influence potential					
Stakeholder empowerment					
Collective audience					
II: Environmental Performance					
Environmental practices					
Pollution controls remediation					
End-of-pipe controls					
Pollution prevention					
Process adoption					
Environmental management					
Environmental establishment					
Environmental identification					
Environmental planning					
Environmental assignment					
Checking and evaluation					
Environmental performance					
Resource consumption					
Consumption of raw material					
Consumption of water					
Pollution production					
Production of polluting agents					
Production of toxic products					
Production of waste					
Post use					
Recyclability					
Re Design					
Compliance with cost (Economic Performance)					
Improvement in cost reduction of output					
Improvement in Product and process Quality					
Better Service capability					
Innovativeness in process operation					
New use of technologies					
Less order fills lead time					
High Order fill rate					

continued on following page

Table 12. Continued

Item	Scale				
Question (Social Criterion)	**1**	**2**	**3**	**4**	**5**
High Order fulfillment cost					
Less Product defect rate of coal					
Less Transportation cost per unit					
High Productivity of plant					
Market value of product					
Better Profitability ratio					
Reliable cost of goods sold					
Return on working capital					
Degree of vertical integration					
Depth of supplier pool					
Breadth of customer base					
Increase in Market share					
Certified by Regulatory compliance (e.g. Consumer Product Safety Act)					
Certified by ISO 9000 certification					
Quality Management System in use					
High Cost investment on the project					
High operating cost					
Achieve economic conditions					
Realized economic profits					
Investment for sustainability					
Cost Investment in IT for sustainability					
Reduce energy usage during production					
Cost investment for waste treatment					
Cost paid as fine for environmental accidents					
Cost investment for disposal of hazardous wastes					
Engage in environmentally friendly disposal					

APPENDIX 2

Table 13 shows the questionnaire about enablers of Sustainable Supply chain management of Thermal power industries.

Table 13. Questionnaire about enablers of Sustainable Supply chain management of Thermal power industries

N°	Taking Into Account Sustainability Drivers	1	2	3	4	5
1	Top management vision and relation					
2	Customer's expectations					
3.	Competitor's actions					
4.	Nature of the business					
5.	Government regulatory requirements					
6.	Supplier's green initiatives					
7.	Other stakeholders (such as NGOs)					
8.	Investors pressure					
9.	Media pressure					
10.	Pressure from local community					
11.	Pressure from labor union					
12.	Employee pressure					
13.	Corporate Social Responsibility					
14.	Eco-literacy amongst Supply Chain Partners					
15.	Voice of customers					
16.	Availability of advanced technology					
17.	Government rewards and incentives					
18.	Pollution regulations followed to reduce global warming					
19.	Maximize ash utilization, ash disposal, to improve situation of water scarcity					
20.	Compliance of emission, ash utilization, and regulatory norms					
21.	Metrics to quantify sustainability benefits in a supply chain					
22.	Support to partners in the supply chain					
23.	Technology management					
24.	Awareness about sustainable practices in a supply chain					
25	Plant affrostation is done to reduce pollution					

Compilation of References

Aagaard, J., & Storr-Paulsen, A. (1995). A comparative study of three different kinds of school furniture. *Ergonomics, 38*(5), 1025–1035. doi:10.1080/00140139508925169 PMID:29105614

Abreu, L., Cid, A., Herrera, G., Lara, J., Laviada, R., Rodríguez, C., & Sánchez, J. (2008). *Perfil por competencias del médico general mexicano.* AMFEM.

Acero Mora, J. (2011). *Los sistemas de bicicleta pública vistos desde la relación servicio - producto.* Bogota: UNC.

Aceves-Gonzalez, C., Prado-León, L. R., & Avila-Chaurand, R. (2010). Factores ergonómicos en el diseño de indicadores visuales estáticos. *Revista de Diseño y Comunicación Visual, Septiembre* (6), 53-59.

Ackoff, R. L. (1981). *Creating the corporate future: Plan or be planned for.* University of Texas Press.

Adham, K. N., Siwar, C., Bhuiyan, M., & Hossain, A. (2015). *An overview of Malaysian government initiatives on sustainable consumption and production practices.* Academic Press.

AgeLab. M. (2017). *Agnes (Age Gain Now Empathy System).* Retrieved August 20, 2017, from: http://agelab.mit.edu/agnes-age-gain-now-empathy-system

Ageron, B., Gunasekaran, A., & Spalanzani, A. (2012). Sustainable supply management: An empirical study. *International Journal of Production Economics, 140*(1), 168–182. doi:10.1016/j.ijpe.2011.04.007

Aguilar-Salinas, C. A. (2013). Comentarios a la ENSANUT 2012. *Salud Pública de México, 55*(2), 347–350.

Ahmed, H., Qureshi, O. M., & Khan, A. A. (2015). Reviving a ghost in the history of technology: The social construction of the recumbent bicycle. *Social Studies of Science, 45*(1), 130–136. doi:10.1177/0306312714560640 PMID:25803920

Alexander, P., Allen, S., & Dutta, D. (1998). Part orientation and build cost determination in layered manufacturing. *Computer Aided Design, 30*(5), 343–356. doi:10.1016/S0010-4485(97)00083-3

Alferez-Padron, C., Maldonado-Macías, A. A., García-Alcaraz, J., Avelar-Sosa, L., & Realyvasquez-Vargas, A. (2017). Workload Assessment and Human Error Identification During the Task of Taking a Plain Abdominal Radiograph: A Case Study. In *International Conference on Applied Human Factors and Ergonomics* (pp. 108–119). Springer.

Allen, D., Karanasios, S., & Slavova, M. (2011). Working with activity theory: Context, technology, and information behavior. *Journal of the Association for Information Science and Technology, 62*(4), 776–788. doi:10.1002/asi.21441

Alonso, L. (2015). The Proust Effect. The Senses as Doorways to Lost Memories. *Revista investigación y Ciencia, 70.* Retrieved from http://www.investigacionyciencia.es/revistas/mente-y-cerebro/pensamiento-creativo-619/memoria-sensorial-12753

Altinkemer, K., Ozcelik, Y., & Ozdemir, Z. (2011). Productivity and performance effects of business process reengineering; a firm level analysis. *Journal of Management Information Systems*, *27*(4), 129–162. doi:10.2753/MIS0742-1222270405

Altshuller, G. (1984). *Creativity as an exact science : the theory of the solution of inventive problems*. New York: Gordon and Breach Science Publishers. Retrieved from http://cds.cern.ch/record/450367

Altshuller, G. (2002). *40 principles : TRIZ keys to innovation* (L. Shulyak & S. Rodman, Eds.; 3rd ed.). Worcester, MA: Technical Innovation Center. Retrieved from https://books.google.com.mx/books?id=mqlGEZgn5cwC&dq=triz&lr=&hl=es&source=gbs_navlinks_s

Ambrose, G., & Harris, P. (2009). *Design thinking*. London, UK: AVA Publishing.

Anaya Echeverría, N. (2017). *Modelo Industrial de Escritorio Escolar Ergonómico. MX/f/2017/003341*. Zapopan, Jalisco: Instituto Mexicano de la Propiedad Industrial.

Anaya Echeverría, N. (2017). *Modelo Industrial de Silla Escolar Ergonómica. MX/f/2017/080312*. Zapopan, Jalisco: Instituto Mexicano de la Propiedad Industrial.

Ankrum, D. R. (1999). Visual ergonomics in the office — Guidelines. *Occupational Health & Safety (Waco, Tex.)*, *68*(7), 64–74. PMID:10081834

Anshel, J. R. (2006). Visual ergonomics in the workplace. *Professional Safety*, *51*(8), 20.

ANSI 2535. (1987). *Criteria for safety symbols*. American National Standards Institute.

Anwar, R.M., Masud, A.K.M., Abedin, F., & Hossain, M.E. (2010). QFD for utility services: A case study of electricity distribution company DESCO. *Int. J. Quality and Innovation*, *1*(2), 184–195.

Apple. (2017). *AirPods*. Retrieved from https://www.apple.com/mx/shop/product/MMEF2BE/A/airpods

APR. (2013). "Dinks", la nueva clase de parejas mexicanas. *El economista*. Retrieved from http://eleconomista.com.mx/finanzas-personales/2013/07/16/dinks-nueva-clase-parejas-mexicanas

Arias, J. (2004). *Propedéutica quirúrgica: preoperatorio, operatorio, postoperatorio*. Editorial Tebar.

Armenta- Hernández. O. D., Maldonado-Macías, A. A., García-Alcaraz, J. L., & Avelar-Sosa, L. (2016). Analysis of mental workload and human error in the task of watching a movie using a home entertainment device. In Ergonomía Ocupacional. Investigaciones y soluciones (9th ed.; pp. 42–52). SEMAC.

Arnette, A. N., Brewer, B. L., & Choal, T. (2014). Design for sustainability (DFS): The intersection of supply chain and environment. *Journal of Cleaner Production*, *83*, 374–390. doi:10.1016/j.jclepro.2014.07.021

Assunção, A., Carnide, F., Vieira, F., Silva, S., & Araújo, J. (2013). Mismatch of school furniture and back pain in adolescents with different maturation levels. *Int. J. of Human Factors and Ergonomics*, *2*(1), 66–81.

Astrom, K. J., Klein, R. E., & Lennartsson, A. (2005). Bicycle dynamics and control: Adapted bicycles for education and research. *IEEE Control Systems*, *25*(4), 26–47. doi:10.1109/MCS.2005.1499389

Attorney, B. Y. (1934). *Automatic Anthropometer*. Retrieved from https://www.google.com/patents/US1946075

Ávila Chaurand, R., Prado León, L., & González Muñoz, E. (2007). *Anthropometric dimensions of Latin American population* (2nd ed.). Universidad de Guadalajara.

Avila-Chaurand, R. (2013). *La usabilidad en el diseño de productos de consumo: un punto de vista desde la ergonomía*. Retrieved from https://es.scribd.com/document/172195315/LA-USABILIDAD-EN-EL-DISENO

Avila-Chaurand, R., Prado-León, L. R., & González, Muñoz, E. L. (2007). *Dimensiones Antropométricas de Población Latinomericana* (2nd ed.). Guadalajara, Mexico: Universidad de Guadalajara.

Ávila–Chaurand, R., Prado–León, L. R., & González–Muñoz, E. L. (2007). Dimensiones antropométricas de la población latinoamericana: México, Cuba, Colombia, Chile (2nd ed.). Guadalajara, México: Universidad de Guadalajara.

Ávila-Chaurand, R., Prado-León, L. R., González-Muñoz, E. L., & Muñoz, G. (2007). *Dimensiones antropométricas de población latinoamericana.* Guadalajara: Universidad de Guadalajara. Retrieved from https://www.researchgate.net/profile/Rosalio_Avila-Chaurand/publication/31722433_Dimensiones_antropometricas_de_la_poblacion_latinoamericana_Mexico_Cuba_Colombia_Chile_R_Avila_Chaurand_LR_Prado_Leon_EL_Gonzalez_Munoz/links/55490e8d0cf2ebfd8e3ad6bc/Dimensi

Ayhan, M. B. (2013). A fuzzy ahp approach for supplier selection problem: A case study in a gearmotor company. *International Journal of Managing Value and Supply Chains, 4*(3), 11–23. doi:10.5121/ijmvsc.2013.4302

Azadeh, A., Fam, I., & Garakani, M. (2007). A Total Ergonomic Design Approach to Enhance the Productivity in a Complicated Control System. *Information Technology Journal, 6*(7), 1036–1042. doi:10.3923/itj.2007.1036.1042

Baber, C., & Stanton, N. A. (1996). Human error identification techniques applied to public technology: Predictions compared with observed use. *Applied Ergonomics, 27*(2), 119–131. doi:10.1016/0003-6870(95)00067-4 PMID:15677051

Bang, I. C., & Heo, G. (2009). An axiomatic design approach in development of nanofluid coolants. *Applied Thermal Engineering, 29*(1), 75–90. doi:10.1016/j.applthermaleng.2008.02.004

Bao, S., Howard, N., Spielholz, P., & Silverstein, B. (2007). Two posture analysis approaches and their application in a modified Rapid Upper Limb Assessment evaluation. *Ergonomics, 50*(12), 2118–2136. doi:10.1080/00140130701458230 PMID:17852373

Bao, S., Howard, N., Spielholz, P., Silverstein, B., & Polissar, N. (2009). Interrater Reliability of Posture Observations. *Human Factors: The Journal of the Human Factors and of Ergonomics Society, 51*(3), 292–309. doi:10.1177/0018720809340273 PMID:19750793

Barab, S. A., Barnett, M., Yamagata-Lynch, L., Squire, K., & Keating, T. (2002). Using activity theory to understand the systemic tensions characterizing a technology-rich introductory astronomy course. *Mind, Culture, and Activity, 9*(2), 76–107. doi:10.1207/S15327884MCA0902_02

Barros, R. Q., Marçal, M. A., & Soares, M. M. (2015). Ergonomic Analysis of the job of assembly and maintenance in an electronic equipment company. *Procedia Manufacturing, 3*, 6542–6549. doi:10.1016/j.promfg.2015.07.953

Beard, J., Officer, A., Cassels, A., Bustreo, F., Worning, A. M., & Asamoa-Baah, A. (2015). Informe mundial sobre el envejecimiento y la salud ed. WHO.

Bedny, G. Z., & Harris, S. R. (2005). The systemic-structural theory of activity: Applications to the study of human work. *Mind, Culture, and Activity, 12*(2), 128–147. doi:10.120715327884mca1202_4

Bejarano, M. (2011). Evaluación cuantitativa de la eficiencia en las salas de cirugía. *Revista Colombiana de Cirugía, 26*(4).

Bendix, T. (1986). Sitting postures-A review of biomechanics and ergonomic aspects. *Man and Medicine, 2*, 77–81.

Bendix, T., & Bloch, I. (1986). How Should a seated workplace with a tiltable chair be adjusted? *Applied Ergonomics, 17*(2), 127–135. doi:10.1016/0003-6870(86)90250-4 PMID:15676578

Bernabeu, X. P., Sabater, J. P. G., & Maheut, J. (2010). Arquitectura de un modelo de simulación basado en sistemas multi-agente de una red de suministro en el sector del automóvil. *4th International Conference On Industrial Engineering and Industrial Management*, 3–14.

Berrueta, É. A. R., Malagón, G. V., Alvarado, B. M., Fernández, M. del C. A., & Sosa, L. E. O. M. V. C. (2008). Tiempo perdido entre cirugías y su repercusión en el cumplimiento de la programación quirúrgica del Hospital General de Querétaro. *Cirujano General, 30*(4), 220–223.

Bevilacqua, M., Ciarapica, F. E., Mazzuto, G., & Paciarotti, C. (2013). Visual management implementation and evaluation through mental workload analysis. *IFAC Proceedings Volumes, 46*(7), 294-299.

Bhattacharyya, S. C. (2010). Shaping a sustainable energy future for India: Management challenges. *Energy Policy, 38*(8), 4173–4185. doi:10.1016/j.enpol.2010.03.045

Bilalis, N., Scroubelos, G., Antoniadis, A., Emiris, D., & Koulouriotis, D. (2002). Visual factory: Basic principles and the 'zoning' approach. *International Journal of Production Research, 40*(15), 3575–3588. doi:10.1080/00207540210140031

Bitner, M., & Sierak, J. (2017). Institute of Economic Research Working Papers Kristóf Gyódi. In *9th International Conference on Applied Economics Contemporary Issues in Economy, Institute of Economic Research, Polish Economic Society Branch in Torun, Faculty of Economic Sciences and Management, Nicolaus Copernicus University* (pp. 22–23). Torun, Poland: Institute of Economic Research.

Blackler, F. (1993). Knowledge and the theory of organizations: Organizations as activity systems and the reframing of management. *Journal of Management Studies, 30*(6), 863–884. doi:10.1111/j.1467-6486.1993.tb00470.x

Blain, C., Levy, S. E., & Brent, J. R. (2005). Destination Branding: Insights and Practices from Destination Management Organizations. *Journal of Travel Research, 43*(4), 328–338. doi:10.1177/0047287505274646

Blessing, L., & Chakrabarti, A. (2009). *DRM, a design research methodology*. London, UK: Springer. doi:10.1007/978-1-84882-587-1

Bligård, L. O., & Osvalder, A. L. (2013). Enhanced cognitive walkthrough: Development of the cognitive walkthrough method to better predict, identify, and present usability problems. *Advances in Human-Computer Interaction, 2013*, 9. doi:10.1155/2013/931698

Blum. (2017). *Age explorer®: Preparing for the future by choosing quality hardware.* Retrieved August 20, 2017, from: https://www.blum.com/us/en/03/30/10/

Bonnici, P. (1999). *Lenguaje Visual, el Medio Oculto de la Comunicación.* Luis Hernando Ramirez C.

Borgianni, Y., & Matt, D. T. (2016). Applications of TRIZ and Axiomatic Design: A Comparison to Deduce Best Practices in Industry. In Procedia CIRP (Vol. 39). Academic Press. doi:10.1016/j.procir.2016.01.171

Borkan, G. A., Hults, D. E., & Glynn, R. J. (1983). Role of Longitudinal Change and Secular Trend in Age Differences in Male Body Dimensions. *Human Biology, 55*, 629–641. doi:10.2307/41464710 PMID:6642483

Borshchev, A., & Filippov, A. (2004). From system dynamics and discrete event to practical agent based modeling: reasons, techniques, tools. *Proceedings of the 22nd international conference of the system dynamics society*, 22.

Brady. (2012). *Visual work handbook.* Brady Inc.

Brewer, B., & Arnette, A. N. (2017). Design for procurement: What procurement driven design initiatives result in environmental and economic performance improvement? *Journal of Purchasing and Supply Management, 23*(1), 28–39. doi:10.1016/j.pursup.2016.06.003

Bronfenbrenner, U. (1995). The bioecological model from a life course perspective: Reflections of a participant observer. In P. Moen & G. H. Elder Jr., (Eds.), *Examining Lives in Context: Perspectives on the Ecology of Human Development* (pp. 519–618). Washington, DC: American Psychological Association. doi:10.1037/10176-017

Brown, N., & Scurr, J. (2012). The need for a standardized anthropometric protocol for objective assessment of pre-and postoperative breast surgery. *Gland Surgery*, *1*(27), 1–4. doi:10.3978/j.issn.2227-684X.2012.10.01 PMID:25083437

Bruseberg, A., & McDonagh-Philp, D. (2002). Focus groups to support the industrial/product designer: A review based on current literature and designers' feedback. *Applied Ergonomics*, *32*(1), 27–38. doi:10.1016/S0003-6870(01)00053-9 PMID:11827136

Bubb, H., Engstler, F., Fritzsche, F., Mergl, C., Sabbah, O., Schaefer, P., & Zacher, I. (2006). The development of RAMSIS in past and future as an example for the cooperation between industry and university. *International Journal of Human Factors Modelling and Simulation*, *1*(1), 140–157. doi:10.1504/IJHFMS.2006.011686

Buchanan, R. (1992). Wicked problems in design thinking. *Design Issues*, *8*(2), 5–21. doi:10.2307/1511637

Burdek, B. (1994). History, theory and practice of industrial design. Gustavo Gili.

Burt, C., & Benbow, M. (2008). *Children and handwriting ergonomics. In Ergonomics for Children: Designing products and places for toddlers to teens* (pp. 689–720). London: Taylor & Francis.

Cacha, C. (1999). *Ergonomics and Safety in Hand Tools Design*. Boca Raton, FL: CRC Press.

Cajaiba-Santana, G. (2014). Social innovation: Moving the field forward. a conceptual framework. *Technological Forecasting and Social Change*, *82*, 42–51. doi:10.1016/j.techfore.2013.05.008

Calderoni, D. R., Motta, M. M., & Kharmandayan, P. (2016). Development and implementation of an anthropometric protocol to evaluate results of otoplasty. *Revista Brasileira de Cirurgia Plástica (RBCP) –. British Journal of Plastic Surgery*, *31*(1), 66–73. doi:10.5935/2177-1235.2016RBCP0010

Cañas, J., & Waern, Y. (2001). *Ergonomía cognitiva: aspectos psicológicos de la interacción de las personas con la tecnología de la información*. Granada.

Cardemil, C., Latorre, M., Filp, J., & Galvez, G. (1991). Factors affecting the improvement of learning in basic education. *Journal of Education*, 208.

Carmeli, E., Patish, H., & Coleman, R. (2003). The aging hand. *The Journals of Gerontology. Series A, Biological Sciences and Medical Sciences*, *58*(2), 146–152. doi:10.1093/gerona/58.2.M146 PMID:12586852

Carmichael, C., & Burke, E. (1997). *BICICLETA. Salud y ejercicio*. Ed Paidotribo.

Carse, B., Thompson, A., & Stansfield, B. (2011). A novel device for evaluating forces during the dynamic jar opening action—Do older and younger adults do things differently? *Medical Engineering & Physics*, *33*(4), 521–525. doi:10.1016/j.medengphy.2010.11.017 PMID:21212013

Casado, C., Hernández, V., Jiménez, R., Fernández, C., Carrasco, P., & Palacios, D. (2015). Actividad física en población adulta trabajadora española: Resultados de la Encuesta Europea de Salud en España. *Atencion Primaria*, *1*(5), 10.

Castellucci, H., Arezes, P., & Viviani, C. (2010). Mismatch Between classroom furniture and anthropometric Measures in Chilean schools. *Applied Ergonomics*, *41*(4), 563–568. doi:10.1016/j.apergo.2009.12.001 PMID:20031115

Castellucci, I., Gonçalves, M. A., & Arezes, P. (2009). Ergonomic considerations classrooms in school portuguesas first cycle. *Science. Job*, *11*(34), 184–187.

Cavassa, D. (2006). *Ergonomía y Productividad*. LIMUSA.

Cawthray, R. & Denison, E. (1999). *Packaging*. McGraw Hill.

CBS. (2014). *Ford car markers wear 'age suits' to design for older drivers*. Retrieved August 20, 2017, from: http://www.cbc.ca/news/canada/windsor/ford-car-makers-wear-age-suits-to-design-for-older-drivers-1.2762417

Celik, M. (2009a). A hybrid design methodology for structuring an Integrated Environmental Management System (IEMS) for shipping business. *Journal of Environmental Management*, *90*(3), 1469–1475. doi:10.1016/j.jenvman.2008.10.005 PMID:19038488

Celik, M. (2009b). Establishing an Integrated Process Management System (IPMS) in ship management companies. *Expert Systems with Applications*, *36*(4), 8152–8171. doi:10.1016/j.eswa.2008.10.022

Celik, M., Cebi, S., Kahraman, C., & Er, I. D. (2009). An integrated fuzzy QFD model proposal on routing of shipping investment decisions in crude oil tanker market. *Expert Systems with Applications*, *36*(3), 6227–6235. doi:10.1016/j.eswa.2008.07.031

Celório, C. (1993). *Diseño del embalaje para exportación*. IMPEE.

Ceschin, F., & Gaziulusoy, I. (2016). Evolution of design for sustainability: From product design to design for system innovations and transitions. *Design Studies*, *47*, 118–163. doi:10.1016/j.destud.2016.09.002

CEVECE. (2014). *Factores De Riesgo De Enfermedad Cardiovascular En Adultos Mayores*. Retrieved MARZO, 2015, from http://salud.edomex.gob.mx/cevece/doc/Documentos/Facriesgo_cardio_amayores.pdf

Chang, S. W., & Wang, M. J. J. (2007). Digital human modeling and workplace evaluation: Using an automobile assembly task as an example. *Human Factors and Ergonomics in Manufacturing & Service Industries*, *17*(5), 445–455. doi:10.1002/hfm.20085

Charles, T. S., Micheline, C. D., & Jennifer, D. (2012). Hazard perception in older drivers. *Int. J. Human Factors and Ergonomics*, *1*(3), 221–233. doi:10.1504/IJHFE.2012.050869

Chedi, J. M. (2015). The Need for the Incorporation of Ergonomics into Curriculum of the Technical and Vocational Education and Training for Teacher/Trainer Professionalization. *3rd UPI International Conference on Technical and Vocational Education and Training*, 105-109.

Cheng, W., Fuh, J. Y. H., Nee, A. Y. C., Wong, Y. S., Loh, H. T., & Miyajawa, T. (1995). Multi-objective optimization of part-building orientation in stereolithography. *Rapid Prototyping Journal*, *1*(4), 12–23. doi:10.1108/13552549510104429

Chen, Y., & Wang, R. (2016). Are Humans Rational? Exploring Factors Influencing Impulse Buying Intention and Continous Buying Intention. *Journal of Consumer Behaviour*, *15*(2), 186–197. doi:10.1002/cb.1563

Chiodo, J. D., & Ijomah, W. L. (2014). Use of active disassembly technology to improve remanufacturing productivity: Automotive application. *International Journal of Computer Integrated Manufacturing*, *27*(4), 361–371. doi:10.1080/0951192X.2012.667151

Chomistek, A., Manson, J., Stefanick, M., Lu, B., Sands-Lincoln, M., Going, S., & Eaton, C. (2013). Relationship of Sedentary Behavior and Physical Activity to Incident Cardiovascular Disease. *Journal of the American College of Cardiology*, *61*(23), 9. doi:10.1016/j.jacc.2013.03.031 PMID:23583242

Christensen, C. M., & Raynor, M. E. (2003). *The innovator's solution: Creating and sustaining successful growth*. Boston, MA: Harvard Business Press.

Chung, J., & Wong, T. (2007). Anthropometric evaluation for primary school furniture design. *Ergonomics, 50*(3), 323–334. doi:10.1080/00140130600842328 PMID:17536771

Churchman, C. W. (1967). Guest editorial: Wicked problems. *Management Science, 14*(4), B141–B142.

Civitci, S. (2004). An ergonomic garment design for elderly Turkish men. *Applied Ergonomics, 35*(3), 243–251. doi:10.1016/j.apergo.2004.02.001 PMID:15145287

Clement, F. J. (1974). Longitudinal and Cross-Sectional Assessments of Age Changes in Physical Strength as Related to Sex, Social Class, and Mental Ability. *Journal of Gerontology, 29*(4), 423–429. doi:10.1093/geronj/29.4.423 PMID:4833758

Cochran, D. J., Stentz, T. L., Stonecipher, B. l., & Hallbeck, M. S. (1999). Guide for Videotaping and Gathering Data on Jobs Analysis of Risks of Musculoskeletal Disorders. In W. Karwoski & W. S. Marras (Eds.), *The Occupational Ergonomics Handbook* (pp. 511–524). Boca Ratón, FL: University of Nebraska-Lincoln.

Cochran, D. S., Eversheim, W., Kubin, G., & Sesterhenn, M. L. (2000). The application of axiomatic design and lean management principles in the scope of production system segmentation. *International Journal of Production Research, 38*(6), 1377–1396. doi:10.1080/002075400188906

Conklin, J. (2005). Wicked problems and social complexity. In *Dialogue mapping: Building shared understanding of wicked problems* (pp. 1–25). Chichester, UK: John Wiley & Sons.

Conway, L. (1998). *Diseño de empaque*. Editorial Somohano.

Cooper, R. G. (1994). New products: The factors that drive success. *International Marketing Review, 11*(1), 60–76. doi:10.1108/02651339410057527

Cooper, R. G., & Kleinschmidt, E. J. (1986). An investigation into the new product process: Steps, deficiencies, and impact. *Journal of Product Innovation Management, 3*(2), 71–85. doi:10.1016/0737-6782(86)90030-5

Corkery, J., Mikalsen, M., & Allan, K. (2017). *Corporate social responsibility: The good corporation*. Centre for Commercial Law.

Corlett, E. N., & Bishop, R. P. (1976). A Technique for Assessing Postural Discomfort. *Ergonomics, 19*(2), 175–182. doi:10.1080/00140137608931530 PMID:1278144

Corlett, N. (1995). The evaluation of posture and its effects. In J. r. Wilson & E. N. Corlett (Eds.), *Evaluation of human work. A practical ergonomics methodology* (2nd ed.; pp. 662–713). Nottingham, UK: Academic Press.

Cornell, P. (2002). The impact of changes in teaching and learning on furniture and the learning environment. *New Directions for Teaching and Learning, 2002*(92), 33–42. doi:10.1002/tl.77

Corona, G. P. (2014). *Propuesta de mejora en el ajuste de prendas de vestir para dama por análisis de elementos antropométricos de la Zona Metropolitana de Guadalajara*. University of Guadalajara.

Costa, A., Schoolmeester, D., Dekker, M., & Jongen, W. (2003). Exploring the use of consumer collages in product design. *Trends in Food Science & Technology, 14*(1-2), 17–31. doi:10.1016/S0924-2244(02)00242-X

Costa, J. (1998). *La esquemática: visualizar la información*. Buenos Aires, Argentina: Paidós.

Costa, J. (1999). *Diseñar para los ojos*. Barcelona: Gustavo Gilli.

Coughlin, J. F. (2013). *Agnes & livable communities*. Retrieved August 3, 2017, from: http://agelab.mit.edu/files/final-reports/MITR22-3FP.pdf

Cox, G. (2005, May). *Cox review of creativity in business: Building on the UK's strengths.* London, UK: The Stationary Office.

Crespo, J. J., Delgado, J. L., Blanco, O., & Aldecoa, S. (2015). Guía básica de detección del sedentarismo y recomendaciones de actividad física en atención primaria. *Atencion Primaria, 47*(3), 9.

Creswell, J. W. (1998). *Qualitative Inquiry and Research Design: Choosing among Five Traditions.* Thousand Oaks, CA: SAGE Publications.

Crilly, N., Moultrie, J., & Clarkson, P. J. (2009). Shaping things: Intended consumer response and the other determinants of product form. *Design Studies, 30*(3), 224–254. doi:10.1016/j.destud.2008.08.001

Criollo, M., Romero, M., & Ruiz, T. F. (2017). University students' self-efficacy for learning how to conduct research. *Psicología Educativa, 23*, 63-72. Retrieved 04 04, 2017, from http://web.a.ebscohost.com

Cross, N. (1993). Science and design methodology: A review. *Research in Engineering Design, 5*(2), 63–69. doi:10.1007/BF02032575

Cross, N. (2004). Expertise in design: An overview. *Design Studies, 25*(5), 427–441. doi:10.1016/j.destud.2004.06.002

Cross, N. (2006). *Designerly ways of knowing.* Basel, Switzerland: Birkhäuser.

Crul, M., & Diehl, J. (2006). *Design for Sustainability: A Practical Approach for Developing Economies.* Paris: UNEP.

Cruz, A., & Garnica, A. (2007). Ergonomía. *Evaluación E Investigación, 1*(2), 96–97. Retrieved from http://www.saber.ula.ve/bitstream/123456789/24086/2/resena2.pdf

CSO. (2007). *Ageing in Ireland.* Stationery Office, Dublin, Ireland. Retrieved July 31, 2017, from: http://www.cso.ie/en/media/csoie/releasespublications/documents/otherreleases/2007/ageinginireland.pdf

Cuñat, R. J. (2007). Aplicación de la teoría fundamentada (Grounded Theory) al estudio del proceso de creación de empresa. *Decisiones basadas en el conocimiento y en el papel social de la empresa: XX Congreso anual de AEDEM.* Palma de Mallorca: Asociación Española de Dirección y Economía de la Empresa (AEDEM). Retrieved 03 17, 2015, from Dialnet: https://dialnet.unirioja.es/servlet/articulo?codigo=2499458

Cushman, W. H., & Rosenberg, D. J. (1991). Human factors in product design. *Advances in human factors/ergonomics, 14.*

Cutkosky, M. (1989). On grasp choice, grasp models and the design of hands for manufacturing tasks. *IEEE Transactions on Robotics and Automation, 5*(3), 269–279. doi:10.1109/70.34763

da Costa, B. R., & Ramos-Vieira, E. (2009). Risk factors for work-related musculoskeletal disorders: A systematic review of recent longitudinal studies. *American Journal of Industrial Medicine, 53*(3), 285–323. doi:10.1002/ajim.20750 PMID:19753591

Daalhuizen, J. J. (2014). *Method usage in design: How methods function as mental tools for designers* (Unpublished PhD dissertation). Delft University of Technology.

Datschefki, E. (2002). *El Re-Diseño de Productos. Productos Sustentables, el regreso a los ciclos naturales.* McGraw-Hill.

David, G. (2005). Ergonomic methods for assessing exposure to risk factors for work-related musculoskeletal disorders. *Occupational Medicine, 55*(3), 190–199. doi:10.1093/occmed/kqi082 PMID:15857898

De la Cruz, L., & Mejía, G. M. (2017). Reflective didactic strategy to integrate semiotic theory and creative practice in graphic design education. Art. *Design & Communication in Higher Education, 16*(1), 83–97. doi:10.1386/adch.16.1.83_1

De la Fuente, J., Gustafson, S., Twoney, C., & Bix, L. (2015). An affordance-based methodology for package design. *Packaging Technology & Science*, *28*(2), 157–17. doi:10.1002/pts.2087

De la Torre, G. (2000). *El lenguaje de los símbolos gráficos*. Limusa.

De los Rios, I. C., & Charnley, F. J. (2016). Skills and capabilities for a sustainable and circular economy: The changing role of design. *Journal of Cleaner Production*.

Delgado, J., & Saraiva, P. A., & Almeida, A.T.de. (2007). Electrical power delivery improvement in Portugal through quality function deployment. *Ninth International Conference, Electrical* 10.1109/EPQU.2007.4424223

Delgado, S. (1990). Predictive values for grip and torque strengths using isometric strength testing unit (ISTU) hand attachments. *Journal of Disability*, *1*, 73–78.

Deneçli, S. (2015). The Effect of Consumers' Attitudes towards Logos or Emblems on Brand Preference. *Journal of Marmara University Social Sciences Institute/Öneri*, *11*(44), 319-335. doi:.10.14783/od.v11i44.5000080017

Designkit. (n.d.). *Methods*. Retrieved 05 24, 2016, from http://www.designkit.org/methods/25

Det, M. J., Meijerink, W. J. H. J., Hoff, C., Totte, E. R., & Pierie, J. P. E. N. (2009). Optimal ergonomics for laparoscopic surgery in minimally invasive surgery suites: A review and guidelines. *Surgical Endoscopy*, *23*(6), 1279–1285. doi:10.100700464-008-0148-x PMID:18830751

Diabat, A., Kannan, D., & Mathiyazhagan, K. (2014). Analysis of enablers for implementation of sustainable supply chain management–A textile case. *Journal of Cleaner Production*, *83*, 391–403. doi:10.1016/j.jclepro.2014.06.081

Dianat, I., Karimi, M. A., Asl Hashemi, A., & Bahrampour, S. (2013). Classroom furniture and anthropometric characteristics of Iranian high school students: Proposed dimensions based on anthropometric data. *Applied Ergonomics*, *44*(1), 101–108. doi:10.1016/j.apergo.2012.05.004 PMID:22695080

Dickinson, A. L. (2006). Integrating Axiomatic Design Into a Design for Six Sigma Deployment (DFSS). *Design*, 2–7.

Dickson, E. W., Singh, S., Cheung, D. S., Wyatt, C. C., & Nugent, A. S. (2009). Application of lean manufacturing techniques in the emergency department. *The Journal of Emergency Medicine*, *37*(2), 177–182. doi:10.1016/j.jemermed.2007.11.108 PMID:18722732

DiDomenico, A. T. (2003). *An investigation on subjective assessment of workload and postural stability under conditions of joint mental and physical demands*. Virginia Polytechnic Institute and State University.

Diego-Mas, J.-A., Alcaide-Marzal, J., & Poveda-Bautista, R. (2017). Errors Using Observational Methods for Ergonomics Assessment in Real Practice. *Human Factors*, (0). doi: 10.1177/0018720817723496

Diener, A. (2010). *Afterlife: An Essential Guide to Design for Disassembly*. Retrieved from http://www.core77.com/posts/15799/afterlife-an-essential-guide-to-design-for-disassembly-by-alex-diener-15799

Díez de Castro, E. C., Landa, F. J., & Navarro, A. (2006). *Merchandising: Teoría y práctica*. Pirámide.

Djellal, F., & Gallouj, F. (2012). Social innovation and service innovation. In H.-W. Franz, J. Hochgerner, & J. Howaldt (Eds.), *Challenge social innovation: Potentials for business, social entrepreneurship, welfare and civil society* (pp. 119–137). Berlin, Germany: Springer-Verlag. doi:10.1007/978-3-642-32879-4_8

Domínguez-Sosa, G., Zavala-González, M. A., De la Cruz-Méndez, D. del C., & Ramírez-Ramírez, M. O. (2010). Síndrome de sobrecarga en cuidadores primarios de adultos mayores de Cárdenas, Tabasco, México. Enero a mayo de 2008. *Revista Médicas UIS*, *23*(1), 28–37.

Dorst, K. (2008). Design research: A revolution-waiting-to-happen. *Design Studies, 29*(1), 4–11. doi:10.1016/j.destud.2007.12.001

Drury, C. G., Atiles, M., Chaitanya, M., Lin, J.-F., Marin, C., Nasarwanji, M., ... Sunm, M. (2006). Vicarious perception of postural discomfort and exertion. *Ergonomics, 49*(14), 1470–1485. doi:10.1080/00140130600774372 PMID:17050388

Dubey, R., Gunasekaran, A., Wamba, S. F., & Bag, S. (2015). Building Theory of Green Supply Chain Management using Total Interpretive Structural Modeling. *IFAC-Papers Online, 48*(3), 1688–1694. doi:10.1016/j.ifacol.2015.06.329

Dumas, J. S., & Redish, J. (1999). *A practical guide to usability testing*. Intellect Books.

Dun, J., & Neumann, W. P. (2005). *Ergonomics Contribution to Company Strategies. 10th International Conference on Human Aspects of Advanced Manufacturing: Agility and Hybrid Automation (HAAMAHA)*, San Diego, CA.

Duran, L. (2004). *La salud del adulto mayor en el Instituto Mexicano del Seguro Social.* DF, México.

Durivage, M. A. (2016). *Practical Process Validation.* ASQ Quality Press.

Durmusoglu, M., & Kulak, O. (2008). A methodology for the design of office cells using axiomatic design principles☆. *Omega, 36*(4), 633–652. doi:10.1016/j.omega.2005.10.007

Ebay. (2016). *Precor Icarian Super Banco | eBay*. Retrieved July 20, 2017, from http://www.ebay.com/itm/Precor-Icarian-Super-Banco-/131881903124?_ul=AR

Eduardo Wainstein, G. (2014). Patogénesis de la artrosis. *Revista Médica Clínica Las Condes, 25*(5), 723–727. doi:10.1016/S0716-8640(14)70098-7

Efer, O. O. (2017). 8. Industrial Design: The Roles and Factors of Aesthetics, Modeling, Styling, Product Brand and Branding in Design/Design Education. *Review of Artistic Education, 14*(1), 186–199. doi:10.1515/rae-2017-0024

Egaña, J. (2016). *¿Qué significa Gay Friendly? Empresas Gay Friendly y las lógicas susceptibilidades.* Retrieved from https://www.congresoegf.com/que-significa-gay-friendly-empresas-gay-friendly-y-las-logicas-suceptibilidades/

Embrey, D. (2005). Understanding human behaviour and error. *Human Reliability Associates, 1*, 1–10.

Engeström, Y. (1987). *Learning by expanding: An activity-theoretical approach to developmental research.* Helsinki: Orienta-Konsuit.

Engeström, Y. (2000). Activity Theory as a framework for analyzing and redesign work. *Ergonomics, 43*(7), 960–974. doi:10.1080/001401300409143 PMID:10929830

Engeström, Y. (2001). Expansive learning at work: Toward an activity theoretical reconceptualization. *Journal of Education and Work, 14*(1), 133–156. doi:10.1080/13639080020028747

Entwistle, J. (2000). *The Fashioned Body: fashion, dress and modern social theory.* Cambridge, UK: Polity Press.

Ergonautas.upv.es. (2017). *Método REBA Rapid Entire Body Assessment.* Retrieved January 4, 2017, from http://www.ergonautas.upv.es/metodos/reba/reba-ayuda.php

Estrada-López, U., Vilchis-Martínez, S., Gutiérrez-Osorio, A., Arellano-Carmona, A., & Borja-Ramírez, V. (2013). Desarrollo de un Equipo Auxiliar para Rehabilitación de Pacientes con Lesiones en las Rodillas de bajo Costo. In *Memorias del XIX Congreso Internacional Anual de la Somim* (pp. 139–148). Pachuca, Mexico: SOMIM. Retrieved from https://www.researchgate.net/profile/Sean_Rodolfo_Vilchis_Martinez/publication/273772276_Desarrollo_de_un_equipo_auxiliar_para_rehabilitacion_de_pacientes_con_lesiones_en_las_rodillas_de_bajo_costo/links/550c53230cf2ac2905a3cc8b/desarrollo-de-un-equipo-au

Evans, W., Courtney, A., & Fok, K. (1988). The design of school furniture for Hong Kong schoolchildren: An anthropometric case study. *Applied Ergonomics, 19*(2), 122–134. doi:10.1016/0003-6870(88)90005-1 PMID:15676654

Faisal, M. N. (2010). Sustainable supply chains: A study of interaction among the enablers. *Business Process Management Journal, 16*(3), 508–529. doi:10.1108/14637151011049476

Farrell, R., & Hooker, C. (2013). Design, science and wicked problems. *Design Studies, 34*(6), 681–705. doi:10.1016/j.destud.2013.05.001

Faulhaber, J. (1976). *Investigación Longitudinal del Crecimiento.* Instituto Nacional de Antropología e Historia. Departamento de Antropología Física.

Faulkner, J., Larkin, L., Claflin, D., & Brooks, S. (2007). Age-related changes in the structure and function of skeletal muscles. *Clinical and Experimental Pharmacology & Physiology, 34*(11), 1091–1096. doi:10.1111/j.1440-1681.2007.04752.x PMID:17880359

Faulkner, L. (2003). Beyond the five-user assumption: Benefits of increased sample sizes in usability testing. *Behavior Research Methods, 35*(3), 379–383. doi:10.3758/BF03195514 PMID:14587545

Fausset, C., Baranak, A., & Farmer, S. (2014). Developing a scoring matrix to evaluate the usability of consumer packaging: A pilot study. *Proceedings of the Human Factors and Ergonomics Society 58th Annual Meeting 2014*, 1802 – 1806. 10.1177/1541931214581377

Feld, W. M. (2000). *Lean manufacturing: Tools, techniques, and how to use them.* CRC Press. doi:10.1201/9781420025538

Fernández Núñez, L. (2006). Fichas para investigadores ¿Cómo analizar datos cualitativos? *Bulletí LaRecerca.* : 1886-1946 / Depósito legal: B.20973-2006

Ferreiro Gravié, R. (2014). *New ways of learning and teaching, cooperative learning: theory of cooperative work, development of critical and creative thinking, strategies and teaching functions.* Academic Press.

Ferrer, I., Rios, J., & Ciurana, J. (2009). An approach to integrate manufacturing process information in part design phases. *Journal of Materials Processing Technology, 209*(4), 2085–2091. doi:10.1016/j.jmatprotec.2008.05.009

Filz, S. A. (2008). *"Instant aging"-selbsterfahrung des alterns (Doctoral dissertation).* Universität Würzburg, Medizinische Fakultät.

Flores-González, I., & Calonje, D. H. (2017). *Telemedicina para detección de enfermedades oculares con potencial de ceguera en México.* Revista Mexicana de Oftalmología; doi:10.1016/j.mexoft.2016.12.002

Forlizzi, J. (2007). The product ecology: Understanding social product use and supporting design culture. *International Journal of Design, 2*(1), 11–20.

Fraj, E., & Martínez, E. (2007). Ecological consumer behavior: An empirical analysis. *International Journal of Consumer Studies, 31*(1), 26–33. doi:10.1111/j.1470-6431.2006.00565.x

Franck, K., & von Sommaruga Howard, T. (2010). *Design through dialogue: A guide for architects and clients.* Chichester, UK: John Wiley & Sons.

Frank, D., & Fadel, G. (1994). Expert system based selection of the preferred direction of build for rapid prototyping. *Journal of Intelligent Manufacturing, 6*(5), 334–339.

Franz, H.-W., Hochgerner, J., & Howaldt, J. (Eds.). (2012). *Challenge social innovation: Potentials for business, social entrepreneurship, welfare and civil society.* Berlin, Germany: Springer-Verlag. doi:10.1007/978-3-642-32879-4

Fraval, J. T., & Godfrey, M. T. (2013). United States Patent, 1(12), 5. 10.1126/science.Liquids

Freivalds, A. (2011). *Biomechanics of the Upper Limbs*. Boca Raton, FL: CRC Press.

Friedlaender, J. S., Costa, P. T. Jr, Bosse, R., Ellis, E., Rhoads, J. G., & Stoudt, H. W. (1977). Longitudinal Physique Changes Among Healthy White Veterans at Boston. *Human Biology*, *49*, 541–558. doi:10.2307/41464475 PMID:590953

Galindo, R. V. (2009). Creatividad en el diseño industrial. Experiencias con dos técnicas creativas. *Revista Digital Universitaria, 10*(12), 1-14. Retrieved 04 20, 2016, from http://www.revista.unam.mx/vol.10/num12/art83/int83.htm

Gallagher, S., Pollard, J., & Porter, W. L. (2011). Electromyography of the thigh muscles during lifting tasks in kneeling and squatting postures. *Ergonomics*, *54*(1), 91–102. doi:10.1080/00140139.2010.535025 PMID:21181592

Gangopadhyay, S., Ghosh, T., Das, T., Ghoshal, G., & Das, B. (2010). Effect of Working Posture on Occurrence of Musculoskeletal Disorders among the Sand Core Making Workers of West Bengal. *Central European Journal of Public Health*, *18*(1), 38–42. Retrieved from http://search.proquest.com/openview/6db3fd86bba430f5c0e2054d5b580110/1?pq-origsite=gscholar&cbl=54116 PMID:20586229

García Criollo, R. (2000). *Estudio del trabajo: ingeniería de métodos*. Editorial McGraw-Hill.

Garcia-Acosta, G., & Lange-Morales, K. (2007). Definition of sizes for the design of school furniture for schools Bogota based on anthropometric criteria. *Ergonomics*, *50*(10), 1626–1642. doi:10.1080/00140130701587541 PMID:17917903

García, B. (2008). *Ecodiseño, Nueva Herramienta para la Sustentabilidad*. Designo.

Garcia, C., & Page, A. (1992). *Guide recommendations for designing ergonomic furniture* (1st ed.). Valencia Institute of Biomechanics of Valencia.

Garcia-Escutia, M. (2004). *Manual de procedimientos: protocolos de prevención de riesgos laborales*. Valencia.

Garnaik, A., Kumar, S., & Satapathy, S. (2016). A Case Study of Furniture: Optimizing Product Design by implying Analytic Hierarchy Process and Quality Function Deployment. *International Journal of Engineering Science and Management, 2*(1), 1-7.

Gastélum Gonzalez, H. (2004). Anthropometry in the classrooms of primary education in the State of Sonora. In *Sixth International Congress on Ergonomics*. Guanajuato: Society Ergonomists Mexico.

Gatewood, R., Feild, H. S., & Barrick, M. (2015). *Human resource selection. Nelson Education*. Nelson Education.

GERT. (2017). *Gert simulador de edad*. Retrieved August 20, 2017, from: http://www.simulador-de-edad.com

Ghasemi, M., Nasleseraji, J., Hoseinabadi, S., & Zare, M. (2013). Application of SHERPA to identify and prevent human errors in control units of petrochemical industry. *International Journal of Occupational Safety and Ergonomics*, *19*(2), 203–209. doi:10.1080/10803548.2013.11076979 PMID:23759191

Gilbert, N. (2004). *Agent-based social simulation: dealing with complexity*. Guildford, UK: Centre for Research on Social Simulation, University of Surrey.

Giles, C. (2004). *Que es el packaging?* Barcelona: Editorial GG.

Gobe, M. (2008). *Branding Emocional*. Barcelona: Divine Egg Publicaciones.

Goepp, V., Zwolinski, P., & Caillaud, E. (2014). Design process and data models to support the design of sustainable remanufactured products. *Computers in Industry*, *65*(3), 480–490. doi:10.1016/j.compind.2014.02.002

Gómez-Bull, K. G., Hernández-Arellano, J. L., & Ibarra-Mejía, G. (2015). A proposed methodology for task analysis in ergonomic evaluations. *Procedia Manufacturing*, *3*, 4756–4760. doi:10.1016/j.promfg.2015.07.573

González Cano, M., Aguayo González, F., Lama Ruiz, J. R., & Pérez Gutiérrez, J. R. (2009). Ingeniería Kansei para un diseño de productos centrado en los usuarios. *Técnica Industrial*, *280*(Marzo-Abril), 7.

González, J. M. (2003). *Actividad física, deporte y vida: beneficios, perjuicios y sentido de la actividad física y del deporte. Cuerpo y espíritu.* Basauri, España: Etor-Ostoa.

González-Rey, F. (2010). (n.d.). Las categorías de sentido, -. *Google Académico.*

Gouvali, M., & Boudolos, K. (2006). Match between school furniture dimensions and children's anthropometry. *Applied Ergonomics*, *37*(6), 765–773. doi:10.1016/j.apergo.2005.11.009 PMID:16442494

Govindan, K., Khodaverdi, R., & Jafarian, A. (2013). A fuzzy multi criteria approach for measuring sustainability performance of a supplier based on triple bottom line approach. *Journal of Cleaner Production*, *47*, 345–354. doi:10.1016/j.jclepro.2012.04.014

Goyal, A., Sudha, A., Kumar, A., & Rao, P. V. M. (2000). Optimum part orientation for rapid prototyping based on estimated number of adaptive slices. *Proceedings of International Conference on E-manufacturing,* 247–253.

Graif, M. (1991). *The visual factory*. Portland, OR: Productivity Press.

Grimm, R., Fox, C., Baines, S., & Albertson, K. (2013). Social innovation, an answer to contemporary societal challenges? Locating the concept in theory and practice. *Innovation (Abingdon)*, *26*(4), 436–455. doi:10.1080/13511610.2013.848163

Groover, M. (2007). Fundamentos de Manufactura moderna: materiales, procesos y sistemas. Pearson Prentice-Hall.

Groza, H. L., Sebesi, S. B., & Mandru, D. S. (2017). Age simulation suits for training, research anddevelopment. In S. Vlad, & N. M. Roman (Eds.), *International conference on advancements of medicine and health care through technology* (pp. 77-80). Cham: Springer International Publishing.

Grzybowska, K. (2012). Sustainability in the supply chain: Analyzing the enablers. In Environmental Issues in Supply Chain Management (pp. 25-40). Springer-Verlag.

Guba, E. G., & Lincoln, Y. S. (1994). Competing paradigms in qualitative research. In N. K. Denzin & Y. S. Lincoln (Eds.), *The Handbook of Qualitative Research* (pp. 105–117). Thousand Oaks, CA: Sage.

Guide, V. D. R. Jr. (2000). Production planning and control for remanufacturing: Industry practice and research needs. *Journal of Operations Management*, *18*(4), 467–483. doi:10.1016/S0272-6963(00)00034-6

Gumina, S., Carbone, S., Campagna, V., Candela, V., Sacchetti, F. M., & Giannicola, G. (2013). The impact of aging on rotator cuff tear size. *Musculoskeletal Surgery*, *97*(1), 69–72. doi:10.100712306-013-0263-2 PMID:23588834

Gumus, B., Ertas, A., Tate, D., & Cicek, I. (2008). The Transdisciplinary Product Development Lifecycle model. *Journal of Engineering Design*, *19*(3), 185–200. doi:10.1080/09544820701232436

Gupta, D., Das, A., Jain, V., Soneja, T., Dang, P., & Srivastava, P. (2013). Design of Ergonomic Garment for Elderly Indian Women For Authors Design of Ergonomic Garment for Elderly Indian Women. *Research Journal of Textile and Apparel*, *17*(3), 118–12743. 10.1108/RJTA-17-03-2013-B012

Gustafsson, K., Andersson, I., Andersson, J., Fjellström, C., & Sidenvall, B. (2003). Older women's perceptions of independence versus dependence in food-related work. *Public Health Nursing (Boston, Mass.)*, *20*(3), 237–247. doi:10.1046/j.0737-1209.2003.20311.x PMID:12716404

Gutierrez, A. (1945). *Un aparato antropometro escolar*. Retrieved from http://lp.espacenet.com/publicationDetails/originalDocument?CC=ES&NR=170025A2&KC=A2&FT=D&ND=4&date=19450701&DB=lp.espacenet.com&locale=es_LP

Gutierrez, M., & Morgado, P. (2001). Guide recommendations for the design of school furniture. Santiago de Chile: Ministry of Education and UNESCO.

Gutiérrez, R. (1945). *Un Aparato Antropometro Escolar*. Academic Press.

Gutiérrez-Robledo, L. M., & Arrieta-Cruz, I. (2015). Demencias en México: La necesidad de un plan de acción. *Gaceta Médica de México*, *151*, 667–673. PMID:26526483

Hallgrimsson, B. (2013). *Diseño de producto, maquetas y prototipos*. Promopress.

Hallo, J., & Manning, R. (2009). Transportation and recreation: A case study of visitors driving for pleasure at Acadia National Park. *Journal of Transport Geography*, *17*(6), 9. doi:10.1016/j.jtrangeo.2008.10.001

Hall, T. C., Nixon, M. F., Dias, J. J., Graham, T., & Cook, S. (2010). How accurately does a simulation glove reflect function compared to rheumatoid arthritis suffers? *Annals of the Royal College of Surgeons*, *92*(7), 605–609. doi:10.1308/rcsann.2010.92.7.605 PMID:20529519

Handcock, H. E., Rogers, W. A., Schroeder, D., & Fisk, A. D. (2004). Safety symbol comprehension: Effects of symbol type, familiarity, and age. Human Factors. *The Journal of the Human Factors and Ergonomics Society*, *46*(2), 183–195. doi:10.1518/hfes.46.2.183.37344

Handfield, R., Walton, S. V., Sroufe, R., & Melnyk, S. A. (2002). Applying environmental criteria to supplier assessment: A study in the application of the Analytical Hierarchy Process. *European Journal of Operational Research*, *141*(1), 70–87. doi:10.1016/S0377-2217(01)00261-2

Hannam, K., Sheller, M., & Urry, J. (2006). *Mobilities*. Ed M.i.a. moorings.

Harris, D., Stanton, N. A., Marshall, A., Young, M. S., Demagalski, J., & Salmon, P. (2005). Using SHERPA to predict design-induced error on the flight deck. *Aerospace Science and Technology*, *9*(6), 525–532. doi:10.1016/j.ast.2005.04.002

Harrison, T., & Aurisicchio, M. (2011). Understanding the front end of design. In *Proceedings of the International Conference on Engineering Design, ICED 2011*. Technical University of Denmark.

Hart, S. G. (2006). NASA-task load index (NASA-TLX); 20 years later. *Proceedings of the Human Factors and Ergonomics Society Annual Meeting*, *50*(9), 904–908. doi:10.1177/154193120605000909

Hart, S. G., & Staveland, L. E. (1988). Development of NASA-TLX (Task Load Index): Results of Empirical and Theoretical Research. *Advances in Psychology*, *52*(C), 139–183. doi:10.1016/S0166-4115(08)62386-9

Hassan, Z. (2014). *The social labs revolution: A new approach to solving our most complex challenges*. Oakland, CA: Berrett-Koehler Publishers. doi:10.9774/GLEAF.9781783531455_5

Hatcher, G. D., Ijomah, W. L., & Windmill, J. F. C. (2014). A network model to assist 'design for remanufacture' integration into the design process. *Journal of Cleaner Production*, *64*, 244–253. doi:10.1016/j.jclepro.2013.09.021

Helander, M. G. (2007). Using design equations to identify sources of complexity in human–machine interaction. *Theoretical Issues in Ergonomics Science*, *8*(2), 123–146. doi:10.1080/14639220601092442

Helio de Oliveira, S. (1987). *Antropometro Peniano*. Retrieved from http://lp.espacenet.com/publicationDetails/originalDocument?locale=en_LP&CC=BR&NR=8506503A&ND=4&KC=A&DB=lp.espacenet.com&FT=D&date=19870630

Hendrick, H. W. (2008). Applying ergonomics to systems: Some documented "lessons learned". *Applied Ergonomics, 39*(4), 418–427. doi:10.1016/j.apergo.2008.02.006 PMID:18374303

Hensler, S., Herren, D. B., & Marks, M. (2015). New technical design of food packaging makes the opening process easier for patients with hand disorders. *Applied Ergonomics, 50*, 1–7. doi:10.1016/j.apergo.2015.02.002 PMID:25959312

Herman, F., Van Gorp, A., Simon, F. and Depaepe, M. (2011). The school desk: from concept to object. *History of Education, 40*(1), 97-117.

Hernández-Arellano, J. L., Aguilar-Duque, J. I., & Gómez-Bull, K. G. (2018). Methodology to Determine Product Dimensions Based on User Anthropometric Data. In J. L. García-Alcaraz, G. Alor-Hernández, A. A. Maldonado-Macías, & C. Sánchez-Ramírez (Eds.), New Perspectives on Applied Industrial Tools and Techniques (pp. 373–385). Cham, Switzerland: Springer International Publishing. doi:10.1007/978-3-319-56871-3_18

Hernandez-Arellano, J. L., Talavera-Aguirre, G., Serratos-Perez, J. N., Maldonado-Macias, A. A., & Garcia-Alcaraz, J. L. (2016). Anthropometrics of University Students in Northern Mexico. *Open Journal of Safety Science and Technology, 6*(4), 143–155. doi:10.4236/ojsst.2016.64011

Herrera-Ruiz, M. (2006). *Silla de Ruedas Anfibia para Hidroterapia*. Instituto Politécnico Nacional. Retrieved from http://itzamna.bnct.ipn.mx/bitstream/handle/123456789/13911/Silla de ruedas anfibia para hidroterapia.pdf?sequence=1

Heyward, V. (2008). *Evaluación de la aptitud física y prescripción del ejercicio* (Vol. 5). Madrid: Médica Panamericana.

Hignett, S., & McAtamney, L. (2000). Rapid Entire Body Assessment (REBA). *Applied Ergonomics, 31*(2), 201–205. doi:10.1016/S0003-6870(99)00039-3 PMID:10711982

Hira, D. (1980). An ergonomic appraisal of educational desks. *Ergonomics, 23*(3), 213–221. doi:10.1080/00140138008924735

Hirani, H., & Suh, N. (2005). Journal bearing design using multiobjective genetic algorithm and axiomatic design approaches. *Tribology International, 38*(5), 481–491. doi:10.1016/j.triboint.2004.10.008

Horne, S. (2002). The classroom environments and its effects on the practice of teachers. *Journal of Environmental Psychology, 22*(1-2), 139–156. doi:10.1006/jevp.2001.0239

Hornsby, P. (2010). *Hierarchical task analysis*. Retrieved February 8, 2017, from https://www.uxmatters.com/mt/archives/2010/02/hierarchical-task-analysis.php

Hsu, C.-H. (2008). Applying a Bust-To-Waist Girth Ratio Approach To Develop Body Measurement Charts for Improving Female Clothing Manufacture. *J Chin Inst Industrial Eng, 25*(3), 215–222. doi:10.1080/10170660809509085

Hurlock, E. (2010). Interés y actividades recreativas. Cambios en los intereses. In Paidos (Ed.), Interés y actividades recreativas (pp. 51 - 55). Academic Press.

Hutchins, E. L., Hollan, J. D., & Norman, D. A. (1985). Direct manipulation interfaces. *Human-Computer Interaction, 1*(4), 311–338. doi:10.1207153270511hci0104_2

IEA. (2008). *EQUID Design process guidelines*. IEA Press.

IEA. (2017). *What Is Ergonomics?* Retrieved from http://iea.cc/whats/index.html

Ijomah, W. L., McMahon, C. A., Hammond, G. P., & Newman, S. T. (2007). Development of robust design-for-remanufacturing guidelines to further the aims of sustainable development. *International Journal of Production Research, 45*(18-19), 4513–4536. doi:10.1080/00207540701450138

INAPAM. (2016). *Vinculación productiva para personas adultas mayores.* Retrieved August 20, 2017, from: http://www.inapam.gob.mx/es/INAPAM/Empleo_para_adultos_mayores

Incel, N., Ceceli, E., Durukan, P., Erdem, R., & Yorgancioglu, R. (2002). Grip strength: Effect of hand dominance. *Singapore Medical Journal, 43*, 234–237. PMID:12188074

INEGI. (2010a). *Censo de población y vivienda.* INEGI.

INEGI. (2010b). *Perfil sociodemográfico de adultos mayores.* INEGI.

INEGI. (2014). *Módulo de Práctica Deportiva y Ejercicio Físico.* Instituto Nacional de Estadística y Geografía. Retrieved Abril, 2015, from http://www.inegi.org.mx/est/contenidos/proyectos/accesomicrodatos/encuestas/hogares/modulos/mopradef/?_file=/est/contenidos/proyectos/encuestas/hogares/modulos/mopradef/doc/resultados_mopradef_feb_2014.pdf

INEGI. (2016). *Estadísticas a propósito del día internacional de las personas de edad (1 de octubre).* Aguascalientes. Retrieved July 31, 2017from http://www.inegi.org.mx/saladeprensa/aproposito/2016/edad2016_0.pdf

International Organisation for Standardization. (2010). *Ergonomics of human-system interaction: Part 210: Human-centred design for interactive systems.* Geneva, Switzerland: ISO.

International Standards Organization (ISO). (1984). *International standard for safety colours and safety signs: ISO 3864.* Geneva, Switzerland: Author.

Iqbal, N. J., Vivek, K., & Amithabha, D. (2015). Manual material handling and occupational low back disorders: A narrative literature review emphasising maximum acceptable weight of load. *Int. J. of Human Factors and Ergonomics, 3*(3/4), 376–392. doi:10.1504/IJHFE.2015.073010

ISO 11228-3. (2007). *Ergonomics -Manual handling- Part 3: Handling of low loads at high frequency* (Vol. 11228).

ISO. (2002). *ISO/TR 14062: Environmental Management - Integrating Environmental Aspects into Product Design and Development.* ISO.

Izquierdo, L. R., Galán, J. M., Santos, J. I., & Del Olmo, R. (2008). Modelado de sistemas complejos mediante simulación basada en agentes y mediante dinámica de sistemas. *EMPIRIA. Revista de Metodología de Las Ciencias Sociales*, (16).

James, D. A. (1941, November 4). *Vehicle steering and stabilizing mechanism.* Google Patents.

Jayaram, J., & Avittathur, B. (2015). Green supply chains: A perspective from an emerging economy. *International Journal of Production Economics, 164*, 234–244. doi:10.1016/j.ijpe.2014.12.003

Jesús Hernández Guzmán, R., & Vásquez Quiroga, J. (2009). Analysis of anthropometry-dimensions Pupitres Relationship, likely to cause fatigue of Students of Industrial Engineering. *Research, 4*(1), 10–16.

Jiamsanguanwong, A., & Umemuro, H. (2014). Influence of affective states on comprehension and hazard perception of warning pictorials. *Applied Ergonomics, 45*(5), 1362–1366. doi:10.1016/j.apergo.2013.08.006 PMID:24045098

Jiun, G. W., & Feng, C. J., Bin Kamaruddin, S., & Binti Muhammad, N. A. (2014). A visual board to facilitate production flow regulation at sequential processes. *International Journal of Industrial Engineering and Management, 5*(2), 85–94.

Johnson, R. A., & Wichern, D. W. (2002). *Applied Multivariate Statistical Analysis.* Upper Saddle River, NJ: Prentice-Hall.

Jokela, T., Iivari, N., Matero, J., & Karukka, M. (2003). The standard of user-centered design and the standard definition of usability: analyzing ISO 13407 against ISO 9241-11. In *Proceedings of the Latin American conference on Human-computer interaction* (pp. 53-60). ACM. 10.1145/944519.944525

Jonassen, D. (2000). Designing constructivist learning environments. In C. Reigeluth (Ed.), *Instructional design theories and models. A new paradigm of instructional theory Part I* (1st ed., pp. 225–249). Madrid: Mc Graw Hill.

Jordan, P. W. (2002). *Designing pleasurable products: An introduction to the new human factors*. CRC Press.

Kai, C. C., & Fai, L. K. (1997). *Rapid Prototyping, Principles and Applications in Manufacturing. John Wiley & Sons* .

Kallman, D., Plato, C., & Tobin, J. (1990). The Role of Muscle Loss in the Age-Related Decline of Grip Strength: Cross-sectional and Longitudinal Perspectives. *The Journals of Gerontology. Series A, Biological Sciences and Medical Sciences*, *45*, 82–88. PMID:2335723

Kaptelinin, V., Nardi, B. A., & Macaulay, C. (1999). Methods & tools: The activity checklist: a tool for representing the "space" of context. *Interactions, 6*(4), 27-39.

Kaptelinin, V. (2005). The object of activity: Making sense of the sense-maker. *Mind, Culture, and Activity*, *12*(1), 4–18. doi:10.120715327884mca1201_2

Kaptelinin, V., & Nardi, B. A. (2006). *Acting with technology: Activity theory and interaction design*. MIT Press.

Kasuo, S., & Kagekuni, A. (2002a). Diagnostic Network Processing System, Bill Collecting System and Physical Distribution System. *Japan Platform for Patent Information*. Retrieved from https://www19.j-platpat.inpit.go.jp/PA1/cgi-bin/PA1DETAIL?MaxCount=1000&PageCount=1000&SearchType=0&TempName=wuZalqa&MaxPage=1&DispPage=1+1000&HitCount=1&ResultId=I00727005602&CookieId=2&DetailPage=1&Language=ENG&Reserve1=DetailPaging&Reserve2=do8yRaE58_c

Kasuo, S., & Kagekuni, A. (2002b). *Diagnostic Network Processing System, Bill Collecting System and Physical Distribution System*. Japon.

Kato, M., Tsutsumi, T., Yamaguchi, T., Kurakane, S., & Chang, H. (2011). Characteristics of maximum performance of pedaling exercise in recumbent and supine positions. *Journal of Sports Science & Medicine*, *10*(3), 491. PMID:24150623

Kavanagh, M., & Johnson, R. (2017). *Human resource information systems: Basics, applications, and future directions*. S. Publications.

Keinonen, T. K., & Takala, R. (2010). *Product Concept Design: A Review of the Conceptual Design of Products in Industry*. Springer Science & Business Media. Retrieved 04 10, 2017, from https://books.google.com.mx

Kelly, M. (2017). Analysing the Complex Relationship between Logo and Brand. *Place Branding and Public Diplomacy*, *13*(1), 18–33. doi:10.1057/pb.2016.3

Kessler, E. H., & Chakrabarti, A. K. (1996). Innovation speed: A conceptual model of context, antecedents, and outcomes. *Academy of Management Review*, *21*(4), 1143–1191.

Khurana, A., & Rosenthal, S. R. (1998). Towards holistic "Front ends" in new product development. *Journal of Product Innovation Management*, *15*(1), 57–74. doi:10.1016/S0737-6782(97)00066-0

Kim, J., & Wilemon, D. (2002). Focusing the fuzzy front-end in new product development. *R & D Management*, *22*(4), 269–279. doi:10.1111/1467-9310.00259

Kim, J., & Wilemon, D. (2003). Sources and assessment of complexity in npd projects. *R & D Management*, *33*(1), 15–30. doi:10.1111/1467-9310.00278

Kim, Y.-S., & Cochran, D. S. (2000). Reviewing TRIZ from the perspective of Axiomatic Design. *Journal of Engineering Design*, *11*(1), 79–94. doi:10.1080/095448200261199

Kirkendall, D., & Garrett, W. (1998). The effects of aging and training on skeletal muscle. *The American Journal of Sports Medicine*, *26*(4), 598–602. doi:10.1177/03635465980260042401 PMID:9689386

Knight, G., & Noyes, J. (1999). Children's Behavior and the design of school furniture. *Ergonomics*, *42*(5), 747–760. doi:10.1080/001401399185423 PMID:10327894

Koen, P. A., Ajamian, G., Burkart, R., Clamen, A., Davidson, J., D'Amore, R., & Wagner, K. (2001). Providing clarity and a common language to the "Fuzzy front end". *Research Technology Management*, *44*(2), 46–55. doi:10.1080/08956308.2001.11671418

Kolb, D. A., Boyatzis, R. E., Mainemelis, C., & ... (2001). Experiential learning theory: Previous research and new directions. *Perspectives on Thinking, Learning, and Cognitive Styles*, *1*(8), 227–247.

Konz, S., & Johnson, S. (2000). *Work Design: Industrial Ergonomics* (5th ed.). Holcomb Hathaway Publishers.

Kopec, D. (2012). *Environmental psychology for design* (O. T. Kontzias, Ed.). Fairchild Books, Inc.

Kremer, G., Chiu, M.-C., Lin, C.-Y., Gupta, S., Claudio, D., & Thevenot, H. (2012). Application of axiomatic design, TRIZ, and mixed integer programming to develop innovative designs: A locomotive ballast arrangement case study. *International Journal of Advanced Manufacturing Technology*, *61*(5-8), 827–842. doi:10.100700170-011-3752-1

Kulak, O., Cebi, S., & Kahraman, C. (2010). Applications of axiomatic design principles : A literature review. *Expert Systems with Applications*, *37*(9), 6705–6717. doi:10.1016/j.eswa.2010.03.061

Kullman, K. (2016). Prototyping bodies: A post-phenomenology of wearable simulations. *Design Studies*, *47*, 73–90. doi:10.1016/j.destud.2016.08.004

Kushwaha, D. K., & Prasad, V. K. (2016). Ergonomic assessment and workstation design of shipping crane cabin in steel industry. *International Journal of Industrial Ergonomics*, *52*, 29–39. doi:10.1016/j.ergon.2015.08.003

Labuschagne, C., Brent, A. C., & Claasen, S. J. (2005a). Environmental and social impact considerations for sustainable project life cycle management in the process industry. *Corporate Social Responsibility and Environmental Management*, *12*(1), 38–54. doi:10.1002/csr.76

Labuschagne, C., Brent, A. C., & Van Erck, R. P. G. (2005). Assessing the sustainability performances of industries. *Journal of Cleaner Production*, *13*(4), 373–385. doi:10.1016/j.jclepro.2003.10.007

Laios, L., & Giannatsis, J. (2010). Ergonomic evaluation and redesign of children bicycles based on anthropometric data. *Applied Ergonomics*, *41*(3), 428–435. doi:10.1016/j.apergo.2009.09.006 PMID:19837401

Laitala, K., Klepp, G., & Benedicte, H. (2011). Materialised Ideals: Sizes and Beauty. *Culture Unbound*, *3*(3), 19–41. doi:10.3384/cu.2000.1525.11319

Landsmeer, J. (1962). Power grip and precision handling. *Annals of the Rheumatic Diseases*, *21*(2), 164–170. doi:10.1136/ard.21.2.164 PMID:14461996

Lan, P.-T., Chou, S.-Y., Chen, L.-L., & Gemmill, D. (1997). Determining fabrication orientation for rapid prototyping with stereolithography apparatus. *Computer Aided Design*, *29*(1), 53–62. doi:10.1016/S0010-4485(96)00049-8

Laughery, K. R. (2006). Safety communications: Warnings. *Applied Ergonomics*, *37*(4), 467–478. doi:10.1016/j.apergo.2006.04.020 PMID:16759630

Lavalle-Herrera, Y. D. (2014). Diseño y ergonomía para la tercera edad. Aplicación al diseño de calzado. Coordinación de Estudios de Posgrado y el Programa de Maestría en Diseño Industrial de la Universidad Nacional Autónoma de México.

Lawson, B. (1994). *Design in mind*. Amsterdam: Architectural Press.

Lawson, B. (2005). *How designers think: The design process demystified*. Amsterdam: Architectural Press.

Lawson, B., & Dorst, K. (2009). *Design expertise*. Amsterdam: Architectural Press.

Lawton, M. P., & Nahemow, L. (1973). Ecology and the aging process. In C. Eisdorfer (Ed.), The psychology of adult development and aging (Vol. 6, pp. 619-674). Washington DC: American Psychological Association. doi:10.1037/10044-020

Lee, D., & Ferreira, J. (2003). *Reliability and usability evaluation of the Manual handling Assessment Charts (MAC) for use by non-regulatory professionals*. Health & safety laboratory.

Lee, I., Shirona, E., Lobelo, F., Puska, P., Blair, S., & Katzmaryk, P. (2012). Effect of physical inactivity on major non-communicable diseases worldwide: An analysis of burden of disease and life expectancy. *Lancet*, *380*(9838), 10. doi:10.1016/S0140-6736(12)61031-9 PMID:22818936

Lee, J. S., & Frongilo, E. A. Jr. (2001). Factors Associated With Food Insecurity Among U.S. Elderly Persons: Importance of Functional Impairments. *The Journals of Gerontology. Series B, Psychological Sciences and Social Sciences*, *56B*(2), S94–S99. doi:10.1093/geronb/56.2.S94 PMID:11245369

Lee, J. Y., Istook, C. L., Nam, Y. J., & Park, S. M. (2007). International Journal of Clothing Science and Technology. *International Journal of Clothing Science and Technology International Journal of Clothing Science and Technology Journal of Fashion Marketing and Management: An International Journal*, *19*(3), 374–391. doi:10.1108/09556220710819555

Lee, S., Yu, S., Choi, J., & Han, C. (2011). A methodology to quantitatively evaluate the safety of a glazing robot. *Applied Ergonomics*, *42*(3), 445–454. doi:10.1016/j.apergo.2010.09.002 PMID:20870215

Lee, Y.-S. (2000). 3-d anthropometric information for designing the wearable. *Proceedings of the Human Factors and Ergonomics Society Annual Meeting*, *6*(38), 751–755. doi:10.1177/154193120004403818

Lehto, M. R., & Landry, S. J. (2012). *Introduction to human factors and ergonomics for engineers*. CRC Press.

Lenz, R. K., & Cochran, D. S. (2000). The application of axiomatic design to the design of the product development organization. Academic Press.

Leont'ev, A. N. (1978). *Activity, consciousness, and personality*. Englewood Cliffs, NJ: Prentice Hall.

Li, G., & Buckle, P. (1999). Current techniques for assessing physical exposure to work-related musculoskeletal risks, with emphasis on posture-based methods. *Ergonomics*, *42*(5), 674–695. doi:10.1080/001401399185388 PMID:10327891

Li, K. W., & Yu, R. (2011). Assessment of grip force and subjective hand force exertion under handedness and postural conditions. *Applied Ergonomics*, *42*(6), 929–933. doi:10.1016/j.apergo.2011.03.001 PMID:21439549

Lindkvist, L., & SÖDerberg, R. (2003). Computer-aided tolerance chain and stability analysis. *Journal of Engineering Design*, *14*(1), 17–39. doi:10.1080/0954482031000078117

Lindlof, L., & Soderberg, B. (2011). Pros and cons of lean visual planning: experiences from four product development organisations. *International Journal of Technology Intelligence and Planning*, *7*(3), 269-279.

Lindlöf, L. (2014). *Visual management-on communication in product development organizations*. Chalmers University of Technology.

Lindroos, O. (2009). Relationships between observed and perceived deviations from normative work procedures. *Ergonomics*, *52*(12), 1487–1500. doi:10.1080/00140130903197461 PMID:19941182

Linton, S., Hellsing, A., Halme, T., & Åkerstedt, K. (1994). The effects of ergonomically designed furniture school on Pupils' attitudes, symptoms and Behavior. *Applied Ergonomics*, *25*(5), 299–304. doi:10.1016/0003-6870(94)90044-2 PMID:15676981

Lippincott, J. (2009). Learning Spaces: Involving Faculty to Improve Pedagogy. *EDUCAUSE Review*, *44*(2), 16.

Liu, C. J., Kemper, S., & McDowd, J. (2009). The use of illustration to improve older adults' comprehension of health-related information: Is it helpful? *Patient Education and Counseling*, *76*(2), 283–288. doi:10.1016/j.pec.2009.01.013 PMID:19286343

Liu, P., & Li, Z. (2014). Comparison between conventional and digital nuclear power plant main control rooms: A task complexity perspective. *International Journal of Industrial Ergonomics*, *51*, 10–20. doi:10.1016/j.ergon.2014.06.011

Liu, W. C. V., Sanchez-Monroy, D., & Parga, G. (1999). Anthropometry of female maquiladora workers. *International Journal of Industrial Ergonomics*, *24*(3), 273–280. doi:10.1016/S0169-8141(98)00036-5

Loker, S., Ashdown, S., & Schoenfelder, K. (2005). Size-specific analysis of body scan data to improve apparel fit. Journal of Textile and Apparel. *Journal of Textile and Apparel, Technology and Management*, *4*(3), 1–15.

Lom, H., Macías, M., Madrid, S., Martínez, T., & Soto, N. (2009). *Áreas integrales de diseño industrial: Percepción, metodologías, manufactura y ergonomía*. Ciudad Juárez, Chihuahua, México: Universidad Autónoma de Ciudad Juárez.

Long, P., & Holeton, R. (2009). Signposts of the Revolution? What We Talk About When We Talk about Learning Spaces. *EDUCAUSE Review*, *44*(2), 36.

Lo, S., & Helander, M. G. (2007). Use of axiomatic design principles for analysing the complexity of human–machine systems. *Theoretical Issues in Ergonomics Science*, *8*(2), 147–169. doi:10.1080/14639220601092475

Macarro, J., Romero, C., & Torres, J. (2010). Motivos de abandono de la práctica de actividad físico-deportiva en los estudiantes de Bachillerato de la provincia de Granada. *Review of Education*, *353*, 25.

Madriz-Quirós, C., Ramírez-Coretti, A., & Serrano-Montero, R. (2008). Estudio antropométrico para el diseño de mobiliario para niños de edad escolar en Costa Rica. *Tecnología en Marcha*, *21*(4), 17–28. Retrieved from http://revistas.tec.ac.cr/index.php/tec_marcha/article/viewFile/221/218

Magdaleno-Adame, S., Olivares-Galvan, J., & Fofana, I. (2014). Electric Field Distribution to Analyze Possible Manufacturing Defects of a Large Preventive Autotransformer. In *International Conference on High Voltage Engineering and Application* (p. 4). Retrieved from http://s3.amazonaws.com/academia.edu.documents/34736938/final_tornillo_dielectrico_09_June_2014.pdf?AWSAccessKeyId=AKIAIWOWYYGZ2Y53UL3A&Expires=1498110339&Signature=emThzCWMeIUSTqLpL9GpmoiGWnA%3D&response-content-disposition=inline%3Bfilename%3DElectri

Magnier, L., & Schoormans, J. (2015). Consumer reactions to sustainable packaging: The interplay of visual appearance, verbal claim and environmental concern. *Journal of Environmental Psychology*, *44*, 53–62. doi:10.1016/j.jenvp.2015.09.005

Maguire, M., & Bevan, N. (2002). User Requirements Analysis: A Review of Supporting Methods. In *17th World Computer Congress* (pp. 25-30). Montreal, Canada: Kluwer Academic Publisher. Retrieved 05 24, 2016, from http://www.usabilitynet.org/papers/WCC_UserRequirements.pdf

Maldonado-Macías, A., Realyvásquez, A., Hernández, J., & García-Alcaraz, J. (2015). Ergonomic assessment for the task of repairing computers in a manufacturing company: A case study. *Work (Reading, Mass.)*, *52*(2), 393–405. doi:10.3233/WOR-152118 PMID:26409363

Malina, R. (1995). Antropometría. *PubliCE Standard, 1.16*. Retrieved from http://g-se.com/es/articulos/articulo/pdf/antropometria-718

Mandal, A. (1982). The correct height of school furniture. *Human Factors: The Journal of the Human Factors and Ergonomics Society*, *24*(3), 257–269. doi:10.1177/001872088202400301

Manrique-Espinoza, B., Andersson, I., & Sidenvall, B. (2001). Case studies of food shopping, cooking and eating habits in older women with Parkinson's disease. *Journal of Advanced Nursing*, *35*(1), 69–78. doi:10.1046/j.1365-2648.2001.01823.x PMID:11442684

Manrique-Espinoza, B., Salinas-Rodríguez, A., Moreno-Tamayo, K. M., Acosta-Castillo, I., Sosa-Ortiz, A. L., Gutiérrez-Robledo, L. M., & Téllez-Rojo, M. M. (2013). Condiciones de salud y estado funcional de los adultos mayores en México. *Salud Pública de México*, *55*(Supl.2), S323–S331. doi:10.21149pm.v55s2.5131 PMID:24626711

Manzini, E. (2014). Making things happen: Social innovation and design. *Design Issues*, *30*(1), 57–66. doi:10.1162/DESI_a_00248

Manzini, E., & Coad, R. (2015). *Design, when everybody designs: An introduction to design for social innovation*. Cambridge, MA: MIT Press.

Margolin, V., & Margolin, S. (2002). A "social model" of design: Issues of practice and research. *Design Issues*, *18*(4), 24–30. doi:10.1162/074793602320827406

Márquez, S. (2013). *Actividad física y salud* (E. D. d. Santos, Ed.). Ediciones Díaz de Santos.

Martha, V. (2007). Significado de la calidad de vida del adulto mayor para sí mismo y para su familia. *Scielo*, *68*, 284–290.

Martínez-Nopala, S. (2015). *Diseño y Construcción de un Vehículo Terrrestre para la Expulsión de Fauna Voladora en los Aeropuertos*. Escuela Superior de Ingeniería Mecánica y Eléctrica.

Martinho, G., Pires, A., Portela, G., & Fonseca, M. (2015). Factors affecting consumer choices concerning sustainable packaging during product purchase and recycling. *Resources, Conservation and Recycling*, *103*, 58–68. doi:10.1016/j.resconrec.2015.07.012

Masood, S. H., Rattanawong, W., & Iovenitti, P. (2000). Part build orientations based on volumetric error in fused deposition modeling. *International Journal of Advanced Manufacturing Technology*, *16*(3), 162–168. doi:10.1007001700050022

Masood, S. H., Rattanawong, W., & Iovenitti, P. (2001). A volumetric approach to part-built orientations in rapid prototyping. *Journal of Materials Processing Technology*, *119*(1-3), 348–353. doi:10.1016/S0924-0136(01)00924-4

Masood, S. H., Rattanawong, W., & Iovenitti, P. (2003). A generic algorithm for part orientation system for complex parts in rapid prototyping. *Journal of Materials Processing Technology*, *139*(1–3), 110–116. doi:10.1016/S0924-0136(03)00190-0

Mathiowetz, V., Kashmann, N., Volland, G., Weber, K., Dowe, M., & Rogers, S. (1985). Grip and pinch strength: Normative data for adults. *Archives of Physical Medicine and Rehabilitation*, *66*, 69–74. PMID:3970660

Mathiyazhagan, K., Govindan, K., & Noorul Haq, A. (2014). Pressure analysis for green supply chain management implementation in Indian industries using analytic hierarchy process. *International Journal of Production Research*, *52*(1), 188–202. doi:10.1080/00207543.2013.831190

Matsumoto, M., Yang, S., Martinsen, K., & Kainuma, Y. (2016). Trends and research challenges in remanufacturing. *International Journal of Precision Engineering and Manufacturing-Green Technology, 3*(1), 129-142.

Maya-Rivero, A. (2012). *Gerontodiseño: Conceptualización y formulación de sus estrategias para una mejor calidad de vida del adulto mayor. Caso de estudio: Problemas del adulto mayor mexicano en la preparación de alimentos (Master Degree Dissertation)*. Toluca, México: Universidad Autónoma del Estado de México.

Mayer, D. L., & Laux, L. F. (1989, October). Recognizability and effectiveness of warning symbols and pictorials. In *Proceedings of the Human Factors Society Annual Meeting* (Vol. 33, No. 15, pp. 984-988). Los Angeles, CA: SAGE Publications. 10.1177/154193128903301518

McClurkin, J. E., & Rosen, D. W. (1998). Computer-aided build style decision support for stereolithography. *Rapid Prototyping Journal*, *4*(1), 4–13. doi:10.1108/13552549810197505

McDonagh, D., Bruseberg, A., & Haslam, C. (2002). Visual Product Evaluation: Exploring Users´ Emotional Relationships with Products. *Applied Ergonomics*, *33*(3), 231–240. doi:10.1016/S0003-6870(02)00008-X PMID:12164507

McGraw, K. O., & Wong, S. P. (1996). Forming Inferences About Some Intraclass Correlation Coefficients. *Psychological Methods*, *1*(1), 30–46. doi:10.1037/1082-989X.1.1.30

Medina Ramírez, R. (2013). *Main Factors of Surgical Deferment in Elective Surgery: A Systems Dynamics Analysis* (Unpublished Doctoral Dissertation). University of Guadalajara, Jalisco, México.

Medina, A. (2006). Ergonomía cognitiva y usabilidad.

Melton, T. (2005). The benefits of lean manufacturing: What lean thinking has to offer the process industries. *Chemical Engineering Research & Design*, *83*(6), 662–673. doi:10.1205/cherd.04351

Menéndez, J., Guevara, A., Arcia, N., León Díaz, E. M., Marín, C., & Alfonso, J. C. (2005). Enfermedades crónicas y limitación funcional en adultos mayores: Estudio comparativo en siete ciudades de América Latina y el Caribe. *Revista Panamericana de Salud Pública*, *17*(5-6), 353–361. doi:10.1590/S1020-49892005000500007 PMID:16053645

Merí, Á. (2005). Fundamentos de fisiología de la actividad física y el deporte (E. M. Panamericana, Ed.). Academic Press.

Merlino, A., & Roqué, G. (2004). *Los nuevos jóvenes. Un estudio psicográfico de sus actitudes y estilos de vida*. Ed Brujas. Córdova.

Meuter, M. L., Ostrom, A. L., Bitner, M. J., & Roundtree, R. (2003). The influence of technology anxiety on consumer use and experiences with self-service technologies. *Journal of Business Research*, *56*(11), 899–906. doi:10.1016/S0148-2963(01)00276-4

Mighell, R. (2009, February 10). *Tilting wheeled vehicle*. Google Patents.

Milton, A., & Rodgers, P. (2013). *Research methods for product design*. Laurence King Publishing Limited.

Mircheski, I., Kandikjan, T., & Sidorenko, S. (2014). Comfort analysis of vehicle driver's seat through simulation of the sitting process. *Tehnicki Vjesnik (Strojarski Fakultet)*, *21*(2), 291–298.

Mishra, A. K., & Thirumavalavan, S. (2014). A Study of Part Orientation in Rapid Prototyping. *Middle East Journal of Scientific Research*, *20*(9), 1197–1201.

Mishra, J. L., Allen, D. K., & Pearman, A. D. (2011). Information sharing during multi-agency major incidents. *Proceedings of the Association for Information Science and Technology*, *48*(1), 1–10. doi:10.1002/meet.2011.14504801039

Mital, A., Kilbom, A., & Kumar, S. (2000). *Ergonomics Guidelines and Problem Solving*. Amsterdam: Elsevier.

Mittal, V., Ross, W. T., & Baldasare, P. M. (1998). The asymmetric impact of negative and positive attribute-level on overall satisfaction and repurchase intentions. *Journal of Marketing*, *62*(3), 33–48. doi:10.2307/1251801

Mohamed, O. A., Masood, S. H., & Bhowmik, J. L. (2015). Optimization of fused deposition modeling process parameters: A review of current research and future prospects. *Advances in Manufacturing*, *3*(1), 42–53. doi:10.100740436-014-0097-7

Mohamed, O. A., Masood, S. H., & Bhowmik, J. L. (2016). Optimization of fused deposition modeling process parameters for dimensional accuracy using I-optimality criterion. *Measurements*, *81*, 174–196.

Mohammed, H. (2011). *Modelling the enablers and alternatives for sustainable supply chain management* (Ph.D. Thesis). Concordia Institute for Information Systems Engineering, Canada.

Mohanty, P. P., & Mahapatra, S. S. (2014). A Compromise Solution by VIKOR Method for Ergonomically Designed Product with Optimal Set of Design Characteristics. *Procedia Materials Science, 6*, 633-640.

Mokdad, M., & Al-Ansari, M. (2009). Anthropometrics for the design of Bahraini school furniture. *International Journal of Industrial Ergonomics*, *39*(5), 728–735. doi:10.1016/j.ergon.2009.02.006

Molenbroek, J., Kroon-Ramaekers, Y., & Snijders, C. (2003). Revision of the design of a standard for the dimensions of school furniture. *Ergonomics*, *46*(7), 681–694. doi:10.1080/0014013031000085635 PMID:12745681

Moles, A. (2007). *La imagen comunicación funcional.* Trillas.

Moore, P., & Paul, C. (1985). *Disguised: A true story.* W Publishing Group.

Morency, R. R., Rooney, E. F., & Foerster, D. R. (2003). A Methodology To Implement and Validate Ergonomic Improvements To Computer Workstations At LL Bean. *Advances In Industrial Ergonomics And Safety*, *V*, 319.

Muduli, K., & Barve, A. (2013). Developing a Framework for Study of GSCM Criteria in Indian Mining Industries. *APCBEE Procedia*, *5*, 22–26. doi:10.1016/j.apcbee.2013.05.005

Mulet, M. T., Domínguez, M. M., & Flores, E. F. (2015). Didactic Strategy with the Support of the ICT for the development of professional Skills in the Licentiate in Accounting and Finance in the Model Semiprescencial. *Pedagogía Universitaria, 20*(1), 71-81. Retrieved 06 04, 2017, from http://web.a.ebscohost.com

Mulgan, G. (2007). *Social innovation: What it is, why it matters and how it can be accelerated.* London, UK: The Young Foundation.

Muñoz, I., & Salgado, P. (2011). Ocupaciones de tiempo libre: Una aproximación desde la perspectiva de los jóvenes. *Universidad de Chile*, *1*, 10.

Murphy, S. A., & Kumar, V. (1997). The front end of new product development: A canadian survey. *R & D Management*, *27*(1), 5–15. doi:10.1111/1467-9310.00038

Murray, R., Cualier-Grice, J., & Mulgan, G. (2010). *The open book of social innovation.* London: National Endowment for Science, Technology and the Arts.

Mwanza, D. (2002). *Towards an activity-oriented design method for HCI research and practice* (Doctoral dissertation). Open University.

Nael, M. (2011). IEA EQUID Template for Cooperation between Product Designers and Ergonomists. *Human Factors and Ergonomics in Consumer Product Design: Methods and Techniques*, 261.

Nakao, M., Kobayashi, N., Hamada, K., Totsuka, T., & Yamada, S. (2007). Decoupling Executions in Navigating Manufacturing Processes for Shortening Lead Time and Its Implementation to an Unmanned Machine Shop. *CIRP Annals - Manufacturing Technology, 56*(1), 171–174. 10.1016/j.cirp.2007.05.041

Napier, J. (1956). The prehensile movements of the human hands. *The Journal of Bone and Joint Surgery*, *38*, 902–913. PMID:13376678

Narayanan, A., Mathew, C., & Baby, V.Y. (2013). Improvement of Ergonomic Factors That Affects Employees in a Textile Industry. *International Journal of Engineering Science and Innovative Technology, 2*(1), 276-283.

National Institute for Educational Infrastructure. (2014). Standards and specifications for studies, projects, construction and installation, Vol. 3 Livability and operation Volume III furniture. Author.

Neumann, P. (2006). *Inventory of Tools for Ergonomic Evaluation: Research programme: Strategies and methods for Management of the Work Environment*. Academic Press.

Newcomb, E., & Istook, C. (2011). Confronting stereotypes: Apparel fit preferences of Mexican-American women. *Journal of Fashion Marketing and Management: An International Journal, 15*(4), 389–411. doi:10.1108/13612021111169915

Ng, A. W., Lo, H. W., & Chan, A. H. (2011). Measuring the Usability of Safety Signs: A use of system usability scale (SUS). In *Proceedings of the International MultiConference of Engineers and Computer Scientists* (Vol. 2, pp. 1296-1301). Hong Kong: IAENG.

Niebel, B. W. N., & Benjamin, W. (1996). *Ingeniería industrial: métodos, tiempos y movimientos*. Alfaomega.

Niebel, B. W., Freivalds, A., & Osuna, M. A. G. (2004). *Métodos, estándares y diseño del trabajo*. Alfaomega.

Nielsen, J. (1994). Guerrilla HCI: Using discount usability engineering to penetrate the intimidation barrier. *Cost-Justifying Usability*, 245–272.

Nielsen, J. (2012). How many test users in a usability study. *Nielsen Norman Group, 4*(6).

Nielsen, J. (1993). *Usability engineering*. New York: AP Professional.

NIOSH. (1997). Musculoskeletal Disorders and Workplace Factors: A Critical Review of Epidemiologic Evidence for Work-Related Musculoskeletal Disorders of the Neck, Upper Extremity, and Low Back. Cincinnati, OH: DHHS (NIOSH).

NNDB. (2014). Clarence Saunders. *NNDB, Tracking the entire world.* Retrieved from http://www.nndb.com/people/054/000166553/

Nordmann, L. H., & Luxhoj, J. T. (2000). Neural network forecasting of service problems for aircraft structural component grouping. *Journal of Aircraft, 37*(2), 332–338. doi:10.2514/2.2599

Norman, D. (2004). *Emotional Design: Why we love (or hate) Everyday things*. New York, NY: Basic Books.

Norman, D. (2013). *The design of everyday things: Revised and expanded edition*. Basic Books.

Norman, D. A. (2003). Designing Emotions Pieter Desmet. *The Design Journal, 6*(2), 60–62. doi:10.2752/146069203789355444

Nosova, E., Yen, P., Chong, K., Alley, H., Stock, E., Quinn, A., & Grenon, M. (2014). Short-term physical inactivity impairs vascular function. *The Journal of Surgical Research, 190*(2), 11. doi:10.1016/j.jss.2014.02.001 PMID:24630521

Oatis, C. (2009). *Kinesiology. The Mechanics & Pathomechanics of Human Movement*. Philadelphia: Lippincott Williams & Wilkins.

Olaiz-Fernández, J., Rivera-Dommarco, J., Shamah-Levy, T., Rojas, R., Villalpando-Hernández, S., & Hernández-Ávila, M. (Eds.). (2006). INEG. *Instituto nacional de estadística y geografía*. Retrieved from http://www.inegi.org.mx/

Oliveira, S. (1987). *Antropómetro Peniano*. Brasil.

Ombregt, L. (2013). Applied anatomy of the wrist, thumb and hand. In L. Ombregt (Ed.), A System of Orthopaedic Medicine (pp. 102-111). Elsevier Health Sciences. doi:10.1016/B978-0-7020-3145-8.00072-7

OMS. (2009). Programa Envejecimiento y Ciclo Vital. Envejecimiento activo: un marco político. *Organización Mundial de La Salud*, 74–105.

OMS. (2010). *Inactividad física: un problema de salud pública mundial*. Retrieved MARZO, 2015, from http://www.who.int/dietphysicalactivity/factsheet_inactivity/es/

OMS. (2015). *Estrategia mundial sobre régimen alimentario, actividad física y salud*. Retrieved Abril, 2015, from http://www.who.int/dietphysicalactivity/physical_activity_intensity/es/

Openshaw, S., & Taylor, E. (2006). *Ergonomics and Design: A Reference Guide*. Allsteel.

Oritz Nicolas, J. C., & Aurisicchio, M. (2011). The scenario of user experience. DS 68-7: Proceedings of the 18th International Conference on Engineering Design (ICED 11), Impacting Society through Engineering Design, Vol. 7: Human Behaviour in Design, Lyngby/Copenhagen, Denmark, 15.-19.08. 2011.

Oropeza, S., & Sánchez, A. K. (2004). *Manual de Diseño de Envases*. IMPEE.

Ortíz Nicolás, J.C. (2016a). Innovación social y diseño, una propuesta metodológica. *2° Coloquio Internacional, Las facetas de la evaluación*. Ciudad de México: Universidad Autónoma Metropolitana.

Ortíz Nicolás, J. C. (2016). Diseñando el cambio. La innovación social y sus retos. *Economía Creativa*, *06*, 9–34.

Ortíz Nicolás, J. C., Aurisicchio, M., & Desmet, P. (2013). Designing for anticipation, confidence, and inspiration. In *Proceedings of the 6th international conference on designing pleasurable products and interfaces* (pp. 31–40). ACM. 10.1145/2513506.2513510

Othman, A., & Owen, L. (2001). Adopting and measuring customer service quality (SQ) in Islamic banks: a case study in Kuwait finance house. *International Journal of Islamic Financial Services, 3*(1), 1-26.

Paço, A., & Gouveia, R. (2016). Environmental activism and consumer's perceived responsibility. *International Journal of Consumer Studies*, *40*(4), 466–474. doi:10.1111/ijcs.12272

Panagiotopoulou, G., Christoulas, K., Papanckolaou, A., & Mandroukas, K. (2004). Classroom furniture dimensions and anthropometric measures in primary school. *Applied Ergonomics*, *35*(2), 121–128. doi:10.1016/j.apergo.2003.11.002 PMID:15105073

Papanek, V. (1984). *Design for the real world. Human ecology and social change*. London: Thames and Hudson.

Papanek, V. J. (1985). *Design for the real world: Human ecology and social change*. New York: Academy Chicago Publishing.

Parcells, C., Stommel, M., & Hubbard, R. (1999). Mismatch of classroom furniture and student body dimensions: Empirical findings and health implications. *The Journal of Adolescent Health*, *24*(4), 265–273. doi:10.1016/S1054-139X(98)00113-X PMID:10227346

Parida, R., & Ray, P. K. (2015). Factors influencing construction ergonomic performance in India. 6th International Conference on Applied Human Factors and Ergonomics. *Procedia Manufacturing, 3*, 6587-6592.

Parkes, K. R. P., & Carnell, S. E. (2005). *Farmer Research Report on Musculoskeletal Disorder, Mental Health and Work Environment, Health and safety executive*. University of oxford.

Parmelee, P. A., & Lawton, M. P. (1990). The design of special environment for the aged. In J. E. Birren & K. W. Schaire (Eds.), Handbook of the psychology of aging (pp. 465–489). New York: Academic Press. doi:10.1016/B978-0-12-101280-9.50033-4

Parra–Marujo, J. (2006). *Gerontodesign: A marca de água do design, do design ergoómico, da marca ou das marcas, branca.* Retrieved April 11, 2011, from: JMarujo.Artician website: http://jmarujo.artician.com/portfolio/gerontodesign/

Parry, G. C., & Turner, C. E. (2006). Application of lean visual process management tools. *Production Planning and Control, 17*(1), 77–86. doi:10.1080/09537280500414991

Parvatikar, V., & Mukkannavar, P. (2009). Comparative study of grip strength in different positions of shoulder and elbow with wrist in neutral and extension positions. *Journal of Exercise Science and Physiotherapy, 5*, 67–75.

Patel, A., Bernstein, L., Deka, A., Feigelson, H., Campbell, P., Gapstur, S., & Thun, M. (2010). Leisure Time Spent Sitting in Relation to Total Mortality in a Prospective Cohort of US Adults. *American Journal of Epidemiology, 172*(4), 10. doi:10.1093/aje/kwq155 PMID:20650954

Patiño, A., & Márquez, J. (2009). *Actividad Física y Ejercicio Físico en salud: retos en un contexto globalizado.* Fúnambulos ed.

Patterson, W. B., & Leone, G. L. (2010). The application of handling qualities to bicycle design. In *Proceedings, bicycle and motorcycle dynamics 2010 symposium on the dynamics and control of single track vehicles* (pp. 20–22). Academic Press.

Paul, R., & Anand, S. (2011). Optimal Part Orientation in Rapid Manufacturing Process for Achieving Geometric Tolerances. *Journal of Manufacturing Systems, 30*(4), 214–222. doi:10.1016/j.jmsy.2011.07.010

Pavlovic-Veselinovic, S., Hedge, A., & Veselinovic, M. (2016). An ergonomice xpert system for risk assessment of work-related musculo-skeletal disorders. *International Journal of Industrial Ergonomics, 53*, 130–139. doi:10.1016/j.ergon.2015.11.008

Peebles, L., & Norris, B. (1998). *Adultdata: The handbook of adult anthropometric and strength measurements : data for design safety.* Dept. of Trade and Industry. Retrieved from https://books.google.com.mx/books/about/Adultdata.html?id=ZcdetwAACAAJ&redir_esc=y

Pereda, M., & Zamarreño, J. M. (2015). Modelado Basado en Agentes: un Enfoque desde la Ingeniería de Sistemas. *Revista Iberoamericana de Automática E Informática Industrial RIAI, 12*(3), 304–312.

Perry, T. E., Andersen, T. C., & Kaplan, D. B. (2014). Relocation remembered: Perspectives on senior transitions in the living environment. *The Gerontologist, 54*(1), 75–81. doi:10.1093/geront/gnt070 PMID:23840021

Pham, D. T., & Demov, S. S. (2001). *Rapid Manufacturing: The Technologies and Applications of Rapid Prototyping and Rapid Tooling.* London: Springer-Verlag London Limited. doi:10.1007/978-1-4471-0703-3

Pham, D. T., Demov, S. S., & Gault, R. S. (1999). Part orientation in stereolithography. *International Journal of Advanced Manufacturing Technology, 15*(9), 674–682. doi:10.1007001700050118

Phatak, A. M., & Pande, S. S. (2012). Optimum Part Orientation in Rapid Prototyping Using Genetic Algorithm. *Journal of Manufacturing Systems, 31*(4), 395–402. doi:10.1016/j.jmsy.2012.07.001

Pheasant, S. (1996). *Body Space: Anthropometry, ergonomics and the design of work.* London: Taylor and Francis.

Pheasant, S. (2003). *Bodyspace.* London: Taylor & Francis Ltd.

Pheasant, S., & Haslegrave, C. (2005). *Bodyspace: Anthropometry.* Ergonomics and the Design of Work.

Pheasant, S., & Haslegrave, C. (2016). *Bodyspace: Anthropometry, Ergonomics and the Design of Work* (3rd ed.). Boca Raton, FL: CRC Press.

Pousette, S., Löfgren, M., Nilsson, B., & Gustafsson, A. (2014). An Extended Method to Measure Overall Consumer Satisfaction with Packaging. *Packaging Technology & Science*, *27*(9), 727–738. doi:10.1002/pts.2064

Prado León, L. R., & Avila Chaurand, R. (2010). *Percepción Visual II. Aplicaciones Para el Diseño*. Editorial Universitaria.

Prado, L. L. R., Diaz de Leon, Z. C., Zambrano, P. P.L., Cisneros, H. A. C., & Cardenas, V. S. (2016). *Evaluation of the usability of nutritional recipes: A comparison between a text-based interface and an interface based on ergonomic guidelines*. Paper presented at the The 6th International Conference for Universal Design, International Association for Universal Design, Japan.

Prado-León, L. R., & Cárdenas-Vázquez, S. (2011). Lineamientos ergonómicos en el diseño de páginas Web. *Revista Ação Ergonômica, 2*(2).

Prado-León, L. R. (2014). *Antropometría para el diseño industrial, de espacios y de vestuario: datos y aplicaciones*. Guadalajara, Mexico: Universidad de Guadalajara.

Prado-León, L., & Ávila-Chaurand, R. (2006). *Ergonomía y Diseño de Espacios Habitables*. Guadalajara: Universidad de Guadalajara.

Prebense, N. K. (2007). Explorign Tourists' Images of a Distant destination. *Tourism Management*, *28*(3), 747–756. doi:10.1016/j.tourman.2006.05.005

Punnett, L., Cherniack, M., Henning, R., Morse, T., Faghri, P., & Team, C.-N. R. (2009). A conceptual framework for integrating workplace health promotion and occupational ergonomics programs. *Public Health Reports, 124*(4_suppl1), 16-25.

Qi, H. (2014). *Decision Making Framework For Sustainable Packaging Design Using Life Cycle Assessment* (Doctoral dissertation). Graduate School-New Brunswick Rutgers, The State University of New Jersey.

Quality Schools Program. (2010). Guide to facilitate the inclusion of students with disabilities in schools participating in the PEC. 1st ed. Secretary of Public Education. Curriculum 2011 Teacher's Guide Basic Primary Education First Degree.

Ramirez, C. (2006). Ergonomía y Productividad. LIMUSA.

Ramírez-Ibanez, M. (2004, December 14). *Aplicacion del Nylamid XL en un sistema de potencia*. Universidad de las Américas Puebla. Retrieved from http://catarina.udlap.mx/u_dl_a/tales/documentos/lim/ramirez_i_m/portada.html

Ramos, E. Z., Torres, L. O., Guzmán, J. M., Salazar, M. N., González, R. G., Espinosa, D. H., . . . Aviña, K. P. (2016). Didactic strategies in teaching-learning: In respect to the study of nomenclatureof organic chemistry in students of the Atotonilco Regional High School. *Educación Química, 27*, 43-51. Retrieved 06 05, 2017, from http://www.sciencedirect.com

Ramos, Pedro M. A.& Paulo, A. (2016). Analysis of the return on preventive measures in musculoskeletal disorders through the benefit–cost ratio: A case study in a hospital. *International Journal of Industrial Ergonomics*. doi:10.1016/j.ergon.2015.11.003

Ramos-Galvan, R. (1975). Somatometría pediátrica : Estudio semilongitudinal en ninõs de la ciudad de México. *Archivos de Investigacion Medica*, *6*(1).

Ranganathan, V., Siemionow, V., Sahgal, V., & Yue, G. (2001). Effects of aging and hand function. *Journal of the American Geriatrics Society*, *49*(11), 1478–1484. doi:10.1046/j.1532-5415.2001.4911240.x PMID:11890586

Rankin, J., & Brown, V. (2016). Creative teaching method as a learning strategy for student midwives: A qualitative study. *Nurse Education Today, 38*, 93-100. Retrieved 06 04, 2017, from www.elsevier.com/nedt

Rao, A. B., & Kumar, P. (2014). Cost implications of carbon capture and storage for the coal power plants in India. *Energy Procedia*, *54*, 431–438. doi:10.1016/j.egypro.2014.07.285

Ray, P. K., Parida, R., & Sarkar, S. (2015). Ergonomic analysis of construction jobs in India: A biomechanical modelling approach. 6th International Conference on Applied Human Factors and Ergonomics. *Procedia Manufacturing*, *3*, 4606-4612.

Reijnierse, M., & Fernandes, N. (2013). Hand and wrist. In I. Beggs (Ed.), *Musculoskeletal Ultrasound* (pp. 73–102). Lippincott Williams & Wilkins.

Reiser, R. F. II, Peterson, M. L., & Broker, J. P. (2002). Understanding recumbent cycling: Instrumentation design and biomechanical analysis. *Biomedical Sciences Instrumentation*, *38*, 209–214. PMID:12085603

Reuters. (2008). *Japan aging suit puts car makers in senior circuit.* Retrieved August 20, 2017, from: http://www.reuters.com/article/us-japan-nissan-suit-idUST9224920080417

Ritchey, T. (2011a). About Fritz Zwicky. In *Wicked problems–Social messes*. Berlin, Germany: Springer-Verlag. doi:10.1007/978-3-642-19653-9_9

Ritchey, T. (2011b). Wicked problems and genuine uncertainty. In *Wicked problems–Social messes* (pp. 19–29). Berlin, Germany: Springer-Verlag. doi:10.1007/978-3-642-19653-9_3

Rittel, H. W. J., & Webber, M. M. (1973). Dilemmas in a general theory of planning. *Policy Sciences*, *4*(2), 155–169. doi:10.1007/BF01405730

Rivera, P. B., Cuartas, M. C., & Renteria, Y. P. (2011). *Facultad de Educación y Humanidades: La importancia del uso de estrategias de aprendizaje.* Retrieved 06 04, 2017, from Universidad Católica Luis Amigó: http://www.funlam.edu.co/uploads/facultadeducacion/53_LA_IMPORTANCIA_DEL__USO_ESTRATEGIAS_DE_APRENDIZAJE.pdf

Rivlin, L., & Weinstein, C. (1984). Educational issues, school settings, and environmental psychology. *Journal of Environmental Psychology*, *4*(4), 347–364. doi:10.1016/S0272-4944(84)80005-5

Roberts, N. (2000). Wicked problems and network approaches to resolution. *International Public Management Review*, *1*(1), 1–19.

Robinette, K. (2012). Anthropometry for Product Design. In G. Salvendy (Ed.), *Handbook of Human Factors and Ergonomics* (4th ed.; p. 1752). Wiley & Sons. doi:10.1002/9781118131350.ch11

Robinette, K. M., & Daanen, H. A. M. (2006). Precision of the CAESAR scan-extracted measurements. *Applied Ergonomics*, *37*(3), 259–265. doi:10.1016/j.apergo.2005.07.009 PMID:16202970

Robinson, P. A. (2009). *Writing and Designing Manuals and Warnings 4e*. CRC Press. doi:10.1201/9781420069860

Rodgers, P., & Milton, A. (2011). *Product Design.* London: Laurence King Publishing. Retrieved 04 10, 2017, from https://www.ebscohost.com

Rodríguez, A. V., & Pérez, E. V. (2015). Didactic strategy to stimulate the personality education. *Educación Médica Superior, 29*(2), 340-350. Retrieved 06 04, 2017, from https://www.ebscohost.com

Rodríguez, Y. (2011). ERIN: método práctico para evaluar la exposición a factores de riesgo de desórdenes músculo-esqueléticos (Doctor en Ciencias Técnicas Tesis de Doctorado). Instituto Superior Politécnico "José Antonio Echeverría", La Habana, Cuba.

Rodríguez, J. A. (2006). *Manual de Ingeniería y Diseño en envase y embalaje*. IMPEE.

Rodriguez, J. A. (2008). *Enciclopedia IMPEE, Manual para elaboración de Especificaciones Técnicas para materiales de Envase y Embalaje*. IMPEE.

Rodríguez, Y., & Pérez, E. (2013a). Implementación de un proceso de intervención ergonómica en una imprenta cubana. *Ação Ergonômica. Revista da Associação Brasileira de Ergonomia*, *8*(2), 1–13.

Rodríguez, Y., & Pérez, E. (2013b). Introducción de mejoras ergonómicas. Beneficios para la organización. *Revista Técnica de Ingeniería de la Universidad del Zulia*, *36*(2), 1–10.

Rodríguez, Y., Viña, S., & Montero, R. (2013a). ERIN: A practical tool for assessing work-related musculoskeletal disorders. *Occupational Ergonomics*, *11*(2-3), 15.

Rodríguez, Y., Viña, S., & Montero, R. (2013b). A method for non-experts in assessing exposure to risk factors for work-related musculoskeletal disorders-ERIN. *Industrial Health*, *51*(6), 5. doi:10.2486/indhealth.2013-0008 PMID:24077444

Roebuck, J. A. Jr. (1995). *Anthropometric Methods: Designing to Fit the Human Body*. Santa Monica, CA: Human Factors & Ergonomics Society.

Roebuck, J., Kromer, K., & Thomson, W. (1975). *Engineering Anthropometry Method* (2nd ed.). New York: Wiley.

Roffey, D. M., Wai, E. K., Bishop, P., Kwon, B. K., & Dagenais, S. (2010). Causal assessment of awkward occupational postures and low back pain: Results of a systematic review. *The Spine Journal*, *10*(1), 89–99. doi:10.1016/j.spinee.2009.09.003 PMID:19910263

Rogers, W. A., Lamson, N., & Rousseau, G. K. (2000). Warning research: An integrative perspective. *Human Factors*, *42*(1), 102–139. doi:10.1518/001872000779656624 PMID:10917149

Roqueiro, N., Vieira, R. S., & Faria, M. G. (2010). Tilting control of a three-wheeled vehicle by steering. *CBA 2010—XVIII Cong. Brasileiro de Automatica*, 3464–3471.

Rossi, D., Bertoloni, E., Fenaroli, M., Marciano, F., & Alberti, M. (2013). A multi-criteria ergonomic and performance methodology for evaluating alternatives in "manuable" material handling. *International Journal of Industrial Ergonomics*, *43*(4), 314–327. doi:10.1016/j.ergon.2013.04.009

Rouyendegh, B. D., & Erkan, T. E. (2012). Selection of academic staff using the fuzzy analytic hierarchy process (fahp): A pilot study. *Technical Gazette*, *19*(4), 923-929.

Rubin, J., & Chisnell, D. (2008). *Handbook of usability testing: howto plan, design, and conduct effective tests*. John Wiley & Sons.

Rundcrantz, B. L., Johnsson, B., & Mortiz, U. (1991). Occupational cervico-brachial disorders among dentists: Analysis of ergonomics and locomotor functions. *Swedish Dental Journal*, *15*(3), 105–115. PMID:1876977

Ruskin, J. (1849). *The seven lamps of architecture*. Chichester, UK: John Wiley & Sons.

Rybaczewska-Błażejowska, M., Masternak-Janus, A., & Gierulski, W. (2016). Inverse problem of life cycle assessment (LCA): Its application in designing for environment (DfE). *Management*, *20*(2), 224–241. doi:10.1515/manment-2015-0062

Saaty, T. L. (1970). *Optimization in integers and related external problems*. McGraw-Hill.

Salas Perea, R. S. (2010). Propuesta de estrategia para la evaluación del desempeño laboral de los médicos en Cuba. *Educación Médica Superior*, *24*(3), 387–417.

Salvendy, G. (2006). Handbook of human Factors and Ergonomics. Academic Press. doi:10.1002/0470048204

Sambasivan, M., & Fei, N. Y. (2008). Evaluation of critical success factors of implementation of ISO 14001 using analytic hierarchy process (AHP): A case study from Malaysia. *Journal of Cleaner Production, 16*(13), 1424–1433. doi:10.1016/j.jclepro.2007.08.003

Sanchez-Monroy, D. (2007). Operadores de autotransporte de 18 a 68 años. In R. Avila-Chaurand, L. R. Prado-León, & E. L. González-Muñoz (Eds.), Dimensiones Antropométricas de Población Latinoamericana (2nd ed.). Guadalajara, Mexico: Academic Press.

Sánchez-Sosa, J. J., & González-Celis, A. (2006). Evaluación de la calidad de vida desde la perspectiva psicológica. In *Manual para la evaluación clínica de los trastornos psicológicos: trastornos de la edad adulta e informes psicológicos* (pp. 473–492). Pirámide.

Satapathy, S. (2014). ANN, QFD and ISM approach for framing electricity utility service in India for consumer satisfaction. *International Journal of Services and Operations Management, 18*(4), 404–428. doi:10.1504/IJSOM.2014.063243

Satapathy, S., Patel, S. K., Mahapatra, S. S., & Biswas, A. (2013). A framework for electricity utility service by QFD technique. *International Journal of Services Sciences, 5*(1), 58–73. doi:10.1504/IJSSCI.2013.057659

Satapathy, S., Patel, S. K., & Mishra, P. D. (2012). Discriminate analysis and neural network approach in water utility service. *International Journal of Services and Operations Management, 12*(4), 468–489. doi:10.1504/IJSOM.2012.047954

Savransky, S. (2000). *Engineering of creativity : introduction to TRIZ methodology of inventive problem solving*. Boca Raton, FL: CRC Press. doi:10.1201/9781420038958

Schatzki, T. R. (2010). *Site of the social: A philosophical account of the constitution of social life and change*. Penn State University Press.

Schnetzler, M. J., Sennheiser, A., & Schönsleben, P. (2007). A decomposition-based approach for the development of a supply chain strategy. *International Journal of Production Economics, 105*(1), 21–42. doi:10.1016/j.ijpe.2006.02.004

Schöggl, J. P., Baumgartner, R. J., & Hofer, D. (2017). Improving sustainability performance in early phases of product design: A checklist for sustainable product development tested in the automotive industry. *Journal of Cleaner Production, 140*, 1602–1617. doi:10.1016/j.jclepro.2016.09.195

Scholtz, J. (2004). Usability evaluation. National Institute of Standards and Technology, 1.

Schön, D. A. (1987). *Educating the reflective practitioner: Toward a new design for teaching and learning in the professions*. Chichester, UK: Jossey-Bass.

Schreuders, T., Brandsma, W., & Stam, H. (2014). Functional Anatomy and Biomechanics of the Hand. In M. Tuncay Duruöz (Ed.), Hand Function (pp. 3–21). London: Springer Science+Business Media. doi:10.1007/978-1-4614-9449-2_1

Schumpeter, J. A. (1934). *The theory of economic development*. Cambridge, MA: Harvard University Press.

Schwab, A. L. (2012). Bicycling safety and the lateral stability of the bicycle. *Proceedings, International Cycling Safety Conference*.

Scott, L., & Vigar-Ellis, D. (2014). Consumer understanding, perceptions and behaviors with regard to environmentally friendly packaging in a developing nation. *International Journal of Consumer Studies, 38*(6), 642–649. doi:10.1111/ijcs.12136

SEP. (2002). *Lineamientos Operativos del Sistema de Monitoreo para la Gestión de la Calidad.* Retrieved September 12, 2014, from http://www.salud.gob.mx/unidades/cdi/documentos/DOCSAL7494.pdf

Sevices., D. o. H. H. (2008). *2008 Physical Activity Guidelines for Americans*. Retrieved 04, 20015, from http://www.health.gov/paguidelines/pdf/paguide.pdf

Shaoqing, L. (1999). *Tao*. China: X-Ray Photograph Anthropometer.

Shen, L., Muduli, K., & Barve, A. (2015). Developing a sustainable development framework in the context of mining industries: AHP approach. *Resources Policy*, *46*, 15–26. doi:10.1016/j.resourpol.2013.10.006

Shiffman, L. (1992). Effects of Aging on Adult Hand Function. *The American Journal of Occupational Therapy*, *46*(9), 785–792. doi:10.5014/ajot.46.9.785 PMID:1514564

Shikdar, A., & Sawaqed, N. (2003). Worker productivity, and occupational health and safety issues in selected industries. *Computers & Industrial Engineering*, *45*(4), 563–572. doi:10.1016/S0360-8352(03)00074-3

Shirzaei, M., Mirzaei, R., Khaje-Alizade, A., & Mohammadi, M. (2015). Evaluation of ergonomic factors and postures that cause muscle pains in dentistry students' bodies. *Journal of Clinical and Experimental Dentistry*, *7*(3), e414–e418. doi:10.4317/jced.51909 PMID:26330941

Shrout, P. E., & Fleiss, J. L. (1979). Intraclass Correlations: Uses in Assessing Rater Reliability. *Psychological Bulletin*, *86*(2), 420–428. doi:10.1037/0033-2909.86.2.420 PMID:18839484

Shu, L. H., & Flowers, W. C. (1999). Application of a design-for-remanufacture framework to the selection of product life-cycle fastening and joining methods. *Robotics and Computer-integrated Manufacturing*, *15*(3), 179–190. doi:10.1016/S0736-5845(98)00032-5

Sidenvall, B., Nydahl, M., & Fjellström, C. (2001). Managing food shopping and cooking: The experiences of older Swedish women. *Ageing and Society*, *21*(02), 151–168. doi:10.1017/S0144686X01008121

Silva, B. Y. (2011). Reseña de "¿Qué es el branding?" [Review of the book ¿Qué es el Branding?]. *Revista del Centro de Investigación. Universidad La Salle, 9*(35) 43-44. Retrieved 05 08, 2017 from http://www.redalyc.org/articulo.oa?id=34218346006

Sinclair, M. A. (1995). Subjective assessment. In J. R. Wilson & E. N. Corlett (Eds.), *Evaluation of Human Work. A practical ergonomics methodology* (pp. 69–100). London: Taylor and Francis.

Smith, P., & Apple, D. K. (2007). *Methodology for creating methodologies*. Retrieved 03 17, 2017, from Faculty guidebook: http://www.pcrest3.com/fgb/efgb4/2/2_4_16.htm

Soh, S. L., Ong, S. K., & Nee, A. Y. C. (2016). Design for assembly and disassembly for remanufacturing. *Assembly Automation*, *36*(1), 12–24. doi:10.1108/AA-05-2015-040

Sonnenwald, D. H. (2006). Challenges in Sharing Information Effectively: Examples from Command and Control. *Information Research: an International Electronic Journal*, *11*(4), n4.

Sonnenwald, D. H., & Pierce, L. G. (2000). Information behavior in dynamic group work contexts: Interwoven situational awareness, dense social networks and contested collaboration in command and control. *Information Processing & Management*, *36*(3), 461–479. doi:10.1016/S0306-4573(99)00039-4

Stamatakis, E., Hamer, M., & Dunstan, D. (2011). Screen-Based Entertainment Time, All-Cause Mortality, and Cardiovascular Events: Population-Based Study with Ongoing Mortality and Hospital Events Follow-Up. *Journal of the American College of Cardiology*, *57*(3), 8. doi:10.1016/j.jacc.2010.05.065 PMID:21232666

Stanton & Baber. (2002). Error by Design: Methods for Predicting Device Usability. *Design Studies, 23*(4), 363–84.

Stanton, N. A. (2006). Hierachical Task Analysis. In W. Karwowski (Ed.), International Encycopledia of Ergonomics and Human factors (pp. 3183–3190). Taylor & Francis.

Stanton, N. A., & Baber, C. (2002). Error by design: Methods for predicting device usability. *Design Studies*, *23*(4), 363–384. doi:10.1016/S0142-694X(01)00032-1

Stanton, N. A., Hedge, A., Brookhuis, K., Salas, E., & Hendrick, H. W. (2004). *Handbook of human factors and ergonomics methods*. CRC Press. doi:10.1201/9780203489925

Stanton, N. A., & Young, M. S. (1999). Ergonomics Methods in the Design of Consumer Products. In W. Karwowski & W. S. Marras (Eds.), *The occupational ergonomics handbook* (pp. 741–760). Boca Raton, Florida.

Stanton, N., Hedge, A., Brookhuis, K., Salas, E., & Hendrick, H. (2006). *Handbook of Human Factors and Ergonomics Methods*. Washington, DC: CRC.

Stanton, N., Salmon, P. M., & Rafferty, L. A. (2013). *Human factors methods: a practical guide for engineering and design*. Ashgate Publishing, Ltd. doi:10.1201/9781315587394

Steenkamp, L. P., Hagedorn-Hansen, D., & Oosthuizen, G. A. (2017). Visual Management System to Manage Manufacturing Resources. *Procedia Manufacturing*, *8*, 455–462. doi:10.1016/j.promfg.2017.02.058

Stegink Jansen, C., Niebuhr, B., Coussirat, D., Hawthorne, D., Moreno, L., & Phillip, M. (2008). Hand Force of Men and Women Over 65 Years of Age as Measured by Maximum Pinch and Grip Force. *Journal of Aging and Physical Activity*, *16*(1), 24–41. doi:10.1123/japa.16.1.24 PMID:18212392

Stephens, J.-P., Vos, G. A., Stevens, E. M. Jr, & Steven Moore, J. (2006). Test-retest repeatability of the Strain Index. *Applied Ergonomics*, *37*(3), 275–281. doi:10.1016/j.apergo.2005.07.007 PMID:16182231

Stevens, E. M., Vos, G. A., Stephens, J.-P., & Steven Moore, J. (2004). Inter-Rater Reliability of the Strain Index. *Journal of Occupational and Environmental Hygiene*, *1*(11), 745–751. doi:10.1080/15459620490521142 PMID:15673095

Stoudt, H. W. (1981). The Anthropometry of the Elderly. *Human Factors: The Journal of the Human Factors and Ergonomics Society*, *23*(1), 29–37. doi:10.1177/001872088102300104 PMID:7228046

Straker, L. M. (1999). Body discomfort assessment tools. In W. Karwowski & W. S. Marras (Eds.), *Handbook of Occupational Ergonomics* (pp. 1239–1251). Boca Ratón, FL: Academic Press.

Study Program 2011 Teacher's Guide Basic Education Elementary Second Grade. (2011). DF, Mexico: Ministry of Public Education.

Suh, N. (1990). *The principles of design. Oxford series on advanced manufacturing: 6*. New York: Oxford University Press.

Suh, N. (1998). Axiomatic Design Theory for Systems. *Research in Engineering Design*, *10*(4), 189–209. doi:10.1007001639870001

Suh, N. (2001). *Axiomatic Design: Advances and Applications*. Oxford University Press.

Superiores, I. T. (2006). Movilidad: Una visión estratégica (ITESO Ed. ITESO ed.). Guadalajara, México: ITESO.

Sylla, N., Bonnet, V., Colledani, F., & Fraisse, P. (2014). Ergonomic contribution of ABLE exoskeleton in automotive industry. *International Journal of Industrial Ergonomics*, *44*(4), 475–481. doi:10.1016/j.ergon.2014.03.008

Taguchi, G. (1993). *Taguchi on robust technology development: bringing quality engineering upstream*. American Society of Mechanical Engineers. doi:10.1115/1.800288

Takala, E.-P., Pehkonen, I., Forsman, M., Hansson, G.-Å., Mathiassen, S., Neumann, W., ... Winkel, J. (2010). Systematic evaluation of observational methods assessing biomechanical exposures at work. *Scandinavian Journal of Work, Environment & Health*, *36*(1), 3–24. doi:10.5271jweh.2876 PMID:19953213

Takao, I., Yoshiyuki, K., Toshio, M., & Katsuhiko, H. (1986). *Motor-Driven Anthropometer*. Japón.

Tang, D., Zhang, G., & Dai, S. (2009). Design as integration of axiomatic design and design structure matrix. *Robotics and Computer-integrated Manufacturing*, *25*(3), 610–619. doi:10.1016/j.rcim.2008.04.005

Taylor, C., & Schwarz, R. (1955). The anatomy and mechanics of the human hand. *Artificial Limbs*, 22–35. PMID:13249858

TC^2. (2003). *SizeUSA project notebook*. Author.

TC^2. (2017). *SizeUSA*. Retrieved from http://www.tc2.com/size-usa.html

Technical Innovation Center. (2013). *40 Principles*. Retrieved June 25, 2017, from http://triz.org/index.php/triz/principles

Tero, M. (2012). The Logotype, Fundamental Unit for a Brand's Visual Identity. *Scientific Bulletin of The Petru Maior University of Targu Mures*, *9*(1), 64–68.

Teymourian, K., Seneviratne, D., & Galar, D. (2016). Ergonomics Contribution in Maintainability. In D. Galar & D. Seneviratne (Eds.), *Proceedings of MPMM 2016: 6th International Conference on Maintenance Performance Measurement and Management* (pp. 180–186). Luleå, Sweden: Luleå Tekniska Universitet. Retrieved from http://www.diva-portal.org/smash/record.jsf?pid=diva2%3A1108271&dswid=-2491

Thomas, P., & Macredie, R. D. (2002). Introduction to the new usability. *ACM Transactions on Computer-Human Interaction*, *9*(2), 69–73. doi:10.1145/513665.513666

Tjell, J., & Bosch-Sijtsema, P. M. (2015). Visual management in mid-sized construction design projects. *Procedia Economics and Finance*, *21*, 193–200. doi:10.1016/S2212-5671(15)00167-7

Togay, C., Dogru, A. H., & Tanik, J. U. (2008). Systematic Component-Oriented development with Axiomatic Design. *Journal of Systems and Software*, *81*(11), 1803–1815. doi:10.1016/j.jss.2007.12.746

Tompa, E., Dolinschi, R., de Oliveira, C., Amick, B., & Irvin, E. (2010). A Systematic Review of Workplace Ergonomic Interventions with Economic Analyses. *Journal of Occupational Rehabilitation*, *20*(2), 220–234. doi:10.100710926-009-9210-3 PMID:19890618

Trevino E., & Trevino, G. (2004). Study on educational inequality in Mexico: the incidence of school in the academic performance of students and the role of teachers. *Research Papers Collection,* (5).

Tromp, N. (2013). *Social design* (Unpublished PhD dissertation). Delft University of Technology.

Tromp, N., & Hekkert, P. (2016). Assessing methods for effect-driven design: Evaluation of a social design method. *Design Studies*, *43*, 24–47. doi:10.1016/j.destud.2015.12.002

Tyagi, M., Kumar, P., & Kumar, D. (2015). Analysis of interactions among the drivers of green supply chain management. *International Journal of Business Performance and Supply Chain Modelling*, *7*(1), 92–108. doi:10.1504/IJBPSCM.2015.068137

U.S. EPA Design for Enviroment from United States Environmental Protection Agency. (2017). *Reports and Publications of the Design for the Environment and Safer Choice Program.* Retrieved June 27, 2017 from http://www.epa.gov/saferchoice/design-enviroment-programs-initiatives-and-projects

UNICEF Innocenti Research Centre (2007). *UNICEF: Child poverty in perspective: An overview of child well-being in rich countries*. UNICEF Innocenti Report Card 7.

Urdiain, R. (2006). Las cuatro P. *Revista Entrepreneur.* Retrieved from https://www.entrepreneur.com/article/256375

Uribe, L. (n.d.). *Dispositivo para realizar estudios antropometricos*. Retrieved from http://siga.impi.gob.mx/newSIGA/content/common/busquedaSimple.jsf

Uribe, P. (2008). Dispositivo para realizar estudios antropométricos.

Valero, A., Gómez, M., Gavala, J., & Ruiz, F. (2007). Porque no se realiza actividad fisco-deportivo en el tiempo libre. *FEADEF, 12*(5).

Van der Have, R. P., & Rubalcaba, L. (2016). Social innovation research: An emerging area of innovation studies? *Research Policy, 45*(9), 1923–1935. doi:10.1016/j.respol.2016.06.010

Van Kuijk, J. (2008) Make It Easy On Yourself, An 8-step guide to buying easy to use products. Dutch Ergonomics Magazine, volume 33, issue 1

Vandergrift, J., Gold, J., Hanlon, A., & Punnett, L. (2012). Physical and psychosocial ergonomic risk factors for low back pain in automobile manufacturing workers. *Occupational and Environmental Medicine, 69*(1), 29–34. doi:10.1136/oem.2010.061770 PMID:21586759

Veitch, J., Ball, K., Crawford, D., Abbott, G., & Salmon, J. (2013). Is park visitation associated with leisure-time and transportation physical activity? *Preventive Medicine, 57*(3). PMID:23933224

Vélez, E., Schiefelbein, E., & Valenzuela, J. (1993). *Factors affecting academic achievement in primary education.* Literature Review of Latin America and the Caribbean.

Vidales, M. D. (1999). *El envase en el tiempo: Historia del envase*. Trillas – UAM.

Vygotsky, L. S. (1978). *Mind in society: The development of higher mental process*. Cambridge, MA: Harvard University Press.

Wahl, H.-W., & Weisman, G. D. (2003). Environmental gerontology at the beginning of the new millennium: Reflections on its historical, empirical, and theoretical development. *The Gerontologist, 43*(5), 616–627. doi:10.1093/geront/43.5.616 PMID:14570958

Walker, H., & Jones, N. (2012). Sustainable supply chain management across the UK private sector. *Supply Chain Management, 17*(1), 15–28. doi:10.1108/13598541211212177

Weisner, K., & Deuse, J. (2014). Assessment methodology to design an ergonomic and sustainable order picking system using motion capturing systems. *Procedia CIRP, 17*, 422–427. doi:10.1016/j.procir.2014.01.046

Wendt, T. (2015). *Design for Dasein: Understanding the Design of Experiences*. Thomas Wendt.

Werle, S., Goldhahn, J., Drerup, S., Simmen, B., Sprott, H., & Herren, D. (2009). Age and Gender Specific Normative Data of Grip and Pinch Strenght in a Healthy Adult Swiss Population. *The Journal of Hand Surgery, 34*(1), 76–84. doi:10.1177/1753193408096763 PMID:19129352

West, A. P., Sambu, S. P., & Rosen, D. W. (2001). A process planning method to improve build performance in stereolithography. *Computer Aided Design, 33*(1), 65–79. doi:10.1016/S0010-4485(00)00064-6

Wheelen, T., & Hunger, J. (2011). *Concepts in strategic management and business policy*. Pearson Education India.

Whiteley, N. (1997). *Design for society*. London: Reaktion Books.

Wilson, T. D. (2008). Activity theory and information seeking. *Annual Review of Information Science & Technology, 42*(1), 119–161. doi:10.1002/aris.2008.1440420111

Wogalter, M. S. & Silver, N. C. (1995). Warning signal words: connoted strength and understandability by children, elders, and non-native English speakers. *Ergonomics, 38*, 2188-2206.

Wogalter, M. S., Sojourner, R. J., & Brelsford, J. W. (1997). Comprehension and retention of safety pictorials. *Ergonomics, 40*(5), 531–542. doi:10.1080/001401397188017

Wolff, J. S., & Wogalter, M. S. (1998). Comprehension of pictorial symbols: Effects of context and test method. Human Factors. *The Journal of the Human Factors and Ergonomics Society, 40*(2), 173–186. doi:10.1518/001872098779480433

Won, K., & Joon, Y. (2005, November). Mutual Compensation of TRIZ and Axiomatic Design. *Design*, 1–12.

Wong, R., & Díaz, J. J. (2007). Health care utilization among older Mexicans: Health and socioeconomic inequalities. *Salud Pública de México, 49*, s505–s514. doi:10.1590/S0036-36342007001000010 PMID:17724524

World Economic Forum. (2011). *The global competitiveness report 2011-2012.* 10.4102ajip.v40i1.1117

World Health Organization. (2002). *Active Ageing: A Policy Framework.* Madrid, Spain: WHO.

World Health Organization. (2015). *World report on ageing and health.* WHO.

X-ray photograph anthropometer. (1999). Retrieved from https://www.google.com/patents/CN2347543Y?cl=en

Xu, F., Wong, Y. S., Loh, H. T., Fuh, J. Y. H., & Miyazawa, T. (1997). Optimal orientation with variable slicing in stereolithography. *Rapid Prototyping Journal, 3*(3), 76–88. doi:10.1108/13552549710185644

Yakimishyn, J. E., & Magill-Evans, J. (2002). Comparisons Among tools, surface orientation, and pencil grasp for children 23 months of age. *The American Journal of Occupational Therapy, 56*(5), 564–572. doi:10.5014/ajot.56.5.564 PMID:12269511

Yang, K., & Zhang, H. (2000). A Comparison of TRIZ and Axiomatic Design. In *First International Conference on Axiomatic Design* (pp. 235–243). Cambridge. Retrieved from http://moodle.stoa.usp.br/file.php/1359/TRIZ_AD.pdf

Yang, S. C. (2003). Reconceptualizing think-aloud methodology: Refining the encoding and categorizing techniques via contextualized perspectives. *Computers in Human Behavior, 19*(1), 95–115. doi:10.1016/S0747-5632(02)00011-0

Yi, J.-W., & Park, G.-J. (2005). Development of a design system for EPS cushioning package of a monitor using axiomatic design. *Advances in Engineering Software, 36*(4), 273–284. doi:10.1016/j.advengsoft.2004.06.016

Yilmaz, S., & Seifert, C. M. (2011). Creativity through design heuristics: A case study of expert product design. *Design Studies, 32*(4), 384–415. doi:10.1016/j.destud.2011.01.003

Yokota, M. (2005). Head and facial anthropometry of mixed-race US Army male soldiers for military design and sizing: A pilot study. *Applied Ergonomics, 36*(3), 379–383. doi:10.1016/j.apergo.2005.01.009 PMID:15854582

Young, M. S., & Stanton, N. A. (2005). *Applying interviews to usability assessment. In The handbook of human factors and ergonomics methods* (pp. 29–1). CRC Press LLC.

Zorrilla, M. (2010)., Efficiency performance of pupils in primary schools between attendance and quality and access. In SITEAL (Ed.), Reflections from basic education in Mexico (8th ed.). Academic Press.

Zwicky, F. (1969). *Discovery, Invention, Research - Through the Morphological Approach.* Toronto: The Macmillan Company.

About the Contributors

Juan Luis Hernandez Arellano is a researcher and a full time teacher of the Design Department at the Autonomous University of Ciudad Juarez, Mexico. His research interest are about industrial ergonomics, ergonomic product design, statistics applied to research ergonomics, and biomechanics. Dr. Hernandez have published numerous articles, books and books chapters in indexed Journals and Internationals books. Currently, Dr. Hernandez is the coordinator of the Master of Product Design at the Autonomous University of Ciudad Juarez.

Aide Maldonado-Macías works as a professor-Investigator for the Autonomous University of Juarez Mexico. Dr. Maldonado has published several papers in various international journals and conference proceedings. She has been a member of the programme committee and scientific committee of several national and international conferences, in most cases also serving as referee. She is a certified professional in ergonomics in Mexico and she serves as Member of the Mexican Society on Ergonomics and the Society of Industrial Engineers. She also serves as Associate Editor of the AcademiaJournals.com journals and the Occupational Ergonomics Advances and Applications. She serves as referee for the International Journal of Advanced Manufacturing Technology and the World Conference of Industrial Engineering. She has also been evaluator and reviewer for research projects in Mexico. She has obtained the recognition with her research the best investigation in Industrial Engineering in Mexico for the Technological Institute of Juarez México and has participated in the most prestigious International Conferences and Congresses on Ergonomics and Computers and Industrial Engineering. She has publications on indexed journals in her country and the International Journal of Advanced Manufacturing Technology and the International Journal of Industrial Engineering. Also has contributions in book chapters for prestigious editorials.

Porfirio Peinado Coronado is at Design Department of the Autonomous University of Ciudad Juarez, Mexico, (UACJ). He is a full time teacher and researcher. Human activities and their impact to the environment, as well as green energies' strategies and practices, Leadership in Energy & Environmental Design's (LEED) certification, and geochemistry are his areas of interest. His recent work comprises ecodesign and its environmental quality improvement, life cycle analysis, green building certification, geochemical characterization, and factorial analyses.

***

Carlos Aceves-González is Professor of Ergonomics & Inclusive Design, University of Guadalajara. He has a background in Psychology and completed his PhD at Loughborough University, England.

Julian Aguilar-Duque is a Professor - Researcher of Industrial Engineering, specialized in simulation systems, ergonomics and manufacturing systems.

Guillermo Amaya-Parra is an Industrial engineer from the Autonomous University of Baja California, Master's Degree in Engineering by UABC; Professor Researcher from 2011 to date.

Natalia Anaya is an interior designer with a masters degree in ergonomics from the University of Guadalajara.

Karina Arredondo-Soto is a Full Time Research Professor, Chemical Sciences and Engineering Faculty. Areas of interest are related to Manufacturing, Technological Innovation and design and development of products and processes.

Cesar Balderrama Armendariz is Professor-Researcher in the Institute of Architecture, Design and Arts. Head of prototype laboratory in Industrial design Programs: Undergraduate Program of Industrial Design, Master in Design and Product Development, Master in Studies and Creative Processes in Art and Design.

Nora Bustamante is Graphic Designer by Universidad Autónoma de Cd. Juárez, current student of Master in Product design and development.

David Cortés Sáenz, PhD, is a Researcher focusing on Ecodesign and Life Cycle assessment in product design.

Luis Elias Davila-Leyva got the degrees of Industrial Designer and Master in Product Design in the Autonomous University of Ciudad Juarez, México. Presently, he is a professor of the Industrial Design program at the Autonomous University of Ciudad Juarez. Also, he is the author of two patents.

Carlos Diaz de Leon Zuloaga, M.S. in Ergonomics and B.A. in Industrial Design.

Adrian Durán is Graphic Designer by Universidad Autónoma de Cd. Juárez, current student of Master in Product design and development, and Professor at Universidad Autónoma de Cd. Juárez of web usability.

Adriana Garcia got a master in science and engineering with a specialization related with optimization at the Universidad Politecnica de Aguascalientes, Mexico.

Elvia Luz González-Muñoz has a Ph.D. in Psychology from National Autonomous University of Mexico and Master of Science in Public Health from Regional Institute of Public Health Research. Providing courses at various universities such as FES Aragón de la UNAM; Autonomous University of Nayarit and UAM Xochimilco, as well as acting as an Instructor in refresher courses for different companies. Publications include four books, 24 chapters of books with various editorials, and 15 articles in journals of international circulation.

Thomas Harrison received his M.Eng. in Engineering from Durham University (2000), before working as a Management Consultant specialising in Enabling Technologies, and then Business Transformation. He studied towards a PhD at Imperial College London and has research interests in areas such as: Designerly Uncertainties, Complexities & Ambiguities; Adaptation to the Design Situation; Resolution of Fuzziness & Wickedness; and finally, the development of Tolerance & Coping.

Guadalupe Hernandez-Escobedo received a PhD in Information Behaviour and Cognitive Engineering at the University of Leeds, UK and has more than 20 years of experience in teaching, research and application of Industrial Engineering.

Victor M. Juarez-Luna is a full-time professor at the Autonomous University of Baja California at the Ensenada Engineering College. He is an engineer in control and automation graduated from the National Polytechnic Institute of the City of Mexico. Master's degree from IPN-Mex's digital technology research center. He is currently studying PhD in the Ensenada scientific and higher education research center in the field of renewable energy.

Kaushik Kumar, B.Tech (Mechanical Engineering, REC (Now NIT), Warangal), MBA (Marketing, IGNOU) and Ph.D (Engineering, Jadavpur University), is presently an Associate Professor in the Department of Mechanical Engineering, Birla Institute of Technology, Mesra, Ranchi, India. He has 14 years of Teaching & Research and over 11 years of industrial experience in a manufacturing unit of Global repute. His areas of teaching and research interest are CAD / CAM, Quality Management Systems, Optimization, Non-conventional machining, Rapid Prototyping and Composites. He has 9 Patents, 4 Books, 3 Edited Books, 19 Book Chapters, 118 International Journal publications, 18 International and 8 National Conference publications to his credit. He is on the editorial board and review panel of 14 International and 1 National Journals of repute. He has been felicitated with many awards and honours.

Ercilia Loera Anchondo is a Full-time professor and researcher at the Autonomous University of Ciudad Juarez with 19 years of experience. Her research interests are related to Tourism and the Teaching of English as a Foreign Language.

Annika Maya Rivero studied a degree in Industrial Design, an MBA in Design and currently a PhD in Design. She has taken some short courses on disability and rehabilitation strategies and their implementation, sustainable design, dementia and person-centered model of dementia care. She has focused on research on design for aging, user-centered design, environmental design and design for dementia.

Mario Moreno graduated in industrial design, with master's degree in product design, currently teaching in technology and design subjects.

Alberto Ochoa Ortiz-Zezzatti (Bs'94–Eng.Master'00; PhD'04-Postdoctoral Researcher'06, Industrial Postdoctoral Research'09 & Energy Studies Postdoctoral Research'18). He joined the Juarez City University in 2008. He participates in the organization of several International Conferences. He has review of four important Journals from Elsevier: highlighting Applied Soft Computing and Computers in Human Behavior. His research interests include ubiquitous compute, evolutionary computation, natural processing language, social modeling, anthropometrics characterization and Social Data Mining. In May 2016 begin a Sabbatical internship year in Barcelona Supercomputing Center with a project related with simulation of human stampedes and refugee boat sinking improved with Artificial Intelligence.

Silvia Oropeza Herrera is an Industrial designer from Universidad Autonoma Metropolitana, Campus Xochimilco, with a University Merit medal, Master degree in Engineering and Packaging Design and Master degree in Teaching. Coauthor of Packaging Design Manual, Experiences in industrial design teaching and Creativity development and teaching portfolio. Professional career in packaging, with national and international recognition, Full time teacher at UAM Xochimilco and was guest teacher in Sao Pablo, Brasil University.

Juan Carlos Ortíz Nicolas holds a Ph.D in User Experience from Imperial College London. He is interested in understanding how consumers experience positive encounters with products. He holds a Master's degree in Design for Interaction from Delft University of Technology and a Bachelor's degree in Industrial Design from the National Autonomous University of Mexico. He has been lecturer in User Centred Design and Design for Social Innovation at undergraduate level at the National Autonomous University of Mexico.

Lilia Prado León has a Ph.D. in Health Sciences, M.S. in Educative Psychology and B.S. in Psychology. Full professor and researcher of the Ergonomics Research Center and Coordinator of the Ergonomics Master of the University of Guadalajara. Member of the Mexican Color Research Association. Founding member of the Asociación Jalisciense de Ciencias. Accredited Ergonomist by the Mexican Ergonomics Society. Member of the National System of Researchers.

Andres Quiñones is a Mechanical Engineering Undergraduate Student at the University of Texas at El Paso.

Arturo Realyvásquez has studied Ergonomics since 2008. He developed an expert system to measure ergonomic compatibility of advanced manufacturing technology. In the field of macroergonomics, he developed a macroergonomic compatibility questionnaire, proposed several hypothetical causal model to measure the effects of macroergonomic compatibility on work systems, and he proposed an index to measure macroergonomic compatibility levels on manufacturing systems. Currently, he works as a full time professor at Instituto Tecnológico de Tijuana.

Rosa Reyes-Martínez is a full-time professor in the Department of Postgraduate Studies and Research.

John Rey-Galindo is an Industrial designer by Universidad de Bogotá Jorge Tadeo Lozano, Master Degree in Design with an emphasis in Ergonomics by Universidad Nacional Autonoma de México. Member of Ergonomics Research Center at Universidad de Guadalajara.

Alicia Libertad Rizo-Corona is an Industrial designer by Universidad de Guadalajara, current student of Master in Ergonomics at Universidad de Guadalajara. 5 years experience in product design.

Yordán Rodríguez is a Professor at the National School of Public Health, University of Antioquia, Medellin, Colombia. He was Professor of Industrial Engineering over seven years (2006-2013) at José Antonio Echeverría Polytechnic Institute, Havana, Cuba. He obtained a first degree in industrial engineering with a specialization in management enterprise in 2006 from José Antonio Echeverría Polytechnic Institute. He subsequently holds an M.S. (2010) in Human Resource Management, and a Ph.D. (2011) in Technical Sciences from José Antonio Echeverría Polytechnic Institute. In 2012, he received the Mention to Annual Best Doctoral Research in Technical Sciences award in Cuba. He is recipient of the highest recognition to scientific results in Cuba, National Award of Cuban Academy of Sciences (2011). His research, teaching, and consulting activities focus on Human Factors and Ergonomics, specifically in the prevention of work-related musculoskeletal disorders, assessment exposure, workplace design, and macroergonomics evaluation of work systems. In 2008, he was guest-scientist at Ergonomics Institute of Technische Universität München in Germany, and he also obtained the MAPFRE scholarship in Ergonomics and Applied Psychology in Spain. He has given conferences and courses in the ergonomic fields in Brazil, Colombia, Cuba, Mexico, Peru and Spain. Dr. Rodríguez has presented his work in some peer-reviewed journals, and in national and international congresses. He has also served as professor of the master program in Human Resource Management, and as coordinator of master program in Ergonomics and Occupational Health and Safety at José Antonio Echeverría Polytechnic Institute. He was Fellow of the Technical Committee of Occupational Safety & Health of the Cuban republic (2009-2013). Currently, he serves as professor of the master and specialization programs in Occupational Health at National School of Public Health, University of Antioquia, Colombia; and he is Fellow (full member) of the Human Factors and Ergonomics Society (HFES).

Berthana Salas Dominguez is an Industrial Designer from Universidad Autonoma Metropolitana, Campus Xochimilco, Master degree in Industrial Design and Ergonomics, PhD candidate in architecture from Universidad Nacional Autonoma de Mexico. Full time professor in UAM Xochimilco, Specialist in Ergonomics, Design for Disability and Usability, Guest scholar researcher at the University at Buffalo, New York State University during August 2016- July 2017.

Jaime Sánchez-Leal is a full-time professor in the Department of Postgraduate Studies and Research.

Suchismita Satapathy is currently working as an Associate Professor at the School of Mechanical Sciences, KIIT University. She published more than 25 articles many national and international journals and conferences. She has more than 14 years of teaching and research experience. Her areas of interest are quality management, ergonomics, sustainability, service quality, supply chain management, etc. She completed her PhD from NIT Rourkela and guided many Btech, Mtech and PhD scholars.

Juan Manuel Solórzano is a full-time professor with 14 years of experience of teaching in Industrial Design major. His publications are about visual perception products.

Ludovico Soto Nogueira is a full-time professor with 10 years of experience of teaching in Industrial Design major. Diverse experience in Management and Operations within Manufacturing, Quality, Facility Engineering and Design disciplines.

Ulises Jesús Tamayo Pérez was born in the city of Guadalajara, on February 24, 1977. He studied the Bachelor's Degree in Electronics Engineering at the Faculty of Engineering of the Autonomous University of Baja California (UABC). He completed a Master of science and engineering in nanotechnology in the faculty of science of the UABC with the thesis entitled "Electrical properties of cylindrical packages of multiwall carbon nanotubes, graphite and amorphous carbon depending on their density" obtaining the publication of a scientific article. Also, a patent was submitted to the Mexican Institute of Industrial Property (IMPI) with the title Instrument and method to electrically characterize conductor and semiconductor materials in volumes with different porosity. He obtained the recognition to the 2011 university merit during his master's studies. He completed a Doctorate in science and engineering in nanotechnology at the Faculty of Sciences of the Autonomous University of Baja California with the thesis entitled Nano-structured electrodes for the production of hydrogen through the electrolysis of water based on clean energies, with which developed a pair of prototypes for the exploitation of marine energies. He obtained the recognition to the 2016 university merit during his Ph.D. degree studies. He is currently a professor of alternative energies in the nanotechnology degree at the UNAM and is a full-time professor in the engineering career in Nanotechnology at the School of Architecture and Design Engineering at the UABC.

Diego Tlapa is a professor of industrial engineering in the Faculty of Engineering, Architecture and Design at the Autonomous University of Baja California, Mexico. He is member of the National System of Researchers of the National Council of Science and Technology in Mexico; his research focuses on supply chain management, and processes improvement projects including six sigma and lean manufacturing. He has participated in several research projects related to process improvement and is author/coauthor of 20 journal papers and conference papers.

Ana Tovar-Hernández has a Bachelor degree in Industrial engineering from Instituto Tecnologico de Celaya. Student of Master in Product Design and Develop at Universidad Autónoma de Ciudad Juárez.

Divya Zindani, (BE, Mechanical Engineering, Rajasthan Technical University, Kota), M.E. (Design of Mechanical Equipment, BIT Mesra), presently pursuing PhD (National Institute of Technology, Silchar). He has over 2 years of Industrial experience. His areas of interests are Optimization, Product and Process Design, CAD/CAM/CAE, Rapid prototyping and Material Selection. He has 1 Patent, 2 Books, 1 Book Chapter, 1 Journal (Web of Sc.), 5 Scopus Indexed international journal and 4 International Conference publications to his credit.

Index

V

W

Y

Purchase Print, E-Book, or Print + E-Book

IGI Global books can now be purchased from three unique pricing formats: Print Only, E-Book Only, or Print + E-Book. Shipping fees apply.

www.igi-global.com

Recommended Reference Books

ISBN: 978-1-4666-9975-5
© 2016; 539 pp.
List Price: $225

ISBN: 978-1-4666-9811-6
© 2016; 441 pp.
List Price: $215

ISBN: 978-1-4666-9619-8
© 2016; 1,670 pp.
List Price: $2,250

ISBN: 978-1-4666-8473-7
© 2015; 1,702 pp.
List Price: $2,295

ISBN: 978-1-4666-9911-3
© 2016; 410 pp.
List Price: $310

ISBN: 978-1-4666-8291-7
© 2015; 743 pp.
List Price: $335

Looking for free content, product updates, news, and special offers?
Join IGI Global's mailing list today and start enjoying exclusive perks sent only to IGI Global members.
Add your name to the list at **www.igi-global.com/newsletters.**

Publishing Information Science and Technology Research Since 1988

IGI Global
DISSEMINATOR OF KNOWLEDGE

www.igi-global.com | Sign up at www.igi-global.com/newsletters | facebook.com/igiglobal | twitter.com/igiglobal

Stay Current on the Latest Emerging Research Developments

Become an IGI Global Reviewer for Authored Book Projects

The overall success of an authored book project is dependent on quality and timely reviews.

In this competitive age of scholarly publishing, constructive and timely feedback significantly decreases the turnaround time of manuscripts from submission to acceptance, allowing the publication and discovery of progressive research at a much more expeditious rate. Several IGI Global authored book projects are currently seeking highly qualified experts in the field to fill vacancies on their respective editorial review boards:

Applications may be sent to:
development@igi-global.com

Applicants must have a doctorate (or an equivalent degree) as well as publishing and reviewing experience. Reviewers are asked to write reviews in a timely, collegial, and constructive manner. All reviewers will begin their role on an ad-hoc basis for a period of one year, and upon successful completion of this term can be considered for full editorial review board status, with the potential for a subsequent promotion to Associate Editor.

If you have a colleague that may be interested in this opportunity, we encourage you to share this information with them.

www.igi-global.com

InfoSci®-Books

A Database for Information Science and Technology Research

Maximize Your Library's Book Collection!

Invest in IGI Global's InfoSci®-Books database and gain access to hundreds of reference books at a fraction of their individual list price.

The InfoSci®-Books database offers unlimited simultaneous users the ability to precisely return search results through more than 80,000 full-text chapters from nearly 3,900 reference books in the following academic research areas:

Business & Management Information Science & Technology • Computer Science & Information Technology
Educational Science & Technology • Engineering Science & Technology • Environmental Science & Technology
Government Science & Technology • Library Information Science & Technology • Media & Communication Science & Technology
Medical, Healthcare & Life Science & Technology • Security & Forensic Science & Technology • Social Sciences & Online Behavior

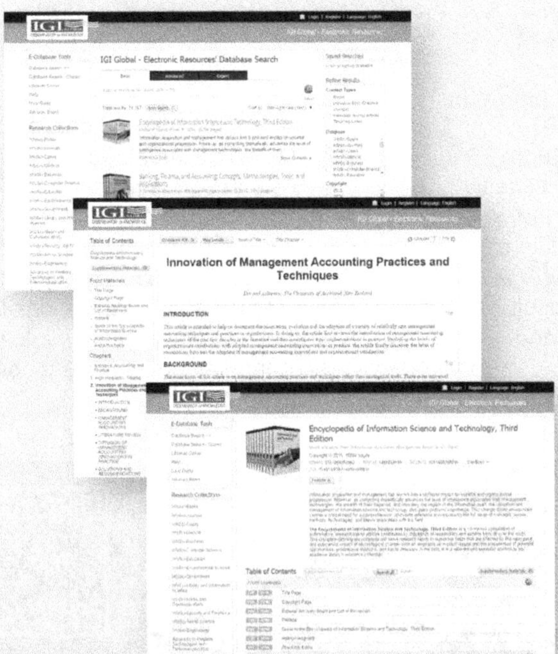

Peer-Reviewed Content:
- Cutting-edge research
- No embargoes
- Scholarly and professional
- Interdisciplinary

Award-Winning Platform:
- Unlimited simultaneous users
- Full-text in XML and PDF
- Advanced search engine
- No DRM

Librarian-Friendly:
- Free MARC records
- Discovery services
- COUNTER4/SUSHI compliant
- Training available

To find out more or request a free trial, visit: www.igi-global.com/eresources

IGI Global
DISSEMINATOR OF KNOWLEDGE
www.igi-global.com

IGI Global
DISSEMINATOR OF KNOWLEDGE
www.igi-global.com

IGI Global Proudly Partners with

eContent Pro International

Expert Copy Editing

eContent Pro International copy editors, with over 70 years of combined experience, will provide complete and comprehensive care for your document by resolving all issues with spelling, punctuation, grammar, terminology, jargon, semantics, syntax, consistency, flow, and more. In addition, they will format your document to the style you specify (APA, Chicago, etc.). All edits will be performed using Microsoft Word's Track Changes feature, which allows for fast and simple review and management of edits.

Additional Services

eContent Pro International also offers fast and affordable proofreading to enhance the readability of your document, professional translation in over 100 languages, and market localization services to help businesses and organizations localize their content and grow into new markets around the globe.

IGI Global Authors Save 25% on eContent Pro International's Services!

Scan the QR Code to Receive Your 25% Discount

The 25% discount is applied directly to your eContent Pro International shopping cart when placing an order through IGI Global's referral link. Use the QR code to access this referral link. eContent Pro International has the right to end or modify any promotion at any time.

Email: customerservice@econtentpro.com

econtentpro.com

IRMA INTERNATIONAL EST. 1987

Information Resources Management Association

Advancing the Concepts & Practices of Information Resources Management in Modern Organizations

Become an IRMA Member

Members of the **Information Resources Management Association (IRMA)** understand the importance of community within their field of study. The Information Resources Management Association is an ideal venue through which professionals, students, and academicians can convene and share the latest industry innovations and scholarly research that is changing the field of information science and technology. Become a member today and enjoy the benefits of membership as well as the opportunity to collaborate and network with fellow experts in the field.

IRMA Membership Benefits:

- **One FREE Journal Subscription**
- **30% Off Additional Journal Subscriptions**
- **20% Off Book Purchases**
- Updates on the latest events and research on Information Resources Management through the IRMA-L listserv.
- Updates on new open access and downloadable content added to Research IRM.
- A copy of the Information Technology Management Newsletter twice a year.
- A certificate of membership.

IRMA Membership $195

Scan code or visit **irma-international.org** and begin by selecting your free journal subscription.

Membership is good for one full year.

www.irma-international.org

IGI Global
DISSEMINATOR OF KNOWLEDGE
www.igi-global.com

Available to Order Now

Order through www.igi-global.com with **Free Standard Shipping**.

The Premier Reference for Information Science & Information Technology

100% Original Content
Contains 705 new, peer-reviewed articles with color figures covering over 80 categories in 11 subject areas

Diverse Contributions
More than 1,100 experts from 74 unique countries contributed their specialized knowledge

Easy Navigation
Includes two tables of content and a comprehensive index in each volume for the user's convenience

Highly-Cited
Embraces a complete list of references and additional reading sections to allow for further research

Included in:
InfoSci®-Books

Encyclopedia of Information Science and Technology Fourth Edition

A Comprehensive 10-Volume Set

Mehdi Khosrow-Pour, D.B.A. (Information Resources Management Association, USA)
ISBN: 978-1-5225-2255-3; © 2018; Pg: 8,104; Release Date: July 2017

The **Encyclopedia of Information Science and Technology, Fourth Edition** is a 10-volume set which includes 705 original and previously unpublished research articles covering a full range of perspectives, applications, and techniques contributed by thousands of experts and researchers from around the globe. This authoritative encyclopedia is an all-encompassing, well-established reference source that is ideally designed to disseminate the most forward-thinking and diverse research findings. With critical perspectives on the impact of information science management and new technologies in modern settings, including but not limited to computer science, education, healthcare, government, engineering, business, and natural and physical sciences, it is a pivotal and relevant source of knowledge that will benefit every professional within the field of information science and technology and is an invaluable addition to every academic and corporate library.

Scan for Online Bookstore

Pricing Information

Hardcover: **$5,695** E-Book: **$5,695** Hardcover + E-Book: **$6,895**

Recommend this Title to Your Institution's Library: www.igi-global.com/books

www.igi-global.com/infosci-ondemand

InfoSci®-OnDemand

Continuously updated with new material on a weekly basis, InfoSci®-OnDemand offers the ability to search through thousands of quality full-text research papers. Users can narrow each search by identifying key topic areas of interest, then display a complete listing of relevant papers, and purchase materials specific to their research needs.

Comprehensive Service

- Over 81,600+ journal articles, book chapters, and case studies.
- All content is downloadable in PDF format and can be stored locally for future use.

No Subscription Fees

- One time fee of $37.50 per PDF download.

Instant Access

- Receive a download link immediately after order completion!

Database Platform Features:

- Comprehensive Pay-Per-View Service
- Written by Prominent International Experts/Scholars
- Precise Search and Retrieval
- Updated With New Material on a Weekly Basis
- Immediate Access to Full-Text PDFs
- No Subscription Needed
- Purchased Research Can Be Stored Locally for Future Use

"It really provides an excellent entry into the research literature of the field. It presents a manageable number of highly relevant sources on topics of interest to a wide range of researchers. The sources are scholarly, but also accessible to 'practitioners'."

- Lisa Stimatz, MLS, University of North Carolina at Chapel Hill, USA

"It is an excellent and well designed database which will facilitate research, publication and teaching. It is a very very useful tool to have."

- George Ditsa, PhD, University of Wollongong, Australia

"I have accessed the database and find it to be a valuable tool to the IT/IS community. I found valuable articles meeting my search criteria 95% of the time."

- Lynda Louis, Xavier University of Louisiana, USA

Recommended for use by researchers who wish to immediately download PDFs of individual chapters or articles.

www.igi-global.com/e-resources/infosci-ondemand

IGI Global
DISSEMINATOR OF KNOWLEDGE
www.igi-global.com

www.ingramcontent.com/pod-product-compliance
Lightning Source LLC
LaVergne TN
LVHW060354300926
849162LV00006B/1265

* 9 7 8 1 5 2 2 5 5 2 3 4 5 *